# LECTURES
ON
# MATHEMATICAL FINANCE
AND
## RELATED TOPICS

# LECTURES

ON

# MATHEMATICAL FINANCE

AND

# RELATED TOPICS

## YURI KIFER

Hebrew University of Jerusalem, Israel

**World Scientific**

NEW JERSEY · LONDON · SINGAPORE · BEIJING · SHANGHAI · HONG KONG · TAIPEI · CHENNAI · TOKYO

*Published by*

World Scientific Publishing Co. Pte. Ltd.
5 Toh Tuck Link, Singapore 596224
*USA office:* 27 Warren Street, Suite 401-402, Hackensack, NJ 07601
*UK office:* 57 Shelton Street, Covent Garden, London WC2H 9HE

Library of Congress Control Number: 2019052122

**British Library Cataloguing-in-Publication Data**
A catalogue record for this book is available from the British Library.

LECTURES ON MATHEMATICAL FINANCE AND RELATED TOPICS

ISBN 978-981-120-956-7

For any available supplementary material, please visit
https://www.worldscientific.com/worldscibooks/10.1142/11534#t=suppl

Desk Editor: Liu Yumeng

To my family

# Preface

This book can be used as a year long course of lectures on mathematical finance for advanced graduate students in mathematics who mastered both measure theory and probability courses, where the latter based on the former. For instance, each of the books [1] or [7] provides sufficient background both in probability and measure theory. On the other hand, no prior finance or economics knowledge is required. The author taught parts of this course at Hebrew University of Jerusalem and at University of Pennsylvania. There are many books on mathematical finance (other names: financial mathematics or stochastic finance) out there. Those which can be considered as textbooks are usually on a lower level with regard to mathematics than the present book, and so they are not appropriate for an advanced course on the level comparable to algebra and analysis courses studied by graduate students in mathematics at the same time. On the other hand, advanced monographs are usually more suitable to serve as reference books for working mathematicians and lecturers rather than textbooks for students.

One of the main features of the present book is that we keep the exposition self-contained beyond standard measure and probability theory courses. This is especially important when the book is used for self-study so that the reader does not need to look in other sources for proofs of theorems from other fields of mathematics used in mathematical finance such as optimal stopping or functional analysis. All necessary results with complete proofs from the theory of martingales and stochastic analysis are provided here as well. When several proofs of some theorems are available we chose those which can be presented more easily in a self-contained fashion. Referring to proofs in other sources requires the reader to understand new notations and to look through previous statements there, and so concentrating the material in one place when possible makes the reading and learning easier and more productive. Another characteristic feature of this book is the detailed study of Israeli (game) options introduced by the author in [35] which is supplemented by complete proofs of corresponding results on Dynkin's optimal stopping games both in discrete and continuous time. Though game options are not yet traded as such in the financial markets, they and their variations became a popular topic of research. We view mathematical finance as a branch of mathematics with problems motivated by financial markets which does not mean that the study should be restricted to current realities in these markets. The approach here should be similar to the one in mathematical physics where the study of Schrödinger operators, for instance, is not restricted only to those which describe the behavior of real particles.

The book is divided into three parts. The first part deals with the discrete time case and it contains the material most of which can be used as a stand alone one semester course. We start with reviewing the theory of martingales and optimal stopping in Chapter 1 proceeding to problems of mathematical finance itself in Chapter 2, where we discuss the binomial Cox–Ross–Rubinstein model of a financial market and describe pricing of derivatives there. In Chapter 3 we present a somewhat novel exposition of two fundamental theorems of asset pricing. In the main text we give full proofs relying on certain separation statement, then in appendices to the corresponding lecture we fill the gap providing complete proofs, which include even the Hahn–Banach theorem, keeping the exposition self-contained. Other books usually present these fundamental theorems either in simplified form for a finite probability space as in [72] and [69] or providing a proof relying on a series of lemmas and on a specific version of the Hahn–Banach theorem, referring the reader to another source for the proof of the latter as in [23].

Chapter 4 deals with superhedging in incomplete markets. We present the optional decomposition theorem first and then study superhedging for European, American and Israeli contingent claims. In particular, we prove the existence of a superhedging strategy for the latter when the initial capital equals the superhedging price which did not appear with all details in the literature before. In Chapter 5 we study partial hedging, i.e., portfolio strategies where an agent accepts certain risk that the portfolio value may not be sufficient to cover a contract obligation. In particular, we prove the existence of a portfolio strategy which minimizes the shortfall risk for Israeli contingent claims that did not appear in the literature yet in this generality.

Discrete time case provides many of the ideas of mathematical finance and gives a good preparation for the study of similar problems in the continuous time which is more technically involved. It requires the background of continuous time martingales and optimal stopping in continuous time that we will present in Chapter 6, which opens Part II. Another machinery required here is the stochastic analysis which we exhibit in Chapter 7. We provide here the basic material with essentially all proofs keeping the exposition self-contained, but for more comprehensive study of this field the reader should explore other books such as [28] and [44]. In Chapter 8 we apply the machinery of Chapters 6 and 7 in order to study the pricing of European, American and Israeli contingent claims in the Black–Scholes market when the underlying asset evolves according to the geometric Brownian motion.

We divide Parts I and II into lectures, each exhibiting one topic or closely related topics. The length of lectures is not uniform, and so they require different time to study. All chapters in Parts I and II are supplemented by exercises. To all of them we provide solutions at the end of the book. The purpose of exercises here is to learn how the theoretical material can be applied in more specific situations.

Part III of the book differs from the previous parts since there we include various notions and results, mostly without proofs, which are not intended for frontal lectures but can be considered as suggestions for further study or used in the teaching process as topics for self-study by some students who may present them to a class at the end of the course. Mathematical finance became a huge field of research by now and many topics remained outside of the scope of this book, among them there are: more general stock evolutions described, for instance, by Lévy processes, partial differential equations approach to American and Israeli option pricing, backward and forward-backward stochastic differential equations related to the American and Israeli type options, volatility uncertainty, risk measures, insider information and others. For historical reviews on the development of mathematical finance we refer the readers to [31] and [9].

# Contents

# Further Topics 265

# PART 1
# Discrete Time

# Chapter 1

# Martingales and Optimal Stopping

## 1.1 Lecture 1: Conditional expectations and martingales

### 1.1.1 *Conditional expectation*

We always work on a given complete probability space which is a triple $(\Omega, \mathcal{F}, P)$ where $\Omega$ is a (sample) space, $\mathcal{F}$ is a $\sigma$-algebra and $P$ is a probability measure.

**Definition 1.1.1.** Let $X$ be a random variable such that $E|X| < \infty$ (finite expectation). Let $\mathcal{A} \subset \mathcal{F}$ be a $\sigma$-subalgebra. A random variable $Y$ defined up to a probability zero set is called the conditional expectation of $X$ with respect to $\mathcal{A}$ and denoted by $Y = E(X|\mathcal{A})$ if for any $A \in \mathcal{A}$,

$$\int_A Y \, dP = \int_A X \, dP.$$

Such $Y$ always exists since, if we set $\mu(A) = \int_A X \, dP$, this defines a signed measure on $(\Omega, \mathcal{A}, P)$ which is absolutely continuous with respect to $P$ (written $\mu \prec P$), and so by the Radon–Nikodym theorem from the measure theory we obtain the existence and uniqueness (up to a $P$-almost sure equality) of $Y$ satisfying the formula above. Observe that since the conditional expectation is unique only almost surely (a.s.) all equalities and inequalities involving it are understood only a.s., i.e., up to sets of $P$-measure zero.

    **Properties of conditional expectations**: Assuming that all conditional expectations below are defined we have

- a) linearity: $E(aX + bY|\mathcal{A}) = aE(X|\mathcal{A}) + bE(Y|\mathcal{A})$, $a, b$ are constants;
- b) if $\mathcal{B} \subset \mathcal{A}$ then $E(E(X|\mathcal{A})|\mathcal{B}) = E(X|\mathcal{B})$;

- c) if $X$ is $\mathcal{A}$-measurable then $E(X|\mathcal{A}) = X$;
- d) if $X$ is independent of the $\sigma$-algebra $\mathcal{A}$ (meaning any events $\{X \leq a\}$ and $A \in \mathcal{A}$ are independent) then $E(X|\mathcal{A}) = EX$;
- e) generalization of c): if $X$ is $\mathcal{A}$-measurable then $E(XY|\mathcal{A}) = XE(Y|\mathcal{A})$;
- f) if $f = f(x,y)$ is a bounded Borel function, $X$ is $\mathcal{A}$-measurable and $Y$ is independent of $\mathcal{A}$, then $E(f(X,Y)|\mathcal{A}) = h(X)$ where $h(x) = Ef(x,Y)$.

**Lemma 1.1.1. Jensen inequality** *Let $\varphi$ be a convex function and $X$ be a random variable such that $E|X| < \infty$ and $E|\varphi(X)| < \infty$. Then*

$$E(\varphi(X)|\mathcal{A}) \geq \varphi(E(X|\mathcal{A})).$$

**Proof.** Since $\varphi$ is convex then

$$\varphi(x) \geq \varphi(a) + \lambda(a)(x - a)$$

for some function $\lambda$ (slope of the line passing through the point $a$), and so

$$\varphi(X) \geq \varphi(E(X|\mathcal{A})) + \lambda(E(X|\mathcal{A}))(X - E(X|\mathcal{A})).$$

Taking the conditional expectation in this inequality we obtain the result. It is legitimate to take the expectation if $E(X|\mathcal{A})$ is bounded. Otherwise, restrict first the inequality to the set (event) $G_n = \{E(X|\mathcal{A}) \leq n\}$, then take the expectation and, finally, let $n \to \infty$. $\qquad\square$

    **Example** Let $\Omega = \cup_{i=1}^{\infty} A_i$ and this union is disjoint, i.e., $\xi = \{A_i\}_{i=1}^{\infty}$ is a partition of $\Omega$. Let $\mathcal{A}$ be the $\sigma$-algebra generated by $\xi$, i.e., $\mathcal{A}$ is the minimal $\sigma$-algebra which contains each $A_i, i = 1, 2, ...$ Let $X$ be a random variable such that $E|X| < \infty$. Then $E(X|\mathcal{A})(\omega) = a_i$ is constant for all (almost all) $\omega \in A_i$. Hence,

$$a_i P(A_i) = \int_{A_i} E(X|\mathcal{A})(\omega)dP = \int_{A_i} X dP,$$

and so $a_i = \frac{1}{P(A_i)} \int_{A_i} X dP$ provided $P(A_i) \neq 0$.

### 1.1.2    *Martingales, submartingales and supermartingales*

We will often deal with families of random variables (or vectors) denoted by $X_t$ or $X(t)$ and called also stochastic processes where a real or integer parameter $t$ plays the role of time. All definitions and results here can be considered for processes defined on infinite or finite time intervals $t \in \mathbb{R}_+ = [0, \infty)$, $t \in \mathbb{Z}_+ = \{0, 1, 2, ...\}$ or $t \in [0, T]$, $t \in \{0, 1, ..., N\}$ with $T, N < \infty$.

In fact, if $X_t$ is defined for $t \in [0, T]$ or $t \in \{0, 1, ..., N\}$ then we always can set $X_t = X_T$ or $X_t = X_N$ for $t > T$ or $t > N$ to obtain the process defined now on the infinite time interval having the same properties as before. Thus, we can always view processes as defined on an infinite time interval but we may restrict our attention to a finite time interval relevant to a problem under consideration.

**Definition 1.1.2.** Let $M_t$, $t \in \mathbb{R}_+$ or $t \in \mathbb{Z}_+$ be a family of random variables defined on a probability space $(\Omega, \mathcal{F}, P)$ and let $\{\mathcal{F}_t, t \in \mathbb{R}_+$ or $t \in \mathbb{Z}_+\}$ be a filtration of $\sigma$-algebras, i.e., a family of $\sigma$-algebras such that $\mathcal{F}_s \subset \mathcal{F}_t \subset \mathcal{F}$ if $s < t$. The family $\{M_t\}$ is called adapted to the filtration $\{\mathcal{F}_t\}$ if for any $t$ the random variable $M_t$ is $\mathcal{F}_t$-measurable. The family $\{M_t\}$ is called a martingale with respect to the filtration $\{\mathcal{F}_t\}$ if it is adapted to the filtration, $E|M_t| < \infty$ for all $t$ and $E(M_t|\mathcal{F}_s) = M_s$ for all $s < t$. If, instead, we have $E(M_t|\mathcal{F}_s) \geq M_s$ for all $s < t$ ($E(M_t|\mathcal{F}_s) \leq M_s$ for all $s < t$) then the family $\{M_t\}$ is called a submartingale (a supermartingale ). If $M_n$, $n = 0, 1, 2, ...$ is a martingale then $\Delta M_n = M_n - M_{n-1}$, $n = 1, 2, ...$ is called a sequence of martingale differences.

Let $\mathcal{F}_t = \sigma\{M_s, s \leq t\}$, $t \geq 0$ be the filtration generated by random variables $M_s$, $0 \leq s \leq t$, i.e., $\mathcal{F}_t$ is the minimal $\sigma$-algebra (intersection of all $\sigma$-algebras) such that all random variables $M_s$, $0 \leq s \leq t$ are measurable with respect to $\mathcal{F}_t$. If $M_t$, $t \geq 0$ is a martingale (submartingale, supermartingale) with respect to some filtration $\mathcal{G}_t$, $t \geq 0$ then the same holds true also with respect to the above filtration $\mathcal{F}_t$, $t \geq 0$. Indeed, $\mathcal{G}_t \supset \mathcal{F}_t$ for all $t \geq 0$, since by definition all $M_s$, $0 \leq s \leq t$ must be measurable with respect to $\mathcal{G}_t$, and so $E(M_t|\mathcal{F}_s) = E(E(M_t|\mathcal{G}_s)|\mathcal{F}_s)$ if $s \leq t$, which implies the claim. Hence, we can say that $M_t$, $t \geq 0$ is a martingale (submartingale, supermartingale) also without specifying a filtration if we mean that the filtration in question is the one generated by these random variables.
  **Examples**

- 1) Let $X_1, X_2, ...$ be independent identically distributed (i.i.d.) random variables such that $E|X_n| < \infty$ and $EX_n = 0$. Then $M_n = \sum_{j=1}^{n} X_j$ is a martingale with respect to the filtration of $\sigma$-algebras $\mathcal{F}_n = \sigma\{X_1, ..., X_n\}$ generated by $X_1, ..., X_n$.
- 2) The sequence $M_n = (X_1 + \cdots + X_n)^2 - nEX_1^2$ is a martingale with respect to the same filtration where $X_1, X_2, ...$ are the same as in 1) with $EX_1^2 < \infty$.

- 3) Let $X_1, X_2, ...$ be i.i.d. such that $X_n > 0$ almost surely (a.s.) and $EX_n = 1$. Then the sequence $M_n = \prod_{j=1}^{n} X_n$ is a martingale with respect to the same filtration as above.
- 4) Let $X$ be an integrable random variable and $\{\mathcal{F}_n, n \geq 0\}$ be a filtration. Then $M_n = E(X|\mathcal{F}_n)$, $n = 0, 1, ...$ is a martingale (of Doob).
- 5) If the sequence $\{M_n\}$ is a martingale and $\varphi$ is a convex function such that $E|\varphi(M_n)| < \infty$ for all $n$, then by Jensen's inequality the sequence $\varphi(M_n)$ is a submartingale. For instance, this is true for $|M_n|^q$ for any $q \geq 1$.

**Remark 1.1.1.** The name "martingale" probably comes from the strategy of game with the same name when the player doubles her/his bet after each loss. There is also martingale which is the strap used to control height of a horse's head.

### 1.1.3  *Supermartingale decomposition*

**Definition 1.1.3.** A sequence of random variables (vectors) $\{A_n\}_{n\geq0}$ is called predictable with respect to a filtration $\{\mathcal{F}_n\}_{n\geq0}$ if $A_n$ is $\mathcal{F}_{n-1}$-measurable for each $n = 1, 2, ...$

**Theorem 1.1.1.** *Let a sequence $\{Y_n\}_{n\geq0}$ be a supermartingale with respect to a filtration $\{\mathcal{F}_n\}_{n\geq0}$. Then there exists a martingale $\{M_n\}_{n\geq0}$ with respect to the same filtration and a predictable non-decreasing process $\{A_n\}_{n\geq0}$ such that $A_0 = 0$, $M_0 = Y_0$,*

$$Y_n = M_n - A_n \quad \text{for all } n \geq 0$$

*and this decomposition is unique. If $\{Z_n\}_{n\geq0}$ is a submartingale then $-Z_n$, $n \geq 0$ is the supermartingale, thus we obtain the submartingale decomposition in the form*

$$Z_n = -M_n + A_n = \hat{M}_n + A_n$$

*where $\hat{M}_n$, $n \geq 1$ is a martingale and $A_n$, $n \geq 0$, $A_0 = 0$ is predictable and non-decreasing.*

**Proof.** Set $M_0 = Y_0$, $A_0 = 0$ and for $n \geq 1$,

$$M_n = Y_0 + \sum_{k=1}^{n} \left(Y_k - E(Y_k|\mathcal{F}_{k-1})\right) \text{ and } A_n = \sum_{k=1}^{n} E\left(Y_{k-1} - Y_k|\mathcal{F}_{k-1}\right)$$

which gives the required decomposition. To prove the uniqueness assume that $Y_n = M'_n - A'_n$ is another decomposition with the same properties. Then $\mathcal{M}_n = M_n - M'_n = A'_n - A_n$, $n \geq 0$ is a martingale such that $\mathcal{M}_n$ is $\mathcal{F}_{n-1}$-measurable and $\mathcal{M}_0 = 0$. But then $\mathcal{M}_{n-1} = E(\mathcal{M}_n | \mathcal{F}_{n-1}) = \mathcal{M}_n$, so $\mathcal{M}_n = \mathcal{M}_{n-1} = \cdots \mathcal{M}_0 = 0$ which yields uniqueness of the above decomposition. $\qquad\square$

### 1.1.4 *Stopping times, local martingales and martingale transform*

**Definition 1.1.4.** Given a filtration $\{\mathcal{F}_t, t \geq 0\}$ (or $\{\mathcal{F}_n, n \in \mathbb{Z}_+\}$), a nonnegative random variable $\tau$ is called a stopping time if $\{\tau \leq t\} \in \mathcal{F}_t$ for any $t \geq 0$. In the discrete time case this is equivalent to $\{\tau = n\} \in \mathcal{F}_n$ for any $n \geq 0$.

If $\tau_1$ and $\tau_2$ are stopping times, so are $\tau_1 \vee \tau_2 = \max(\tau_1, \tau_2)$ and $\tau_1 \wedge \tau_2 = \min(\tau_1, \tau_2)$.

 **Example.** $\tau_B = \min\{n \geq 0 : X_n \in B\}$ is a stopping time for any Borel set $B$ provided $X_n$ is $\mathcal{F}_n$-measurable for each $n$.

 Stopping times enable us to extend the notion of martingales (sub-, supermartingales) which leads to the following useful notion.

**Definition 1.1.5.** A stochastic process (family of random variables) $Z = Z_t$, $t \geq 0$ adapted to a filtration $\{\mathcal{F}_t, t \geq 0\}$ (both for discrete and continuous time $t$) is called a local martingale if there exists a sequence of stopping times $\tau_n \uparrow \infty$ as $n \uparrow \infty$, which is called a localizing sequence, such that for each $n$ the stopped process $Z^{\tau_n}(t) = Z(t \wedge \tau_n)$, $t \geq 0$ is a martingale. Local sub- and supermartingales are defined similarly.

In applications to the mathematical finance the following notion plays a useful role.

**Definition 1.1.6.** Let $M = \{M_n, n \geq 0\}$ be a martingale with respect to a filtration $\{\mathcal{F}_n, n \geq 0\}$ so that $\Delta M_n = M_n - M_{n-1}$, $n \geq 1$ is a sequence of martingale differences. Let $\gamma = \{\gamma_n, n \geq 0\}$ be a predictable sequence of random variables (with respect to the same filtration as the martingale $M$). The sequence $Z_n$, $n \geq 0$, where $Z_0$ is an $\mathcal{F}_0$-measurable random variable and $Z_n = Z_0 + \sum_{k=1}^n \gamma_k \Delta M_k$, $n = 1, 2, ...$, is called a martingale transform of $M$ by means of the sequence $\gamma$.

The following useful result relates the notions of a martingale, a local martingale and a martingale transform.

**Proposition 1.1.1.** *Let $X = \{X_n, \, n \geq 0\}$ be a discrete time stochastic process adapted to a filtration $\mathcal{F} = \{\mathcal{F}_n, \, n \geq 0\}$ and such that $E|X_0| < \infty$. The time parameter $n$ can run in the statements below either over all nonnegative integers $\mathbb{Z}_+ = \{0, 1, ...\}$ or over a bounded set $\{0, 1, ..., N\}$ with $N < \infty$.*

*(i) If $X$ is a martingale transform with respect to the filtration $\mathcal{F}$ then it is a local martingale with respect to the same filtration;*

*(ii) If $X$ is a local martingale with respect to the filtration $\mathcal{F}$ then it is a martingale if and only if $E|X_n| < \infty$ for all $n \geq 0$;*

*(iii) If $X$ is a local martingale with respect to the filtration $\mathcal{F}$ and for all $n = 0, 1, ...,$*

$$\text{either} \quad EX_n^- < \infty \quad \text{or} \quad EX_n^+ < \infty, \tag{1.1.1}$$

*where $x^- = -(x \wedge 0)$ and $x^+ = x \vee 0$, then $X$ is a martingale;*

*(iv) Let $X$ be a local martingale with respect to the filtration $\mathcal{F}$ without assuming additionally that $E|X_0| < \infty$. Suppose that for some $N < \infty$,*

$$\text{either} \quad EX_N^- < \infty \quad \text{or} \quad EX_N^+ < \infty.$$

*Then $\{X_n, \, n = 0, 1, ..., N\}$ is a martingale.*

**Proof.** (i) Let $X_n = X_0 + \sum_{k=1}^n \gamma_k \Delta M_k$, $n = 1, 2, ...$ where $\{M_k\}$ is a martingale and $\{\gamma_k\}$ is a predictable sequence. Define $\tau_m = \min\{k - 1 : |\gamma_k| > m\}$ where we set $\tau_m = \infty$ if the event in braces never occurs. Since $\gamma_k$ is $\mathcal{F}_{k-1}$-measurable,

$$\{\tau_m > k\} = \cap_{j=0}^{k+1} \{|\gamma_j| \leq m\} \in \mathcal{F}_k,$$

so $\tau_m$, $m = 1, 2, ...$ are stopping times and, clearly, $\tau_m \uparrow \infty$ as $m \uparrow \infty$ a.s. Then

$$X_n^{\tau_m} = X_{n \wedge \tau_m} = X_0 + \sum_{k=1}^{n \wedge \tau_m} \gamma_k \Delta M_k = X_0 + \sum_{k=1}^n \gamma_k^{(m)} \Delta M_k$$

where $\gamma_k^{(m)} = \gamma_k \mathbb{I}_{\tau_m > k-1}$ and $\mathbb{I}_\Gamma = 1$ if an event $\Gamma$ occurs and $\mathbb{I}_\Gamma = 0$ otherwise. Hence, $X^{\tau_m}$ is a martingale transform by means of a predictable sequence $\gamma^{(m)}$ bounded (in absolute value) by $m$, and since $\gamma_k^{(m)} \Delta M_k$ is integrable now (as $M$ is a martingale) we obtain that for any $n \geq 1$,

$$E(X_n^{\tau_m} | \mathcal{F}_{n-1}) = X_{n-1}^{\tau_m} + \gamma_n^{(m)} E(\Delta M_n | \mathcal{F}_{n-1}) = X_{n-1}^{\tau_m}.$$

Together with our assumption that $X_0$ is integrable this yields that $X_n^{\tau_m}$, $n \geq 0$ is a martingale. Hence, $X$ is a local martingale as claimed.

(ii) Any martingale is a local martingale, and if $\{X_n, \, n \geq 0\}$ is a martingale, then by definition each $X_n$ is integrable. Now, suppose that $X$ is a local martingale and each $X_n$, $n \geq 0$ is integrable. Let $\tau_m$, $m \geq 1$ be a localizing sequence for $X$. Since

$$|X_{(n+1)\wedge\tau_m}| \leq \sum_{j=0}^{n+1} |X_j| \text{ and } E\sum_{j=0}^{n+1} |X_j| = \sum_{j=0}^{n+1} E|X_j| < \infty$$

we can apply the Lebesgue dominated convergence theorem (for conditional expectations) letting $m \uparrow \infty$ in the equality $E(X_{(n+1)\wedge\tau_m}|\mathcal{F}_n) = X_{n\wedge\tau_m}$ to obtain that $E(X_{n+1}|\mathcal{F}_n) = X_n$, proving that $X$ is a martingale.

(iii) Let $X$ be a local martingale with a localizing sequence $\tau_m \uparrow \infty$. By the Fatou lemma

$$EX_n^+ = E \liminf_{m\uparrow\infty} X_{n\wedge\tau_m}^+ \leq \liminf_{m\uparrow\infty} EX_{n\wedge\tau_m}^+$$

$$= \liminf_{m\uparrow\infty}(EX_{n\wedge\tau_m} + EX_{n\wedge\tau_m}^-) = EX_0 + \liminf_{m\uparrow\infty} EX_{n\wedge\tau_m}^-$$

$$\leq E|X_0| + \sum_{j=0}^n EX_j^- .$$

If $EX_j^- < \infty$ for all $j$ then $EX_n^+ < \infty$, thus $E|X_n| < \infty$ for all $n$ which implies by (ii) that $X$ is a martingale.

(iv) Let $X$ be a local martingale with $EX_N^+ < \infty$ for some $N < \infty$ and a localizing sequence $\tau_m \uparrow \infty$. Since $f(x) = x^+$ is a convex function, it follows by the Jensen inequality that $X_j^+$, $j = 0, 1, ..., N$ is a local submartingale, i.e.,

$$E(X_{n\wedge\tau_m}^+|\mathcal{F}_{n-1}) \geq X_{(n-1)\wedge\tau_m}^+$$

whenever $N \geq n \geq 1$ and $m \geq 1$. Thus, if $EX_n^+ < \infty$ then

$$E(X_n^+\mathbb{I}_{\tau_m>n-1}|\mathcal{F}_{n-1}) = E(X_{n\wedge\tau_m}^+\mathbb{I}_{\tau_m>n-1}|\mathcal{F}_{n-1})$$

$$= \mathbb{I}_{\tau_m>n-1}E(X_{n\wedge\tau_m}^+|\mathcal{F}_{n-1}) \geq \mathbb{I}_{\tau_m>n-1}X_{(n-1)\wedge\tau_m}^+.$$

Hence,

$$EX_n^+ \geq E(X_n^+\mathbb{I}_{\tau_m>n-1}) \geq E(X_{(n-1)\wedge\tau_m}^+\mathbb{I}_{\tau_m>n-1}).$$

Assuming $EX_n^+ < \infty$ we obtain by the Fatou lemma that

$$\infty > \liminf_{m\uparrow\infty} E(X_{(n-1)\wedge\tau_m}^+\mathbb{I}_{\tau_m>n-1})$$

$$\geq E(\liminf_{m\uparrow\infty} X_{(n-1)\wedge\tau_m}^+\mathbb{I}_{\tau_m>n-1}) = EX_{n-1}^+.$$

Since $EX_N^+ < \infty$ we can repeat this argument for $n = N, N-1, ..., 1$ to obtain that $EX_n^+ < \infty$ for all $n = 0, 1, ..., N$. It follows now by (iii) that $X = \{X_0, X_1, ..., X_N\}$ is a martingale. We obtain the same conclusion under $EX_N^- < \infty$ by replacing $X$ with $-X$ which exchanges $X^+$ and $X^-$. $\qquad\square$

## 1.2 Lecture 2: Optional sampling, martingale inequalities and upcrossings

### 1.2.1 *Optional sampling (stopping) theorem*

We start with the following result showing that martingale (sub-, super-martingale) properties are preserved under stopping.

**Lemma 1.2.1.** *Let $\{M_n\}_{n\geq 0}$ be a martingale (sub-, supermartingale) with respect to a filtration $\{\mathcal{F}_n\}$ and $\tau$ be a stopping time. Then $M_n^\tau = M_{n\wedge\tau}$ is a martingale (sub-, supermartingale).*

**Proof.** We write

$$M_{n\wedge\tau} = \sum_{m=0}^{n-1} M_m \mathbb{I}_{\tau=m} + M_n \mathbb{I}_{\tau\geq n}$$

where $\mathbb{I}_\Gamma = 1$ if an event $\Gamma$ occurs and $\mathbb{I}_\Gamma = 0$ otherwise. Hence,

$$M_{(n+1)\wedge\tau} - M_{n\wedge\tau} = \mathbb{I}_{\tau>n}(M_{n+1} - M_n),$$

and so $E(M_{(n+1)\wedge\tau} - M_{n\wedge\tau}|\mathcal{F}_n) = 0$ in the martingale case or $E(M_{(n+1)\wedge\tau} - M_{n\wedge\tau}|\mathcal{F}_n) \geq 0$ in the submartingale case (or $\leq 0$ for a supermartingale). $\square$

For any stopping time $\tau$ with respect to a filtration $\{\mathcal{F}_t\}_{t\geq 0}$ set $\mathcal{F}_\tau = \{A \in \mathcal{F} : A \cap \{\tau \leq t\} \in \mathcal{F}_t, \forall t\}$. Clearly, $\tau$ is $\mathcal{F}_\tau$-measurable.

**Theorem 1.2.1.** *(Optional sampling (or stopping) theorem) Let $X = (X_n, \mathcal{F}_n)_{n\geq 0}$ (where $X_n, n \geq 0$ is a sequence of random variables and $\{\mathcal{F}_n\}_{n\geq 0}$ is a filtration) be a martingale (submartingale). Let $\tau_1 \leq \tau_2$ be stopping times such that $E|X_{\tau_i}| < \infty$, $i = 1, 2$ and*

$$\liminf_{n\to\infty} \int_{\{\tau_2>n\}} |X_n| dP = 0.$$

*Then $E(X_{\tau_2}|\mathcal{F}_{\tau_1}) = X_{\tau_1}$ in the martingale case and $E(X_{\tau_2}|\mathcal{F}_{\tau_1}) \geq X_{\tau_1}$ in the submartingale case.*

**Proof.** It suffices to prove that for any $A \in \mathcal{F}_{\tau_1}$,

$$\int_A X_{\tau_2} dP = \int_A X_{\tau_1} dP$$

in the martingale case and change $=$ to $\geq$ in the submartingale case. Then it suffices to prove that for any $n \in \mathbb{Z}_+$,

$$\int_{A\cap\{\tau_1=n\}} X_{\tau_2} dP = \int_{A\cap\{\tau_1=n\}} X_n dP$$

in the martingale case and change $=$ to $\geq$ in the submartingale case. Now,

$$\int_{A\cap\{\tau_1=n\}} X_n dP = \int_{A\cap\{\tau_1=n\}\cap\{\tau_2\geq n\}} X_n dP$$

$$= \int_{A\cap\{\tau_1=n\}\cap\{\tau_2=n\}} X_n dP + \int_{A\cap\{\tau_1=n\}\cap\{\tau_2>n\}} X_n dP$$

$$= (\text{or} \leq \text{in the submartingale case}) \int_{A\cap\{\tau_1=n\}\cap\{\tau_2=n\}} X_{\tau_2} dP$$

$$+ \int_{A\cap\{\tau_1=n\}\cap\{\tau_2>n\}} E(X_{n+1}|\mathcal{F}_n)dP = \int_{A\cap\{\tau_1=n\}\cap\{\tau_2=n\}} X_{\tau_2} dP$$

$$+ \int_{A\cap\{\tau_1=n\}\cap\{\tau_2\geq n+1\}} X_{n+1} dP = \cdots = (\text{or} \leq \cdots \leq \text{for submartingales})$$

$$\int_{A\cap\{\tau_1=n\}\cap\{n\leq\tau_2\leq m\}} X_{\tau_2} dP + \int_{A\cap\{\tau_1=n\}\cap\{\tau_2>m\}} X_m dP.$$

Letting $m \to \infty$ and using the conditions we derive the result. $\qquad\square$

**Definition 1.2.1.** A sequence $\{X_n\}$ of random variables is uniformly integrable if

$$\lim_{a\to\infty} \sup_n \int_{|X_n|>a} |X_n| dP = 0.$$

**Corollary 1.2.1.** *If $\{X_n\}$ is a uniformly integrable submartingale then the conditions of the above theorem are satisfied, and so its result holds true as well. In particular, this is true when $|X_n| \leq C$ for all $n$ and some constant $C$.*

**Proof.** First, observe that $X_n^+ = \max(X_n, 0)$, $n \geq 0$ is also a uniformly integrable submartingale, and so $\sup_n EX_n^+ < \infty$. Let $\tau < \infty$ a.s. be a stopping time, by the above theorem $EX_{\tau\wedge n}^+ \leq EX_n^+$, so

$$\sup_n EX_{\tau\wedge n}^+ \leq \sup_n EX_n^+ < \infty.$$

Clearly, $X_\tau^+ = \lim_{n\to\infty} X_{\tau\wedge n}^+$ a.s., and by the Fatou lemma,

$$EX_\tau^+ = E\liminf_{n\to\infty} X_{\tau\wedge n}^+ \leq \liminf_{n\to\infty} EX_{\tau\wedge n}^+ \leq \sup_n EX_n^+ < \infty.$$

Since $X_n$, $n \geq 0$ is a submartingale, $EX_n^- = EX_n^+ - EX_n \leq EX_n^+ - EX_0$, where $X_n^- = \max(-X_n, 0)$, and so

$$E|X_\tau| = EX_\tau^+ + EX_\tau^- \leq 2EX_\tau^+ - EX_0 < \infty.$$

In order to check the remaining condition of the above theorem we write

$$\int_{\{\tau>n\}} |X_n| dP \leq \int_{\{|X_n|>a\}} |X_n| dP + \int_{\{\tau>n\}\cap\{|X_n|\leq a\}} |X_n| dP$$

$$\leq \sup_n \int_{\{|X_n|>a\}} |X_n| dP + aP\{\tau > n\}.$$

Letting first $n \to \infty$ and then $a \to \infty$ we obtain that $\int_{\{\tau>n\}} |X_n| dP \to 0$ as $n \to \infty$, completing the proof of this corollary. $\qquad\square$

**Corollary 1.2.2.** *If $0 \leq \tau_i \leq N$, $i = 1,2$ and $X_n$, $n \geq 0$ is a martingale then*

$$EX_0 = EX_{\tau_1} = EX_{\tau_2} = EX_N.$$

### 1.2.2    *Martingale inequalities*

**Theorem 1.2.2.** *([28] and [67]) Let $X_t, t \geq 0$ be a submartingale and $0 \leq t_1 < t_2 < ... < t_n$.*

*(i) For any $a > 0$,*

$$aP\{\max_{0\leq i\leq n} X_{t_i} \geq a\} \leq E(X_{t_n}\mathbb{I}_{\{X\geq a\}}) \leq EX_{t_n}^+ \leq E|X_{t_n}|.$$

*In particular, if $X_t$ is a martingale and $p \geq 1$ then*

$$P\{\max_{0\leq i\leq n} |X_{t_i}| \geq a\} \leq \frac{E|X_{t_n}|^p}{a^p}.$$

*(ii) For any $a > 0$,*

$$aP\{\min_{0\leq i\leq n} X_{t_i} \leq -a\} \leq -EX_0 + E(X_{t_n}\mathbb{I}_{\{\min_i X_{t_i} > -a\}}) \leq E|X_0| + E|X_{t_n}|.$$

*(iii) Suppose that $X_t \geq 0$ is a submartingale and $p > 1$. Then*

$$E(\max_{0\leq i\leq n} X_{t_i}^p) \leq (\frac{p}{p-1})^p EX_{t_n}^p.$$

*In particular, if $X_t$ is a martingale (not assuming nonnegativity) then*

$$E(\max_{0\leq i\leq n} |X_{t_i}|^p) \leq (\frac{p}{p-1})^p E|X_{t_n}|^p.$$

*(iv) More generally, let $\tau < \infty$ be a stopping time and $0 \leq t_1 < t_2 < ...$ be an infinite sequence. Then (i)–(iii) remain true with $\max_{i: t_i \leq \tau} X_{t_i}^p$ and $|X_\tau|$ in place of $\max_{0\leq i\leq n} X_{t_i}^p$ and $|X_{t_n}|$, respectively.*

**Proof.** (i) Define the stopping time

$$\tau = \begin{cases} \min\{t_i : X_{t_i} \geq a\} & \text{if an } i \text{ exists that the event in brackets occurs;} \\ t_n & \text{otherwise.} \end{cases}$$

Then, setting $X = \max_{0\leq i\leq n} X_{t_i}$,

$$EX_{t_n} \geq EX_\tau = E(X_\tau \mathbb{I}_{X\geq a}) + E(X_{t_n}\mathbb{I}_{X<a}) \geq aP\{X \geq a\} + E(X_{t_n}\mathbb{I}_{X<a}),$$

and so

$$E(X_{t_n}\mathbb{I}_{X\geq a}) \geq aP\{X \geq a\}$$

implying the first inequality in the assertion (i). If $X_t$ is a martingale, then $|X_t|$ is a submartingale. Applying Chebyshev's inequality to the above we obtain the second inequality in (i).

(ii) Define the stopping time

$$\tau = \begin{cases} \min\{t_i : X_{t_i} \leq -a\} & \text{if an } i \text{ exists that the event in brackets occurs;} \\ t_n & \text{otherwise.} \end{cases}$$

Then

$$EX_0 \le EX_\tau = E(X_\tau \mathbb{I}_{\min_i X_{t_i} \le -a}) + E(X_{t_n} \mathbb{I}_{\min_i X_{t_i} > -a})$$
$$\le -aP\{\min_i X_{t_i} \le -a\} + E(X_{t_n} \mathbb{I}_{\min_i X_{t_i} > -a})$$

and the required inequality follows.

(iii) We have

$$EX^p = \int_\Omega dP \int_0^X p\lambda^{p-1} d\lambda = \int_\Omega dP \int_0^\infty \mathbb{I}_{\{X \ge \lambda\}} p\lambda^{p-1} d\lambda$$
$$= p\int_0^\infty \lambda^{p-1} P\{X \ge \lambda\} d\lambda \le p\int_0^\infty \lambda^{p-2}(EX_{t_n} \mathbb{I}_{\{X \ge \lambda\}}) d\lambda$$
$$= p\int_0^\infty \int_\Omega \lambda^{p-2} \mathbb{I}_{\{X \ge \lambda\}} X_{t_n} dP d\lambda = \frac{p}{p-1} \int_\Omega X^{p-1} X_{t_n} dP$$

where we used (i) and integrated

$$\int_0^\infty \lambda^{p-2} \mathbb{I}_{\{X \ge \lambda\}} d\lambda = \int_0^X \lambda^{p-2} d\lambda = \frac{X^{p-1}}{p-1}.$$

By the Hölder inequality $(E|YZ| \le (E|Y|^p)^{1/p}(E|Y|^q)^{1/q}, p,q > 0, 1/p + 1/q = 1)$,

$$EX^{p-1} X_{t_n} \le (EX^p)^{\frac{p-1}{p}} (EX_{t_n}^p)^{1/p}.$$

It follows that

$$EX^p \le \frac{p}{p-1}(EX^p)^{\frac{p-1}{p}} (EX_{t_n}^p)^{1/p},$$

and so

$$(\frac{p}{p-1})^p EX_{t_n}^p \ge EX^p.$$

(iv) Consider the stopped submartingale $X_{t_i \wedge \tau}$, $i = 1, 2, ...$ Use the inequalities (i)–(iii) for $i = 1, ..., n$ and pass to the limit as $n \to \infty$ using the optional sampling theorem if $\min(\tau, \lim_{i \to \infty} t_i) < \tau$, completing the proof. □

### 1.2.3 *Martingale convergence theorem*

Another important result about submartingales is the Doob upcrossing inequality which yields the (sub)martingale convergence theorem. Let $[a, b]$, $a < b$ be an interval and $X_1, X_2, ...$ be a sequence of random variables. The number $N_n(a, b) = N_n(a, b)(\omega)$ of upcrossings of $[a, b]$ by $X_1(\omega), ..., X_n(\omega)$ is the number of times this $n$-sequence passes from $(-\infty, a]$ to $[b, \infty)$. Namely, define the following random variables

$$\tau_0 = 0, \ \tau_1 = \min\{k > 0 : X_k \le a\}, \ \tau_2 = \min\{k > \tau_1 : X_k \ge b\}, ...,$$
$$\tau_{2m-1} = \min\{k > \tau_{2m-2} : X_k \ge b\},$$

where $\tau_k = \infty$ if the corresponding set in brackets is empty. Then $N_n(a,b) = 0$ if $\tau_2 > n$ and $N_n(a,b) = \max\{m : \tau_{2m} \leq n\}$ for otherwise.

Suppose that the sequence $X_1, X_2, \ldots$ is adapted to a filtration $\mathcal{F}_1 \subset \mathcal{F}_2 \subset \cdots \subset \mathcal{F}$. Then $\tau_1, \tau_2, \ldots$ are stopping times. Indeed, $\tau_0$ is the stopping time and if $k$ is odd and $\tau_{k-1}$ is a stopping time then

$$\{\tau_k = l\} = \cup_{j=0}^{l-1}\{\tau_{k-1} = j, \, X_{j+1} > a, \ldots, X_{l-1} > a, \, X_l \leq a\} \in \mathcal{F}_l.$$

Thus $\tau_1$ is a stopping time, and if $k$ is even and $\tau_{k-1}$ is a stopping time then

$$\{\tau_k = l\} = \cup_{j=0}^{l-1}\{\tau_{k-1} = j, \, X_{j+1} < b, \ldots, X_{l-1} < b, \, X_l \geq b\} \in \mathcal{F}_l.$$

**Theorem 1.2.3.** *Let $X_1, X_2, \ldots$ be a submartingale with respect to a filtration $\mathcal{F}_1 \subset \mathcal{F}_2 \subset \cdots \subset \mathcal{F}$. Then for any $n \geq 1$ the corresponding upcrossing number $N_n(a,b)$ satisfies*

$$EN_n(a,b) \leq \frac{E(X_n - a)^+}{b-a}$$

*where $c^+ = \max(0, c)$.*

**Proof.** Set $Y_n = (X_n - a)^+$, $n \geq 1$. Observe that

$$E((X_n - a)^+|\mathcal{F}_{n-1}) \geq \mathbb{I}_{X_{n-1} \geq a} E(X_n - a|\mathcal{F}_{n-1})$$
$$\geq \mathbb{I}_{X_{n-1} \geq a}(X_{n-1} - a) = (X_{n-1} - a)^+,$$

and so $Y_n$, $n \geq 1$ is a submartingale. The number of upcrossings by $X_1, \ldots, X_n$ of $[a,b]$ is equal to the number of upcrossings by $Y_1, \ldots, Y_n$ of $[0, b-a]$, thus without loss of generality we can assume that $X_n$, $n \geq 1$ is a nonnegative submartingale and that $a = 0$, so it suffices to show that

$$EN_n(0,b) \leq \frac{EX_n}{b}.$$

Set $X_0$, $\mathcal{F}_0 = \{\emptyset, \Omega\}$, and for $i = 1, 2, \ldots$ define

$$\varphi_i = \begin{cases} 1 & \text{if } \tau_m < i \leq \tau_{m+1} \text{ for some odd } m \\ 0 & \text{if } \tau_m < i \leq \tau_{m+1} \text{ for some even } m. \end{cases}$$

Clearly,

$$bN_n(0,b) \leq \sum_{i=1}^{n}(X_i - X_{i-1})\varphi_i$$

and

$$\{\varphi_i = 1\} = \cup_{\text{odd } m}(\{\tau_m < i\} \setminus \{\tau_{m+1} < i\}) \in \mathcal{F}_{i-1}.$$

Hence,

$$bEN_n(0,b) \leq E \sum_{i=1}^{n}(X_i - X_{i-1})\varphi_i = \sum_{i=1}^{n}\int_{\{\varphi_i=1\}}(X_i - X_{i-1})dP$$

$$= \sum_{i=1}^{n}\int_{\{\varphi_i=1\}} E(X_i - X_{i-1}|\mathcal{F}_{i-1})dP$$

$$= \sum_{i=1}^{n}\int_{\{\varphi_i=1\}}\left(E(X_i|\mathcal{F}_{i-1}) - X_{i-1}\right)dP$$

$$\leq \sum_{i=1}^{n}\int_{\Omega}\left(E(X_i|\mathcal{F}_{i-1}) - X_{i-1}\right)dP = EX_n$$

completing the proof. □

Applying this result we obtain the following Doob's submartingale convergence theorem.

**Corollary 1.2.3.** *Let $X_1, X_2, \ldots$ be a submartingale with respect to a filtration $\mathcal{F}_1 \subset \mathcal{F}_2 \subset \cdots \subset \mathcal{F}$ such that*

$$\sup_n EX_n^+ < \infty.$$

*Then, the limit $\lim_{n\to\infty} X_n = X_\infty$ exists with probability one and $E|X_\infty| < \infty$.*

**Proof.** Set $\Gamma = \{\omega : \limsup_{n\to\infty} X_n(\omega) > \liminf_{n\to\infty} X_n(\omega)\}$ and $\Gamma_{a,b} = \{\omega : \limsup_{n\to\infty} X_n(\omega) > b > a > \liminf_{n\to\infty} X_n(\omega)\}$. Then

$$\Gamma = \cup_{a<b,\, a,b \text{ are rational}}\Gamma_{a,b}.$$

It suffices to show that $P(\Gamma_{a,b}) = 0$ for any $a < b$. Let $N_n(a,b)$ be the number of upcrossings by $X_1, \ldots, X_n$ of the interval $[a,b]$. Then by the above theorem

$$EN_n(a,b) \leq \frac{E(X_n - a)^+}{b-a} \leq \frac{EX_n^+ + |a|}{b-a} \leq \frac{E|X_n| + |a|}{b-a},$$

and by the monotone convergence theorem

$$EN_\infty(a,b) = \lim_{n\to\infty} EN_n(a,b) \leq \frac{\sup_n E|X_n| + |a|}{b-a} < \infty.$$

In particular, it follows that $N_\infty(a,b) < \infty$ a.s. which implies that $P(\Gamma_{a,b}) = 0$ for any $a < b$. Hence, $\lim_{n\to\infty} X_n = X_\infty$ exists with probability one and by Fatou's lemma

$$E|X_\infty| = E\liminf_{n\to\infty}|X_n| \leq \liminf_{n\to\infty} E|X_n|$$

$$\leq \sup_n EX_n^+ + \sup_n EX_n^- \leq 2\sup_n EX_n^+ - EX_0 < \infty,$$

where we used that $EX_n^- = EX_n^+ - EX_n \leq EX_n^+ - EX_0$ since $X_n, n \geq 0$ is a submartingale, completing the proof. □

## 1.3  Lecture 3: Optimal stopping

### 1.3.1  *Single player*

A person observes a reward process $R_n$, $n \geq 0$ and can stop it at any time $n \leq N$ receiving then the amount $R_n$. The goal is to maximize the average reward using only stopping times as strategies which means that only current information is available and no clairvoyance is permitted. The following result describes the problem and its solution in more precise terms.

**Theorem 1.3.1.** *(see [59] or [62]) Let $R_n$, $n \geq 0$ be a sequence of random variables adapted to a filtration $\{\mathcal{F}_n, \, n \geq 0\}$ and such that $E|R_n| < \infty$ for all $n$. Let $N < \infty$ be fixed and denote by $\mathcal{T}_{nN}$ the set of all stopping times $\tau$ such that $N \geq \tau \geq n$. Then*

$$L_n = \text{ess sup}_{\tau \in \mathcal{T}_{nN}} E(R_\tau | \mathcal{F}_n) \qquad (1.3.1)$$

*satisfies for $n = 1, ..., N-1$ the backward induction relation*

$$L_n = \max(R_n, E(L_{n+1} | \mathcal{F}_n)) \quad a.s.. \qquad (1.3.2)$$

*The process $\{L_n, \, n \geq 0\}$ is the smallest supermartingale dominating the sequence $R_n, n \geq 0$, i.e., satisfying $L_n \geq R_n$ (called Snell's envelope). Furthermore,*

$$EL_n = \sup_{\tau \in \mathcal{T}_{nN}} ER_\tau.$$

*If $E \sup_{n \geq 0} |R_n| < \infty$ then the results hold true for $N = \infty$ as well.*

**Remark 1.3.1.** Recall that if $q_\alpha$ is a family of random variables then $Q = \text{ess sup}_\alpha q_\alpha$ is a random variable such that $Q \geq q_\alpha$ a.s. for each $\alpha$, and if $\tilde{Q}$ also satisfies the latter property then $Q \leq \tilde{Q}$ (see [58]). When $q_\alpha$ is a countable family then $\sup_\alpha q_\alpha$ is a random variable and ess sup is not needed but if the family is uncountable then $\sup_\alpha q_\alpha$ may not be measurable. When we will deal with a finite probability space, there will be only finitely many stopping times bounded by $N < \infty$ and we can take max in place of ess sup in the above theorem.

**Proof.** Let $\tau_1, \tau_2 \in \mathcal{T}_{nN}$ and define

$$\tau = \mathbb{I}_{\Omega \backslash \Gamma} \tau_1 + \mathbb{I}_\Gamma \tau_2 \text{ where } \Gamma = \{\omega : E(R_{\tau_1} | \mathcal{F}_n) < E(R_{\tau_2} | \mathcal{F}_n)\}.$$

Then $\tau$ is a stopping time and

$$E(R_\tau | \mathcal{F}_n) = \mathbb{I}_{\Omega \backslash \Gamma} E(R_{\tau_1} | \mathcal{F}_n) + \mathbb{I}_\Gamma E(R_{\tau_2} | \mathcal{F}_n) = \max(E(R_{\tau_1} | \mathcal{F}_n), E(R_{\tau_2} | \mathcal{F}_n))$$

(which means that the set $\{E(R_\tau|\mathcal{F}_n), \tau \in \mathcal{T}_{nN}\}$ is a lattice, i.e., with each two elements it contains their maximum). By the property of ess sup (see, for instance, [58]) there exists a sequence $\tau_k \in \mathcal{T}_{nN}$, $\tau_0 \equiv n$ such that $\sup_k E(R_{\tau_k}|\mathcal{F}_n) = L_n$ a.s., and by the above argument we can choose this sequence so that this supremum is obtained by a monotonic non-decreasing convergence, i.e.,

$$E(R_{\tau_k}|\mathcal{F}_n) \uparrow L_n \text{ as } k \uparrow \infty \text{ a.s.}$$

Now

$$L_n \geq E(R_{\tau_k}|\mathcal{F}_n) \geq E(R_{\tau_0}|\mathcal{F}_n) = R_n.$$

Since for any $\tau \in \mathcal{T}_{nN}$, $|R_\tau| \leq \max_{n \leq m \leq N} |R_m| \leq \sum_{m=0}^{N} |R_m|$ and the right hand side here is integrable, we can apply the dominated convergence theorem to obtain from the above that

$$E(L_n|\mathcal{F}_{n-1}) = \lim_{k \uparrow \infty} \uparrow E(R_{\tau_k}|\mathcal{F}_{n-1}) \leq L_{n-1}$$

as $L_{n-1} = \text{ess sup}_{\tau \in \mathcal{T}_{n-1}} E(R_\tau|\mathcal{F}_{n-1})$ and $\mathcal{T}_{n-1,N} \supset \mathcal{T}_{nN}$. Since, clearly, $L_{n-1} \geq R_{n-1}$ we obtain

$$L_{n-1} \geq \max\left(R_{n-1}, E(L_n|\mathcal{F}_{n-1})\right),$$

and so $L_n$, $n \geq 0$ is a supermartingale.

In order to prove that, in fact, the above inequality is an equality we have to derive the inequality in the other direction. We write

$$R_\tau = R_n \mathbb{I}_{\{\tau=n\}} + R_{\tau \vee (n+1)} \mathbb{I}_{\{\tau>n\}}$$

for $\tau \in \mathcal{T}_{nN}$ and since $\tau \vee (n+1) \in \mathcal{T}_{n+1,N}$, it follows that

$$E(R_{\tau \vee (n+1)}|\mathcal{F}_{n+1}) \leq L_{n+1}.$$

Hence, for any $\tau \in \mathcal{T}_{nN}$,

$$E(R_\tau|\mathcal{F}_n) = R_n \mathbb{I}_{\{\tau=n\}} + E(R_{\tau \vee (n+1)}|\mathcal{F}_n)\mathbb{I}_{\{\tau>n\}}$$
$$\leq R_n \mathbb{I}_{\{\tau=n\}} + E(L_{n+1}|\mathcal{F}_n)\mathbb{I}_{\{\tau>n\}} \leq \max(R_n, E(L_{n+1}|\mathcal{F}_n)),$$

and so

$$L_n \leq \max(R_n, E(L_{n+1}|\mathcal{F}_n)),$$

proving that, in fact, we have here the equality with probability one.

Next,

$$E\big(\text{ess sup}_{\tau \in \mathcal{T}_{nN}} E(R_\tau|\mathcal{F}_n)\big) \geq \sup_{\tau \in \mathcal{T}_{nN}} E(E(R_\tau|\mathcal{F}_n)) = \sup_{\tau \in \mathcal{T}_{nN}} E(R_\tau).$$

On the other hand,

$$E(\text{ess sup}_{\tau \in \mathcal{T}_{nN}} E(R_\tau | \mathcal{F}_n)) = \lim_{k \uparrow \infty} \uparrow E(E(R_{\tau_k} | \mathcal{F}_n))$$
$$= \lim_{k \uparrow \infty} \uparrow E(R_{\tau_k}) \leq \sup_{\tau \in \mathcal{T}_{nN}} E(R_\tau).$$

Hence,

$$E(\text{ess sup}_{\tau \in \mathcal{T}_{nN}} E(R_\tau | \mathcal{F}_n)) = \sup_{\tau \in \mathcal{T}_{nN}} E(R_\tau),$$

i.e.,

$$EL_n = \sup_{\tau \in \mathcal{T}_{nN}} E(R_\tau).$$

If $\tilde{L}_n, n \geq 0$ is another supermartingale which dominates $R_n, n \geq 0$, i.e., $\tilde{L}_n \geq R_n$ a.s. then for any $\tau \in \mathcal{T}_{nN}$ by the optional stopping theorem

$$\tilde{L}_n \geq E(\tilde{L}_\tau | \mathcal{F}_n) \geq E(R_\tau | \mathcal{F}_n),$$

and so $\tilde{L}_n \geq L_n$.

If $N = \infty$ and $E \sup_{n \geq 0} |R_n| < \infty$, we first use the above results for $N < \infty$ and then let $N \to \infty$ relying on the dominated convergence theorem for Lebesgue integrals. □

**Remark 1.3.2.** In Ch. I of [62] the process $L_n$, $n \geq 0$ is defined directly by the backward induction (dynamical programming) formula (1.3.2) and then this process is shown to be the smallest supermartingale dominating the process $R_n$, $n \geq 0$. In this way it is possible to avoid talking about ess sup definition (1.3.1) here. Still, (1.3.1) justifies more directly the optimal stopping name of this topic and in the continuous time case it is the only way to introduce the corresponding Snell's envelopes.

Assume that $\mathcal{F}_0$ is the trivial $\sigma$-field. If we interpret $R_n$ as a reward a player receives upon stopping a game at time $n$ then $L_0 = \sup_{0 \leq \tau \leq N} ER_\tau$ is interpreted as the maximal possible average reward in such a game. We assume that only stopping times can be used since this means that only the information available up to the present time $n$ which is carried by the $\sigma$-algebra $\mathcal{F}_n$ is used to decide whether to stop or not (no clairvoyance!). The backward induction (dynamical programming ) procedure described in the above theorem allows to compute this maximal reward (game value). The following result describes the optimal stopping time to achieve this reward.

**Theorem 1.3.2.** *In notations and conditions of the previous theorem define*

$$\tau_0 = \begin{cases} \min\{n : R_n = L_n\} & \text{if for an } n \leq N \text{ the event in brackets occurs,} \\ N & \text{otherwise.} \end{cases}$$

*Then $L_0 = ER_{\tau_0}$.*

**Proof.** On the event $\{\tau_0 > n\}$ we have that $L_n > R_n$ a.s., and so by the previous theorem on this event $L_n = E(L_{n+1}|\mathcal{F}_n)$. Hence,

$$E(L_{\tau_0 \wedge (n+1)}|\mathcal{F}_n) = L_{\tau_0}\mathbb{I}_{\{\tau_0 \leq n\}} + E(L_{n+1}|\mathcal{F}_n)\mathbb{I}_{\{\tau_0 > n\}}$$
$$= L_{\tau_0}\mathbb{I}_{\{\tau_0 \leq n\}} + L_n\mathbb{I}_{\{\tau_0 > n\}} = L_{\tau_0 \wedge n}.$$

Hence, the sequence $\{L_{\tau_0 \wedge n}, n \geq 0\}$ is a martingale, and so $E(L_{\tau_0 \wedge n}) = EL_0$ for all $n$. In particular, for $n = N$ we have $EL_{\tau_0} = EL_0 = L_0$ since we assumed that $\mathcal{F}_0$ is trivial, and so $L_0$ is a constant. Since $L_{\tau_0} = R_{\tau_0}$ we obtain from here that $L_0 = ER_{\tau_0}$ completing the proof. $\square$

**Remark 1.3.3.** Some information about the history of optimal stopping problems in discrete time can be found in [5].

### 1.3.2 *Dynkin stopping games*

The modern setup for a Dynkin's game (in discrete time) consists of two players and two $\{\mathcal{F}_n\}$-adapted stochastic processes $Y_n$ and $Z_n$, so that when the first player stops the game at time $m$ and the second one stops at time $n$, the former pays to the latter the amount

$$R(m, n) = Y_m\mathbb{I}_{\{m < n\}} + Z_n\mathbb{I}_{\{m \geq n\}}.$$

The time $n$ runs up to some horizon $N < \infty$ when the game is automatically stopped. If it was not stopped before and then the first player pays to the second one the amount

$$Y_N = Z_N.$$

Next, assume that for any $n \leq N$,

$$Z_n \leq Y_n \quad \mathbb{P}\text{-almost surely (a.s)} \quad \text{and}$$

$$E(|Y_n| + |Z_n|) < \infty.$$

It is also possible to consider additional payoff process $U_n$ so that $Z_n \leq U_n \leq Y_n$, and the first player pays $U_n$ if both players stop at the same time. There are also perpetual versions when $N = \infty$ but we will not consider these cases here. Denote by $\mathcal{T}_{mn}$, $m \leq n$ the collection of all stopping times $\tau$ with values between $m$ and $n$. Introduce the upper and the lower values of the game starting at time $n \leq N$ by

$$\bar{V}_n = \text{ess inf}_{\sigma \in \mathcal{T}_{nN}} \text{ess sup}_{\tau \in \mathcal{T}_{nN}} E\big(R(\sigma,\tau)|\mathcal{F}_n\big) \text{ and}$$

$$\underline{V}_n = \text{ess sup}_{\tau \in \mathcal{T}_{nN}} \text{ess inf}_{\sigma \in \mathcal{T}_{nN}} E\big(R(\sigma,\tau)|\mathcal{F}_n\big).$$

**Theorem 1.3.3.** *(see [59]) Under the above conditions $V_\tau \stackrel{def}{=} \bar{V}_\tau = \underline{V}_\tau$ a.s. for any stopping time $\tau \in \mathcal{T}_{0N}$ and, in particular, the Dynkin's game has a value*

$$V = V_0 = \bar{V}_0 = \underline{V}_0.$$

*Furthermore, the stopping times*

$$\sigma_0 = \min\{n \leq N : V_n = Y_n\} \text{ and } \tau_0 = \min\{n \leq N : V_n = Z_n\}$$

*are optimal (saddle point), i.e., for any $\sigma, \tau \in \mathcal{T}_{0N}$,*

$$E\big(R(\sigma_0,\tau)\big) \leq E\big(R(\sigma_0,\tau_0)\big) \leq E\big(R(\sigma,\tau_0)\big).$$

*We have also the following backward recursive (dynamical programming) relation*

$$V_n = \min\big(Y_n, \max(Z_n, E(V_{n+1}|\mathcal{F}_n))\big).$$

**Proof.** It is clear that $Z_n \leq \bar{V}_n \leq Y_n$ since the first and the second players can always use the stopping times $\sigma \equiv n$ and $\tau \equiv n$, respectively. Let $\sigma \in \mathcal{T}_{nN}$. Set

$$\tilde{\sigma} = \begin{cases} \sigma & \text{if } \sigma > n \\ N & \text{if } \sigma = n \end{cases}$$

which is a stopping time. Then

$$\text{ess sup}_{\tau \in \mathcal{T}_{nN}} E(R(\sigma,\tau)|\mathcal{F}_n)$$
$$= \text{ess sup}_{\tau \in \mathcal{T}_{nN}} E(R(\tilde{\sigma},\tau)|\mathcal{F}_n)\mathbb{I}_{\{\sigma>n\}} + Y_n \mathbb{I}_{\{\sigma=n\}}.$$

Hence,

$$\text{ess sup}_{\tau \in \mathcal{T}_{nN}} E(R(\sigma,\tau)|\mathcal{F}_n) \geq \min(Y_n, \text{ess sup}_{\tau \in \mathcal{T}_{nN}} E(R(\tilde{\sigma},\tau)|\mathcal{F}_n))$$
$$\geq \min(Y_n, E(\bar{V}_{n+1}|\mathcal{F}_n))$$

where we used that $\mathcal{T}_{n+1,N} \subset \mathcal{T}_{nN}$ and that

$$E(\bar{V}_{n+1}|\mathcal{F}_n) \leq \text{ess sup}_{\tau \in \mathcal{T}_{n+1,N}} E(R(\sigma,\tau)|\mathcal{F}_n).$$

The latter follows since for any $\sigma \in \mathcal{T}_{n+1,N}$,

$$E(\text{ess sup}_{\tau \in \mathcal{T}_{n+1,N}} E(R(\sigma,\tau)|\mathcal{F}_{n+1})|\mathcal{F}_n) = \text{ess sup}_{\tau \in \mathcal{T}_{n+1,N}} E(R(\sigma,\tau)|\mathcal{F}_n)$$

which is proved similarly to the single player case by obtaining ess sup as a monotonic increasing limit and using the monotone convergence theorem, i.e.,

$$E(\lim_{k \uparrow \infty} \uparrow E(R(\sigma,\tau_k)|\mathcal{F}_{n+1})|\mathcal{F}_n) = \lim_{k \uparrow \infty} \uparrow E(R(\sigma,\tau_k)|\mathcal{F}_n).$$

Thus,

$$\bar{V}_n = \text{ess inf}_{\sigma \in \mathcal{T}_{nN}} \text{ess sup}_{\tau \in \mathcal{T}_{nN}} E(R(\sigma,\tau)|\mathcal{F}_n) \geq \min(Y_n, E(\bar{V}_{n+1}|\mathcal{F}_n)).$$

Next, define

$$\tilde{\tau} = \begin{cases} \tau & \text{if } \tau > n \\ N & \text{if } \tau = n \end{cases}$$

which is a stopping time. Then for any $\sigma \in \mathcal{T}_{nN}$,

$$E(R(\sigma,\tau)|\mathcal{F}_n) = E(R(\sigma,\tilde{\tau})|\mathcal{F}_n)\mathbb{I}_{\{\tau>n\}} + Z_n\mathbb{I}_{\{\tau=n\}}.$$

Hence for any $\sigma \in \mathcal{T}_{nN}$,

$$\text{ess sup}_{\tau \in \mathcal{T}_{nN}} E(R(\sigma,\tau)|\mathcal{F}_n) \leq \max(Z_n, \text{ess sup}_{\tau \in \mathcal{T}_{n+1,N}} E(R(\sigma,\tilde{\tau})|\mathcal{F}_n)).$$

Therefore,

$$\bar{V}_n \leq \text{ess inf}_{\sigma \in \mathcal{T}_{n+1,N}} \text{ess sup}_{\tau \in \mathcal{T}_{n,N}} E(R(\sigma,\tau)|\mathcal{F}_n)$$
$$\leq \max(Z_n, \text{ess inf}_{\sigma \in \mathcal{T}_{n+1,N}} \text{ess sup}_{\tau \in \mathcal{T}_{n+1,N}} E(R(\sigma,\tau)|\mathcal{F}_n))$$
$$= \max(Z_n, E(\bar{V}_{n+1}|\mathcal{F}_n))$$

where we used that

$$\text{ess inf}_{\sigma \in \mathcal{T}_{n+1,N}} \text{ess sup}_{\tau \in \mathcal{T}_{n+1,N}} E(R(\sigma,\tau)|\mathcal{F}_n)$$
$$= E(\text{ess inf}_{\sigma \in \mathcal{T}_{n+1,N}} \text{ess sup}_{\tau \in \mathcal{T}_{n+1,N}} E(R(\sigma,\tau)|\mathcal{F}_{n+1})|\mathcal{F}_n)$$

which is proved similarly to the above by obtaining ess sup and ess inf as monotonic limits and using the monotone convergence theorem.

It follows that

$$\bar{V}_n = \min(Y_n, \max(Z_n, E(\bar{V}_{n+1}|\mathcal{F}_n))).$$

In the same way we prove that

$$\underline{V}_n = \min(Y_n, \max(Z_n, E(\underline{V}_{n+1}|\mathcal{F}_n))).$$

On the other hand,

$$\bar{V}_N = \underline{V}_N = Y_N = Z_N,$$

and so by the backward induction we obtain that $\bar{V}_n = \underline{V}_n$ for all $n = N, N-1, N-2, ..., 1, 0$.

It remains only to show that $\sigma_0$ and $\tau_0$ is the optimal (saddle point) pair of stopping times. Indeed, we have from the above that $Z_n \leq V_n \leq Y_n$. Now, on the event $\{\sigma_0 > n\}$ we have that $V_n < Y_n$, and so on this event by above

$$V_n = \max(Z_n, E(V_{n+1}|\mathcal{F}_n)) \geq E(V_{n+1}|\mathcal{F}_n).$$

It follows that

$$E(V_{\sigma_0 \wedge (n+1)}|\mathcal{F}_n) = V_{\sigma_0}\mathbb{I}_{\{\sigma_0 \leq n\}} + E(V_{n+1}|\mathcal{F}_n)\mathbb{I}_{\{\sigma_0 > n\}}$$
$$\leq V_{\sigma_0}\mathbb{I}_{\{\sigma_0 \leq n\}} + V_n\mathbb{I}_{\{\sigma_0 > n\}} = V_{\sigma_0 \wedge n}.$$

Hence, the sequence $V_{\sigma_0 \wedge n}$ is a supermartingale, and so

$$V_0 \geq EV_{\sigma_0 \wedge \tau} \geq ER(\sigma_0, \tau).$$

The latter inequality holds true since on the event $\{\sigma_0 < \tau\}$ we have $V_{\sigma_0 \wedge \tau} = V_{\sigma_0} = Y_{\sigma_0} = R(\sigma_0, \tau)$ (where the second equality follows by the definition of $\sigma_0$) while on the event $\{\sigma_0 \geq \tau\}$ we have $V_{\sigma_0 \wedge \tau} = V_\tau \geq Z_\tau = R(\sigma_0, \tau)$ since always $V_n \geq Z_n$.

Similarly, on the event $\{\tau_0 > n\}$ we have that $V_n > Z_n$, and so on this event by above we have

$$V_n = \min(Y_n, E(V_{n+1}|\mathcal{F}_n)) \leq E(V_{n+1}|\mathcal{F}_n).$$

It follows that

$$E(V_{(n+1) \wedge \tau_0}|\mathcal{F}_n) = V_{\tau_0}\mathbb{I}_{\{\tau_0 \leq n\}} + E(V_{n+1}|\mathcal{F}_n)\mathbb{I}_{\{\tau_0 > n\}}$$
$$\geq V_{\tau_0}\mathbb{I}_{\{\tau_0 \leq n\}} + V_n\mathbb{I}_{\{\tau_0 > n\}} = V_{n \wedge \tau_0}.$$

Hence, the sequence $V_{n \wedge \tau_0}$ is a submartingale, and so

$$V_0 \leq EV_{\sigma \wedge \tau_0} \leq ER(\sigma, \tau_0).$$

The latter inequality holds true since on the event $\tau_0 \leq \sigma$ we have $V_{\sigma \wedge \tau_0} = V_{\tau_0} = Z_{\tau_0} = R(\sigma, \tau_0)$, while on the event $\{\sigma < \tau_0\}$ we have $V_{\sigma \wedge \tau_0} = V_\sigma \leq Y_\sigma = R(\sigma, \tau_0)$ since always $V_n \leq Y_n$.

□

**Remark 1.3.4.** The original setup of optimal stopping games appeared in [8] consisted of a probability space $(\Omega, \mathcal{F}, \mathbb{P})$, of a filtration of $\sigma$-algebras $\{\mathcal{F}_t\}$, $\mathcal{F}_t \subset \mathcal{F}$ with either $t \in \mathbb{N} = \{0, 1, 2, ...\}$ (discrete time case) or $t \in \mathbb{R}_+ = [0, \infty)$ (continuous time case), of $\{\mathcal{F}_t\}$-adapted payoff process $\{X_t\}$ and of a pair of $\{\mathcal{F}_t\}$-adapted 0–1 valued "permission" processes $\varphi_t^{(i)}$, $i =$

$1, 2$ such that the player $i$ is allowed to stop the game at time $t$ if and only if $\varphi_t^{(i)} = 1$. If the game is stopped at time $t$ then the first player pays to the second one the sum $X_t$. Clearly, if $\varphi_t^1 \equiv 1$ and $\varphi_t^{(2)} \equiv 0$ we arrive back at the usual (one player) optimal stopping problem. The useful generalization of this setup which is described above was suggested in [59] and it turned out to be more convenient both for further study and for applications. In this setup with two payoff adapted processes $Y_t \geq Z_t$ we can have virtual "permission" processes not by direct regulations but by a "market economy" tools. Namely, in order to accomplish this it suffices to prescribe very high payment $Y_t$ or zero payment $Z_t$ where we "forbid" to stop the game by the first player or by the second one, respectively. The continuous time version of the original setup of optimal stopping games was studied in [34]. The two payoffs continuous time setup will be discussed later in Lecture 12 following mostly [51].

## 1.4 Exercises

1) Let $X_1, X_2, \ldots$ be independent random variables such that $X_i$ has the Gaussian distribution with $EX_i = 0$ and $EX_i^2 = \sigma_i^2 > 0$ for all $i$. Let $\mathcal{F}_n = \sigma\{X_1, X_2, \ldots, X_n\}$, $\mathcal{F}_0 = \{\emptyset, \Omega\}$ (i.e., $\mathcal{F}_n$ is generated by $X_1, \ldots, X_n$ meaning that it is the smallest $\sigma$-algebra with respect to which $X_1, \ldots, X_n$ are measurable). Prove that $M_n = \exp(\sum_{k=1}^n \alpha_k X_k - \frac{1}{2} \sum_{k=1}^n \alpha_k^2 \sigma_k^2)$, $n = 1, 2, \ldots$; $M_0 = 1$ is a martingale with respect to the filtration $\{\mathcal{F}_n\}$ where $\alpha_1, \alpha_2, \ldots$ are arbitrary numbers.

2) Let $X_1, X_2, \ldots$ be independent random variables such that each $X_i$ has a symmetric distribution (i.e., $X_i$ and $-X_i$ have the same distribution) and assume that $E|X_i|^3 < \infty$ for all $i$. Prove that $L_n = (\sum_{k=1}^n X_k)^3 - 3(\sum_{k=1}^n X_k)(\sum_{k=1}^n \sigma_k^2)$, $L_0 = 0$ is a martingale with respect to the same filtration as in 1).

3) (a) What is the connection between 2) and 1)?

(b) How to generalize 1) to independent random variables with arbitrary distributions (having moment generating functions) so that a similar exponential expression will be a martingale there?

4) Let $X_1, X_2, \ldots$ be a sequence of random variables and $\mathcal{F}_n = \sigma\{X_1, X_2, \ldots, X_n\}$, $\mathcal{F}_0 = \{\emptyset, \Omega\}$. Show that $\tau_1 = \min\{n : \max_{1 \leq i \leq n} X_i \in (a, b)\}$ and $\tau_2 = \min\{n : \min_{1 \leq i \leq n} X_i \in (a, b)\}$ (where $a < b$ are numbers) are stopping times with respect to the above filtration but that $\tau_3 = \min\{n : \max_{n \leq i \leq 2n} X_i \in (a, b)\}$ is not necessarily a stopping time.

5) (Secretary problem) A manager chooses one secretary out of $N$

candidates which arrive at random. The candidates can be ranked in a linear order so that there are no equal candidates. The candidates arrive one in a day in consecutive days and once rejected they do not appear again. The manager wants to maximize the expectation of the rank of the chosen secretary. Formulate the corresponding optimal stopping problem, find the optimal stopping time and the corresponding maximal expectation of the rank.

6) (Stopping game version of the secretary problem) In addition to the manager of the company A described above, there is a manager of a rival company B who wants the manager of the company A to choose as bad a secretary as possible (with the lowest possible rank). To do this he has a possibility to stop future candidates from arriving (starting from tomorrow) for interview to the company A but he has to pay some penalty for doing that. If nobody arrives for interview then A hires the candidate who appeared last for the interview. Formulate the corresponding optimal stopping (Dynkin's) game and describe optimal stopping times of the companies A and B together with the equilibrium price of the game (expectation of the rank of a chosen candidate).

# Chapter 2

# Derivatives in General and Binomial Markets

## 2.1 Lecture 4: Derivatives in discrete time markets

### 2.1.1 *General financial market in discrete time*

We start with a probability space $(\Omega, \mathcal{F}, P)$ together with a filtration $\{\mathcal{F}_n\}_{n \geq 0}$ and with two types of securities:

    a) a bond $B = \{B_n\}_{n \geq 0}$ such that its price $B_n > 0$ at time $n \geq 1$ is a $\mathcal{F}_{n-1}$-measurable (i.e., predictable) random variable and

    b) $d$-stocks $S = (S_n^1, S_n^2, ..., S_n^d)$ where the price $S_n^i > 0$ of $i$-th stock at time $n$ is an $\mathcal{F}_n$-measurable random variable. We assume that $\mathcal{F}_0$ is the trivial $\sigma$-algebra and that $B_0, B_1$ and $S_0$ are constants.

    We write $B_n = (1 + r_n)B_{n-1}$ and $\Delta B_n = B_n - B_{n-1}$ so that $r_n = \frac{\Delta B_n}{B_{n-1}} > -1$ is $\mathcal{F}_{n-1}$-measurable. Similarly, we have $S_n^i = (1 + \rho_n^i)S_{n-1}^i$, $\Delta S_n^i = S_{n+1}^i - S_n^i$ so that $\rho_n^i = \frac{\Delta S_n^i}{S_{n-1}^i}$ is $\mathcal{F}_n$-measurable. Thus,

$$B_n = B_0 \prod_{1 \leq k \leq n} (1 + r_k), \quad S_n^i = S_0^i \prod_{1 \leq k \leq n} (1 + \rho_k^i).$$

**Definition 2.1.1.** A pair $\pi = (\beta, \gamma)$ of predictable sequences of random variables $\beta = \{\beta_n\}_{n \geq 1}$ and $\gamma = \{\gamma_n^1, ..., \gamma_n^d\}_{n \geq 1}$ is called a trading strategy (thus, $\beta_n, \gamma_n^i$ are $\mathcal{F}_{n-1}$-measurable). The value of the corresponding investment portfolio at time $n \geq 1$ is given by

$$X_n^\pi = \beta_n B_n + \sum_{i=1}^d \gamma_n^i S_n^i = \beta_n B_n + (\gamma_n, S_n)$$

where $(\cdot, \cdot)$ denotes the inner product. We assume that the investment portfolio value $X_0^\pi$ at time 0 is also defined and it is a constant. The investment portfolio (and the corresponding trading strategy) is called self-financing if

$$X^\pi_{n-1} = \beta_n B_{n-1} + \sum_{i=1}^{d} \gamma^i_n S^i_{n-1}$$

or equivalently

$$\Delta X^\pi_n = X^\pi_n - X^\pi_{n-1} = \beta_n \Delta B_n + \sum_{i=1}^{d} \gamma^i_n \Delta S^i_n = \beta_n \Delta B_n + (\gamma_n, \Delta S_n).$$

If $\pi$ is self-financing then we can write

$$\frac{X^\pi_n}{B_n} = \beta_n + \sum_{i=1}^{d} \gamma^i_n \frac{S^i_n}{B_n}$$

and

$$\Delta\left(\frac{X^\pi_n}{B_n}\right) = \frac{X^\pi_n}{B_n} - \frac{X^\pi_{n-1}}{B_{n-1}} = \sum_{i=1}^{d} \gamma^i_n \left(\frac{S^i_n}{B_n} - \frac{S^i_{n-1}}{B_{n-1}}\right)$$
$$= \sum_{i=1}^{d} \gamma^i_n \Delta \frac{S^i_n}{B_n} = (\gamma_n, \Delta \frac{S_n}{B_n}).$$

Summing this equality in $n$ we obtain

$$\frac{X^\pi_n}{B_n} = \frac{X^\pi_0}{B_0} + \sum_{k=1}^{n} (\gamma_k, \Delta \frac{S_k}{B_k}).$$

It follows from here that

**Lemma 2.1.1.** *If $\{\frac{S^i_n}{B_n}\}_{n\geq 0}$ is a martingale for each $i = 1,...,d$ (equivalently $\{\frac{S_n}{B_n}\}_{n\geq 0}$ is a d-dimensional martingale) and $\pi = (\beta_n, \gamma_n)_{n\geq 1}$ is a self-financing trading strategy then $\{\frac{X^\pi_n}{B_n}\}_{n\geq 0}$ is a local martingale.*

**Proof.** Under the conditions of the lemma $\{\frac{X^\pi_n}{B_n}\}_{n\geq 0}$ is a martingale transform of $\{\frac{S_n}{B_n}\}_{n\geq 0}$ by means of the predictable sequence $\{\gamma_n\}_{n\geq 0}$, and so the assertion follows from Proposition 1.1.1 in Lecture 1. $\square$

The ratios $\frac{S^i_n}{B_n}$ and $\frac{X^\pi_n}{B_n}$ are called discounted (or adjusted ) values of the corresponding quantities. In this case $B_n$ is called also numeraire since we measure other quantities in units of $B_n$. Thus, the above lemma says that if discounted stock prices form a martingale then discounted portfolio values (of a self-financing trading strategy) form a local martingale. Considering discounted values usually enables us to study only the case when the bond price is constant (interest rate is zero), i.e., above $r_n \equiv 0$ for all $n$. Observe that in a finite market model, i.e., when the probability $P$ is concentrated on a finitely many atoms of the sample space $\Omega$, all random variables are bounded. Hence, the discounted portfolio value $\frac{X_n^{\pi}}{B_n}$ is bounded as well, and if $\{\frac{X_n^{\pi}}{B_n}\}_{n\geq 0}$ forms a local martingale then by Proposition 1.1.1 it forms, in fact, a martingale. Later on in this lecture we will see that the same happens in a general discrete time markets for self-financing portfolio values which are relevant to pricing of derivative securities.

### 2.1.2 Derivative securities

One of the main questions of financial mathematics is the fair pricing of derivative securities, also called contingent claims, which are contracts whose monetary settlements depend on the evolution of values of underlying risky assets or indexes such as stock prices, foreign currency exchange rates, commodity prices, etc. The original use of derivatives was a protection against risk. For instance, a car company which produces cars in Europe and has expenses in euro plans to ship to USA certain number of cars in, say, three months. It sets the price in US dollars and in order to protect itself from exchange rate fluctuations it buys an appropriate derivative security which ensures that the company will have an opportunity to buy back euros for its dollars according to certain specified exchange rate. For a more detailed description of derivative securities traded in financial markets we refer the readers to [57].

We will consider three types of derivative securities with a payoff process $R$.

**Definition 2.1.2.** (i) A European contingent claim (option or other derivative) is a contract between its buyer and seller so that the latter accepts an obligation to pay an amount $R = R_N$ at a fixed time $N$ (called the expiration time or horizon) to the buyer where $R \geq 0$ is a $\mathcal{F}_N$-measurable random variable.

(ii) An American contingent claim (option or other derivative) is a contract between its buyer and a seller who accepts an obligation to pay an amount $R_n$ to the buyer provided the latter exercises the contract at time $n$. Here $\{R_n\}_{n\geq 0}$ is a nonnegative sequence adapted to the filtration $\{\mathcal{F}_n\}_{n\geq 0}$ (i.e., $R_n \geq 0$ is $\mathcal{F}_n$-measurable) and there is a horizon $N$ so that the contract is stopped automatically at time $N$ if it was not exercised before in which case the seller pays to the buyer the amount $R_N$.

(iii) An Israeli (game) contingent claim (option or other derivative) introduced in [35] is a contract between its buyer and seller so that the latter accepts an obligation to pay an amount $R(m,n) = Y_m \mathbb{I}_{m<n} + Z_n \mathbb{I}_{m \geq n}$ to the buyer provided the seller cancels the contract at time $m$ and the buyer exercises it at time $n$. Here, $\{Y_n\}_{n \geq 0}$ and $\{Z_n\}_{n \geq 0}$ are nonnegative sequences adapted to the filtration $\{\mathcal{F}_n\}_{n \geq 0}$ such that $Y_n \geq Z_n$ a.s. for all $n$. Again, there is a horizon $N$ so that the contract is stopped automatically at time $N$ if it was not cancelled or exercised before in which case the seller pays to the buyer the amount $R(N,N) = Y_N = Z_N$. The amount $\delta_n = Y_n - Z_n \geq 0$ is interpreted as a penalty the seller has to pay to the buyer for cancellation of the contract at time $n$.

**Example 2.1.** The payoffs $R_n = (K - S_n)^+$ and $R_n = (S_n - K)^+$ (where $x^+ = \max(x,0)$) correspond to put and call options, respectively. Here $K$ is called the strike price. A put (call) option enables (but not obliges) its owner to sell (buy) a stock for the price $K$ at a fixed expiration time $N$ (in the European option case) or at any time $n$ up to the expiration time (in the American option case). The amount $R_n$ is considered as a payoff by the seller of the option to its buyer since, if $K > S_n$ in the put option case, the seller of the contract has to pay to its buyer the amount $K$ for the stock while she/he can sell the stock on the market for the price $S_n$. Thus, the actual payment amounts to $K - S_n$. If $K \leq S_n$ then it does not make sense for the buyer to use the contract, and so the payment by the seller is zero. Similar arguments are applied in the call option case to justify the above payoff formula. In the Israeli option case we take $R(m,n) = (R_m + \delta_m)\mathbb{I}_{m<n} + R_n \mathbb{I}_{m \geq n}$ where $R_n$ as above and $\delta_n$ is a cancellation penalty.

European and American types of contingent claims are traded on financial markets such as Chicago Mercantile Exchange. In [42] callable convertible bonds were considered as financial derivatives of the game type. The latter is a security where the holder can convert it into a specified number of shares of an underlying asset (such as a stock) while the issuer of the bond can recall the bond paying some compensation.

**Definition 2.1.3.** (Hedging) A self-financing trading strategy $\pi = (\beta, \gamma)$ is called a hedging strategy or a hedge (and the corresponding portfolio is called a hedging portfolio) in the European and the American contingent claims cases if the portfolio value $X^\pi$ is kept sufficient a.s. to cover the payment obligation of the seller according to the corresponding contingent claim. This means:

(i) In the European contingent claim case $X_N^\pi \geq R_N$ a.s.;

(ii) In the American contingent claim case $X_n^\pi \geq R_n$ a.s. for all $n = 0, 1, ..., N$;

**Definition 2.1.4.** An investment strategy of the seller in the game contingent claim case is a pair $(\pi, \sigma)$ where $\pi = (\beta, \gamma)$ is a trading strategy and $\sigma$ is the cancellation of the contract stopping time. Such an investment strategy $(\pi, \sigma)$ is called hedging if $\pi$ is self-financing and $X_{\sigma \wedge n}^\pi \geq R(\sigma, n)$ a.s. for all $n = 0, 1, ..., N$.

**Definition 2.1.5.** The fair price $V$ of a contingent claim is the infimum of initial capitals of all hedging (self-financing) portfolios, i.e., $V = \inf\{x :$ there exists a hedging trading strategy $\pi$ (hedging investment strategy $(\pi, \sigma)$ in the game options case) such that $X_0^\pi = x\}$.

Sometimes this notion is called the fair price from seller's point of view while the fair price from buyer's point of view will be described later in Lecture 8.

**Definition 2.1.6.** A probability measure $Q$ on the probability space $(\Omega, \mathcal{F}, P)$ is called a martingale measure if it is equivalent to $P$ (both measures have the same sets of zero measure), written $Q \sim P$, and the discounted stock prices $\{\frac{S_n}{B_n}\}_{n \geq 0}$ form a martingale with respect to $Q$.

**Theorem 2.1.1.** *(Lower bound of the fair price) If $Q$ is a martingale measure then the fair price $V$ of a contingent claim with a payoff process $R$ satisfies*
*(i) in the European case:*

$$V \geq B_0 E_Q(\frac{R_N}{B_N}),$$

*where $E_Q$ denotes the expectation with respect to $Q$;*
*(ii) in the American case:*

$$V \geq \sup_{0 \leq \tau \leq N} B_0 E_Q(\frac{R_\tau}{B_\tau}),$$

*where the supremum is taken over the stopping times;*
*(iii) in the Israeli (game) case*

$$V \geq \inf_{0 \leq \sigma \leq N} \sup_{0 \leq \tau \leq N} B_0 E_Q(\frac{R(\sigma, \tau)}{B_{\sigma \wedge \tau}}),$$

*where both the infimum and the supremum are taken over the stopping times.*

**Proof.** (i) Let $\pi$ be a hedging trading strategy. Then by the lemma above $\{\frac{X_n^\pi}{B_n}\}_{n\geq 0}$ is a local martingale with respect to the measure $Q$ such that $\frac{X_N^\pi}{B_N} \geq \frac{R_N}{B_N} \geq 0$, i.e., $(\frac{X_N^\pi}{B_N})^- = 0$. It follows from Proposition 1.1.1 that $\{\frac{X_n^\pi}{B_n}\}_{n\geq 0}$ is a martingale with respect to $Q$, and so

$$\frac{X_0^\pi}{B_0} = E_Q(\frac{X_N^\pi}{B_N}) \geq E_Q(\frac{R_N}{B_N})$$

where the inequality above follows by the definition of hedging, proving (i).

(ii) Let $\pi$ be a hedging trading strategy. Then by the lemma above $\{\frac{X_n^\pi}{B_n}\}_{n\geq 0}$ is a local martingale with respect to the measure $Q$ such that $\frac{X_n^\pi}{B_n} \geq \frac{R_n}{B_n} \geq 0$ for $n = 0, 1, ..., N$. By Proposition 1.1.1 $\{\frac{X_n^\pi}{B_n}\}_{n\geq 0}$ is a martingale with respect to $Q$, and so by the optional stopping theorem for any stopping time $\tau \leq N$,

$$\frac{X_0^\pi}{B_0} = E_Q(\frac{X_\tau^\pi}{B_\tau}) \geq E_Q(\frac{R_\tau}{B_\tau})$$

where the last inequality holds true by the definition of hedging. This inequality holds true for any stopping time $\tau \leq N$, and so we can take the supremum in stopping times $\tau \leq N$ in the last expression, proving (ii).

(iii) Let $(\pi, \sigma)$ be a hedging investment strategy. Then, again, $\{\frac{X_n^\pi}{B_n}\}_{n\geq 0}$ is a local martingale with respect to the measure $Q$ such that $\frac{X_{\sigma\wedge n}^\pi}{B_{\sigma\wedge n}} \geq \frac{R(\sigma,n)}{B_{\sigma\wedge n}} \geq 0$ for $n = 0, 1, ..., N$. It follows from Proposition 1.1.1 and the optional stopping theorem that $\{\frac{X_{\sigma\wedge n}^\pi}{B_{\sigma\wedge n}}, n = 0, 1, ..., N\}$ is a martingale, and so for any stopping time $\tau \leq N$ by the optional stopping theorem and the definition of hedging

$$\frac{X_0^\pi}{B_0} = E_Q(\frac{X_{\sigma\wedge\tau}^\pi}{B_{\sigma\wedge\tau}}) \geq E_Q(\frac{R(\sigma,\tau)}{B_{\sigma\wedge\tau}}).$$

The inequality remains true after taking the supremum in $\tau$. The definition of the fair price requires to take the infimum over initial capitals which allow construction of a hedging pair $(\pi, \sigma)$, so taking the infimum in $\sigma$ we justify the assertion (iii). $\qquad\square$

## 2.2   Lecture 5: Pricing derivatives in binomial markets

### 2.2.1   *Cox–Ross–Rubinstein (CRR) (binomial) market model*

In this model the market acts on a probability space $(\Omega, \mathcal{F}, P)$ and it consists of two securities:

a) a bond with the price evolution $B_n = B_0(1+r)^n$ where $r \geq 0$ is a constant (interest rate) and

b) a stock with the price evolution $S_n = S_0 \prod_{1 \leq k \leq n}(1 + \rho_k)$ where $\rho_1, \rho_2, \ldots$ are i.i.d. random variables taking on only two values so that

$$\rho_k = \begin{cases} b & \text{with probability } p \\ a & \text{with probability } 1-p \end{cases}$$

where $0 < p < 1$. In addition, we assume that $b > r > a > -1$ since if this does not hold true then the model becomes trivial and not interesting. Indeed, if $r \geq b$ then it does not make sense to buy a stock since investing all money in a bond yields the riskless maximal profit. If $a \geq r$ then it is best to invest all money in the stock and the bond becomes useless. We usually assume also that $a < 0$ so that investing in the stock may yield both a profit and a loss. The assumption $a > -1$ means that the stock price remains positive though it may become arbitrarily small.

It is assumed also that there is a horizon $N$ so that the market is active at times $n = 0, 1, \ldots, N$. We consider the filtration $\{\mathcal{F}_n\}_{0 \leq n \leq N}$ generated by the stock prices process $\mathcal{F}_n = \sigma\{S_0, S_1, \ldots, S_n\}$ where $\mathcal{F}_0 = \{\Omega, \emptyset\}$ is the trivial $\sigma$-algebra, and so $S_0$ is a constant, while we assume that $\mathcal{F}_N = \mathcal{F}$, i.e., all our information comes from the evolution of stock prices and $\mathcal{F}_n$ is interpreted as information market participants have up to time $n$ (inclusive). Set $\varepsilon_n = (2\rho_n - a - b)(b - a)^{-1}$. Then $\varepsilon_1, \varepsilon_2, \ldots, \varepsilon_N$ are i.i.d. random variables and

$$\varepsilon_k = \begin{cases} 1 & \text{with probability } p \\ -1 & \text{with probability } 1-p. \end{cases}$$

Then $\rho_n = \frac{1}{2}(a+b) + \frac{1}{2}(b-a)\varepsilon_n$, and so each sequence $\varepsilon_1, \ldots, \varepsilon_n, \rho_1, \ldots, \rho_n$ and $S_1, \ldots, S_n$ determines each other uniquely. It follows that in order to describe randomness generated by the stock evolution it suffices to consider the product space $(\Omega, \mathcal{F}, P)$ where $\Omega = \{-1, 1\}^N = \{\omega = (\omega_1, \omega_2, \ldots, \omega_N), \omega_i = 1 \text{ or } = -1\}$ and $P = \{p, 1-p\}^N$, so that for $\omega = (\omega_1, \ldots, \omega_N)$,

$$P(\omega) = p^{\frac{1}{2}\sum_{i=1}^N(1+\omega_i)}(1-p)^{\frac{1}{2}\sum_{i=1}^N(1-\omega_i)}.$$

The $\sigma$-algebra $\mathcal{F}$ here is just the (finite) collection of all subsets of the space (of sequences) $\Omega$.

**Theorem 2.2.1.** *In the market model above there exists a unique martingale measure* $P^* = \{p^*, 1-p^*\}^N$ *where* $p^* = \frac{r-a}{b-a}$.

**Proof.** Denote by $E^*$ the expectation with respect to the probability $P^*$ then

$$E^*\left(\frac{S_n}{B_n}|\mathcal{F}_{n-1}\right) = \frac{S_{n-1}}{B_{n-1}}E^*\frac{1+\rho_n}{1+r}$$

$$= \frac{S_{n-1}}{B_{n-1}}(1+r)^{-1}(p^*(1+b)+(1-p^*)(1+a)) = \frac{S_{n-1}}{B_{n-1}}$$

where we use that $S_{n-1}$ is $\mathcal{F}_{n-1}$-measurable while $\rho_n$ is independent of the $\sigma$-algebra $\mathcal{F}_{n-1}$. Hence, the sequence $\{\frac{S_n}{B_n}, n = 0, 1, ..., N\}$ is a martingale with respect to the probability measure $P^*$, and so the latter is a martingale measure. In fact, this is true also when $N = \infty$ taking $P^*$ to be the product measure on the space of infinite sequences defined on cylinder sets determined by finite sequences. Nevertheless, we will consider usually $N < \infty$.

Next, we will prove the uniqueness. Let $Q$ be another martingale probability measure on $(\Omega, \mathcal{F})$ with $\mathcal{F} = \mathcal{F}_N$ and let $E_Q$ be the expectation with respect to $Q$. If

$$E_Q\left(\frac{S_n}{B_n}|\mathcal{F}_{n-1}\right) = \frac{S_{n-1}}{B_{n-1}}$$

then

$$E_Q\left(\frac{1+\rho_n}{1+r}|\mathcal{F}_{n-1}\right) = 1 \text{ and so } E_Q(\rho_n|\mathcal{F}_{n-1}) = r.$$

It follows that

$$Q\{\rho_n = a|\mathcal{F}_{n-1}\}a + Q\{\rho_n = b|\mathcal{F}_{n-1}\}b = r.$$

Since $Q\{\rho_n = a|\mathcal{F}_{n-1}\} + Q\{\rho_n = b|\mathcal{F}_{n-1}\} = 1$ we obtain that

$$Q\{\rho_n = b|\mathcal{F}_{n-1}\} = \frac{r-a}{b-a} \text{ and } Q\{\rho_n = a|\mathcal{F}_{n-1}\} = \frac{b-r}{b-a}.$$

Hence, we have a random variable $X = \rho_n$ such that for any Borel set $\Gamma$ the conditional probability

$$Q\{X \in \Gamma|\mathcal{G}\} = E_Q(\mathbb{I}_{\{X \in \Gamma\}}|\mathcal{G}), \ \mathcal{G} = \mathcal{F}_{n-1}$$

is constant $Q$-almost surely. Thus,

$$E_Q(\mathbb{I}_{\{X \in \Gamma\}}|\mathcal{G}) = Q\{X \in \Gamma|\mathcal{G}\} = Q\{X \in \Gamma\} = E_Q(\mathbb{I}_{\{X \in \Gamma\}}) \ \ Q\text{-a.s.},$$

and so for any $A \in \mathcal{G}$,

$$Q(A \cap \{X \in \Gamma\}) = \int_A \mathbb{I}_\Gamma(X)dQ = \int_A E_Q(\mathbb{I}_{\{X \in \Gamma\}}|\mathcal{G})dQ$$

$$= \int_A Q\{X \in \Gamma\}dQ = Q(A)Q\{X \in \Gamma\}.$$

Hence, $A$ and $\{X \in \Gamma\}$ are independent and this being true for all $A \in \mathcal{G}$ and any Borel $\Gamma$ means that $X$ is independent of the $\sigma$-algebra $\mathcal{G}$.

Applying this to our situation we conclude that $\rho_1, \rho_2, ..., \rho_N$ are independent with respect to $Q$ (since each $\mathcal{F}_k$ is generated by $\rho_1, ..., \rho_k$ by definition) and $Q\{\rho_k = b\} = \frac{b-r}{b-a} = 1 - Q\{\rho_k = a\}$ for all $k = 1, ..., N$. It follows that on the sequence space $\Omega$ described above $Q$ coincides with $P^*$. $\square$

It is easy to see with essentially the same proof that if we extend slightly our model considering independent two-valued but not identically distributed $\rho_k$, $k = 1, 2, ...$, then all results discussed here will go through. Namely, let

$$\rho_k = \begin{cases} b_k & \text{with probability } p_k \\ a_k & \text{with probability } 1 - p_k, \end{cases}$$

with $b_k > r_k > a_k > -1$, $B_n = B_0 \prod_{k=1}^{n}(1+r_k)$ and $S_n = S_0 \prod_{k=1}^{n}(1+\rho_k)$. Then setting $p_k^* = \frac{r_k - a_k}{b_k - a_k}$ we will obtain that the product measure $P^* = \prod_{k=1}^{N}\{p_k^*, 1 - p_k^*\}$ given by

$$P^*(\omega) = \prod_{k=1}^{N}(p_k^*)^{\frac{1}{2}(1+\omega_k)}(1 - p_k^*)^{\frac{1}{2}(1-\omega_k)}, \quad \omega = (\omega_1, ..., \omega_N) \in (-1,1)^N$$

is the unique martingale measure in this setup.

It will be important to understand for what follows that if we consider a multinomial instead of binomial model as there are already infinitely many martingale measures. Namely, let $\rho_1, \rho_2, ..., \rho_N$ be i.i.d. random variables such that $\rho_1 = a_j$ with probability $p_j$, $j = 1, 2, ..., m$ with $m \geq 3$, $a_1 < a_2 < ... < a_N$, $p_j \geq 0$ and $p_1 + p_2 + \cdots + p_m = 1$. Now, we consider the product (sequence) space $(\Omega, \mathcal{F}, P)$ where $\Omega = \{1, 2, ..., m\}^N = \{\omega = (\omega_1, \omega_2, ..., \omega_N)$ where each $\omega_j$ takes values $1, 2, ..., m\}$ and $P = \{p_1, p_2, ..., p_m\}^N$. Then $P$ will be a martingale measure if and only if

$$E_P \frac{1 + \rho_1}{1 + r} = 1, \text{ i.e., } E_P \rho_1 = r$$

which means that

$$\sum_{k=1}^{m} p_k a_k = r.$$

Here we have one equation for $m$ variables $p_1, ..., p_m$ to which we have to add another equation $p_1 + \cdots + p_m = 1$ and the condition $p_j \geq 0$ for all $j = 1, ..., m$. It is easy to see that if $\min_j a_j < r < \max_j a_j$ these equations have infinitely many nonnegative solutions in $p_1, ..., p_m$.

## 2.2.2 *A martingale representation lemma*

Consider again the product probability space $(\Omega, \mathcal{F}, P)$ where $\Omega = \{-1, 1\}^N = \{\omega = (\omega_1, \omega_2, ..., \omega_N), \omega_i = 1 \text{ or } = -1\}$ and $P = \{p, 1 - p\}^N$ together with the i.i.d. random variables $\varepsilon_1, \varepsilon_2, ..., \varepsilon_N$ given by $\varepsilon_k(\omega) = \omega_k$

for each $\omega = (\omega_1, \omega_2, ..., \omega_N)$ so that $\varepsilon_k = 1$ with probability $p$ and $\varepsilon_k = -1$ with probability $1 - p$. As before we consider also the filtration $\{\mathcal{F}_k\}_{0 \le k \le N}$ where $\mathcal{F}_k = \sigma\{\varepsilon_1, ..., \varepsilon_k\}$ for $k \ge 1$, $\mathcal{F}_N = \mathcal{F}$ and $\mathcal{F}_0$ is a trivial $\sigma$-algebra.

**Lemma 2.2.1.** *Let $M = \{M_n\}_{0 \le n \le N}$ be a martingale on the probability space $(\Omega, \mathcal{F}, P)$ with respect to the filtration $\{\mathcal{F}_n\}_{0 \le n \le N}$. Then there exists a unique predictable sequence $\{H_n\}_{1 \le n \le N}$ such that*

$$M_n = M_0 + \sum_{k=1}^{n} H_k(\varepsilon_k - 2p + 1) \quad for \ n = 1, 2, ..., N.$$

**Proof.** Since $M_n$ is $\mathcal{F}_n$-measurable and the latter is generated by $\varepsilon_1, ..., \varepsilon_n$ then $M_n(\omega) = f_n(\varepsilon_1(\omega), ..., \varepsilon_n(\omega)) = f_n(\omega_1, ..., \omega_n)$ for some function $f_n : \{-1, 1\}^n \to \mathbb{R}$. (This holds true also in general, i.e., if $X$ is a random vector and $Y$ is a random variable measurable with respect to the $\sigma$-algebra generated by $X$, then $Y = f(X)$ for some Borel function $f$.)

Now, $M_k, k \ge 0$ is a martingale, and so

$$0 = E(M_n - M_{n-1}|\mathcal{F}_{n-1})(\omega)$$
$$= pf_n(\omega_1, ..., \omega_{n-1}, 1) + (1 - p)f_n(\omega_1, ..., \omega_{n-1}, -1) - f_{n-1}(\omega_1, ..., \omega_{n-1})$$

where we used that $\varepsilon_n$ is independent of $\mathcal{F}_{n-1}$ while $\varepsilon_1, ..., \varepsilon_{n-1}$ are measurable with respect to it. In view of this equality we can define

$$H_n(\omega) = \frac{f_n(\omega_1, ..., \omega_{n-1}, 1) - f_{n-1}(\omega_1, ..., \omega_{n-1})}{2(1-p)}$$
$$= \frac{f_{n-1}(\omega_1, ..., \omega_{n-1}) - f_n(\omega_1, ..., \omega_{n-1}, -1)}{2p}.$$

Then $H_n$ is $\mathcal{F}_{n-1}$-measurable and we have to show that

$$M_n = M_0 + \sum_{k=1}^{n} H_k(\varepsilon_k - 2p + 1).$$

This equality holds true trivially for $n = 0$. Suppose that it holds true for all $n$ up to $m - 1$. Then

$$M_m(\omega) = f_m(\omega_1, ..., \omega_m)$$

$$= \begin{cases} \dfrac{f_m(\omega_1, ..., \omega_{m-1}, 1) - f_{m-1}(\omega_1, ..., \omega_{m-1})}{2(1-p)}(\omega_m - 2p + 1) \\ + f_{m-1}(\omega_1, ..., \omega_{m-1}) \quad \text{if } \omega_m = 1 \\ -\dfrac{f_m(\omega_1, ..., \omega_{m-1}, -1) - f_{m-1}(\omega_1, ..., \omega_{m-1})}{2p}(\omega_m - 2p + 1) \\ + f_{m-1}(\omega_1, ..., \omega_{m-1}) \quad \text{if } \omega_m = -1. \end{cases}$$

$$= H_m(\omega)(\varepsilon_m(\omega) - 2p + 1) + M_{m-1}(\omega)$$

completing the proof by induction.

It remains only to establish the uniqueness. Suppose that the representation holds true also for a predictable sequence $H'_k, k = 1, ..., N$, then

$$\sum_{k=1}^{n}(H_k - H'_k)(\varepsilon_k - 2p + 1) = 0.$$

Thus

$$0 = E\Big(\sum_{k=1}^{n}(H_k - H'_k)(\varepsilon_k - 2p + 1)|\mathcal{F}_1\Big) = (H_1 - H'_1)(\varepsilon_1 - 2p + 1)$$

implying that $H_1 = H'_1$ since $\varepsilon_1 - 2p + 1 \neq 0$ and we used that $E\big((H_k - H'_k)(\varepsilon_k - 2p + 1)|\mathcal{F}_{k-1}\big) = 0$ for each $k$. Now assume that $H_k = H'_k$ for all $k \leq m - 1$. Then

$$0 = E\Big(\sum_{k=m}^{n}(H_k - H'_k)(\varepsilon_k - 2p + 1)|\mathcal{F}_m\Big) = (H_m - H'_m)(\varepsilon_m - 2p + 1)$$

implying that $H_m = H'_m$ completing the proof by induction. □

### 2.2.3 *Fair price of options in CRR market*

**Theorem 2.2.2.** *Let $P^* = \{p^*, 1 - p*\}^N$, $p^* = \frac{r-a}{b-a}$ be the martingale measure in the CRR market and denote by $E^*$ the expectation with respect to $P^*$. Then the fair price $V$ of a contingent claim with a payoff process $R$ is given by*

*(i) in the European case:*

$$V = B_0 E^*\Big(\frac{R_N}{B_N}\Big);$$

*(ii) in the American case:*

$$V = \sup_{0 \leq \tau \leq N} B_0 E^*\Big(\frac{R_\tau}{B_\tau}\Big),$$

*where the supremum is taken over the stopping times;*

*(iii) in the Israeli (game) case:*

$$V = \inf_{0 \leq \sigma \leq N} \sup_{0 \leq \tau \leq N} B_0 E^*\Big(\frac{R(\sigma, \tau)}{B_{\sigma \wedge \tau}}\Big),$$

*where both the infimum and the supremum are taken over the stopping times. (In fact, we can write* max *and* min *in (ii) and (iii) since there exist only finitely many stopping times taking on values between 0 and N on the finite sample space $\Omega$.)*

**Proof.** (i) We already obtained the appropriate lower bound even in the case of a more general discrete time market, and so it remains only to show that there exists a hedging self-financing portfolio strategy with the initial capital

$$x^* = B_0 E^*(\frac{R_N}{B_N}).$$

Introduce the martingale

$$M_n = E^*(B_0 \frac{R_N}{B_N}|\mathcal{F}_n),\ n = 0, 1, 2, ....$$

Since $\varepsilon_k - 2p^* + 1 = 2(\rho_k - r)(b - a)^{-1}$ we obtain by the martingale representation lemma for the CRR market that

$$M_n = M_0 + \sum_{k=1}^{n}(1+r)^{-k}\gamma_k S_{k-1}\frac{1}{2}(b-a)(\varepsilon_k - 2p^* + 1)$$
$$= M_0 + \sum_{k=1}^{n}(1+r)^{-k}\gamma_k S_{k-1}(\rho_k - r)$$

where $(1+r)^{-k} = B_0 B_k^{-1}$ and $\gamma_k$, $k \geq 1$ is a predictable sequence obtained from the martingale representation lemma so that $\gamma_k = 2H_k(1+r)^k((b-a)S_{k-1})^{-1}$. Set $X_n = (1+r)^n M_n$ and $\beta_n = (X_{n-1} - \gamma_n S_{n-1})B_{n-1}^{-1}$ so that

$$X_{n-1} = \beta_n B_{n-1} + \gamma_n S_{n-1}.$$

Observe that $(\beta_n, \gamma_n)$, $n \geq 1$ is a predictable sequence and in order to prove that it is a self-financing portfolio strategy we have to show that

$$X_n = \beta_n B_n + \gamma_n S_n,\ n = 1, 2, ....$$

By the martingale representation and the formula for $X_{n-1}$ we obtain

$$X_n = (1+r)^n M_n = (1+r)^n M_{n-1} + \gamma_n S_{n-1}(\rho_n - r)$$
$$= (1+r)X_{n-1} + \gamma_n S_{n-1}(\rho_n - r) = (1+r)(\beta_n B_{n-1} + \gamma_n S_{n-1})$$
$$+\gamma_n S_{n-1}(\rho_n - r) = \beta_n B_n + \gamma_n(1 + \rho_n)S_{n-1} = \beta_n B_n + \gamma_n S_n,$$

and so $\pi = (\beta_n, \gamma_n)$, $n \geq 1$ is a self-financing portfolio strategy. Finally, $X_0 = M_0 = x^*$ and $X_N = (1+r)^N = R_N$, so the strategy $\pi$ with the initial capital $x^*$ is hedging. It follows that the fair price $V$ should not be bigger than $x^*$ which together with the estimate in the other direction obtained earlier yields that $V = x^*$.

(ii) In the American contingent claim case we define

$$U_n = \max_{n \leq \tau \leq N} E^*(\frac{B_0}{B_\tau}R_\tau|\mathcal{F}_n)$$

(we can take max here since there are only finitely many stopping times between 0 and $N$ on a finite probability space $\Omega$). As we proved it in the optimal stopping section the sequence $\{U_n, \, n = 0, 1, ..., N\}$ is a super-martingale and by the Doob supermartingale decomposition theorem

$$U_n = M_n - A_n, \ n = 0, 1, ..., N$$

where $M_n$, $n \geq 0$ is a martingale, $M_0 = U_0$ and $A_n$, $n \geq 0$, $A_0 = 0$ is a non-decreasing predictable process.

Again, we use the martingale representation lemma for the CRR market to obtain

$$M_n = M_0 + \sum_{k=1}^{n}(1+r)^{-k}\gamma_k S_{k-1}\tfrac{1}{2}(b-a)(\varepsilon_k - 2p^* + 1)$$
$$= M_0 + \sum_{k=1}^{n}(1+r)^{-k}\gamma_k S_{k-1}(\rho_k - r)$$

where $\gamma_k$, $k \geq 1$ is a predictable sequence. We set again $X_n = (1+r)^n M_n$ and $\beta_n = (X_{n-1} - \gamma_n S_{n-1})B_{n-1}^{-1}$ so that

$$X_{n-1} = \beta_n B_{n-1} + \gamma_n S_{n-1}.$$

In the same way as in (i) we see that $\pi = (\beta_n, \gamma_n)$, $n \geq 1$ is a self-financing strategy, i.e., that we also have here $X_n = \beta_n B_n + \gamma_n S_n$, $n = 1, 2, ....$ Now, $X_0 = M_0 = U_0 = V$ and

$$X_n = (1+r)^n M_n = (1+r)^n (U_n + A_n) \geq (1+r)^n U_n \geq R_n$$

where the last inequality follows since $U_n$ is a supremum over all stopping times $\tau$ greater or equal $n$ and we can always take $\tau \equiv n$. Hence, $\pi$ is hedging with the initial capital $V$ and we conclude again that the fair price should not be bigger than $V$ which together with the estimate in the other direction obtained earlier yields that, in fact, it equals $V$.

(iii) In the game contingent claim case let the payoff function be given by

$$R(m,n) = Y_m \mathbb{I}_{m<n} + Z_n \mathbb{I}_{m \geq n}$$

where $Y_n \geq Z_n$ are processes adapted to the filtration $\{\mathcal{F}_n, \, n \geq 0\}$. Now, we fix a stopping time $\sigma$ and define $Q_k^\sigma = B_0 B_{\sigma \wedge k}^{-1} R(\sigma, k)$, $k = 0, 1, ..., N$ and

$$U_n^\sigma = \max_{n \leq \tau \leq N} E^*(Q_\tau^\sigma | \mathcal{F}_n).$$

Observe that

$$B_{\sigma \wedge k}^{-1} R(\sigma, k) = B_k^{-1} Z_k \mathbb{I}_{\sigma \geq k} + \sum_{l=0}^{k-1} B_l^{-1} Y_l \mathbb{I}_{\sigma = l},$$

and so $Q_k^\sigma$ is $\mathcal{F}_k$-measurable. As we proved it in the optimal stopping section the sequence $\{U_n^\sigma,\, n = 0, 1, ..., N\}$ is a supermartingale and by the Doob supermartingale decomposition theorem

$$U_n^\sigma = M_n^\sigma - A_n^\sigma,\; n = 0, 1, ..., N$$

where $M_n^\sigma$, $n \geq 0$ is a martingale, $M_0^\sigma = U_0^\sigma$ and $A_n^\sigma$, $n \geq 0$, $A_0^\sigma = 0$ is a non-decreasing predictable process.

Again, we use the martingale representation lemma for the CRR market to obtain

$$M_n^\sigma = M_0^\sigma + \sum_{k=1}^n (1+r)^{-k} \gamma_k^\sigma S_{k-1} \tfrac{1}{2}(b-a)(\varepsilon_k - 2p^* + 1)$$
$$= M_0^\sigma + \sum_{k=1}^n (1+r)^{-k} \gamma_k^\sigma S_{k-1}(\rho_k - r)$$

where $\gamma_k^\sigma$, $k \geq 1$ is a predictable sequence. We set again $X_n^\sigma = (1+r)^n M_n^\sigma$ and $\beta_n^\sigma = (X_{n-1}^\sigma - \gamma_n^\sigma S_{n-1})B_{n-1}^{-1}$ so that

$$X_{n-1}^\sigma = \beta_n^\sigma B_{n-1} + \gamma_n^\sigma S_{n-1}.$$

In the same way as in (i) we see that $\pi^\sigma = (\beta_n^\sigma, \gamma_n^\sigma)$, $n \geq 1$ is a self-financing strategy, i.e., that we also have here $X_n^\sigma = \beta_n^\sigma B_n + \gamma_n^\sigma S_n$, $n = 1, 2, ....$ Now,

$$X_n^\sigma = (1+r)^n M_n^\sigma = (1+r)^n (U_n^\sigma + A_n^\sigma) \geq (1+r)^n U_n^\sigma \geq R(\sigma, n),$$

and so $\pi^\sigma$ is a hedging strategy with the initial capital

$$X_0^\sigma = M_0^\sigma = U_0^\sigma = \max_{0 \leq \tau \leq N} E^*(Z_\tau^\sigma).$$

Take $\sigma^* = \min\{n : Y_n(1+r)^{-n} = V_n\}$ where

$$V_n = \min_{n \leq \sigma \leq N} \max_{n \leq \tau \leq N} E^*(Q_\tau^\sigma | \mathcal{F}_n).$$

It follows from the above formulas and the theorem about Dynkin's games proved in Lecture 3 that $X_0^{\sigma^*} = V$, thus $\pi^{\sigma^*}$ is a hedging strategy with the initial capital $V$. Relying on the same concluding argument as in (i) and (ii) we complete the proof of (iii) and of the whole theorem. $\quad\square$

From the backward induction (dynamical programming) formulas derived for the single player and the game versions of the optimal stopping problems in Lecture 3 we obtain the following.

**Corollary 2.2.1.** *(i) The fair price $V$ of an American contingent claim can be obtained by the backward induction as $V = V_0$ where $V_N = \frac{B_0 R_N}{B_N}$ and*

$$V_n = \max(B_0 \frac{R_n}{B_n}, E^*(V_{n+1} | \mathcal{F}_n))$$

*for* $n = N - 1, N - 2, ..., 1, 0$;

(ii) The fair price $V$ of a game contingent claim can be obtained by the backward induction as $V = V_0$ where $V_N = \frac{B_0 Y_N}{B_N}$ and

$$V_n = \min \left( B_0 \frac{Y_n}{B_n}, \ \max(B_0 \frac{Z_n}{B_n}, \ E^*(V_{n+1}|\mathcal{F}_n))) \right)$$

for $n = N - 1, N - 2, ..., 1, 0$ where $R(m, n) = Y_m \mathbb{I}_{m<n} + Z_n \mathbb{I}_{m\geq n}$.

## 2.3 Exercises

1) Consider an American contingent claim in the CRR binomial market with the horizon $N = 3$, the constant bond price $B_n \equiv 1$ and the stock price given by $S_n = \prod_{i=1}^{n}(1+\rho_i)$, $n = 1, 2, ...$ and $S_0 = 1$. Here $\rho_1, \rho_2, ...$ are i.i.d. random variables such that $\rho_i = 1/2$ or $\rho_i = -1/2$ both with probability $1/2$. The payoff process is given by $R_n = (1 + \rho_n)^2$, $n = 1, 2, 3$ and $R_0 = 1$. Find the fair price $V$ of this contingent claim, a hedging portfolio strategy of the seller with the initial capital $V$ and an optimal exercise time of the buyer.

2) Consider the following example of a game contingent claim in the CRR binomial market model where we assume that the horizon $N = 3$, the bond and the stock prices at time $n$ are given by $B_n = 2^n$ and $S_n = \prod_{i=1}^{n}(1 + \rho_i)$ respectively, where $\rho_1, \rho_2, ...$ are i.i.d. random variables such that $\rho_i = 2$ with probability $1/4$ and $\rho_i = -\frac{1}{2}$ with probability $3/4$. The payoff function is given by

$$R(m, n) = Y_m \mathbb{I}_{m<n} + Z_n \mathbb{I}_{n\leq m}$$

where $Y_n = Z_n + 1$ and $Z_n = \rho_n^2$. Find the fair price of this contingent claim and an optimal (rational) investment strategy of the seller and an optimal exercise time of the buyer.

# Chapter 3

# Fundamental Theorems of Asset Pricing

## 3.1 Lecture 6: Arbitrage and completeness

### 3.1.1 *Arbitrage*

We return to the general discrete time financial market described in Lecture 4 which consists of a probability space with a filtration $(\Omega, \mathcal{F}, \{\mathcal{F}_n\}_{n\geq 0}, P)$ and of $d$-stocks $S = (S^1, ..., S^d)$, $S^i = (S^i_n)_{n\geq 0}$ and a bond $B = (B_n)_{n\geq 0}$ which will be called a general $(B, S)$ financial market. Here

$$B_n = B_0 \prod_{1\leq k\leq n} (1 + r_k) \text{ and } S^i_n = S^i_0 \prod_{1\leq k\leq n} (1 + \rho^i_k)$$

with a predictable sequence $r_k \geq 0$, $k = 1, 2, ...$ and adapted sequences $\rho^i_k$, $-1 < \rho^i_k$, $k = 1, 2, ...$ We recall that a pair $\pi = (\beta, \gamma)$ of predictable sequences of random variables $\beta = \{\beta_n\}_{n\geq 0}$ and $\gamma = \{\gamma^1_n, ..., \gamma^d_n\}_{n\geq 0}$ is called a self-financing trading strategy if

$$\Delta X^\pi_n = X^\pi_n - X^\pi_{n-1} = \beta_n \Delta B_n + \sum_{i=1}^{d} \gamma^i_n \Delta S^i_n = \beta_n \Delta B_n + (\gamma_n, \Delta S_n),$$

where $X^\pi_n = \beta_n B_n + \sum_{i=1}^{d} \gamma^i_n S^i_n = \beta_n B_n + (\gamma_n, S_n)$ is the portfolio value at time $n$ corresponding to the strategy $\pi$.

**Definition 3.1.1.** (arbitrage) A self-financing trading strategy $\pi$ provides an arbitrage opportunity at the time $N$ if $X^\pi_0 = 0$, $X^\pi_N \geq 0$ a.s. and $P\{X^\pi_N > 0\} > 0$ (equivalently, $EX^\pi_N > 0$). A financial market has no arbitrage opportunities
    1) if $X^\pi_0 = 0$ and $X^\pi_N \geq 0$ implies that $X^\pi_N = 0$ a.s.;
    2) (in the weak sense) if $X^\pi_0 = 0$ and $X^\pi_n \geq 0$ for all $n = 1, 2, ..., N$ implies that $X^\pi_N = 0$ a.s.;
    3) (in the strong sense) if $X^\pi_0 = 0$ and $X^\pi_N \geq 0$ implies that $X^\pi_n = 0$ a.s. for all $n = 1, 2, ..., N$.

**Remark 3.1.1.** For studying arbitrage related issues we are talking about events of the form $\{X_n^\pi > 0\}$, $\{X_n^\pi \geq 0\}$ and $\{X_n^\pi = 0\}$, and so we can deal instead with adjusted (discounted) quantities $\tilde{X}_n^\pi = \frac{X_n^\pi}{B_n}$, $\tilde{S}_n = \frac{S_n}{B_n}$ and $\tilde{B}_n = 1$ (i.e., in fact, we can assume that the bond interest rate is zero).

**Theorem 3.1.1.** *(First fundamental theorem of asset pricing) The general $(B, S)$ financial market defined above has no arbitrage opportunities (in the sense of 1)) if and only if there exists a martingale measure (i.e., a measure $\tilde{P}$ equivalent to $P$ such that $\tilde{S}_n = \frac{S_n}{B_n}$, $n \geq 0$ is a martingale with respect to it).*

**Proof.** a) Suppose that a martingale measure $\tilde{P}$ exists, $\pi$ is a self-financing strategy and $X_n^\pi$ is the value of the corresponding portfolio. Then by Proposition 1.1.1 from Lecture 1, $\tilde{X}_n^\pi = \frac{X_n^\pi}{B_n}$, $n \geq 0$ is a local martingale with respect to $\tilde{P}$. If $\tilde{X}_N^\pi \geq 0$ a.s. then by Proposition 1.1.1 $\tilde{X}_n^\pi = \frac{X_n^\pi}{B_n}$, $n = 0, 1, ..., N$ is a martingale. Hence, if $X_0^\pi = 0$ and $\tilde{X}_N^\pi \geq 0$ a.s. then $0 = E\tilde{X}_0^\pi = E\tilde{X}_N^\pi$, and so $\tilde{X}_N^\pi = 0$ a.s., i.e., there are no arbitrage opportunities.

b) The other direction: "no arbitrage opportunities implies the existence of a martingale measure", is more difficult to prove. The proof here will follow mostly [44] (see also [10], [23] and [66]). We can assume here that the initial value $X_0^\pi$ of the portfolio is zero and then the adjusted value of the portfolio $\tilde{X}_n^\pi = \frac{X_n^\pi}{B_n}$ at time $n$ is given by the formula

$$\tilde{X}_n^\pi = \sum_{k=1}^{n} (\gamma_k, \Delta \tilde{S}_k)$$

where $\pi = (\beta_k, \gamma_k)_{k=1}^N$ is a self-financing trading strategy and $\tilde{S}_k = \frac{S_k}{B_k}$. No arbitrage means that if $\tilde{X}_N^\pi \geq 0$ a.s. then $\tilde{X}_N^\pi = 0$ a.s.

Let $M_N$ be the set of all random variables $\xi$ having the form

$$\xi = \sum_{k=1}^{N} (\gamma_k, \Delta \tilde{S}_k)$$

for all predictable $d$-dimensional sequences $\gamma_k$, $k = 1, ..., N$ and denote by $L_+^0$ the set of all nonnegative $\mathcal{F}_N$-measurable random variables. Then the no arbitrage assumption is equivalent to

$$M_N \cap L_+^0 = \{0\}. \tag{3.1.1}$$

Indeed, clearly, (3.1.1) implies that there is no arbitrage. In the other direction, if there is no arbitrage then there exists no self-financing trading strategy $\pi = (\beta_k, \gamma_k)_{k=1}^N$ such that $\tilde{X}_N^\pi \geq 0$ and $P\{\tilde{X}_N^\pi > 0\} > 0$ since, recall, $\tilde{X}_0^\pi = 0$. But if (3.1.1) does not hold true then there exists a predictable sequence $\gamma_k$, $k = 1, ..., N$ such that

$$\xi = \sum_{k=1}^N (\gamma_k, \Delta\tilde{S}_k) \geq 0 \text{ and } P\{\xi > 0\} > 0.$$

Now define $\beta_1 = -(\gamma_1, \tilde{S}_0)$ and recursively,

$$\beta_{k+1} = \beta_k + (\gamma_k, \tilde{S}_k) - (\gamma_{k+1}, \tilde{S}_k), \ k = 1, ..., N - 1.$$

Then $\pi = (\beta_k, \gamma_k)_{k=1}^N$ is a self-financing trading strategy and $\tilde{X}_N^\pi = \xi$ which contradicts the no arbitrage assumption.

Set

$$K_N = M_N - L_+^0 = \{\eta = \xi - \zeta : \xi \in M_N, \zeta \in L_+^0\}$$

which is a convex cone. Then (3.1.1) is equivalent to

$$K_N \cap L_+^0 = \{0\}.$$

We will show in Appendix to this lecture that $K_N$ is closed with respect to the convergence in probability which is the same as to be closed with respect to the almost sure convergence since any convergent in probability sequence has a subsequence converging almost surely.

Set $K_N^1 = K_N \cap L^1$ where $L^1 = L^1(\Omega, \mathcal{F}_N, P)$ is the space of all integrable $\mathcal{F}_N$-measurable random variables. Then $K_N^1$ is a convex cone in $L^1$. Since, as will be shown in Appendix, $K_N$ is closed with respect to the convergence in probability, it is also closed with respect to the stronger convergence in $L^1$, and so $K_N^1$ is closed with respect to the latter convergence. Then relying on the Hahn–Banach theorem and related functional analysis results we will derive in Appendix that any $x \neq 0$ in the set $L_+^1 = L_+^1(\Omega, \mathcal{F}_N, P)$ of all nonnegative integrable $\mathcal{F}_N$-measurable random variables can be separated from $K_N^1$ in the sense that for any $x \in L_+^1$, $x \neq 0$ there exists $z_x \in L^\infty = L^\infty(\Omega, \mathcal{F}_N, P)$ such that

$$E z_x \xi < E z_x x \tag{3.1.2}$$

for all $\xi \in K_N^1$.

Relying on the above assertion we observe that $E z_x x > 0$ for any $x \in L_+^1$, $x \neq 0$ since $0 \in K_N^1$, and since $K_N^1$ is a cone, i.e., (3.1.2) being true for some $\xi \in K_N^1$ must remain true for any $a\xi$ where $a \geq 0$ is a constant, we obtain that $E z_x \xi \leq 0$. Moreover, since $K_N^1$ contains all integrable negative random variables, $z_x \geq 0$ a.s. as for otherwise we could take a random variable $\xi$, which is negative precisely where $z_x$ is negative being zero elsewhere, which would give $E z_x \xi > 0$ leading to the contradiction. By a version of the Halmos-Savage theorem, whose proof is provided below, the family of measures $\{z_x P\}$ contains a countable equivalent subfamily $\{z_{x_i} P, \ i \in \mathbb{N}\}$ (i.e., both families have the same null sets). Set $\rho = \sum_i 2^{-i} \frac{z_{x_i}}{\|z_{x_i}\|_\infty}$ and $\tilde{x} = \mathbb{I}_{\{\rho = 0\}}$. Then $E z_{x_i} \tilde{x} = 0$ for all $i$, and so $E z_x \tilde{x} = 0$ for all $x \in L_+^1$. Hence, $\tilde{x} = 0$ a.s. since for otherwise, $E z_{\tilde{x}} \tilde{x} > 0$ by the above. It follows that $\rho > 0$ a.s. and the probability measure $\tilde{P} = c\rho P$ with $c = 1/E\rho$ is equivalent to $P$, $\frac{d\tilde{P}}{dP} \in L^\infty$ and $E_{\tilde{P}} \xi \leq 0$ for all $\xi \in K_N$.

Replacing $P$ if necessary, by the equivalent probability

$$P' = C e^{-\sum_{1 \leq k \leq N} |S_k|} P$$

we can assume without loss of generality that $S_1, S_2, ..., S_N$ are $P$-integrable, and so they are also $\tilde{P}$-integrable. Taking $\xi = \xi_k = \pm (\gamma_k, \Delta \tilde{S}_k)$ with bounded predictable $\gamma_1, ..., \gamma_N$ we conclude that $E_{\tilde{P}} \xi \leq 0$ implies that, in fact, $E_{\tilde{P}} \xi = 0$, and so

$$0 = E_{\tilde{P}} (\gamma_k, E(\Delta \tilde{S}_k | \mathcal{F}_{k-1})).$$

Since this holds true for any bounded $\mathcal{F}_{k-1}$-measurable $d$-dimensional random vector $\gamma_k$ then $E_{\tilde{P}}(\Delta \tilde{S}_k | \mathcal{F}_{k-1}) = 0$ a.s. Thus, $\tilde{P}$ is a martingale measure.

It remains to provide the Halmos-Savage theorem argument while the more technically heavy closedness and separation arguments needed for (3.1.2) we leave for Appendix. Consider the family $\{yP\}$ where $y$'s are finite convex combinations of $z_x$'s. As always, ess $\sup_y \mathbb{I}_{\{y > 0\}}$ can be attained on a sequence $\mathbb{I}_{\{\tilde{y}_k > 0\}}$, $k = 1, 2, ...$ and taking $y_k = \frac{1}{k} \sum_{i=1}^k \tilde{y}_i$ we obtain that this essential supremum is attained on the increasing sequence $\mathbb{I}_{\{y_k > 0\}}$, $k = 1, 2, ...$ Clearly, $\{y_k P\}$ is a countable equivalent subfamily of $\{yP\}$ and each $y_k$ has the form $y_k = m_k^{-1} \sum_{j=1}^{l_k} q_{j,k} z_{x_{j,k}}$ where $x_{j,k}$ are all different, $q_{j,k} \geq 0$ and $\sum_{j=1}^{l_k} q_{j,k} = m_k$. Now, the countable family $\{z_{x_{j,k}} P\}$ is an equivalent subfamily of $\{z_x P\}$, as required. This completes the proof of both Halmos-Savage theorem and the first fundamental theorem of asset pricing, remembering that the proof of the separation assertion (3.1.2) will be given in Appendix. $\qquad \square$

### 3.1.2   *Counterexamples*

We proved the first fundamental theorems of asset pricing assuming that the market contains finitely many securities (stocks) $d < \infty$ and that it acts during a limited time, i.e., that the horizon is finite $N < \infty$. It turns out that if $d = \infty$ (infinitely many stocks) or when the horizon is infinite $N = \infty$ then there exist markets without arbitrage opportunities and without martingale measures. If $N = \infty$ (infinite horizon, perpetual market) then there also exists a market where arbitrage opportunities coexist with martingale measures.

The counterexample with infinitely many stocks which appeared in [66] (and modified in [68]) consists of the probability space $(\Omega, \mathcal{F}, P)$ where $\Omega = \{1, 2, 3, ...\}$, $P = \sum_{k=1}^{\infty} 2^{-k} \delta_k$, i.e., $P\{k\} = 2^{-k}$, and $\mathcal{F}$ is the $\sigma$-algebra of all subsets of $\Omega$. Let $\mathcal{F}_0 = \{\emptyset, \Omega\}$ be the trivial $\sigma$-algebra, $\mathcal{F}_1 = \mathcal{F}$ and assume that the market acts only one unit of time, i.e., the horizon $N = 1$. Let the prices of the bond be constant $B_0 = B_1 = 1$ and the prices of infinitely many stocks $S_n = (S_n^1, S_n^2, S_n^3, ...)$, $n = 0, 1$ be given by $S_0^i = 1$ for all $i = 1, 2, 3, ...$ and

$$\Delta S_1^i(\omega) = S_1^i(\omega) - 1 = \begin{cases} 1 & \text{if } \omega = i \\ -1 & \text{if } \omega = i + 1 \\ 0 & \text{otherwise.} \end{cases}$$

Then the value of a portfolio $X^\pi$ at the time 1 can be represented in the form

$$X_1^\pi = c_0 + \sum_{i=1}^{\infty} c_i S_1^i = X_0^\pi + \sum_{i=1}^{\infty} c_i \Delta S_1^i$$

where $X_0^\pi = c_0 + \sum_{i=1}^{\infty} c_i$ is the value of the portfolio at time 0 which keeps the self-financing condition and we assume that $\sum_i |c_i| < \infty$. Then, if $X_0^\pi = c_0 + \sum_{i=1}^{\infty} c_i = 0$ and $X_1^\pi \geq 0$, we obtain $X_1^\pi(1) = c_1 \geq 0$ and $X_1^\pi(i) = c_i - c_{i-1} \geq 0$ for all $i \geq 2$, which yields that $c_i = 0$ for all $i \geq 1$. So $X_1^\pi = 0$ which means that the market is without arbitrage. On the other hand, if $\tilde{P} \sim P$ is a probability measure and $S$ is a martingale with respect to $\tilde{P}$ then we must have $E_{\tilde{P}} \Delta S_1^i = 0$ for all $i \geq 1$, which yields that $\tilde{P}\{i\} = \tilde{P}\{i + 1\}$ for $i = 1, 2, ...$ which is impossible. So there are no martingale measures for this market.

In the perpetual market, i.e., $N = \infty$, with stock prices $S_n$, $n \geq 0$ the natural "no arbitrage" condition should say that a sure gain is impossible on each step. Namely, if $X_n^\pi = X_0^\pi + \sum_{k=1}^{n} \gamma_k \Delta S_k$ is a value of a self-financing portfolio at time $n$, where $\{\gamma_k, k \geq 1\}$ is a predictable sequence, then $X_n^\pi - X_{n-1}^\pi = \gamma_n \Delta S_n \geq 0$ a.s. should imply that $\gamma_n \Delta S_n = 0$ a.s. Next, we provide an example (similar to the one appeared in [3]) showing that when $N = \infty$ this condition does not guarantee the existence of a martingale measure. Namely, consider the perpetual binomial market where the interest rate $r = 0$ and the stock price at time $n$ is equal to $S_n = S_0 \prod_{1 \leq k \leq n} (1 + \rho_k)$ where $\rho_1, \rho_2, \ldots$ are i.i.d. random variables such that $P\{\rho_k = a\} = 1 - P\{\rho_k = -a\} = p$ where $0 < a < 1$, $0 < p < 1$, $p \neq \frac{1}{2}$ and, as before, $\mathcal{F}_n = \sigma\{\rho_1, \ldots, \rho_n\}$. We showed in Lecture 5 that there exists a unique probability measure $P^*$ such that $S_n$, $n \geq 0$ becomes a martingale with respect to it and $\rho_1, \rho_2, \ldots$ becomes an i.i.d. sequence of random variables under $P^*$ such that $P^*\{\rho_k = a\} = 1 - P\{\rho_k = -a\} = \frac{1}{2}$. Set $\Gamma = \{\omega : \lim_{n \to \infty} \frac{1}{n} \sum_{1 \leq k \leq n} \rho_k(\omega) = 0\}$. Then by the strong law of large numbers $P^*(\Gamma) = 1$ while $P(\Gamma) = 0$ since $\lim_{n \to \infty} \frac{1}{n} \sum_{1 \leq k \leq n} \rho_k = a(2p - 1) \neq 0$ $P$-a.s. Thus $P$ and $P^*$ are singular, and so there is no martingale measure in this market. Nevertheless, the above "no arbitrage" condition holds true here. Indeed, if $\gamma_n$ is an $\mathcal{F}_{n-1}$-measurable random variable such that $\gamma_n \Delta S_n = \gamma_n S_{n-1} \rho_n \geq 0$ $P$-a.s. then we must have $\gamma_n = 0$ $P$-a.s. since $\gamma_n$ and $\rho_n$ are independent, and so

$$0 = P\{\gamma_n S_{n-1} \rho_n < 0\} = pP\{\gamma_n < 0\} + (1 - p)P\{\gamma_n > 0\}$$

which means that $P\{\gamma_n < 0\} = P\{\gamma_n > 0\} = 0$.

Next, we describe a counterexample with the infinite horizon $N = \infty$ as appeared in [68] where a martingale measure coexist with an arbitrage. Let $\xi_1, \xi_2, \ldots$ be a sequence of i.i.d. random variables on a probability space $(\Omega, \mathcal{F}, P)$ such that $P\{\xi_n = 1\} = P\{\xi_n = -1\} = \frac{1}{2}$. The price of the bond will remain constant $B_n \equiv 1$ while the price of the (one) stock at time $n \geq 1$ will be given here by $S_n = \xi_1 + \cdots + \xi_n$ with the zero price at time 0, $S_0 = 0$. Clearly, $S_n$, $n \geq 0$ is a martingale. Consider the (self-financing) strategy $\pi = (\beta_k, \gamma_k)$, $k \geq 1$ where $\beta_k \equiv 0$ and $\gamma_k = 2^{k-1}$ if $\xi_1 = \cdots = \xi_{k-1} = -1$ while $\gamma_k = 0$ for otherwise. Clearly, $X_n^\pi$ can be viewed as the gain of a gambler playing the "martingale", i.e., doubling the stake after each loss. Let $\tau = \min\{k : X_k^\pi = 1\}$. Since $P\{\tau = k\} = 2^{-k}$, thus $P\{\tau < \infty\} = 1$, and this gives an opportunity for the arbitrage since $X_0^\pi = 0$ and $P\{X_\tau^\pi = 1\} = 1$. Of course, in order to adopt such strategy an investor must be infinitely rich or should have an unlimited credit line which is impossible in practice.

### 3.1.3    *Complete and incomplete markets*

**Definition 3.1.2.** A financial market defined above on a probability space $(\Omega, \mathcal{F}, P)$ with a filtration $\{\mathcal{F}_n\}_{n \geq 0}$ and a horizon $N$ is called complete (or $N$-complete) if for any $\mathcal{F}_N$-measurable payoff function (claim) $f_N = f_N(\omega)$

there exists a self-financing trading strategy $\pi$ with an initial capital $x$ so that the corresponding portfolio value satisfies $X_0^\pi = x$ and $X_N^\pi = f_N$, i.e., any (European) contingent claim is replicable (or attainable). Otherwise, the market is called incomplete.

In the CRR binomial market we saw that a martingale measure exists and it is unique so this market is without arbitrage and it is complete, while in multinomial markets there exist infinitely many martingale measures, and so such markets have no arbitrage but they are incomplete.

Denote by $\mathcal{P}_N(P)$ the set of all martingale measures for the above market with a horizon $N$.

**Theorem 3.1.2.** *(Second fundamental theorem of asset pricing) A financial market defined above without opportunity of arbitrage and with $d, N < \infty$ is complete if and only if $\mathcal{P}_N(P)$ consists of one measure only.*

**Proof.** a) Assume that the market is complete. Let $\Gamma \in \mathcal{F}_N$ and define $f_N(\omega) = \mathbb{I}_\Gamma(\omega)$. By completeness of the market there exists a self-financing trading strategy $\pi$ and an initial capital $x$ such that $X_0^\pi = x$ and $X_N^\pi = f_N$. If there exist two martingale measures $P_1$ and $P_2$ then $\frac{X_n^\pi}{B_n}$, $n \geq 0$ is a local martingale with respect to both $P_1$ and $P_2$. Since $\frac{X_N^\pi}{B_N} = \frac{f_N}{B_N} \geq 0$ it follows by Proposition 1.1.1 that $\frac{X_n^\pi}{B_n}$, $n \geq 0$ is, in fact, a martingale with respect to both $P_1$ and $P_2$. Then for $i = 1, 2$,

$$\frac{x}{B_0} = \frac{X_0^\pi}{B_0} = E_{P_i}\frac{X_N^\pi}{B_N} = E_{P_i}\frac{\mathbb{I}_\Gamma}{B_N} = \int_\Gamma B_N^{-1} dP_i.$$

Hence, $\int_\Gamma B_N^{-1} dP_1 = \int_\Gamma B_N^{-1} dP_2$ for any $\Gamma \in \mathcal{F}_N$ and since $B_N > 0$ we obtain that $P_1 = P_2$. Indeed, if $\mu(\Gamma) = \int_\Gamma B_N^{-1} dP_1 = \int_\Gamma B_N^{-1} dP_2$ then $P_1(\Gamma) = \int_\Gamma B_N d\mu = P_2(\Gamma)$ for any $\Gamma \in \mathcal{F}_N$.

b) It is more difficult to prove the other direction: if there exists only one martingale measure then any payoff function (claim) is attainable or, equivalently, if some $\mathcal{F}_N$-measurable claim cannot be replicated by a self-financing trading strategy then there exist at least two martingale measures. Since the market is without arbitrage then by the first fundamental theorem of asset pricing at least one martingale measure $Q$ exists.

Let $L_N$ be the set of all random variables $\xi$ having the form

$$\xi = \lambda + \sum_{k=1}^{N}(\gamma_k, \Delta \tilde{S}_k)$$

where $\lambda$ is a real number and $\gamma_k$, $k = 1, ..., N$ is a predictable $d$-dimensional sequence. If some $\mathcal{F}_N$-measurable claim cannot be replicated then there exists an $\mathcal{F}_N$-measurable random variable $H_N \notin L_N$. As in the proof of the first fundamental theorem of asset pricing we can assume without loss of generality that $H_N$ and $\tilde{S}_k$, $k = 0, 1, ..., N$ are integrable with respect to $P$ since for otherwise we can take the equivalent measure $e^{-|H_N| + \sum_{1 \leq k \leq N} |\tilde{S}_k|} P$ and proceed with the proof for the latter measure. By the construction in the proof of the first fundamental theorem of asset pricing we may assume that the Radon–Nikodym derivative $\frac{dQ}{dP}$ is bounded, and so $H_N$ and $\tilde{S}_k$, $k = 0, 1, ..., N$ are integrable with respect to $Q$ as well.

We will see in Appendix that $L_N$ is closed with respect to the convergence in probability, and so it is also closed with respect to the $L^1$-convergence. By the above we can assume without loss of generality that $L_N \subset L^1(\Omega, \mathcal{F}_N, Q)$ and since $H_N \notin L_N$ is also assumed to be $Q$-integrable it will follow by the separation theorem in Appendix that there exists $z \in L^\infty(\Omega, \mathcal{F}_N, Q)$ which separates $L_N$ and $H_N$, i.e.,

$$E_Q z H_N > E_Q z \ell \tag{3.1.3}$$

for any $\ell \in L_N$. Since $L_N$ is a subspace it follows, in particular, that if $\ell \in L_N$ then $a\ell \in L_N$ for any constant $a \in \mathbb{R}$, and so we must have $E_Q z \ell = 0$ for all $\ell \in L_N$.

Since $\gamma_k \equiv 0$ for all $k$ is a predictable sequence then $1 \in L_N$, and so

$$E_Q z \cdot 1 = E_Q z = \int z dQ = 0.$$

Set

$$g = 1 + \frac{z}{2\|z\|_\infty} \geq \frac{1}{2} > 0,$$

where $\| \cdot \|_\infty$ is the $L^\infty$-norm, and

$$\tilde{Q}(\Gamma) = \int_\Gamma g dQ$$

for any $\Gamma \in \mathcal{F}_N$. The Radon–Nikodym derivative $g = \frac{d\tilde{Q}}{dQ}$ is bounded away from zero and infinity, and so integrable random variables with respect to $Q$ and $\tilde{Q}$ are the same.

Next,

$$\tilde{Q}(\Omega) = E_Q 1 + \frac{E_Q z}{2\|z\|_\infty} = 1,$$

and so $\tilde{Q}$ is an equivalent probability measure. Observe that taking $\lambda = 0$ we see that

$$\sum_{k=1}^N (\gamma_k, \Delta \tilde{S}_k) \in L_N$$

for any predictable sequence $\gamma_k$, $k = 1, ..., N$. Hence, if $\gamma_k$, $k = 1, ..., N$ is predictable and bounded then

$$E_{\tilde{Q}}\Big(\sum_{k=1}^{N}(\gamma_k, \Delta\tilde{S}_k)\Big) = E_Q\Big((1 + \tfrac{z}{2\|z\|_\infty})\sum_{k=1}^{N}(\gamma_k, \Delta\tilde{S}_k)\Big)$$

$$= E_Q\Big(\sum_{k=1}^{N}(\gamma_k, \Delta\tilde{S}_k)\Big) = \sum_{k=1}^{N}E_Q(\gamma_k, \Delta\tilde{S}_k).$$

Since $Q$ is a martingale measure,

$$E_Q(\gamma_k, \Delta\tilde{S}_k) = E_Q(\gamma_k, E_Q(\Delta\tilde{S}_k|\mathcal{F}_{k-1})) = 0.$$

Thus,

$$E_{\tilde{Q}}\Big(\sum_{k=1}^{N}(\gamma_k, \Delta\tilde{S}_k)\Big) = 0$$

for any bounded predictable sequence $\gamma_k = (\gamma_k^{(1)}, ..., \gamma_k^{(d)})$, $k = 1, ..., N$. Taking $\gamma_k^{(i)} = \mathbb{I}_\Gamma$ for an arbitrary $\Gamma \in \mathcal{F}_{k-1}$, $\gamma_k^{(j)} = 0$ for $j \neq i$ and $\gamma_l \equiv 0$ for all $l \neq k$ we obtain

$$E_{\tilde{Q}}(\mathbb{I}_\Gamma \Delta\tilde{S}_k^{(i)}) = \int_\Gamma (\tilde{S}_k^{(i)} - \tilde{S}_{k-1}^{(i)})d\tilde{Q} = 0.$$

Since this holds true for any $\Gamma \in \mathcal{F}_{k-1}$ and $i = 1, ..., d$ we obtain by the definition of the conditional expectation that

$$E_{\tilde{Q}}(\tilde{S}_k^{(i)}|\mathcal{F}_{k-1}) = \tilde{S}_{k-1}^{(i)}$$

for all $i = 1, ..., d$. This being true for all $k = 1, ..., N$ yields that $\tilde{Q}$ is a martingale measure different from $Q$. This completes the proof of the second fundamental theorem of asset pricing, remembering that the separation assertion (3.1.3) will be established in Appendices below. $\square$

## 3.2 Appendices to Lecture 6: Finite sample space, general closedness and separation theorems

### 3.2.1 *Appendix 1: Finite sample space*

Before considering the general case we will deal with the simpler situation where $\Omega = \{\omega_1, ..., \omega_m\}$ is a finite sample space, and so each random variable $\xi$ on $\Omega$ can be described by the vector $(\xi(\omega_1), ..., \xi(\omega_m))$. Then the separation results we relied upon in the first and the second fundamental theorems of asset pricing can be reduced to the following separation hyperplane theorem.

**Theorem 3.2.1.** *Let $F \subset \mathbb{R}^m$ be a non-empty compact convex set and $L$ be a non-empty linear subspace of $\mathbb{R}^m$ such that $L \cap F = \emptyset$. Then there exists an $(m-1)$-dimensional hyperplane $H \supset L$ such that for some $z \in \mathbb{R}^m \setminus \{0\}$,*

$$H = \{x \in \mathbb{R}^m : \langle x, z \rangle = 0\} \quad \text{and} \quad \langle y, z \rangle > 0 \quad \forall y \in F$$

*where $\langle x, z \rangle = \sum_{i=1}^m x_i z_i$ for $x = (x_1, ..., x_m)$ and $z = (z_1, ..., z_m)$.*

**Proof.** Set

$$G = F - L = \{x \in \mathbb{R}^m : x = f - \ell, \, f \in F, \, \ell \in L\}.$$

Then $0 \notin G$, $G \neq \emptyset$ and it is easy to see that $G$ is a closed convex set. Let $B = \{x \in \mathbb{R}^m : |x| \leq r\}$ be the ball centered at $0$ with the radius $r$ large enough so that $B \cap G \neq \emptyset$. Then $B \cap G$ is a closed and bounded subset of $\mathbb{R}^m$, and so it is compact. Thus, $g(x) = |x|$ attains its minimum in $B \cap G$ at a point $z \in B \cap G$. By construction, $z \neq 0$ and $|z| \leq r$. Clearly, $|x| \geq |z|$ for any $z \in G$ since $|x| = g(x) \geq g(z) = |z|$ for any $x \in B \cap G$ and $|x| > r$ when $x \in G \setminus B$. Since $\lambda x + (1 - \lambda)z \in G$ for any $\lambda \in (0, 1)$ and $x \in G$ by convexity of $G$ we obtain

$$|\lambda x + (1 - \lambda)z|^2 \geq |z|^2 \quad \forall x \in G, \, \lambda \in (0, 1).$$

Hence, $2(1 - \lambda)\langle x, z \rangle - 2|z|^2 + \lambda(|x|^2 + |z|^2) \geq 0$. Letting $\lambda \to 0$ we obtain $\langle x, z \rangle \geq |z|^2$ which means $\langle f - \ell, z \rangle \geq |z|^2$ for all $f \in F$ and $\ell \in L$, and so $\langle \ell, z \rangle \leq \langle f, z \rangle - |z|^2$. Since this holds true for any vector $\ell$ in a linear space $L$, in particular, for $a\ell$, $a \in \mathbb{R}$, we must have $\langle \ell, z \rangle = 0$ for each $\ell \in L$ and $\langle f, z \rangle \geq |z|^2 > 0$. This completes the proof of the theorem defining $H = \{x \in \mathbb{R}^m : \langle x, z \rangle = 0\}$. $\qquad \square$

In the application of this result to the first fundamental theorem of asset pricing we recall that the adjusted stock values $\tilde{S}_n = \frac{S_n}{B_n}$ and the coefficients $\gamma_n$, $n = 1, ..., N$ are $d$-dimensional random vectors. Since $\Omega = \{\omega_1, ..., \omega_m\}$ we can view now the set $M_N$ defined at the beginning of the proof as a linear subspace of $\mathbb{R}^m$ containing (non-random) vectors of the form $\xi = \sum_{k=1}^N (\gamma_k, \Delta \tilde{S}_k)$ where $\xi = (\xi(\omega_1), ..., \xi(\omega_m))$ and $(\gamma_k, \Delta \tilde{S}_k) = \big((\gamma_k(\omega_1), \Delta \tilde{S}_k(\omega_1)), ..., (\gamma_k(\omega_m), \Delta \tilde{S}_k(\omega_m))\big)$. Let

$$F = \{\zeta = (\zeta_1, ..., \zeta_m) \in \mathbb{R}^m : \zeta_i \geq 0, \sum_{i=1}^m \zeta_i = 1\}.$$

Then assuming no arbitrage, $M_N \cap F = 0$. Hence, by the separation hyperplane theorem there exists $z \in \mathbb{R}^m \setminus \{0\}$ such that $\langle \xi, z \rangle = 0$ for any $\xi \in M_N$ and $\langle \zeta, z \rangle > 0$ for any $\zeta \in F$, where, again, $\langle x, y \rangle = \sum_{i=1}^m x_i y_i$ for

$x = (x_1, ..., x_m)$, $y = (y_1, ..., y_m) \in \mathbb{R}^m$. Since $\zeta^{(i)} = (\zeta_1, ..., \zeta_m) \in F$ when $\zeta_i = 1$ and $\zeta_j = 0$ for $j \neq i$, we obtain that $z = (z_1, ..., z_m)$ has $z_j > 0$ for all $j = 1, ..., m$. Set

$$\tilde{P}(\{\omega + i\}) = \frac{z_i}{\sum_{j=1}^{m} z_j}, \quad i = 1, ..., m.$$

Then for any $\xi \in M_N$,

$$E_{\tilde{P}}\xi = (\sum_{i=1}^{m} z_i)^{-1} \langle \xi, z \rangle = 0.$$

In particular, this holds true for any $\xi = (\gamma_k, \Delta \tilde{S}_k)$ with an arbitrary bounded $\gamma_k$ measurable with respect to $\mathcal{F}_{k-1}$, and so

$$0 = E_{\tilde{P}}(\gamma_k, \Delta \tilde{S}_k) = E_{\tilde{P}}(\gamma_k, E_{\tilde{P}}(\Delta \tilde{S}_k | \mathcal{F}_{k-1})).$$

It follows that $E_{\tilde{P}}(\Delta \tilde{S}_k | \mathcal{F}_{k-1}) = 0$ for $k = 1, 2, ..., N$, implying that $\tilde{P}$ is a martingale measure, completing the proof of the first fundamental theorem of asset pricing for the case of a finite sample space $\Omega$.

For the second fundamental theorem of asset pricing in this case we apply the separating hyperplane theorem directly by taking $L$ as a subspace of $\mathbb{R}^m$ containing vectors of the form $\{\lambda + \sum_{k=1}^{N}(\gamma_k, \Delta \tilde{S}_k)(\omega_i), i = 1, ..., m\}$ with $\lambda \in \mathbb{R}$ and predictable sequences $\{\gamma_k, k = 1, ..., N\}$. Choosing $F$ to be the single vector $H_N = (H_N(\omega_1), ..., H_N(\omega_m))$ which is a non-replicable claim, we find $z \in \mathbb{R}^m$ such that $\langle z, H_N \rangle > \langle z, \ell \rangle$ for any $\ell \in L$ and proceed with the proof as in the main part of the lecture.

### 3.2.2  *Appendix 2: A closedness theorem*

**Theorem 3.2.2.** *(i) The set of random variables* $M_N = \{\xi : \xi = \sum_{k=1}^{N}(\gamma_k, \Delta \tilde{S}_k), \gamma_k$ *is* $\mathcal{F}_{k-1}$*-measurable,* $k = 1, ..., N\}$ *is closed with respect to convergence in probability;*

*(ii) The set of random variables* $L_N = \{\lambda + \xi : \xi \in M_N, \lambda \in \mathbb{R}\}$ *is closed with respect to convergence in probability;*

*(iii) Under the no arbitrage condition the set of random variables* $K_N = M_N - L_+^0 = \{\eta = \xi - \zeta : \xi \in M_N, \zeta \in L_+^0\}$, *where* $L_+^0$ *is the set of all nonnegative* $\mathcal{F}_N$*-measurable random variables, is closed with respect to convergence in probability.*

**Proof.** We will first prove (iii) following [44], indicating in appropriate places that (i) follows by simplifying the same proof. At the end we will add arguments which will yield (ii). Observe that, if $\Omega$ is a finite sample space then $M_N$ and $L_N$ can be viewed as linear subspaces of a finite dimensional Euclidean space, and so they are closed, while $K_N$ is a convex cone there which, as can be easily seen, is also closed. For the proof of the general case we begin with the following result from [44].

**Lemma 3.2.1.** *For any sequence $\eta^n$, $n \geq 1$ of random d-dimensional vectors such that $\eta = \liminf_{n\to\infty} |\eta^n| < \infty$, we can choose another sequence $\tilde{\eta}^k$, $k \geq 1$ of random d-dimensional vectors such that for any $\omega$ the sequence $\tilde{\eta}^k(\omega)$, $k \geq 1$ is a convergent subsequence of $\eta^n(\omega)$, $n \geq 1$ (though for different $\omega$'s the subsequences may be different), and if $k(\omega) = n$ whenever $\tilde{\eta}^{k(\omega)}(\omega) = \eta^n(\omega)$ then $k = k(\omega)$ is measurable.*

**Proof.** Set $\tau_0 = 0$ and recursively,

$$\tau_k = \min\{n > \tau_{k-1} : ||\eta^n| - \eta| \leq k^{-1}\}.$$

Let $\tilde{\eta}_0^k = \eta^{\tau_k}$, then $\sup_k |\tilde{\eta}_0^k| < \infty$. Next, let $\tilde{\eta}_0^k = (\tilde{\eta}_{0,1}^k, \tilde{\eta}_{0,2}^k, ..., \tilde{\eta}_{0,d}^k)$ and set $\tilde{\eta}_{0,1} = \liminf_{k\to\infty} \tilde{\eta}_{0,1}^k$. Define $\tau_0' = 0$ and recursively, $\tau_k' = \min\{n > \tau_{k-1}' : |\tilde{\eta}_{0,1} - \tilde{\eta}_{0,1}^n| \leq k^{-1}\}$. Now, set $\tilde{\eta}_1^k = \tilde{\eta}_0^{\tau_k'}$. Clearly, $\tilde{\eta}_1^k(\omega)$, $k \geq 1$ is a subsequence of $\eta^n(\omega)$, $n \geq 1$ such that its first component $\tilde{\eta}_{1,1}^k(\omega)$ converges as $k \to \infty$. Furthermore, if $k(\omega) = n$ when $\tilde{\eta}_1^{k(\omega)}(\omega) = \eta^n(\omega)$ then $k = k(\omega)$ is measurable. Next, we apply this procedure to the second component of $\tilde{\eta}_1^k$ by choosing a subsequence $\tilde{\eta}_2^k$ where both first and second components converge as $k \to \infty$. Repeating this $d$ times in total we obtain a sequence $\tilde{\eta}^k$, $k \geq 1$ with required properties. $\qquad\square$

Next, we return to the proof of closedness of $K_N$ with respect to convergence in probability assuming that there is no arbitrage. We proceed by induction taking $N = 1$ first. Since from every convergent in probability sequence we can choose an a.s. convergent subsequence, we can take

$$(\gamma_1^n, \Delta\tilde{S}_1) - r^n \to \zeta \quad \text{a.s. as } n \to \infty$$

where $\gamma_1^n$ is $\mathcal{F}_0$-measurable and $r^n \in L_+^0$. It suffices to find $\mathcal{F}_0$-measurable random variables $\tilde{\gamma}_1^k$ and $\tilde{r}^k \in L_+^0$ such that $\tilde{\gamma}_1^k \to \tilde{\gamma}_1$ a.s. and $(\tilde{\gamma}_1^k, \Delta\tilde{S}_1) - \tilde{r}^k \to \zeta$ a.s. as $k \to \infty$, since then $\tilde{r}^k \to (\tilde{\gamma}_1, \Delta\tilde{S}_1) - \zeta \in L_+^0$, and so $\zeta \in K_1$. Clearly, if $\Omega_1, ..., \Omega_l \in \mathcal{F}_0$ is a finite partition of $\Omega$ then it suffices to construct the required sequences $\tilde{\gamma}_1^k$ and $\tilde{r}^k$ on each $\Omega_j$ separately. Observe that in proving closedness of $M_N$ we have $r^n \equiv 0$ for all $n$, and so we can view this as $r^n \to 0$ as $n \to \infty$ which holds true automatically, i.e., in this case we do not have to deal with $\tilde{r}^k$'s at all.

Let $\gamma_1 = \liminf_{n \to \infty} |\gamma_1^n|$. On the set $\Omega_1 = \{\gamma_1 < \infty\}$ we can use the above lemma to construct $\mathcal{F}_0$-measurable $\tilde{\gamma}_1^k$ such that $\tilde{\gamma}_1^k(\omega)$ is a convergent subsequence of $\gamma_1^n(\omega)$ for every $\omega \in \Omega_1$, and if $\tilde{\gamma}_1^k(\omega) = \gamma_1^n(\omega)$ we take $\tilde{r}^k(\omega) = r^n(\omega)$. If $P(\Omega_1) = 1$ then we are done; if not we consider $\Omega_2 = \{\gamma_1 = \infty\}$. On $\Omega_2$ set $\eta_1^n = \frac{\gamma_1^n}{|\gamma_1^n|}$ and $h_1^n = \frac{r_1^n}{|\gamma_1^n|}$ and observe that $(\eta_1^n, \Delta \tilde{S}_1) - h_1^n \to 0$ a.s. as $n \to \infty$. By the above lemma we obtain an $\mathcal{F}_0$-measurable $\tilde{\eta}_1^k$ such that $\tilde{\eta}_1^k(\omega)$ is a convergent subsequence of $\eta_1^n(\omega)$ for any $\omega$. Denoting the limit by $\tilde{\eta}_1$, we obtain that $(\tilde{\eta}_1, \Delta \tilde{S}_1) = \tilde{h}_1$ where $\tilde{h}_1$ is nonnegative, and so by the no arbitrage assumption $K_N \cap L_+^0 = \{0\}$ we have $(\tilde{\eta}_1, \Delta \tilde{S}_1) = 0$. Observe that, since $r_1^n = 0$ for all $n$ in the proof of closedness of $M_N$ we obtain $(\tilde{\eta}_1, \Delta \tilde{S}_1) = 0$ automatically without any need in the no arbitrage assumption.

Next, if $d = 1$ then $\Delta \tilde{S}_1 = 0$ a.s. as $|\tilde{\eta}_1| = 1$. But in this case $-r^n \to \zeta$ a.s. as $n \to \infty$, and so $-\zeta \in L_+^0$, whence $\zeta = 0 - (-\zeta) \in K_1$. Now, suppose that $d > 1$, then we proceed by induction decreasing the dimension on each step. Suppose that we constructed the required sequences when the dimension does not exceed $d - 1$. Now let our random vectors have dimension $d$. Since $|\tilde{\eta}_1| = 1$ we can partition $\Omega_2$ into $d$ disjoint subsets $\Omega_2^i \in \mathcal{F}_0$, $i = 1, ..., d$ such that the $i$-th component $\tilde{\eta}_1^{(i)}$ of $\tilde{\eta}_1$ does not vanish on $\Omega_2^i$. For each $i$ define $\bar{\gamma}_1^n$ on $\Omega_2^i$ by $\bar{\gamma}_1^n = \gamma_1^n - \frac{\gamma_1^{ni}}{\tilde{\eta}_1^{(i)}} \tilde{\eta}_1$ where $\gamma_1^{ni}$ is the $i$-th component of the $d$-dimensional random vector $\gamma_1^n$. Then

$$(\bar{\gamma}_1^n, \Delta \tilde{S}_1) = \sum_{j=1}^{d} (\gamma_1^{nj} - \frac{\tilde{\eta}_1^{(j)} \gamma_1^{ni}}{\tilde{\eta}_1^{(i)}}) \Delta \tilde{S}_1^j = (\gamma_1^n, \Delta \tilde{S}_1)$$

since $(\tilde{\eta}_1, \Delta \tilde{S}_1) = 0$ as proved above. Now on each $\Omega_2^i$ we have, essentially, $(d-1)$-dimensional vectors $\bar{\gamma}_1^n$ and $\Delta \tilde{S}_1$ since $\bar{\gamma}_1^{ni} = 0$ on $\Omega_2^i$, and so the $i$-th component $\Delta \tilde{S}_1^i$ of $\Delta \tilde{S}_1$ plays no role in the scalar product $(\bar{\gamma}_1^n, \Delta \tilde{S}_1)$ as well. In view of the above equality

$$(\bar{\gamma}_1^n, \Delta \tilde{S}_1) - r^n \to \zeta \quad \text{a.s. as } n \to \infty,$$

thus we decreased the dimension accomplishing the induction step.

Thus, we derived the closedness assertion for $M_N$ and $K_N$ when $N = 1$; and in order to complete the proof for any $N$ we assume the validity of the assertion for $N - 1$ and show that it remains true for $N$. Thus, let

$$\sum_{j=1}^{N-1} (\gamma_j^n, \Delta \tilde{S}_j) - r^n \to \zeta \quad \text{a.s. as } n \to \infty$$

where $\gamma_j^n$ are $\mathcal{F}_{j-1}$-measurable and $r^n \in L_+^0$. Consider $(\gamma_N^n, \Delta \tilde{S}_N)$ and set $\gamma_N = \liminf_{n \to \infty} |\gamma_N^n|$. On the set $\Omega_1 = \{\gamma_N < \infty\}$ by the above lemma we can choose $\mathcal{F}_{N-1}$-measurable $\tilde{\gamma}_N^k$ such that $\tilde{\gamma}_N^k(\omega)$ is a convergent subsequence of $\gamma_N^n(\omega)$ for every $\omega$. Then $\tilde{\gamma}_N^k \to \eta$ a.s. as $k \to \infty$ where $\eta$ is $\mathcal{F}_{N-1}$-measurable, and so

$$\sum_{j=1}^{N-1} (\gamma_j^k, \Delta \tilde{S}_j) + (\tilde{\gamma}_N^k, \Delta \tilde{S}_N) \to \zeta + (\eta, \Delta S_N) \text{ a.s. as } k \to \infty.$$

Observe that $(\eta, \Delta \tilde{S}_N) \in K_N$ since we can take zero coefficients in $\Delta \tilde{S}_j$ for $j = 1, ..., N - 1$. Furthermore, $\zeta \in K_N$ since $\zeta \in K_{N-1}$ by the induction hypothesis and $\zeta = \zeta + (0, \Delta \tilde{S}_N) \in K_N$. Hence, $\zeta + (\eta, \Delta \tilde{S}_N) \in K_N$.

On $\Omega_2 = \{\gamma_N = \infty\}$ we proceed as in the case $N = 1$ replacing $\gamma_N^n$ by $\bar{\gamma}_N^n$, which is a random vector having less nonzero components than $\gamma_N^n$ on elements of a finite partition of $\Omega_2$. Proceeding in the same way we eliminate $\gamma_N^n$ completely on elements of a certain finite partition where we reduce the situation to the sum $\sum_{j=1}^{N-1} (\gamma_j^n, \Delta \tilde{S}_j)$, which completes the proof of the assertions (i) and (iii) of the theorem in view of the induction hypothesis.

It remains to provide arguments which yield the assertion (ii) of the theorem. Let

$$\lambda_n + \xi_n \to \zeta \text{ a.s. as } n \to \infty$$

where $\lambda_n \in \mathbb{R}$ and $\xi_n \in M_N$, $n = 1, 2, ...$ If $1 \in M_N$ then $\lambda_n + \xi_n \in M_N$, and so $\zeta \in M_N$ since we proved that $M_N$ is closed. Now suppose that $1 \notin M_N$. If $\liminf_{n \to \infty} |\lambda_n| = 0$ we can choose a subsequence $\lambda_{n_i} \to 0$ as $i \to \infty$ and then $\xi_{n_i} \to \zeta$ as $i \to \infty$, which means that $\zeta \in L_N$ since $M_N$ is closed. Now, suppose that $\liminf_{n \to \infty} |\lambda_n| > 0$. Then there exist $a > 0$ and $n_0$ such that $|\lambda_n| \geq a$ when $n \geq n_0$. Now, for any $n \geq n_0$ we define $\beta_n = -\frac{\xi_n}{\lambda_n}$, and since $M_N$ is a linear space, $\beta_n \in M_N$.

For each $\varepsilon > 0$ set $\Gamma_{n,\varepsilon} = \{\omega : |\beta_n(\omega) - 1| \geq \varepsilon\}$. If $P(\Gamma_{n,\varepsilon}) \to 0$ we can choose a subsequence $n_i$ such that $\sum_i P(\Gamma_{n_i,\varepsilon}) < \infty$, and so by the Borel–Cantelli lemma there exists a random $N_\varepsilon < \infty$ a.s. such that $|\beta_n(\omega) - 1| < \varepsilon$ for all $n \geq n_\varepsilon(\omega)$. If this holds true for any $\varepsilon > 0$, by taking a sequence $\varepsilon_k \to 0$ as $k \to \infty$ we can use the diagonal procedure to obtain a subsequence $n_j$ such that $\beta_{n_j} \to 1$ a.s. But $\beta_{n_j} \in M_N$, $1 \notin M_N$ and $M_N$ is closed, whence this cannot happen. It follows that there exist $\varepsilon, \delta > 0$ and a subsequence $n_j \to \infty$ as $j \to \infty$ such that

$$P(\Gamma_{n_j,\varepsilon}) \geq \delta \text{ for all } j.$$

Set $\Omega_D = \{\omega : |\zeta(\omega)| \leq D\}$ and $\Omega_{D,m} = \{\omega : |\lambda_n + \xi_n(\omega)| \leq 2D \text{ for all } n \geq m\}$. Then we can choose $D$ and $m$ so that

$$P(\Omega_{D,m}) > 1 - \delta.$$

But for all $n_j \geq m$ and $\omega \in \Omega_{D,m} \cap \Gamma_{n_j,\varepsilon} \neq \emptyset$,

$$2D \geq |\lambda_{n_j} + \xi_{n_j}(\omega)| = |\lambda_{n_j}||1 - \beta_{n_j}(\omega)| \geq \varepsilon|\lambda_{n_j}|,$$

i.e., $|\lambda_{n_j}| \leq 2D\varepsilon^{-1}$ for all $n_j \geq m$. Hence, we can choose a converging subsequence $\lambda_{n_{j_i}} \to \lambda$ as $i \to \infty$ and then $\xi_{n_{j_i}} \to \xi = \zeta - \lambda \in M_N$ as $i \to \infty$ since $M_N$ is closed. Hence, $\zeta = \lambda + \xi \in L_N$, and so $L_N$ is closed as well. $\qquad\square$

### 3.2.3   *Appendix 3: Separation*

The result we relied upon in the proof of both fundamental theorems of asset pricing is the following.

**Theorem 3.2.3.** *(Separation theorem) Let $M$ be a closed convex set in $L^1 = L^1(\Omega, \mathcal{F}, P)$. Then for any $x \in L^1 \setminus M$ there exists $z_x \in L^\infty = L^\infty(\Omega, \mathcal{F}, P)$ such that*

$$Ez_x\xi < Ez_x x$$

*for all $\xi \in M$.*

**Proof.** First, observe that, when $\Omega$ is a finite sample space the proof reduces to more elementary arguments about a separating hyperplane in the finite dimensional Euclidean space described in Appendix 1 above. For the general case, first recall that the dual space to $L^1(\Omega, \mathcal{F}, P)$ can be identified with $L^\infty(\Omega, \mathcal{F}, P)$. Indeed, if $z \in L^\infty(\Omega, \mathcal{F}, P)$ then

$$\ell(\zeta) = Ez\zeta, \ \zeta \in L^1(\Omega, \mathcal{F}, P)$$

that defines a bounded linear functional on $L^1(\Omega, \mathcal{F}, P)$.

On the other hand, if $\ell$ is a bounded linear functional on $L^1(\Omega, \mathcal{F}, P)$, then for any $\Gamma \in \mathcal{F}$ we can set

$$\mu(\Gamma) = \ell(\mathbb{I}_\Gamma)$$

that defines a finitely additive measure on $\mathcal{F}$, which, in fact, is $\sigma$-additive since $\ell$ is bounded, and it is also absolutely continuous with respect to $P$. Indeed, if $P(\Gamma) = 0$ then by boundedness of $\ell$,

$$\ell(\mathbb{I}_\Gamma) \leq \|\ell\| \int \mathbb{I}_\Gamma dP = 0.$$

Hence, by the Radon–Nikodym theorem there exists a function $g \in L^1$ such that for any $\Gamma \in \mathcal{F}$, $\ell(\mathbb{I}_\Gamma) = \int_\Gamma g dP$. Then

$$\left| \int_\Gamma g dP \right| \leq \|\ell\| P(\Gamma),$$

and so $g \leq \|\ell\|$ with probability one. It follows that $\ell(\zeta) = \int \zeta g dP$ for any $\zeta \in L^1(\Omega, \mathcal{F}, P)$ with $g \in L^\infty(\Omega, \mathcal{F}, P)$, as required.

Thus, we conclude that in order to prove the separation theorem, we have to show that for any $x \in L^1 \backslash M$ there exists a bounded linear functional $\ell_x$ on $L^1$ such that

$$\ell_x(\xi) < \ell_x(x) \tag{3.2.1}$$

for all $\xi \in M$. Without loss of generality, we can assume that $0 \in M$ since for otherwise we can take $M - y$ in place of $M$ and $x - y$ in place of $x$ for some $y \in M$. Let $d_x > 0$ be the distance between $x$ and $M$ and $U_0(r)$ be the open ball centered at 0 with the radius $r < \frac{1}{2} d_x$. Then $U = M + U_0(r)$ and $V_x = x + U_0(r)$ are disjoint open convex sets (neighborhoods) containing $M$ and $x$, respectively.

Recall that, the Minkovski functional of $U$ is a function on our space $L^1$ defined by

$$p_U(y) = \inf\{r > 0 : \frac{y}{r} \in U\}.$$

Let $x \notin M$ and $U, V_x$ be as above then the distance between $x$ and $\bar{U}$ is positive, and so it is easy to see that $p_U(x) > 1$. On the one-dimensional space $L_0 = \{\alpha x, \alpha \in \mathbb{R}\}$ define the linear functional

$$\ell_0(\alpha x) = \alpha p_U(x)$$

which satisfies

$$\ell_0(\alpha x) \leq p_U(\alpha x) \quad \text{for all } \alpha \in \mathbb{R}$$

since

$$p_U(\alpha x) = \alpha p_U(x) \text{ if } \alpha \geq 0 \text{ and } \ell_0(\alpha x) = \alpha \ell_0(x) < 0 \leq p_U(\alpha x) \text{ if } \alpha < 0.$$

It is easy to see that the Minkovski functional is nonnegative, convex and positively homogeneous, i.e.,

$$p_U \geq 0, \ p_U(y + z) \leq p_U(y) + p_U(z) \text{ and } p_U(\alpha y) = \alpha p_U(y) \text{ for all } \alpha > 0.$$

Now we can apply the Hahn–Banach theorem (which will be discussed below) in order to extend the functional $\ell_0$ to a linear functional $\ell_x$ on the whole space $L^1$ so that

$$\ell(\alpha x) = \ell_0(\alpha x) \text{ for all } \alpha \in \mathbb{R} \text{ and } \ell_x(y) \leq p_U(y) \text{ for all } y \in L^1.$$

Then $\ell_x(y) \leq p_U(y)$ for $y \in U \supset M$ and $\ell_x(x) = p_U(x) > 1$, and so (3.2.1) holds true. $\qquad \square$

Next, for the readers' convenience we will formulate and prove the Hahn–Banach theorem.

**Theorem 3.2.4.** *(Hahn–Banach) Let $p$ be a nonnegative, convex and positively homogeneous real valued functional defined on a real linear space $L$ and $L_0$ be a linear subspace of $L$. If $\ell_0$ is a linear functional on $L_0$ satisfying $\ell_0(x) \leq p(x)$ for all $x \in L_0$ then $\ell_0$ can be extended to a linear functional $\ell$ on $L$ such that*

$$\ell(y) = \ell_0(y) \text{ for all } y \in L_0 \text{ and } \ell(y) \leq p(y) \text{ for all } y \in L.$$

**Proof.** If $L_0 = L$ there is nothing to prove; if $L_0 \neq L$ then there exists $z \in L \setminus L_0$, $z \neq 0$. Let $L_1$ be the linear space generated by $L_0$ and $z$, i.e., $L_1 = \{tz+y : t \in \mathbb{R}, y \in L_0\}$. Set $\ell_1(tz+y) = t\ell_1(z)+\ell_0(y)$, $y \in L_0$ which is a linear functional on $L_1$, and we only have to specify $\ell_1(z)$ so that

$$t\ell_1(z) + \ell_0(y) \leq p(tz + y). \tag{3.2.2}$$

Since $p$ is positively homogeneous this is equivalent to

$$\ell_1(z) \leq p(\tfrac{y}{t} + z) - \ell_0(\tfrac{y}{t}) \text{ if } t > 0 \text{ and } \ell_1(z) \geq -p(-\tfrac{y}{t} - z) - \ell_0(\tfrac{y}{t}) \text{ if } t < 0.$$

Since by linearity of $\ell_0$ and by convexity of $p$ for any $y_1, y_2 \in L_0$,

$$\ell_0(y_2) - \ell_0(y_1) \leq p(y_2 - y_1) \leq p(y_2 + z) + p(-y_1 - z),$$

we obtain

$$-\ell_0(y_2) + p(y_2 + z) \geq -\ell_0(y_1) - p(-y_1 - z).$$

Set

$$c_1 = \sup_{y \in L_0} (-\ell_0(y) - p(-y - z)) \text{ and } c_2 = \inf_{y \in L_0} (-\ell_0(y) + p(y + z)).$$

By the above, $c_2 \geq c_1$, and so we can set $\ell_1(z) = c$ for some $c$ satisfying $c_2 \geq c \geq c_1$. Then (3.2.2) will hold true.

If the probability space $(\Omega, \mathcal{F}, P)$ is separable, i.e., the $\sigma$-algebra $\mathcal{F}$ is generated by a countable collection of sets, then $L^1(\Omega, \mathcal{F}, P)$ is a separable space. In general, if $L$ is separable we can choose a countable collection $z_1, z_2, \dots \in L$ which generates $L$ (i.e., $L$ is the minimal linear space containing $z_1, z_2, \dots$), and then we can construct the required functional $\ell$ by extending it as above, successively to the increasing sequence of subspaces $L^{(1)} = \{L_0, z\}$, $L^{(2)} = \{L^{(1)}, z_2\}, \dots$ where $L^{(k)}$ is the minimal linear subspace of $L$ containing $L^{(k-1)}$ and $z_k$, $k = 1, 2, \dots$ Then any $y \in L$ belongs to some $L^{(k)}$, and so $\ell$ will be extended to the whole $L$ with the condition $\ell(y) \leq p(y)$ preserved.

In the non separable case we have to consider the set $\mathcal{T}$ of all extensions $\ell$ satisfying $\ell \leq p$ with the partial order determined by inclusion of the corresponding linear subspaces where the extensions are defined. Each linearly ordered subset $\mathcal{T}_0$ of $\mathcal{T}$ has an upper bound taking the union of the increasing sequence of corresponding subspaces. By the Zorn lemma there exists a maximal element in $\mathcal{T}$ which is the required extension $\ell$ of $\ell_0$ to the whole $L$.          $\square$

## 3.3    Exercises

1) Assume that a discrete time financial market on a probability space $(\Omega, \mathcal{F}, P)$ admits at least two (different) martingale measures. Prove that there exist infinitely (continuum) many martingale measures then.

2) Let $(\Omega, \mathcal{F}, P)$ be a probability space and $\rho_1, \rho_2, \ldots$ be i.i.d. random variables, such that $\rho_i > -1$ with probability one and there is a number $a > 0$ such that

$$P\{\rho_1 < 0\} > 0, \ P\{0 \leq \rho_1 < a\} > 0 \text{ and } P\{\rho_1 \geq a\} > 0.$$

Let the interest rate be zero $r = 0$, i.e., $B_n \equiv B_0$ for all $n$ and the stock evolution is given by $S_n = S_0 \prod_{i=1}^{n}(1+\rho_i)$. Prove that this market with a finite horizon $N$ provides no arbitrage opportunities, but it is incomplete, and so by the previous exercise it admits infinitely many martingale measures.

# Chapter 4

# Superhedging

## 4.1 Lecture 7: Optional decomposition

### 4.1.1 *Optional decomposition theorem*

We start again with a probability space $(\Omega, \mathcal{F}, P)$ with a filtration $\{\mathcal{F}_n\}_{0 \leq n \leq N}$, $\mathcal{F}_0 = \{\emptyset, \Omega\}$. Let $X = \{X_n\}_{0 \leq n \leq N}$ be a (real valued) stochastic process and $S = \{S_n\}_{0 \leq n \leq N}$, $S_n = (S_n^1, ..., S_n^d)$ be a $\mathbb{R}^d$-valued ($d$-dimensional) stochastic process, both processes adapted to the filtration $\{\mathcal{F}_n\}_{0 \leq n \leq N}$. We interpret $S_n$ as adjusted values at time $n$ of $d$-stocks (so that a bond value does not appear here as it is assumed now to be equal to one).

Let $\mathcal{P}(P)$ be the set of all martingale measures, i.e., the set of all probability measures equivalent to $P$ and such that $\{S_n\}_{0 \leq n \leq N}$ is a martingale with respect to each $\tilde{P} \in \mathcal{P}(P)$. Assume that $\mathcal{P}(P) \neq \emptyset$. Suppose that $X = \{X_n\}_{0 \leq n \leq N}$ is a supermartingale with respect to each measure $\tilde{P} \in \mathcal{P}(P)$. If we use the Doob supermartingale decomposition then we obtain that $X_n = X_0 + M_n^{(\tilde{P})} - C_n^{(\tilde{P})}$ where $M = \{M_n^{(\tilde{P})}\}_{0 \leq n \leq N}$, $M_0^{(\tilde{P})} = 0$ is a martingale and $C = \{C_n^{(\tilde{P})}\}_{0 \leq n \leq N}$, $C_0^{(\tilde{P})} = 0$ is a predictable non-decreasing process. This decomposition will depend on $\tilde{P}$ and by this reason it cannot be used for pricing of derivatives if there is more than one martingale measure. In addition, to use methods we discussed for the CRR market we would need also a martingale representation theorem which holds true only in very restricted circumstances.

**Theorem 4.1.1.** *(Optional decomposition theorem) Suppose that $X = \{X_n\}_{0 \leq n \leq N}$ is a supermartingale with respect to each measure $\tilde{P} \in \mathcal{P}(P) \neq$*

$\emptyset$. *Then there exists an optional decomposition*

$$X_n = X_0 + \sum_{k=1}^{n} (\gamma_k, \Delta S_k) - C_n, \ n = 1, 2, ..., N$$

*such that* $\gamma = \{\gamma_k\}_{0 \le k \le N}$ *is a predictable sequence of d-dimensional random vectors and* $C = \{C_k\}_{0 \le k \le N}$ *is an adapted (not predictable!) non-decreasing process with* $C_0 = 0$. *There is no uniqueness, in general, of such a representation.*

**Proof.** We will adopt the proof from Section 7.2 of [23]. We will show that there exist random $\mathcal{F}_{n-1}$-measurable vectors $\gamma_n \in \mathbb{R}^d$ such that

$$\Delta X_n - (\gamma_n, \Delta S_n) \le 0.$$

Then we can define

$$\Delta C_n = -(\Delta X_n - (\gamma_n, \Delta S_n)) \text{ and } C_n = \sum_{k=1}^{n} \Delta C_k.$$

Denote by $\hat{M}_n$ the set of all random variables $\xi$ having the form $\xi = (\gamma_n, \Delta S_n)$ for all $\mathcal{F}_{n-1}$-measurable $d$-dimensional random vectors $\gamma_n$. The representation

$$\Delta X_n = (\gamma_n, \Delta S_n) - \Delta C_n$$

means that

$$\Delta X_n \in \hat{K}_n = \hat{M}_n - L^0_+(\Omega, \mathcal{F}_n, P) = \{\eta = \xi - \zeta : \xi \in \hat{M}_n, \zeta \in L^0_+\}$$

where $L^0_+ = L^0_+(\Omega, \mathcal{F}_n, P)$ is the set of all nonnegative $\mathcal{F}_n$-measurable random variables.

Observe that it follows from the beginning of the proof of the closedness theorem in Appendix 2 to Lecture 6 that $\hat{K}_n$ is closed with respect to the almost sure convergence of random variables. Indeed, in the proof there for $N = 1$ the $\sigma$-algebras $\mathcal{F}_0 \subset \mathcal{F}_1$ where arbitrary, and so the argument works with $\mathcal{F}_{n-1} \subset \mathcal{F}_n$ as well. The only point to check is the no arbitrage assumption employed there, namely, we have to check that in our case

$$\hat{K}_n \cap L^0_+ = \{0\}.$$

Since the set of martingale measures $\mathcal{P}(P)$ is not empty, there is no arbitrage in the market with discounted stock prices $S_0, S_1, ..., S_n$ above in view of the first fundamental theorem of asset pricing. It follows, as we saw in the proof of the latter theorem, that $K_n \cap L^0_+ = \{0\}$, where recall

$$K_n = \{\xi : \xi = \sum_{k=1}^{n} (\gamma_k, \Delta S_k), \gamma_k \text{ is } \mathcal{F}_{k-1} - \text{measurable}, k = 1, ..., n\}.$$

Clearly, $\hat{K}_n \subset K_n$ since we always can take $\gamma_k = 0$ for $k = 1, ..., n-1$, and so our claim follows.

Next, assume by contradiction that $\Delta X_n \notin \hat{K}_n$. Observe that, without loss of generality we can assume from the beginning that $P$ itself is a martingale measure. By definition that any supermartingale is integrable, hence $\Delta X_n \in L^1(\Omega, \mathcal{F}_n, P)$. Thus, our "reductio ad absurdum" assumption reads

$$\Delta X_n \notin \mathcal{C}_n = \hat{K}_n \cap L^1(\Omega, \mathcal{F}_n, P).$$

Recall, that convergence in $L^1(\Omega, \mathcal{F}_n, P)$ implies the existence of an almost surely convergent subsequence, so $\mathcal{C}_n$ is closed in $L^1(\Omega, \mathcal{F}_n, P)$.

Clearly, $\mathcal{C}_n$ is also convex, and we can apply the separation theorem from Appendix 3 to Lecture 6 which yields the existence of a bounded $\mathcal{F}_n$-measurable random variable $Z_n$ such that

$$E Z_n \xi < E(Z_n \Delta X_n) < \infty$$

for all $\xi \in \mathcal{C}_n$. Since $\mathcal{C}_n$ is a cone the above inequality must hold true for any $a\xi$, $a > 0$, and so it follows that $E Z_n \xi \leq 0$ for any $\xi \in \mathcal{C}_n$. Furthermore, $E(Z_n \Delta X_n) > 0$ since $0 \in \mathcal{C}_n$ and $Z_n \geq 0$ a.s. The latter follows as for otherwise, taking into account that $\mathcal{C}_n$ contains all integrable nonpositive $\mathcal{F}_n$-measurable random variables, we could choose a random variable $\xi \in \mathcal{C}_n$ which is negative exactly where $Z_n$ is negative and equals zero everywhere else. Then we would obtain $E Z_n \xi > 0$ in contradiction to the argument above. Since $\pm(\gamma_n, \Delta S_n) \in \mathcal{C}_n$ for any bounded $\mathcal{F}_{n-1}$-measurable $\gamma_n$ we have that, in fact, $E(Z_n(\gamma_n, \Delta S_n)) = 0$, and so

$$0 = E(\gamma_n, Z_n \Delta S_n) = E(\gamma_n, E(Z_n \Delta S_n | \mathcal{F}_{n-1})).$$

This being true for any bounded $\mathcal{F}_{n-1}$-measurable $\gamma_n$ yields that with probability one,

$$E(Z_n \Delta S_n | \mathcal{F}_{n-1}) = 0. \tag{4.1.1}$$

Next, observe that we can ensure that $Z_n \geq \varepsilon > 0$ for $\varepsilon$ sma.ll enough by replacing, if necessary, $Z_n$ by $Z_n + \varepsilon$. Indeed, if $\xi \in \mathcal{C}_n$ then there exists $\mathcal{F}_{n-1}$-measurable $\gamma_n$ such that $(\gamma_n, \Delta S_n) \geq \xi$, and by Fatou's lemma

$$E\xi \leq E(\gamma_n, \Delta S_n) \leq \liminf_{l \to \infty} E\big(\mathbb{I}_{\{|\gamma_n| \leq l\}}(\gamma_n, \Delta S_n)\big) = 0$$

since

$$E\big(\mathbb{I}_{\{|\gamma_n| \leq l\}}(\gamma_n, \Delta S_n)\big) = E\big(\mathbb{I}_{\{|\gamma_n| \leq l\}}(\gamma_n, E(\Delta S_n | \mathcal{F}_{n-1}))\big) = 0$$

because $\{S_n\}$ is a martingale. Hence, $E(Z_n + \varepsilon)\xi \le 0$, and if $\varepsilon > 0$ is small enough then $E((Z_n + \varepsilon)\Delta X_n)$ is still positive. Thus, we can assume from now on that $Z_n \ge \varepsilon > 0$ a.s. and that $E(Z_n \Delta X_n)$ is positive.

Let

$$V_{n-1} = E(Z_n | \mathcal{F}_{n-1})$$

and define a new measure $\tilde{P} \sim P$ by its Radon–Nikodym derivative

$$\frac{d\tilde{P}}{dP} = \frac{Z_n}{V_{n-1}}$$

which is bounded since $V_{n-1} \ge \varepsilon$. Then

$$\tilde{P}(\Omega) = E_{\tilde{P}} 1 = E_P \frac{Z_n}{V_{n-1}} = E_P(V_{n-1}^{-1} E(Z_n | \mathcal{F}_{n-1})) = 1,$$

and so $\tilde{P}$ is a probability measure. Next, we will employ the following general result called in [68] a generalized Bayes formula .

**Lemma 4.1.1.** *Let $\mathcal{G} \subset \mathcal{H}$ be sub $\sigma$-algebras of $\mathcal{F}$ and $Q, R$ be two equivalent probability measures. Let $Z_{\mathcal{H}} = \frac{dQ}{dR}|_{\mathcal{H}}$ and $Z_{\mathcal{G}} = \frac{dQ}{dR}|_{\mathcal{G}}$ be the Radon–Nikodym derivatives corresponding to the measures $Q$ and $R$ restricted to the $\sigma$-algebras $\mathcal{H}$ and $\mathcal{G}$, i.e., $Z_{\mathcal{H}}$ and $Z_{\mathcal{G}}$ are $\mathcal{H}$ and $\mathcal{G}$ measurable respectively, and $Q(A) = \int_A Z_{\mathcal{H}} dR$, $Q(B) = \int_B Z_{\mathcal{G}} dR$ for any $A \in \mathcal{H}$ and $B \in \mathcal{G}$. Then, $Z_{\mathcal{H}} = E_R(\frac{dQ}{dR}|\mathcal{H})$, $Z_{\mathcal{G}} = E_R(\frac{dQ}{dR}|\mathcal{G})$ $Q(or R)$-a.s., and for any $\mathcal{H}$-measurable $Q$-integrable random variable $Y$,*

$$E_Q(Y|\mathcal{G}) = \frac{1}{Z_{\mathcal{G}}} E_R(Y Z_{\mathcal{H}} | \mathcal{G}) \quad Q(or R)\text{-a.s.}$$

*where $E_Q$ and $E_R$ are expectations with respect to $Q$ and $R$ respectively.*

**Proof.** Since the equalities $Q(A) = \int_A Z_{\mathcal{H}} dR$ and $Q(B) = \int_B Z_{\mathcal{G}} dR$ remain true for any $A \in \mathcal{H}$ and $B \in \mathcal{G}$ for representations of $Z_{\mathcal{H}}$ and $Z_{\mathcal{G}}$, both as restricted Radon–Nikodym derivatives and as conditional expectations, we conclude that both representations amount to the same. Next, for any $\Gamma \in \mathcal{G}$,

$$\int_\Gamma E_Q(Y|\mathcal{G})dQ = \int_\Gamma Y dQ = \int_\Gamma Y \frac{dQ}{dR} dR = \int_\Gamma E_R(Y \frac{dQ}{dR}|\mathcal{H})dR$$
$$= \int_\Gamma Y Z_{\mathcal{H}} dR = \int_\Gamma E_R(Y Z_{\mathcal{H}}|\mathcal{G})dR = \int_\Gamma (\frac{dQ}{dR})^{-1} E_R(Y Z_{\mathcal{H}}|\mathcal{G})dQ$$
$$= \int_\Gamma \frac{1}{Z_{\mathcal{G}}} E_R(Y Z_{\mathcal{H}}|\mathcal{G})dQ$$

and the lemma follows.                                                    □

Next, set

$$W_k = E\left(\frac{Z_n}{V_{n-1}}\Big|\mathcal{F}_k\right), \ k = 0, 1, \dots .$$

If $k \geq n$ then $W_k = \frac{Z_n}{V_{n-1}}$; and if $k < n$ then

$$W_k = E\left(\frac{E(Z_n|\mathcal{F}_{n-1})}{V_{n-1}}\Big|\mathcal{F}_k\right) = 1.$$

Hence, $W_k = W_{k-1}$ whenever $k \neq n$. Observe that $S_k \in L^1(\Omega, \mathcal{F}_k, \tilde{P})$ since the density $d\tilde{P}/dP$ is bounded. It follows from here and the lemma above that for $k \neq n$,

$$E_{\tilde{P}}(\Delta S_k|\mathcal{F}_{k-1}) = \frac{1}{W_{k-1}} E(W_k \Delta S_k|\mathcal{F}_{k-1}) = E(\Delta S_k|\mathcal{F}_{k-1}) = 0$$

since $\{S_k\}$ is a $P$-martingale. For $k = n$ we have by the above lemma

$$E_{\tilde{P}}(\Delta S_n|\mathcal{F}_{n-1}) = \frac{1}{V_{n-1}} E(Z_n \Delta S_n|\mathcal{F}_{n-1}) = 0$$

where the last equality follows from (4.1.1). Hence, $\tilde{P} \in \mathcal{P}(P)$, and so $\{X_k\}$ is a supermartingale with respect to $\tilde{P}$. It follows that

$$0 \geq E_{\tilde{P}}\left(V_{n-1} E_{\tilde{P}}(\Delta X_n|\mathcal{F}_{n-1})\right) = E_{\tilde{P}}(V_{n-1}\Delta X_n) = E(Z_n \Delta X_n).$$

Since the right hand side of this formula is positive as proved above we arrived at a contradiction, completing the proof of the theorem. $\qquad\square$

### 4.1.2 *Supermartingales with respect to all martingale measures*

Let $f = (f_0, f_1, \dots, f_N)$ be an adapted sequence of nonnegative random variables on a probability space with filtration $(\Omega, \mathcal{F}, \{\mathcal{F}_n\}_{0 \leq n \leq N}, P)$, $\mathcal{F}_0 = \{\emptyset, \Omega\}$. Let $\mathcal{P}(P)$ be a nonempty set of martingale measures and assume that $\sup_{\tilde{P}} \int f_k d\tilde{P} < \infty$ for every $k = 1, 2, \dots, N$.

**Theorem 4.1.2.** *(i) Set*

$$Y_n = ess \ sup_{\tilde{P} \in \mathcal{P}(P)} E_{\tilde{P}}(f_N|\mathcal{F}_n), \ n = 0, 1, 2, \dots, N,$$

*where $E_{\tilde{P}}$ is the expectation with respect to $\tilde{P}$. Then $\{Y_n\}_{0 \leq n \leq N}$ is a supermartingale with respect to each $\tilde{P} \in \mathcal{P}(P)$.*

*(ii) Set*

$$Y_n = ess \ sup_{\tilde{P} \in \mathcal{P}(P), \ \tau \in \mathcal{T}_{nN}} E_{\tilde{P}}(f_\tau|\mathcal{F}_n), \ n = 0, 1, 2, \dots, N,$$

*where $\mathcal{T}_{nN}$ is the set of stopping times $\tau$ such that $n \leq \tau \leq N$. Then $\{Y_n\}_{0 \leq n \leq N}$ is a supermartingale with respect to each $\tilde{P} \in \mathcal{P}(P)$.*

**Proof.** We will prove only (ii). The proof of (i) is obtained in the same way as the proof of (ii) just by disregarding arguments concerning stopping times. We will use only that the measures from $\mathcal{P}(P)$ are equivalent to each other and without loss of generality we assume that $P \in \mathcal{P}(P)$. Now there is nothing special about $P$ so we will prove without loss of generality that $Y_n, n \geq 0$ is a supermartingale with respect to $P$.

If $\tilde{P} \in \mathcal{P}(P)$ we can consider the Radon–Nikodym derivatives

$$\tilde{Z}_N = \frac{d\tilde{P}}{dP}, \ \tilde{Z}_n = \frac{d\tilde{P}_n}{dP_n}, \ n = 1, ..., N-1, \ \tilde{Z}_0 = 1$$

where $P_n$ and $\tilde{P}_n$ are restrictions to $\mathcal{F}_n$ of $P$ and $\tilde{P}$ respectively. Since $\tilde{P} \sim P$ then $P\{\tilde{Z}_n > 0\} = \tilde{P}\{\tilde{Z}_n > 0\} = 1$ for all $n = 0, 1, ..., N$, and so we can define $\tilde{\rho}_n = \frac{\tilde{Z}_n}{\tilde{Z}_{n-1}}$. Then

$$\tilde{Z}_n = \prod_{k=1}^{n} \tilde{\rho}_k.$$

Observe, that $\{\tilde{Z}_n\}$, $n = 1, 2, ..., N$ is a martingale with respect to $P$ since for any $\Gamma \in \mathcal{F}_n$,

$$\int_{\Gamma} \tilde{Z}_n dP = \int_{\Gamma} \frac{d\tilde{P}_n}{dP_n} dP_n = \tilde{P}_n(\Gamma) = \int_{\Gamma} E(\frac{d\tilde{P}}{dP}|\mathcal{F}_n)dP = \int_{\Gamma} E(\tilde{Z}_N|\mathcal{F}_n)dP,$$

and so $\tilde{Z}_n = E(\tilde{Z}_N|\mathcal{F}_n)$ a.s. for $n = 1, 2, ..., N$ (which also follows immediately from the generalized Bayes formula). We see now that all measures $\tilde{P} \in \mathcal{P}(P)$ and their restrictions $\tilde{P}_n$ are defined by means of $P$ and of each one of sequences $\{\tilde{Z}_n\}$ and $\{\tilde{\rho}_n\}$.

By the generalized Bayes formula for any stopping time $N \geq \tau \geq n$,

$$E_{\tilde{P}}(f_\tau|\mathcal{F}_n) = \frac{1}{\tilde{Z}_n} E_P(f_\tau \tilde{Z}_\tau|\mathcal{F}_n) = E_P(\tilde{\rho}_{n+1} \cdots \tilde{\rho}_\tau f_\tau|\mathcal{F}_n)$$

$$= E_P(\bar{\rho}_1 \cdots \bar{\rho}_n \bar{\rho}_{n+1} \cdots \bar{\rho}_\tau f_\tau|\mathcal{F}_n) = E_P(f_\tau \bar{Z}_\tau|\mathcal{F}_n)$$

where $\bar{\rho}_1 = \cdots = \bar{\rho}_n = 1$, $\bar{\rho}_k = \tilde{\rho}_k$ for $k > n$ and $\bar{Z}_k = \bar{\rho}_1 \cdots \bar{\rho}_k$. Thus,

$$Y_n = \text{ess sup}_{\tau \in \mathcal{T}_{nN}, \bar{Z} \in \mathcal{Z}_{nN}} E_P(f_\tau \bar{Z}_\tau|\mathcal{F}_n)$$

where $\mathcal{Z}_{nN}$ is the set of positive $P$-martingales (martingales with respect to $P$), $\bar{Z} = \{\bar{Z}_k\}_{0 \leq k \leq N}$ such that $\bar{Z}_0 = \bar{Z}_1 = \cdots = \bar{Z}_n = 1$ and $\tilde{P} = \int Z_N dP \in \mathcal{P}(P)$. Thus we reduced the problem from taking the supremum over a set of measures to taking the supremum over a set of martingales which is a bit more tractable. Observe that

$$\bar{Z}_{nN} \subset \bar{Z}_{n-1,N} \quad \text{and} \quad \mathcal{T}_{nN} \subset \mathcal{T}_{n-1,N}.$$

By the basic result about ess sup (see, for instance, [58]) the above ess sup can be obtained as a supremum over a sequence $\tau^{(i)}$, $\bar{Z}_{\tau^{(i)}}^{(i)}$ as $i \to \infty$. In fact, as will be shown in Lemma 4.1.2 below this supremum can be obtained here along monotone non-decreasing sequence, i.e.,

$$\text{ess sup}_{\tau \in \mathcal{T}_{nN}, \bar{Z} \in \mathcal{Z}_{nN}} E_P(f_\tau \bar{Z}_\tau | \mathcal{F}_n) = \lim_{i \uparrow \infty} \uparrow E_P(f_{\tau^{(i)}} \bar{Z}_{\tau^{(i)}}^{(i)} | \mathcal{F}_n).$$

By the monotone convergence theorem

$$
\begin{aligned}
E_P(Y_n | \mathcal{F}_{n-1}) &= E_P\big(\text{ess sup}_{\tau \in \mathcal{T}_{nN}, \bar{Z} \in \mathcal{Z}_{nN}} E_P(f_\tau \bar{Z}_\tau | \mathcal{F}_n) | \mathcal{F}_{n-1}\big) \\
&= E_P\big(\lim_{i \uparrow \infty} \uparrow E_P(f_{\tau^{(i)}} \bar{Z}_{\tau^{(i)}}^{(i)} | \mathcal{F}_n) | \mathcal{F}_{n-1}\big) \\
&= \lim_{i \uparrow \infty} E_P\big(f_{\tau^{(i)}} \bar{Z}_{\tau^{(i)}}^{(i)} | \mathcal{F}_{n-1}\big) \\
&\leq \text{ess sup}_{\tau \in \mathcal{T}_{nN}, \bar{Z} \in \mathcal{Z}_{nN}} E_P(f_\tau \bar{Z}_\tau | \mathcal{F}_{n-1}) \\
&\leq \text{ess sup}_{\tau \in \mathcal{T}_{n-1,N}, \bar{Z} \in \mathcal{Z}_{n-1,N}} E_P(f_\tau \bar{Z}_\tau | \mathcal{F}_{n-1}) = Y_{n-1},
\end{aligned}
$$

completing the proof that $\{Y_n\}_{0 \leq n \leq N}$ is a supermartingale. $\qquad\square$

**Lemma 4.1.2.** *Let* $0 \leq n \leq N < \infty$, $\Gamma \in \mathcal{F}_n$, $\tau^{(1)}, \tau^{(2)} \in \mathcal{T}_{nN}$ *and* $\bar{Z}^{(1)}, \bar{Z}^{(2)} \in \mathcal{Z}_{nN}$. *Then* $\tau = \tau^{(1)} \mathbb{I}_\Gamma + \tau^{(2)} \mathbb{I}_{\Omega \setminus \Gamma} \in \mathcal{T}_{nN}$ *and* $\bar{Z} = \bar{Z}^{(1)} \mathbb{I}_\Gamma + \bar{Z}^{(2)} \mathbb{I}_{\Omega \setminus \Gamma} \in \mathcal{Z}_{nN}$. *Furthermore, let* $f_n \geq 0$, $n = 0, 1, ..., N$ *be an adapted sequence satisfying* $\max_{0 \leq k \leq N} \sup_{\tilde{P} \in \mathcal{P}(P)} \int f_k d\tilde{P} < \infty$, *then*

$$E(f_\tau \bar{Z}_\tau | \mathcal{F}_n) = \mathbb{I}_\Gamma E(f_{\tau^{(1)}} \bar{Z}_{\tau^{(1)}}^{(1)} | \mathcal{F}_n) + \mathbb{I}_{\Omega \setminus \Gamma} E(f_{\tau^{(2)}} \bar{Z}_{\tau^{(2)}}^{(2)} | \mathcal{F}_n).$$

*In particular, if* $\Gamma$ *above is defined by*

$$\Gamma = \{\omega : E(f_{\tau^{(1)}} \bar{Z}_{\tau^{(1)}}^{(1)} | \mathcal{F}_n)(\omega) > E(f_{\tau^{(2)}} \bar{Z}_{\tau^{(2)}}^{(2)} | \mathcal{F}_n)(\omega)\}$$

*then*

$$E(f_\tau \bar{Z}_\tau | \mathcal{F}_n) = \max(E(f_{\tau^{(1)}} \bar{Z}_{\tau^{(1)}}^{(1)} | \mathcal{F}_n), E(f_{\tau^{(2)}} \bar{Z}_{\tau^{(2)}}^{(2)} | \mathcal{F}_n))$$

*which means that the family* $\{E(f_\tau \bar{Z}_\tau | \mathcal{F}_n), \tau \in \mathcal{T}_{nN}, \bar{Z} \in \mathcal{Z}_{nN}\}$ *is a lattice, and so for each sequence of random variables from this family we can construct a monotone non-decreasing sequence from the same family which converges to the supremum of the former sequence.*

**Proof.** First, for $k \geq n$,

$$\{\tau = k\} = (\Gamma \cap \{\tau^{(1)} = k\}) \cup ((\Omega \setminus \Gamma) \cap \{\tau^{(2)} = k\}) \in \mathcal{F}_k,$$

and so $\tau$ is a stopping time since $\tau \geq n$ taking into account that $\tau^{(1)}, \tau^{(2)} \geq n$. Next, for $k > m \geq n$,

$$
\begin{aligned}
E(f_k \bar{Z}_k | \mathcal{F}_m) &= \mathbb{I}_\Gamma E(f_k \bar{Z}_k^{(1)} | \mathcal{F}_m) + \mathbb{I}_{\Omega \setminus \Gamma} E(f_k \bar{Z}_k^{(2)} | \mathcal{F}_m) \\
&= f_m(\mathbb{I}_\Gamma \bar{Z}_m^{(1)} + \mathbb{I}_{\Omega \setminus \Gamma} \bar{Z}_m^{(2)}) = f_m \bar{Z}_m
\end{aligned}
$$

if $\{f_k \bar{Z}_k^{(1)}\}_{k=1}^N$ and $\{f_k \bar{Z}_k^{(2)}\}_{k=1}^N$ are martingales. If $m < n$ then we write

$$E(f_k \bar{Z}_k | \mathcal{F}_m) = E(E(f_k \bar{Z}_k | \mathcal{F}_n) | \mathcal{F}_m) = E(f_n \bar{Z}_n | \mathcal{F}_m) = E(f_n | \mathcal{F}_m)$$

since $\bar{Z}_n^{(1)} = \bar{Z}_n^{(2)} = 1$, and so $\bar{Z}_n = 1$. If $\{f_k\}_{k=1}^N$ is a martingale, in particular, if $f_k \equiv 1$ for all $k$, then the right hand side above becomes $f_m = f_m \bar{Z}_m$ since $\bar{Z}_m^{(1)} = \bar{Z}_m^{(2)} = 1$ when $m \leq n$, thus $\bar{Z}_m = 1$ as well. In particular, if we set $f_k = S_k$, $k \geq 0$, where $S_k$ is the discounted stock price at the time $k$, which is a $P$-martingale, then we obtain that both $\bar{Z}_k$, $k \geq 0$ and $S_k \bar{Z}_k$, $k \geq 0$ are $P$-martingales, and $\bar{Z} \in \mathcal{Z}_{nN}$. The formulas for $E(f_\tau \bar{Z}_\tau | \mathcal{F}_n)$ in the statement of the lemma follow immediately from the definitions there.

Next, let $\tau^{(1)}, \tau^{(2)}, \ldots \in \mathcal{T}_{nN}$ and $\bar{Z}^{(1)}, \bar{Z}^{(2)}, \ldots \in \mathcal{Z}_{nN}$. Define $\sigma^{(1)} = \tau^{(1)}$, $\bar{U}^{(1)} = \bar{Z}^{(1)}$ and recursively for $i > 1$,

$$\sigma^{(i)} = \mathbb{I}_{\Gamma_i} \tau^{(i)} + \mathbb{I}_{\Omega \backslash \Gamma_i} \sigma^{(i-1)} \text{ and } \bar{U}^{(i)} = \mathbb{I}_{\Gamma_i} \bar{Z}^{(i)} + \mathbb{I}_{\Omega \backslash \Gamma_i} \bar{U}^{(i-1)}$$

where

$$\Gamma_i = \{\omega : E(f_{\tau^{(i)}} \bar{Z}_{\tau^{(i)}}^{(i)} | \mathcal{F}_n)(\omega) > E(f_{\sigma^{(i-1)}} \bar{U}_{\sigma^{(i-1)}}^{(i-1)} | \mathcal{F}_n)(\omega)\}.$$

In the same way as above we conclude that $\sigma^{(i)} \in \mathcal{T}_{nN}$, $\bar{U}^{(i)} \in \mathcal{Z}_{nN}$ and

$$E(f_{\sigma^{(i)}} \bar{U}_{\sigma^{(i)}}^{(i)} | \mathcal{F}_n) = \max_{1 \leq j \leq i} E(f_{\tau^{(j)}} \bar{Z}_{\tau^{(j)}}^{(j)} | \mathcal{F}_n)$$

producing a monotone non-decreasing sequence, as required. $\qquad\square$

## 4.2 Lecture 8: Superhedging in incomplete markets

### 4.2.1 *Superhedging of European contingent claims in incomplete markets*

We defined the fair price (from the seller's point of view) of a European contingent claim with a payoff $f_N$ at the time $N$ by

$$V = V(f_N, P) = \inf\{x : \exists \pi \text{ such that } X_0^\pi = x, \ X_N^\pi \geq f_N \ P\text{-a.s}\}$$

where $\pi$ denotes a self-financing trading strategy and $X^\pi$ is the value of the corresponding portfolio.

**Theorem 4.2.1.** *Let a nonnegative $\mathcal{F}_N$-measurable random variable $f_N$ satisfy*

$$\sup_{\tilde{P} \in \mathcal{P}(P)} E_{\tilde{P}} \frac{f_N}{B_N} < \infty$$

*where $B_N$ is the bond price at the time $N$. Then*

$$V = V(f_N, P) = \sup_{\tilde{P} \in \mathcal{P}(P)} B_0 E_{\tilde{P}} \frac{f_N}{B_N}$$

*and there exists a hedging strategy with the initial capital $V$.*

**Proof.** We already proved for the general case that for any martingale measure $\tilde{P}$ the initial capital of a self-financing hedging portfolio cannot be less than $B_0 E_{\tilde{P}} \frac{f_N}{B_N}$, and so

$$V \geq \sup_{\tilde{P} \in \mathcal{P}(P)} B_0 E_{\tilde{P}} \frac{f_N}{B_N}.$$

In order to obtain the inequality in the other direction set

$$Y_n = \operatorname{ess\,sup}_{\tilde{P} \in \mathcal{P}(P)} E_{\tilde{P}}(\frac{f_N}{B_N}|\mathcal{F}_n),\ n = 0, 1, ..., N$$

which, as we proved, is a supermartingale with respect to any $\tilde{P} \in \mathcal{P}(P)$. Hence, there exists an optional decomposition

$$Y_n = Y_0 + \sum_{k=1}^{n} \gamma_k \Delta(\frac{S_k}{B_k}) - C_n,\ C_0 = 0,$$

where $\{\gamma_k\}_{1 \leq k \leq N}$ is a predictable sequence and $\{C_k\}_{1 \leq k \leq N}$ is an adapted non-decreasing process.
Set

$$X_n^\pi = B_n(Y_0 + \sum_{k=1}^{n} \gamma_k \Delta(\frac{S_k}{B_k})),\ X_0^\pi = B_0 Y_0$$

and

$$\beta_n = \frac{1}{B_n}(X_n^\pi - \gamma_n S_n)$$

so that $X_n^\pi = \beta_n B_n + \gamma_n S_n$ and $\pi = (\beta_n, \gamma_n)_{1 \leq n \leq N}$. From the definition of $X_n^\pi$ above we obtain that

$$\frac{X_n^\pi}{B_n} - \frac{X_{n-1}^\pi}{B_{n-1}} = \gamma_n \Delta(\frac{S_n}{B_n}),$$

and so

$$\frac{X_{n-1}^\pi}{B_{n-1}} = \frac{X_n^\pi}{B_n} - \gamma_n \frac{S_n}{B_n} + \gamma_n \frac{S_{n-1}}{B_{n-1}} = \beta_n + \gamma_n \frac{S_{n-1}}{B_{n-1}}.$$

Hence, $\beta_n = \frac{X_{n-1}^\pi}{B_{n-1}} - \gamma_n \frac{S_{n-1}}{B_{n-1}}$, and so $\{\beta_n\}_{1 \leq n \leq N}$ is a predictable sequence and we also have from here that

$$X_{n-1}^\pi = \beta_n B_{n-1} + \gamma_n S_{n-1},$$

implying that $\pi$ is a self-financing trading strategy. It is also hedging since

$$X_N^\pi = B_N(Y_0 + \sum_{k=1}^{N} \gamma_k \Delta(\frac{S_k}{B_k}))$$
$$\geq B_N(Y_0 + \sum_{k=1}^{N} \gamma_k \Delta(\frac{S_k}{B_k}) - C_N) = B_N Y_N = f_N.$$

The initial capital $X_0^\pi = B_0 Y_0 = \sup_{\tilde{P} \in \mathcal{P}(P)} B_0 E_{\tilde{P}} \frac{f_n}{B_N}$ is as required, so the proof is complete. $\square$

### 4.2.2 *Superhedging of American contingent claims*

We defined the fair price (based on hedging) of an American contingent claim with a payoff process $f = \{f_n\}_{0 \leq n \leq N}$ and a horizon $N < \infty$ by

$$V = V_N(f, P) = \inf\{x : \text{there exists a self-financing strategy } \pi \text{ such that}$$
$$X_0^\pi = x, \; X_n^\pi \geq f_n \; P - \text{a.s.} \; \forall n = 1, 2, ..., N\}.$$

Again, let $\mathcal{T}_{nN}$ denote the set of all stopping times $\tau$ with $0 \leq \tau \leq N$ and $\mathcal{P}(P)$ denote the set of all martingale measures.

**Theorem 4.2.2.** *Assume that $\mathcal{P}(P) \neq \emptyset$ and let a nonnegative adapted payoff process $f = \{f_n\}_{0 \leq n \leq N}$ satisfy*

$$\sup_{\tilde{P} \in \mathcal{P}(P)} E_{\tilde{P}} \frac{f_n}{B_n} < \infty \text{ for all } n = 0, 1, ..., N.$$

*Then*

$$V = V_N(f, P) = \sup_{\tilde{P} \in \mathcal{P}(P), \, \tau \in \mathcal{T}_{0N}} B_0 E_{\tilde{P}} \frac{f_\tau}{B_\tau}$$

*and there exists a hedging strategy with the initial capital $V$.*

**Proof.** We already proved for the general case that for any martingale measure $\tilde{P}$ the initial capital of a self-financing hedging portfolio cannot be less than $B_0 \sup_{\tau \in \mathcal{T}_{0N}} E_{\tilde{P}} \frac{f_\tau}{B_\tau}$, so

$$V \geq \sup_{\tilde{P} \in \mathcal{P}(P), \, \tau \in \mathcal{T}_{0N}} B_0 E_{\tilde{P}} \frac{f_\tau}{B_\tau}.$$

In order to obtain the inequality in the other direction set

$$Y_n = \text{ess sup}_{\tilde{P} \in \mathcal{P}(P), \, \tau \in \mathcal{T}_{nN}} E_{\tilde{P}}\left(\frac{f_\tau}{B_\tau}|\mathcal{F}_n\right), \; n = 0, 1, ..., N$$

which, as we proved, is a supermartingale with respect to any $\tilde{P} \in \mathcal{P}(P)$. Hence, there exists an optional decomposition

$$Y_n = Y_0 + \sum_{k=1}^n \gamma_k \Delta\left(\frac{S_k}{B_k}\right) - C_n, \; C_0 = 0,$$

where $\{\gamma_k\}_{1 \leq k \leq N}$ is a predictable sequence and $\{C_k\}_{1 \leq k \leq N}$ is an adapted non-decreasing process.

Set, again,

$$X_n^\pi = B_n\left(Y_0 + \sum_{k=1}^n \gamma_k \Delta\left(\frac{S_k}{B_k}\right)\right), \; X_0^\pi = B_0 Y_0$$

and

$$\beta_n = \frac{1}{B_n}(X_n^\pi - \gamma_n S_n)$$

so that $X_n^\pi = \beta_n B_n + \gamma_n S_n$ and $\pi = (\beta_n, \gamma_n)_{1 \leq n \leq N}$. The initial capital $X_0^\pi = B_0 Y_0 = V_N(f, P)$ for the strategy $\pi$ is as required. The proof that the strategy $\pi = (\beta_n, \gamma_n)_{0 \leq n \leq N}$ is self-financing, i.e., showing that $X_{n-1}^\pi = \beta_n B_{n-1} + \gamma_n S_{n-1}$, is the same as in the case of European contingent claims above since we have similar definitions here and there. The strategy $\pi$ is also hedging since for all $n = 0, 1, .., N$,

$$X_n^\pi \geq B_n Y_n \geq B_n E_{\tilde{P}}\left(\frac{f_n}{B_n} | \mathcal{F}_n\right) = f_n \text{ a.s.,}$$

where the first inequality follows from the definitions of $Y_n$, $X_n^\pi$ and from the fact that $C_n \geq 0$ while the second inequality follows since we always can take the stopping time $\tau \equiv n$. The proof is complete now. □

If we can obtain a complete description of the set $\mathcal{P}(P)$ of martingale measures, then in some cases the above superhedging price can be computed using the backward induction (dynamical programming) derived for optimal stopping problems.

**Corollary 4.2.1.** *For each martingale measure $\tilde{P}$ set $V_N^{\tilde{P}} = B_0 \frac{f_N}{B_N}$ and*

$$V_n^{\tilde{P}} = \max\left(B_0 \frac{f_0}{B_n}, E_{\tilde{P}}(V_{n+1}^{\tilde{P}} | \mathcal{F}_n)\right)$$

*for $n = N-1, N-2, ..., 1, 0$. Then the superhedging price $V$ is given by the formula*

$$V = \sup_{\tilde{P}} V_0^{\tilde{P}}.$$

### 4.2.3  *Superhedging of Israeli (game) contingent claims*

An Israeli contingent claim is determined by a payoff process $R(n, m) = f_n \mathbb{I}_{n < m} + g_m \mathbb{I}_{m \geq n}$ and a horizon $N < \infty$ where $\{f_n\}_{0 \leq n \leq N}$ and $\{g_n\}_{0 \leq n \leq N}$ are adapted sequences of random variables such that $f_n \geq g_n \geq 0$ for all $n = 0, 1, ..., n$. We defined the fair price (based on hedging) of such Israeli contingent claim by

$$V = V_N(f, g, P) = \inf\{x : \text{there exists a self-financing trading strategy } \pi$$
$$\text{and a cancellation stopping time } \sigma \text{ such that}$$
$$X_0^\pi = x, X_{\sigma \wedge n}^\pi \geq R(\sigma, n) \text{ } P\text{-a.s. } \forall n = 1, 2, ..., N\}.$$

Again, let $\mathcal{T}_{nN}$ denote the set of all stopping times $\tau$ with $0 \leq \tau \leq N$ and $\mathcal{P}(P)$ denote the set of all martingale measures.

**Theorem 4.2.3.** *Assume that $\mathcal{P}(P) \neq \emptyset$ and suppose that the nonnegative adapted process $f = \{f_n\}_{0 \leq n \leq N}$ satisfies*

$$\sup_{\tilde{P} \in \mathcal{P}(P)} E_{\tilde{P}} \frac{f_n}{B_n} < \infty \text{ for all } n = 0, 1, ..., N.$$

*Then*

$$V = V_N(f, g, P) = \inf_{\sigma \in \mathcal{T}_{0N}} \sup_{\tilde{P} \in \mathcal{P}(P), \tau \in \mathcal{T}_{0N}} B_0 E_{\tilde{P}} \frac{R(\sigma, \tau)}{B_{\sigma \wedge \tau}}.$$

**Proof.** We already proved for the general case that for any martingale measure $\tilde{P}$ the initial capital of a self-financing hedging investment strategy $(\pi, \sigma)$ cannot be less than $B_0 \sup_{\tau \in \mathcal{T}_{0N}} E_{\tilde{P}} \frac{R(\sigma, \tau)}{B_{\sigma \wedge \tau}}$, and so

$$V \geq \inf_{\sigma \in \mathcal{T}_{0N}} \sup_{\tilde{P} \in \mathcal{P}(P), \tau \in \mathcal{T}_{0N}} B_0 E_{\tilde{P}} \frac{R(\sigma, \tau)}{B_{\sigma \wedge \tau}}.$$

In order to obtain the inequality in the other direction, for any $\sigma \in \mathcal{T}_{0N}$, set

$$Y_n^\sigma = \text{ess sup}_{\tilde{P} \in \mathcal{P}(P), \tau \in \mathcal{T}_{nN}} E_{\tilde{P}} \left( \frac{R(\sigma, \tau)}{B_{\sigma \wedge \tau}} | \mathcal{F}_n \right), \ n = 0, 1, ..., N$$

which, as we proved, is a supermartingale with respect to any $\tilde{P} \in \mathcal{P}(P)$ (since $\frac{R(\sigma, n)}{B_{\sigma \wedge n}}$ is $\mathcal{F}_n$-measurable as we checked it in Lecture 5). Hence, there exists an optional decomposition

$$Y_n^\sigma = Y_0^\sigma + \sum_{k=1}^n \gamma_k^\sigma \Delta\left(\frac{S_k}{B_k}\right) - C_n^\sigma, \ C_0^\sigma = 0,$$

where $\{\gamma_k^\sigma\}_{1 \leq k \leq N}$ is a predictable sequence and $\{C_k^\sigma\}_{1 \leq k \leq N}$ is an adapted non-decreasing process.

Set

$$X_n^{\pi, \sigma} = B_n \left( Y_0^\sigma + \sum_{k=1}^n \gamma_k^\sigma \Delta\left(\frac{S_k}{B_k}\right) \right), \ X_0^{\pi, \sigma} = B_0 Y_0^\sigma$$

and

$$\beta_n^\sigma = \frac{1}{B_n}(X_n^{\pi, \sigma} - \gamma_n^\sigma S_n)$$

so that $X_n^{\pi, \sigma} = \beta_n^\sigma B_n + \gamma_n^\sigma S_n$ and $\pi = \pi^\sigma = (\beta_n^\sigma, \gamma_n^\sigma)_{1 \leq n \leq N}$. The proof that the strategy $\pi = (\beta_n^\sigma, \gamma_n^\sigma)_{0 \leq n \leq N}$ is self-financing, i.e., showing that $X_{n-1}^{\pi, \sigma} = \beta_n^\sigma B_{n-1} + \gamma_n^\sigma S_{n-1}$, is the same as in the case of European contingent claims above since we have similar definitions here and there. The strategy $(\pi, \sigma)$ is also hedging since for all $n = 0, 1, .., N$,

$$X^{\pi,\sigma}_{\sigma \wedge n} \geq B_{\sigma \wedge n} Y^{\sigma}_{\sigma \wedge n} \geq B_{\sigma \wedge n} E_{\tilde{P}}\left(\frac{R(\sigma, n)}{B_{\sigma \wedge n}}\Big|\mathcal{F}_n\right) = R(\sigma, n) \text{ a.s.}$$

where the first inequality follows from the definitions of $Y^{\sigma}_n$, $X^{\pi,\sigma}_n$ and from the fact that $C^{\sigma}_n \geq 0$ while the second inequality follows since we always can take the stopping time $\tau \equiv n$.

Next, just by the definition of the infimum for each $\varepsilon > 0$ we can choose $\sigma_\varepsilon \in \mathcal{T}_{0N}$ such that for $\pi = \pi^{\sigma_\varepsilon}$,

$$X^{\pi,\sigma_\varepsilon}_0 = B_0 Y^{\sigma_\varepsilon}_0 \leq V_N(f, g, P) + \varepsilon$$

which means that there exists a self-financing hedging strategy with any initial capital larger than $V_N(f, g, P)$, and so by definition the fair price should not be bigger than $V_N(f, g, P)$, completing the proof of the theorem. $\square$

### 4.2.4 *Existence of a superhedging strategy for game options with the initial capital equal to the superhedging price*

We have not proved that there exists a self-financed hedging trading strategy $\pi^*$ and a cancellation (stopping) time $\sigma^*$ with the initial capital $X^{\pi^*,\sigma^*}_0 = V_N(f, g, P)$ which does not follow directly from the theory of optimal stopping (Dynkin's) games and requires an additional argument.

**Theorem 4.2.4.** *A superhedging self-financing strategy with the initial capital equal to the superhedging price exists for any game option satisfying conditions of the previous subsection.*

**Proof.** The proof here will be a combination of the method by which we proved the existence of the equilibrium (saddle point) in Dynkin's games and the technique to obtain supermartingales with respect to all martingale measures above where we replaced the supremum over all martingale measures by the supremum over martingales being corresponding Radon–Nikodym derivatives (cf. the proof of existence of an optimal stopping time for the corresponding American contingent claim case in [48]).

We proved that the superhedging price of a game option equals to

$$V = V(f, g, P) = \text{ess inf}_{\sigma \in \mathcal{T}_{0N}} \text{ess sup}_{\tilde{P} \in \mathcal{P}(P)} \text{ess sup}_{\tau \in \mathcal{T}_{0N}} B_0 E_{\tilde{P}}\left(\frac{R(\sigma, \tau)}{B_{\sigma \wedge \tau}}\right)$$

where $R(\sigma, \tau) = f_\sigma \mathbb{I}_{\sigma < \tau} + g_\tau \mathbb{I}_{\sigma \geq \tau}$. Let $\mathcal{T}_{nN}$, $n \leq N$ be the collection of all stopping times $\tau$ satisfying $n \leq \tau \leq N$ and $\mathcal{Z}_{nN}$ be the set of positive $P$-martingales $\bar{Z} = \{\bar{Z}_k\}_{0 \leq k \leq N}$ such that $\bar{Z}_0 = \bar{Z}_1 = \cdots = \bar{Z}_n = 1$. Then

$$V = \inf_{\sigma \in \mathcal{T}_{0N}} \sup_{\bar{Z} \in \mathcal{Z}_{0N}} \sup_{\tau \in \mathcal{T}_{0N}} B_0 E\big(\bar{Z}_{\sigma \wedge \tau} \frac{R(\sigma, \tau)}{B_{\sigma \wedge \tau}}\big).$$

Set

$$V_n = \text{ess inf}_{\sigma \in \mathcal{T}_{nN}} \text{ess sup}_{\bar{Z} \in \mathcal{Z}_{nN}} \text{ess sup}_{\tau \in \mathcal{T}_{nN}} B_0 E\big(\bar{Z}_{\sigma \wedge \tau} \frac{R(\sigma, \tau)}{B_{\sigma \wedge \tau}} | \mathcal{F}_n\big).$$

For $\sigma \in \mathcal{T}_{nN}$ let

$$\tilde{\sigma} = \begin{cases} \sigma & \text{if } \sigma > n \\ N & \text{if } \sigma = n \end{cases}$$

which is a stopping time. Then taking into account that $\bar{Z}_n = 1$ when $\bar{Z} \in \mathcal{Z}_{nN}$, we obtain

$$\text{ess sup}_{\bar{Z} \in \mathcal{Z}_{nN}} \text{ess sup}_{\tau \in \mathcal{T}_{nN}} B_0 E(\bar{Z}_{\sigma \wedge \tau} \frac{R(\sigma, \tau)}{B_{\sigma \wedge \tau}} | \mathcal{F}_n)$$

$$= \text{ess sup}_{\bar{Z} \in \mathcal{Z}_{nN}} \text{ess sup}_{\tau \in \mathcal{T}_{nN}} B_0 E(\bar{Z}_{\tilde{\sigma} \wedge \tau} \frac{R(\tilde{\sigma}, \tau)}{B_{\tilde{\sigma} \wedge \tau}} | \mathcal{F}_n) \mathbb{I}_{\{\sigma > n\}} + B_0 \frac{f_n}{B_n} \mathbb{I}_{\{\sigma = n\}}.$$

Hence,

$$\text{ess sup}_{\bar{Z} \in \mathcal{Z}_{nN}} \text{ess sup}_{\tau \in \mathcal{T}_{nN}} B_0 E(\bar{Z}_{\sigma \wedge \tau} \frac{R(\sigma, \tau)}{B_{\sigma \wedge \tau}} | \mathcal{F}_n)$$

$$\geq \min(B_0 \frac{f_n}{B_n}, \text{ess sup}_{\bar{Z} \in \mathcal{Z}_{nN}} \text{ess sup}_{\tau \in \mathcal{T}_{nN}} B_0 E(\bar{Z}_{\tilde{\sigma} \wedge \tau} \frac{R(\tilde{\sigma}, \tau)}{B_{\tilde{\sigma} \wedge \tau}} | \mathcal{F}_n))$$

$$\geq \min(B_0 \frac{f_n}{B_n}, B_0 E(V_{n+1} | \mathcal{F}_n))$$

where we used that $\mathcal{T}_{n+1, N} \subset \mathcal{T}_{nN}$, $\mathcal{Z}_{n+1, N} \subset \mathcal{Z}_{nN}$ and that

$$E(V_{n+1} | \mathcal{F}_n) \leq \text{ess sup}_{\bar{Z} \in \mathcal{Z}_{n+1, N}} \text{ess sup}_{\tau \in \mathcal{T}_{n+1, N}} B_0 E(\bar{Z}_{\tilde{\sigma} \wedge \tau} \frac{R(\tilde{\sigma}, \tau)}{B_{\tilde{\sigma} \wedge \tau}} | \mathcal{F}_n).$$

The latter follows since for any $\tilde{\sigma} \in \mathcal{T}_{n+1, N}$,

$$E\big(\text{ess sup}_{\bar{Z} \in \mathcal{Z}_{n+1, N}} \text{ess sup}_{\tau \in \mathcal{T}_{n+1, N}} B_0 E(\bar{Z}_{\tilde{\sigma} \wedge \tau} \frac{R(\tilde{\sigma}, \tau)}{B_{\tilde{\sigma} \wedge \tau}} | \mathcal{F}_{n+1}) | \mathcal{F}_n\big)$$

$$\leq \text{ess sup}_{\bar{Z} \in \mathcal{Z}_{n+1, N}} \text{ess sup}_{\tau \in \mathcal{T}_{n+1, N}} B_0 E(\bar{Z}_{\tilde{\sigma} \wedge \tau} \frac{R(\tilde{\sigma}, \tau)}{B_{\tilde{\sigma} \wedge \tau}} | \mathcal{F}_n).$$

This is proved similarly to the first subsection of Lecture 8 obtaining both ess sup above as limits along non-decreasing sequences of expectations and using the monotone convergence theorem, i.e.,

$$E(\lim_{k \uparrow \infty} \uparrow E(\bar{Z}^{(k)}_{\sigma \wedge \tau_k} \frac{R(\sigma, \tau_k)}{B_{\sigma \wedge \tau_k}} | \mathcal{F}_{n+1}) | \mathcal{F}_n)$$

$$= \lim_{k \uparrow \infty} \uparrow E(\bar{Z}^{(k)}_{\sigma \wedge \tau_k} \frac{R(\sigma, \tau_k)}{B_{\sigma \wedge \tau_k}} | \mathcal{F}_n).$$

Thus,

$$V_n = \text{ess inf}_{\sigma \in \mathcal{T}_{nN}} \text{ ess sup}_{\bar{Z} \in \mathcal{Z}_{nN}} \text{ ess sup}_{\tau \in \mathcal{T}_{nN}} B_0 E(\bar{Z}_{\sigma \wedge \tau} \frac{R(\sigma,\tau)}{B_{\sigma \wedge \tau}} | \mathcal{F}_n)$$
$$\geq \min(B_0 \frac{f_n}{B_n}, E(V_{n+1}|\mathcal{F}_n)).$$

From the formula for $V_n$ we have that $B_0 \frac{g_n}{B_n} \leq V_n \leq B_0 \frac{f_n}{B_n}$ since we always can take there $\sigma = n$ or $\tau = n$. Set

$$\sigma_0 = \min\{n \leq N : V_n = B_0 \frac{f_n}{B_n}\}.$$

On the event $\{\sigma_0 > n\}$ we have that $V_n < B_0 \frac{f_n}{B_n}$, and so on the event above

$$V_n \geq \min(B_0 \frac{f_n}{B_n}, E(V_{n+1}|\mathcal{F}_n)) \geq E(V_{n+1}|\mathcal{F}_n).$$

It follows that

$$E(V_{\sigma_0 \wedge (n+1)}|\mathcal{F}_n) = V_{\sigma_0} \mathbb{I}_{\{\sigma_0 \leq n\}} + E(V_{n+1}|\mathcal{F}_n)\mathbb{I}_{\{\sigma_0 > n\}}$$
$$\leq V_{\sigma_0} \mathbb{I}_{\{\sigma_0 \leq n\}} + V_n \mathbb{I}_{\{\sigma_0 > n\}} = V_{\sigma_0 \wedge n}.$$

Hence, the sequence $V_{\sigma_0 \wedge n}$ is a supermartingale, and so

$$V_0 \geq E V_{\sigma_0 \wedge \tau} \geq B_0 E \frac{R(\sigma_0,\tau)}{B_{\sigma_0 \wedge \tau}}$$

for any stopping time $\tau$ where we take into account that $\bar{Z}_{\sigma_0 \wedge \tau} = 1$ when $\bar{Z} \in \mathcal{Z}_{\sigma_0 \wedge \tau, N}$. The latter inequality holds true since on the event $\{\sigma_0 < \tau\}$ we have

$$V_{\sigma_0 \wedge \tau} = V_{\sigma_0} = B_0 \frac{f_{\sigma_0}}{B_{\sigma_0}} = B_0 \frac{R(\sigma_0,\tau)}{B_{\sigma_0 \wedge \tau}},$$

while on the event $\{\sigma_0 \geq \tau\}$ we have that

$$V_{\sigma_0 \wedge \tau} = V_\tau \geq B_0 \frac{g_\tau}{B_\tau} = B_0 \frac{R(\sigma_0,\tau)}{B_{\sigma_0 \wedge \tau}}$$

as $V_n \geq B_0 \frac{g_n}{B_n}$ always.

Without loss of generality we can assume that $P$ is itself a martingale measure and since no measure from $\mathcal{P}(P)$ plays a special role here, we obtain that

$$V_0 \geq \sup_{\tilde{P} \in \mathcal{P}(P)} \sup_{\tau \leq N} B_0 E_{\tilde{P}}(\frac{R(\sigma_0,\tau)}{B_{\sigma_0 \wedge \tau}}) = X_0^{\pi,\sigma_0},$$

where the portfolio $X^{\pi,\sigma_0}$ was constructed before. This inequality implies that the initial capital of the hedging strategy $(\pi, \sigma_0)$, $\pi = \pi^{\sigma_0}$ does not exceed the superhedging price $V_0$ and since it cannot be smaller than $V_0$ it equals precisely $V_0$. $\square$

### 4.2.5   *Pricing of contingent claims from buyer's point of view*

Define

$$W = W(f_N, P) = \sup\{x : \text{there exists a self-financing strategy } \pi$$
$$\text{such that } X_0^\pi = -x, \ X_N^\pi + f_N \geq 0 \ P\text{-a.s.}\}$$

where $f_N \geq 0$ is $\mathcal{F}_N$-measurable and $f_N > 0$ with positive probability. This quantity arises as a fair price of a European contingent claim from the following point of view. The buyer of the contract takes a bank loan $x$, pays for the contract and builds an investment portfolio $X^\pi$ starting with the initial capital $-x$, so that at time $N$ he/she gets the payoff $f_N$ which should cover the minus (overdraft) in his/her portfolio.

**Theorem 4.2.5.** *Assume that a nonnegative $\mathcal{F}_N$-measurable random variable $f_N$ satisfies*

$$\sup_{\tilde{P} \in \mathcal{P}(P)} E_{\tilde{P}} \frac{f_N}{B_N} < \infty$$

*where $B_N$ is the bond price at time $N$. Then*

$$W = W(f_N, P) = \inf_{\tilde{P} \in \mathcal{P}(P)} B_0 E_{\tilde{P}} \frac{f_N}{B_N}.$$

**Proof.** a) Assume that there exists a self-financing portfolio strategy $\pi$ such that $X_0^\pi = -x$ and $X_N^\pi + f_N \geq 0$ $\tilde{P}$-a.s. Since $\{\frac{X_n}{B_n}\}_{0 \leq n \leq N}$ is a martingale with respect to $\tilde{P} \in \mathcal{P}(P)$ we obtain

$$0 \leq E_{\tilde{P}} \frac{X_N^\pi}{B_N} + E_{\tilde{P}} \frac{f_N}{B_N} = -\frac{x}{B_0} + E_{\tilde{P}} \frac{f_N}{B_N},$$

and so $x \leq B_0 E_{\tilde{P}} \frac{f_N}{B_N}$ for any $\tilde{P} \in \mathcal{P}(P)$ yielding that $x \leq W(f_N, P)$.

b) For the inequality in the other direction set

$$Y_n = \text{ess sup}_{\tilde{P} \in \mathcal{P}} E_{\tilde{P}}(-\frac{f_N}{B_N} | \mathcal{F}_n) = -\text{ess inf}_{\tilde{P} \in \mathcal{P}} E_{\tilde{P}}(\frac{f_N}{B_N} | \mathcal{F}_n).$$

Then

$$Y_0 = -\frac{W(f_N, P)}{B_0} \quad \text{and} \quad Y_N = -\frac{f_N}{B_N}.$$

As we know $\{Y_n\}_{0 \leq n \leq N}$ is a supermartingale with respect to all martingale measures $\tilde{P}$, and so an optional decomposition

$$Y_n = Y_0 + \sum_{k=1}^{n} \gamma_k \Delta \frac{S_k}{B_k} - C_n$$

holds true.

Set

$$X_n^\pi = B_n(Y_0 + \sum_{k=1}^{n} \gamma_k \Delta \frac{S_k}{B_k}) \text{ and } \beta_n = \frac{X_n^\pi}{B_n} - \gamma_n \frac{S_n}{B_n} = \frac{X_{n-1}^\pi}{B_{n-1}} - \gamma_n \frac{S_{n-1}}{B_{n-1}}.$$

Then, as before, the strategy $\pi = (\beta_n, \gamma_n)_{0 \le n \le N}$ is self-financing, $X_0^\pi = B_0 Y_0 = -W(f_N, P)$ and

$$\frac{X_N^\pi}{B_N} \ge Y_N = -\frac{f_N}{B_N}$$

which says that starting a portfolio with the initial capital $-W(f_N, P)$ and obtaining the payoff $f_N$ at time $N$ we are left with a nonnegative sum, meaning that the fair price from this point of view should not be less that $W(f_N, P)$ completing the proof. $\qquad\square$

The interval $(W(f_N, P), V(f_N, P))$ is called the no arbitrage interval. The reason is as follows. If the contract is sold for the price $v > V(f_N, P)$ then the seller can start a hedging portfolio with the initial capital $V(f_N, P)$ and to gain the riskless profit $v - V(f_N, P)$. If the contract is sold for the price $v < W(f_N, P)$ then the buyer can take a loan $W(f_N, P)$, to have zero debt at time $N$ and to gain the riskless profit $W(f_N, P) - v$.

On the other hand, if the selling price of the contract was $v \in (W(f_N, P), V(f_N, P))$ then there exists $\tilde{P} \in \mathcal{P}(P)$ such that

$$v = B_0 E_{\tilde{P}} \frac{f_N}{B_N} > 0.$$

Indeed, for any two martingale measures $\tilde{P}_1, \tilde{P}_2$, any measure $\tilde{P} = \alpha \tilde{P}_1 + \beta \tilde{P}_2$, $\alpha, \beta \ge 0$, $\alpha + \beta = 1$ is also the martingale measure and $E_{\tilde{P}} \frac{f_N}{B_N} = \alpha E_{\tilde{P}_1} \frac{f_N}{B_N} + \beta E_{\tilde{P}_2} \frac{f_N}{B_N}$. If $X^\pi$ is a self-financing hedging investment portfolio, i.e., $X_N^\pi \ge f_N$, then

$$\frac{X_0^\pi}{B_0} = E_{\tilde{P}} \frac{X_N^\pi}{B_N} \ge E_{\tilde{P}} \frac{f_N}{B_N},$$

i.e., $X_0^\pi \ge v$. This means that in order to fulfill his/her obligation, the seller has to invest into the hedging portfolio as initial capital at least the sum $v$ obtained from the buyer and he/she gets no riskless profit. On the buyer's side, if $\frac{X_N^\pi}{B_N} + \frac{f_N}{B_N} \ge 0$ then

$$0 \le E_{\tilde{P}} \frac{X_N^\pi}{B_N} + E_{\tilde{P}} \frac{f_N}{B_N} = \frac{X_0^\pi}{B_0} + E_{\tilde{P}} \frac{f_N}{B_N},$$

and so $X_0^\pi \ge -B_0 E_{\tilde{P}} \frac{f_N}{B_N} = -v$. This means that if the buyer takes a loan $v' > v$ and profits from the difference $v' - v$, then any self-financing portfolio with the initial capital $-v'$ cannot cover his/her debt at time $N$ even together with the payoff $f_N$, so there is no riskless profit for the buyer as well. Hence, $v$ is a no arbitrage price.

Similar results can be obtained for American and Israeli contingent claims. Namely, in the former case define

$$W = W(f, P) = \sup\{x : \text{there exists a self-financing strategy } \pi \text{ and}$$

a stopping time $\tau \leq N$ such that $X_0^\pi = -x$ and $X_\tau^\pi + f_\tau \geq 0$ $P$-a.s.$\}$.

Observe that the buyer not only manages a self-financing portfolio here but also chooses an exercise time.

**Theorem 4.2.6.** *Assume that $f_n \geq 0$ is $\mathcal{F}_n$-measurable, $0 \leq n \leq N$ and*

$$\sup_{\tilde{P} \in \mathcal{P}(P)} E_{\tilde{P}} \frac{f_n}{B_n} < \infty,$$

*where $f_n$ and $B_n$ is the payoff function and the bond price at the time $n$, respectively. Then*

$$W = W(f, P) = \sup_{\tau \in \mathcal{T}_{0N}} \inf_{\tilde{P} \in \mathcal{P}(P)} B_0 E_{\tilde{P}} \frac{f_\tau}{B_\tau}. \tag{4.2.1}$$

**Proof.** If $\pi$ is a self-financing strategy and $\tau \leq N$ is a stopping time such that $X_0^\pi = -x$ and $X_\tau^\pi + f_\tau \geq 0$ $P$-a.s., then by the optional stopping theorem

$$-\frac{x}{B_0} = \frac{X_0^\pi}{B_0} = E_{\tilde{P}} \frac{X_\tau^\pi}{B_\tau} \geq -E_{\tilde{P}} \frac{f_\tau}{B_\tau},$$

i.e., $x \leq B_0 E_{\tilde{P}} \frac{f_\tau}{B_\tau}$ for any $\tilde{P} \in \mathcal{P}(P)$. Hence,

$$W \leq \sup_{\tau \in \mathcal{T}_{0N}} \inf_{\tilde{P} \in \mathcal{P}(P)} B_0 E_{\tilde{P}} \frac{f_\tau}{B_\tau}.$$

In the other direction, let $\tau \in \mathcal{T}_{0N}$ and set

$$Y_n^\tau = \underset{\tilde{P} \in \mathcal{P}(P)}{\text{ess sup}}\, E_{\tilde{P}}(-\frac{f_\tau}{B_\tau}|\mathcal{F}_n) = -\underset{\tilde{P} \in \mathcal{P}(P)}{\text{ess inf}}\, E_{\tilde{P}}(\frac{f_\tau}{B_\tau}|\mathcal{F}_n).$$

Then $\{Y_n^\tau, 0 \leq n \leq N\}$ is a supermartingale with respect to any $\tilde{P} \in \mathcal{P}(P)$ and the optional decomposition theorem yields that

$$Y_n^\tau = Y_0^\tau + \sum_{k=1}^n \gamma_k^\tau \Delta \frac{S_k}{B_k} - C_n^\tau,$$

where $C_n^\tau \geq 0$ $P$-a.s. is a non-decreasing process. Set

$$X_n^{\pi_\tau} = B_0(Y_0^\tau + \sum_{k=1}^n \gamma_k^\tau \Delta \frac{S_k}{B_k}) \text{ and } \beta_n^\tau = \frac{X_n^{\pi_\tau}}{B_n} - \gamma_n^\tau \frac{S_n}{B_n} = \frac{X_{n-1}^{\pi_\tau}}{B_{n-1}} - \gamma_n^\tau \frac{S_{n-1}}{B_{n-1}}.$$

Then, as before, the strategy $\pi_\tau = (\beta_n^\tau, \gamma_n^\tau)_{0 \le n \le N}$ is self-financing,

$$X_0^{\pi_\tau} = B_0 Y_0^\tau = - \inf_{\tilde{P} \in \mathcal{P}(P)} B_0 E_{\tilde{P}}\left(\frac{f_\tau}{B_\tau}\right) \text{ and } \frac{X_\tau^{\pi_\tau}}{B_\tau} \ge Y_\tau^\tau = -\frac{f_\tau}{B_\tau}.$$

Thus, for each $\tau$ there exists a self-financing strategy $\pi_\tau$ satisfying the above inequality and having the initial capital

$$-x_\tau = - \inf_{\tilde{P} \in \mathcal{P}(P)} B_0 E_{\tilde{P}}\left(\frac{f_\tau}{B_\tau}\right).$$

Hence, by definition $W \ge x_\tau$ for any $\tau \in \mathcal{T}_{0N}$, and so

$$W \ge \sup_{\tau \in \mathcal{T}_{0N}} \inf_{\tilde{P} \in \mathcal{P}(P)} B_0 E_{\tilde{P}}\left(\frac{f_\tau}{B_\tau}\right)$$

which together with the inequality in the other direction proves (4.2.1). □

In the game contingent claim case define

$$W = W(f, g, P) = \sup\{x : \text{there exists a self-financing strategy } \pi$$
$$\text{and a stopping time } \tau \le N \text{ such that } X_0^\pi = -x \text{ and}$$
$$X_{m \wedge \tau}^\pi + R(m, \tau) \ge 0 \ P\text{-a.s. for all } m = 0, 1, ..., N\}$$

where, recall, $R(m, n) = f_m \mathbb{I}_{m < n} + g_n \mathbb{I}_{m \ge n}$.

**Theorem 4.2.7.** *Assume that $f_n \ge g_n \ge 0$, $f_n$ and $g_n$ are $\mathcal{F}_n$-measurable, $0 \le n \le N$ and*

$$\sup_{\tilde{P} \in \mathcal{P}(P)} E_{\tilde{P}} \frac{f_n}{B_n} < \infty.$$

*Then*

$$W = W(f, g, P) = \sup_{\tau \in \mathcal{T}_{0N}} \inf_{\sigma \in \mathcal{T}_{0N}} \inf_{\tilde{P} \in \mathcal{P}(P)} B_0 E_{\tilde{P}} \frac{R(\sigma, \tau)}{B_{\sigma \wedge \tau}}. \tag{4.2.2}$$

**Proof.** Let $\pi$ be a self-financing strategy and $\tau \in \mathcal{T}_{0N}$ be a stopping time such that $X_0^\pi = -x$ and $X_{m \wedge \tau}^\pi + R(m, \tau) \ge 0 \ P$-a.s. for all $m = 0, 1, ..., N$. Then by the optional stopping theorem for any $\tilde{P} \in \mathcal{P}(P)$ and $\sigma \in \mathcal{T}_{0N}$,

$$-\frac{x}{B_0} = \frac{X_0^\pi}{B_0} = E_{\tilde{P}} \frac{X_{\sigma \wedge \tau}^\pi}{B_{\sigma \wedge \tau}} \ge -E_{\tilde{P}} \frac{R(\sigma, \tau)}{B_{\sigma \wedge \tau}}.$$

Hence, $x$ must satisfy $x \le B_0 E_{\tilde{P}} \frac{R(\sigma, \tau)}{B_{\sigma \wedge \tau}}$ for any $\tilde{P} \in \mathcal{P}(P)$ and $\sigma \in \mathcal{T}_{0N}$, and so

$$W \le \sup_{\tau \in \mathcal{T}_{0N}} \inf_{\sigma \in \mathcal{T}_{0N}} \inf_{\tilde{P} \in \mathcal{P}(P)} B_0 E_{\tilde{P}} \frac{R(\sigma, \tau)}{B_\tau}.$$

For the inequality in the other direction let $\tau \in \mathcal{T}_{0N}$ and set

$$Y_n^\tau = ess\sup_{\sigma \in \mathcal{T}_{nN}} ess\sup_{\tilde{P} \in \mathcal{P}(P)} E_{\tilde{P}}\left(-\frac{R(\sigma,\tau)}{B_{\sigma \wedge \tau}}\Big| \mathcal{F}_n\right)$$
$$= -ess\inf_{\sigma \in \mathcal{T}_{nN}} ess\inf_{\tilde{P} \in \mathcal{P}(P)} E_{\tilde{P}}\left(\frac{R(\sigma,\tau)}{B_{\sigma \wedge \tau}}\Big| \mathcal{F}_n\right).$$

Then $\{Y_n^\tau, 0 \le n \le N\}$ is a supermartingale with respect to any $\tilde{P} \in \mathcal{P}(P)$ and applying the optional decomposition theorem we obtain that

$$Y_n^\tau = Y_0^\tau + \sum_{k=1}^{n} \gamma_k^\tau \Delta \frac{S_k}{B_k} - C_n^\tau,$$

where $C_n^\tau \ge 0$ $P$-a.s. is a non-decreasing process. Set

$$X_n^{\pi_\tau} = B_0 (Y_0^\tau + \sum_{k=1}^{n} \gamma_k^\tau \Delta \frac{S_k}{B_k}) \text{ and } \beta_n^\tau = \frac{X_n^{\pi_\tau}}{B_n} - \gamma_n^\tau \frac{S_n}{B_n} = \frac{X_{n-1}^{\pi_\tau}}{B_{n-1}} - \gamma_n^\tau \frac{S_{n-1}}{B_{n-1}}.$$

As before, the strategy $\pi_\tau = (\beta_n^\tau, \gamma_n^\tau)_{0 \le n \le N}$ is self-financing,

$$X_0^{\pi_\tau} = B_0 Y_0^\tau = -\inf_{\sigma \in \mathcal{T}_{nN}} \inf_{\tilde{P} \in \mathcal{P}(P)} B_0 E_{\tilde{P}}\left(\frac{R(\sigma,\tau)}{B_{\sigma \wedge \tau}}\right)$$

$$\text{and } \frac{X_{m \wedge \tau}^{\pi_\tau}}{B_{m \wedge \tau}} \ge Y_{m \wedge \tau}^\tau = -\frac{R(m,\tau)}{B_{m \wedge \tau}}$$

for any $m = 0, 1, ..., N$, since $R(m,\tau)$ is $\mathcal{F}_{m \wedge \tau}$-measurable and $m \in \mathcal{T}_{m \wedge \tau, N}$. This together with the definition of $W$ yields that, for any $\tau \in \mathcal{T}_{0N}$,

$$W \ge x_\tau = \inf_{\sigma \in \mathcal{T}_{0N}} \inf_{\tilde{P} \in \mathcal{P}(P)} B_0 E_{\tilde{P}}\left(\frac{R(\sigma,\tau)}{B_{\sigma \wedge \tau}}\right),$$

thus

$$W \ge \sup_{\tau \in \mathcal{T}_{0N}} \inf_{\sigma \in \mathcal{T}_{0N}} \inf_{\tilde{P} \in \mathcal{P}(P)} B_0 E_{\tilde{P}}\left(\frac{R(\sigma,\tau)}{B_{\sigma \wedge \tau}}\right).$$

This together with the inequality in the other direction above proves (4.2.2).
□

**Remark 4.2.1.** It follows from Section 6.5 in [23] that in the case of Theorem 4.2.6, for each $0 \le n \le N$ with probability one,

$$\sup_{\tau \in \mathcal{T}_{nN}} \inf_{\tilde{P} \in \mathcal{P}(P)} E_{\tilde{P}}\left(\frac{f_\tau}{B_\tau}\Big| \mathcal{F}_n\right) = \inf_{\tilde{P} \in \mathcal{P}(P)} \sup_{\tau \in \mathcal{T}_{nN}} E_{\tilde{P}}\left(\frac{f_\tau}{B_\tau}\Big| \mathcal{F}_n\right).$$

In the case of Theorem 4.2.7 it is possible to show in a similar way that, for each $0 \le n \le N$ with probability one,

$$\sup_{\tau \in \mathcal{T}_{nN}} \inf_{\sigma \in \mathcal{T}_{nN}} \inf_{\tilde{P} \in \mathcal{P}(P)} E_{\tilde{P}}\left(\frac{R(\sigma,\tau)}{B_{\sigma \wedge \tau}}\Big| \mathcal{F}_n\right)$$
$$= \inf_{\tilde{P} \in \mathcal{P}(P)} \sup_{\tau \in \mathcal{T}_{nN}} \inf_{\sigma \in \mathcal{T}_{nN}} E_{\tilde{P}}\left(\frac{R(\sigma,\tau)}{B_{\sigma \wedge \tau}}\Big| \mathcal{F}_n\right)$$
$$= \inf_{\tilde{P} \in \mathcal{P}(P)} \inf_{\sigma \in \mathcal{T}_{nN}} \sup_{\tau \in \mathcal{T}_{nN}} E_{\tilde{P}}\left(\frac{R(\sigma,\tau)}{B_{\sigma \wedge \tau}}\Big| \mathcal{F}_n\right)$$

where the last equality follows from the fact that Dynkin's games have values as we proved it in Lecture 3. Actually, all these can also be obtained by the method described at the beginning of this section, by replacing the infimum over the martingale measures by the infimum over corresponding martingales and then passing to appropriate monotone limits.

**Remark 4.2.2.** In this and in previous chapters the non-decreasing process $C_n \geq 0$ (or $C_n^\sigma$ in the game contingent claim case) emerging from optional or supermartingale decompositions did not play an important role in pricing of contingent claims, and only its nonnegativity was used to ensure hedging. On the other hand, it is possible to view $C_n - C_{n-1}$ as a consumption on step $n$. Then definitions of self-financing strategies are modified to allow consumption which becomes a part of a strategy, but no infusion of outside capital is still permitted.

## 4.3 Exercises

Consider the trinomial market where $B_n \equiv 1$ (interest rate is zero), and the price of the stock at time $n$ is given by the formula

$$ S_n = \prod_{k=1}^{n} (1 + \rho_k), \quad S_0 = 1, $$

where $\rho_1, \rho_2, \ldots$ are i.i.d. random variables such that $\rho_i$ takes on values $-1/2$, $1/2$ and $1$, each with probability $1/3$. The market is active only for two stages (days), i.e., the horizon is $N = 2$. We consider two payoffs: a put option payoff $f_n = (2 - S_n)^+$ and a call option payoff $f_n = (S_n - 2)^+$. Find

    1) All martingale measures (the market is incomplete!);

    2) The superhedging prices of the corresponding put and call European options and self-financing hedging portfolio strategies with the initial capital equal to the corresponding superhedging prices.

    3) The superhedging prices of the corresponding put and call American options and self-financing hedging portfolio strategies with the initial capital equal to the corresponding superhedging prices.

    4) The superhedging prices of the corresponding put and call game options where the payoff is as above and the penalty is 1 (i.e., if the seller cancels before the buyer exercises then the seller adds to the above payoff the penalty equal 1).

# Chapter 5

# Hedging with Risk

## 5.1 Lecture 9: Partial hedging or hedging with risk

We discussed already superhedging of derivatives in incomplete markets which is a perfect hedging since it allows the seller to build a portfolio whose value covers his/her obligation with probability one. This requires to sell the derivative at a rather high price which is often not practical. In this lecture we will discuss other possible pricing criteria which relax a perfect hedging requirement but still provide certain partial hedging which allow the seller to stay on the safe side with high probability (quantile hedging) or by minimizing the risk that the portfolio value is not sufficient to cover the seller's obligation (shortfall risk minimization). Our discussion will partially follow Chapter 8 of [23].

### 5.1.1 *Quantile hedging*

Let $(\Omega, \mathcal{F}, P)$ be a complete probability space with a filtration $\{\mathcal{F}_n, n \geq 0\}$. Denote again by $\{X_n^\pi, n = 0, 1, ..., N\}$ a self-financing portfolio corresponding to a trading strategy $\pi$ and let $f_N \geq 0$ be $\mathcal{F}_N$-measurable discounted payoff at time $N$ (the horizon) for the corresponding European contingent claim (i.e., the bond price remains equal to 1 all the time). We assume everywhere that

$$\sup_{\tilde{P} \in \mathcal{P}(P)} E_{\tilde{P}} f_N < \infty$$

where $\mathcal{P}(P)$ denotes the set of all martingale measures. Consider the following optimization problem: find a self-financing trading strategy $\pi$ so that

$$P\{X_N^\pi \geq f_N\} \longrightarrow \max; \text{ under the condition } X_0^\pi \leq x, \qquad (5.1.1)$$

i.e., we want to maximize the probability that the portfolio value will be sufficient to cover the payoff at time $N$, under the constraint that the initial capital of the portfolio does not exceed $x$. The corresponding problem for an American contingent claim with a payoff function $\{f_n\}$ can be written as

$$P\{X_n^\pi \geq f_n \ \forall n \leq N\} \longrightarrow \max; \text{ under the condition } X_0^\pi \leq x. \quad (5.1.2)$$

Another type of relevant optimization problems (which we will not discuss in detail) can be written as

$$x = X_0^\pi \longrightarrow \min; \text{ under the condition } P\{X_N^\pi < f_N\} \leq \varepsilon, \quad (5.1.3)$$

i.e., we want to minimize the initial capital which ensures that probability of shortfall of the portfolio at time $N$ does not exceed $\varepsilon$. The corresponding problem for American contingent claims can be written as

$$x = X_0^\pi \longrightarrow \min; \text{ under the condition } P\{X_n^\pi < f_n \text{ for some } n\} < \varepsilon.$$
$$(5.1.4)$$

We will deal with European contingent claims at the beginning and then consider only strategies $\pi$, called admissible, which do not require debt (bank loan) at the exercise time (admissible strategies), i.e., $X_N \geq 0$ with probability one. The event $\{X_N^\pi \geq f_N\}$ will be called "success event". We consider first the case of a complete market.

**Theorem 5.1.1.** *Let $P^*$ be the unique martingale measure and assume that an event $A^* \in \mathcal{F}_N$ maximizes the probability $P(A)$ over $A \in \mathcal{F}_N$ under the condition*

$$E_{P^*}(f_N \mathbb{I}_A) \leq x. \quad (5.1.5)$$

*Then a replicating strategy $\pi^*$ for the payoff function $f_N^* = f_N \mathbb{I}_{A^*}$ solves the maximization problem (5.1.1) and the success event $\{X_N^{\pi^*} \geq f_N\}$ coincides with $A^*$ up to an event of probability zero (i.e., the symmetric difference of these two sets has probability zero).*

**Proof.** Let $\pi$ will be any admissible self-financing trading strategy such that $0 \leq X_0^\pi \leq x$. Set $A = \{X_N^\pi \geq f_N\}$. Then

$$X_N^\pi \geq f_N \mathbb{I}_A.$$

Recall that we consider discounted quantities, in particular, the portfolio value, so $\{X_n^\pi\}$ is a local martingale with respect to $P^*$, but since $\pi$ is admissible, i.e., $X_N^\pi \geq 0$ a.s., it follows by Proposition 1.1.1 from Lecture

1 that, in fact, $\{X_n^\pi, n = 0, 1, ..., N\}$ is a martingale with respect to $P^*$. Hence,

$$E_{P^*}(f_N \mathbb{I}_A) \leq E_{P^*}(X_N^\pi) = X_0^\pi \leq x,$$

so $A$ satisfies (5.1.5). Thus

$$P(A) \leq P(A^*). \tag{5.1.6}$$

Now, let $\pi^*$ be the replicating strategy for the payoff $f_N^* = f_N \mathbb{I}_{A^*}$. Then $\pi^*$ is admissible since $X_N^{\pi^*} = f_N^* \geq 0$ and its success event satisfies

$$\{X_N^{\pi^*} \geq f_N\} = \{f_N \mathbb{I}_{A^*} \geq f_N\} \supset A^*.$$

Since $x \geq E_{P^*}(f_N \mathbb{I}_{A^*}) = X_0^{\pi^*}$ we can apply (5.1.6) to obtain

$$P\{X_N^{\pi^*} \geq f_N\} \leq P(A^*),$$

and the symmetric difference of $A^*$ and $\{X_N^{\pi^*} \geq f_N\}$ has probability zero (i.e., these events coincide up to a probability zero event). $\qquad\square$

It may not always be possible to find an event $A^*$ satisfying the conditions of the above theorem, so we consider now another (still related) optimization problem.

**Definition 5.1.1.** Let $\pi$ be an admissible self-financing strategy. The success ratio of $\pi$ is defined by

$$\psi_\pi = \mathbb{I}_{\{X_N^\pi \geq f_N\}} + \frac{X_N^\pi}{f_N} \mathbb{I}_{\{X_N^\pi < f_N\}}.$$

Observe that the event $\{\psi_\pi = 1\}$ coincides with the success event $\{X_N^\pi \geq f_N\}$.

Next, we state an optimization result in a general no arbitrage market with the set of martingale measures $\mathcal{P}(P) \neq \emptyset$. Denote by $\Pi_x$ the set of admissible strategies with the initial capital $X_0^\pi \leq x$. We want to find $\pi^* \in \Pi_x$ so that

$$E(\psi_{\pi^*}) = \sup_{\pi \in \Pi_x} E(\psi_\pi) \tag{5.1.7}$$

where $E$ denotes (in this chapter) the expectation with respect to the measure $P$.

**Theorem 5.1.2.** *For any $x > 0$, $x < \sup_{\tilde{P} \in \mathcal{P}(P)} E_{\tilde{P}}(f_N) < \infty$ there exists $\pi^* \in \Pi_x$ which solves the optimization problem (5.1.7) and such that*

$$X_0^{\pi^*} = \sup_{\tilde{P} \in \mathcal{P}(P)} E_{\tilde{P}}(f_N \psi_{\pi^*}) \leq x.$$

*In particular, $\pi^*$ can be chosen as a superhedging strategy with the above initial capital for the payoff of the form $f_N \psi^*$ for certain $\psi^* : \Omega \to [0, 1]$.*

**Proof.** We can always pick up a sequence $\pi_n \in \Pi_x$, $n \geq 1$ such that

$$\lim_{n\to\infty} E\psi_{\pi_n} = \sup_{\pi\in\Pi_c} E\psi_\pi.$$

By the Komlos theorem, a version of which will be formulated and proved below, we can choose convex combinations

$$\Psi_n = \sum_{j=n}^{\infty} p_j^{(n)} \psi_{\pi_j} \in \text{conv}(\psi_{\pi_n}, \psi_{\pi_{n+1}}, ...), \; p_j^{(n)} \geq 0, \; \sum_{j=n}^{\infty} p_j^{(n)} = 1$$

such that $\Psi_n$ converges almost surely as $n \to \infty$ to some random variable $\Psi$, $0 \leq \Psi \leq 1$.

Since, $X_N^\pi \geq 0$ for any $\pi \in \Pi_x$, it follows from Proposition 1.1.1 that $X_n^\pi$, $n \geq 0$ is a martingale with respect to any $\tilde{P} \in \mathcal{P}(P)$, and so

$$E_{\tilde{P}} f_N \psi_\pi = E_{\tilde{P}}(f_N \wedge X_N^\pi) \leq E_{\tilde{P}} X_N^\pi = X_0^\pi \leq x$$

for any $\pi \in \Pi_x$. By the Fatou lemma (though under our assumptions the Lebesgue dominated convergence theorem also works) for any $\tilde{P} \in \mathcal{P}(P)$,

$$E_{\tilde{P}}(f_N\Psi) \leq \liminf_{n\to\infty} E_{\tilde{P}}(f_N\Psi_n) = \liminf_{n\to\infty} \sum_{j=n}^{\infty} p_j^{(n)} E_{\tilde{P}}(f_N\psi_{\pi_j}) \leq x.$$

By Lecture 8 there exists a superhedging strategy $\pi^*$ for the payoff $f_N\Psi$ with the initial capital

$$x_0 = \sup_{\tilde{P}\in\mathcal{P}(P)} E_{\tilde{P}}(f_N\Psi) \leq x.$$

Thus $\pi^* \in \Pi_x$. Since $\pi^*$ is hedging the payoff $f_N\Psi$,

$$X_N^{\pi^*} \geq f_N\Psi,$$

and so

$$f_N\psi_{\pi^*} = f_N \wedge X_N^{\pi^*} \geq f_N \wedge (f_N\Psi) = f_N\Psi.$$

Hence,

$$\psi_{\pi^*} \geq \Psi \quad \text{on the set } \{f_N > 0\}.$$

Since, clearly, $\psi_{\pi^*} = \Psi = 1$ on the set $\{f_N = 0\}$ we obtain

$$\psi_{\pi^*} \geq \Psi \quad \text{a.s.}$$

On the other hand, by the Lebesgue dominated convergence theorem,

$$E\Psi = \lim_{n\to\infty} \sum_{j=n}^{\infty} p_j^{(n)} E\psi_{\pi_j} = \sup_{\pi\in\Pi_x} E\psi_\pi \geq E\psi_{\pi^*},$$

so $\psi_{\pi^*} = \Psi$ a.s. Hence,

$$E\psi_{\pi^*} = E\Psi = \lim_{n\to\infty} E\Psi_n = \lim_{n\to\infty} \sum_{j=n}^{\infty} p_j^{(n)} E\psi_{\pi_j} = \sup_{\pi\in\Pi_x} E\psi_\pi,$$

completing the proof of the theorem by setting $\psi^* = \Psi = \psi_{\pi^*}$. $\square$

### 5.1.2 *Minimizing the shortfall risk*

As we saw the perfect hedging requires large initial portfolio capital

$$V = \sup_{\tilde{P} \in \mathcal{P}(P)} E_{\tilde{P}} f_N$$

where $f_N$ is again $\mathcal{F}_N$-measurable discounted payoff of a European contingent claim. Suppose that an investor decides to start a portfolio with a smaller sum $x \in (0, V)$ and he/she is interested to know what is the risk measured by an average portfolio shortfall defined as $E(f_N - X_N^\pi)^+$ and the problem is to choose a self-financing trading strategy $\pi$ which minimizes this shortfall.

**Definition 5.1.2.** Let $\ell$ be a convex function (loss function), $\ell(x) = 0$ for $x \leq 0$, $\ell$ is increasing and finite on $[0, \infty)$ and $E\ell(f_N) < \infty$. The shortfall risk $r(\pi)$ of a strategy $\pi$ in the European contingent claim case is defined by

$$r(\pi) = E\ell(f_N - X_N^\pi) = E\ell((f_N - X_N^\pi)^+).$$

**Theorem 5.1.3.** Let $\ell$ be convex on $[0, \infty)$ and $\mathcal{P}(P) \neq \emptyset$. Then for any $x > 0$, $x < \sup_{\tilde{P} \in \mathcal{P}(P)} E_{\tilde{P}}(f_N) < \infty$ there exists $\pi^* \in \Pi_x$ such that

$$r(\pi^*) = r_{\min} = \inf_{\pi \in \Pi_x} r(\pi).$$

**Proof.** Observe that if we set $Y_N^\pi = \min(f_N, X_N^\pi)$, then $r(\pi) = E\ell(f_N - Y_N^\pi)$ since negative values of $f_N - X_N^\pi$ do not play a role in the formula. Now, choose a sequence $\pi_n$, $n \geq 1$ of strategies from $\Pi_x$ such that

$$\lim_{n \to \infty} r(\pi_n) = \inf_{\pi \in \Pi_x} r(\pi) = r_{\min}$$

which is always possible to do. Since $0 \leq Y_N^{\pi_n} \leq f_N$ we can apply the Komlos theorem (formulated and proved below) which provides convex combinations

$$Z^{(n)} = \sum_{j=n}^{\infty} p_j^{(n)} Y_N^{\pi_j} \in \mathrm{conv}(Y_N^{\pi_n}, Y_N^{\pi_{n+1}}, \ldots)$$

such that $Z^{(n)}$ converges almost surely as $n \to \infty$ to a random variable $Z$. Since $E\ell(f_N) < \infty$ and $\ell$ is increasing, it follows that $\ell(f_N - Y_N^{\pi_n})$, $n \geq 1$ is uniformly integrable. Since $\lim_{n \to \infty} E\ell(f_N - Y_N^{\pi_n}) = r_{\min}$, we obtain by the convexity (and so, continuity) of $\ell$,

$$E\ell(f_N - Z) = \lim_{n \to \infty} E\ell(f_N - Z^{(n)}) \leq \limsup_{n \to \infty} \sum_{j=n}^{\infty} p_j^{(n)} E\ell(f_N - Y_N^{\pi_j}) \leq r_{\min}.$$

The last inequality follows since for any $\varepsilon > 0$ there exists $n_\varepsilon$ such that whenever $n \geq n_\varepsilon$,

$$E\ell(f_N - Y_N^{\pi_n}) \leq r_{\min} + \varepsilon,$$

and so

$$\limsup_{n \to \infty} \sum_{j=n}^{\infty} p_j^{(n)} E\ell(f_N - Y_N^{\pi_j}) \leq r_{\min} + \varepsilon.$$

Again, relying on Proposition 1.1.1 we conclude that $X_n^{\pi_j}, n \geq 0$ is a martingale with respect to any martingale measure $\tilde{P}$, and so

$$E_{\tilde{P}} Z^{(n)} = \sum_{j=n}^{\infty} p_j^{(n)} E_{\tilde{P}} Y_N^{\pi_j} \leq \sum_{j=n}^{\infty} p_j^{(n)} E_{\tilde{P}} X_N^{\pi_j} = \sum_{j=n}^{\infty} p_j^{(n)} X_0^{\pi_j} \leq x.$$

Taking into account that $0 \leq Z^{(n)} \leq f_N$ we obtain from here by the Lebesgue dominated convergence theorem that

$$E_{\tilde{P}} Z \leq x$$

for any $\tilde{P} \in \mathcal{P}(P)$.

As we proved in Lecture 8 concerning superhedging of European contingent claims in incomplete markets, there exists a self-financing strategy $\pi^*$ with the initial capital

$$x_0 = \sup_{\tilde{P} \in \mathcal{P}(P)} E_{\tilde{P}} Z \leq x$$

which is superhedging for the payoff $Z$, i.e., $X_N^{\pi^*} \geq Z$, thus

$$r_{\min} \geq E\ell(f_N - Z) \geq E\ell(f_N - X_N^{\pi^*})$$

since $\ell$ is increasing. By the definition of $r_{\min}$ it follows that, in fact, $r_{\min} = E\ell(f_N - X_N^{\pi^*})$ which concludes the proof of the theorem. $\qquad \square$

**Definition 5.1.3.** Let $\ell$ be a convex increasing function equal to zero on $\{x \leq 0\}$ and $f_n \geq 0$, $n = 0, 1, 2, \ldots$ be a payoff function satisfying $\max_{0 \leq n \leq N} E\ell(f_n) < \infty$. The shortfall risk $r(\pi)$ for an American contingent claim given a self-financing strategy $\pi$ is defined by

$$r(\pi) = \sup_{\tau \in \mathcal{T}_{0N}} E\ell(f_\tau - X_\tau^\pi) = \sup_{\tau \in \mathcal{T}_{0N}} E\ell((f_\tau - X_\tau^\pi)^+)$$

where $\mathcal{T}_{0N}$ is the set of all stopping times $\tau$ taking values between 0 and $N$.

In the American contingent claim case a self-financing portfolio strategy is called admissible if the corresponding portfolio value $X_n^\pi$ is nonnegative for all $n = 0, 1, ..., N$. The set of admissible strategies with $X_0^\pi \leq x$ will be denoted by $\Pi_x$. The next result provides a shortfall minimizing strategy for American contingent claims.

**Theorem 5.1.4.** *Let $\ell$ be convex on $[0, \infty)$ and $\mathcal{P}(P) \neq \emptyset$. Assume that $0 < x < \sup_{\tilde{P} \in \mathcal{P}(P), \tau \in \mathcal{T}_{0N}} E_{\tilde{P}}(f_\tau) < \infty$. Then there exists $\pi^* \in \Pi_x$ such that*

$$r(\pi^*) = r_{\min} = \inf_{\pi \in \Pi_x} r(\pi).$$

**Proof.** Set $Y_n^\pi = \min(f_n, X_n^\pi)$, $n = 0, 1, ..., N$ and observe that $r(\pi) = \sup_{\tau \in \mathcal{T}_{0N}} E\ell(f_\tau - Y_\tau^\pi)$. Choose a sequence of strategies $\pi_n \in \Pi_x$, $n \geq 1$ such that

$$\lim_{n \to \infty} r(\pi_n) = r_{\min}.$$

By the Komlos theorem (see below) we can choose $p_j^{(n)} \geq 0$, $j = n, n + 1, n + 2, ...$ such that $\sum_{j=n}^\infty p_j^{(n)} = 1$ and that the random vectors $Z^{(n)} = (Z_0^{(n)}, Z_1^{(n)}, ..., Z_N^{(n)})$, where

$$Z_k^{(n)} = \sum_{j=n}^\infty p_j^{(n)} Y_k^{\pi_j}, \ k = 0, 1, ..., N,$$

converge almost surely as $n \to \infty$ to a random vector $Z = (Z_0, Z_1, ..., Z_N)$. Then

$$r_{\min} = \lim_{n \to \infty} \sup_{\tau \in \mathcal{T}_{0N}} E\ell(f_\tau - Y_\tau^{\pi_n}) \geq \sup_{\tau \in \mathcal{T}_{0N}} \limsup_{n \to \infty} E\ell(f_\tau - Y_\tau^{\pi_n}).$$

Set $\rho_\tau = \limsup_{n \to \infty} E\ell(f_\tau - Y_\tau^{\pi_n})$. Then for any $\varepsilon > 0$ there exists $n_{\varepsilon,\tau}$ such that $E\ell(f_\tau - Y_\tau^{\pi_n}) \leq \rho_\tau + \varepsilon$ for all $n \geq n_{\varepsilon,\tau}$, and so by the above for any $\tau \in \mathcal{T}_{0N}$,

$$\limsup_{n \to \infty} \sum_{j=n}^\infty p_j^{(n)} E\ell(f_\tau - Y_\tau^{\pi_j}) \leq \rho_\tau + \varepsilon \leq r_{\min} + \varepsilon.$$

Since $\varepsilon > 0$ is arbitrary it follows, by the dominated convergence theorem and by convexity and continuity of $\ell$, that

$$r_{\min} \geq \sup_{\tau \in \mathcal{T}_{0N}} \limsup_{n \to \infty} \sum_{j=n}^\infty p_j^{(n)} E\ell(f_\tau - Y_\tau^{\pi_j})$$

$$\geq \sup_{\tau \in \mathcal{T}_{0N}} \limsup_{n \to \infty} E\ell(f_\tau - Z_\tau^{(n)}) = \sup_{\tau \in \mathcal{T}_{0N}} E\ell(f_\tau - Z_\tau).$$

As we proved in Lecture 8 concerning superhedging of American contingent claims in incomplete markets there exists a self-financing strategy $\pi^*$ with the initial capital

$$x_0 = \sup_{\tilde{P} \in \mathcal{P}(P),\, \tau \in \mathcal{T}_{0N}} E_{\tilde{P}} Z_\tau$$

which hedges the payoff $Z$, i.e., $X_n^{\pi^*} \geq Z_n$ for all $n = 0, 1, ..., N$. Since $0 \leq Z_n \leq f_n$ we can apply the dominated convergence theorem to obtain that for any $\tilde{P} \in \mathcal{P}(P)$ and $\tau \in \mathcal{T}_{0N}$,

$$E_{\tilde{P}} Z_\tau = \lim_{n \to \infty} \sum_{j=n}^{\infty} p_j^{(n)} E_{\tilde{P}} Y_\tau^{\pi_j}$$

$$\leq \limsup_{n \to \infty} \sum_{j=n}^{\infty} p_j^{(n)} E_{\tilde{P}} X_\tau^{\pi_j} = \limsup_{n \to \infty} \sum_{j=n}^{\infty} p_j^{(n)} X_0^{\pi_j} \leq x$$

where we used that $X_n^\pi$, $n \geq 0$ is a martingale for each self-financing $\pi$ with respect to any $\tilde{P} \in \mathcal{P}(P)$. Hence, $x_0 \leq x$. Since $\ell$ is increasing and $X_n^{\pi^*} \geq Z_n$ a.s.,

$$r_{\min} \geq \sup_{\tau \in \mathcal{T}_{0N}} E\ell(f_\tau - Z_\tau) \geq \sup_{\tau \in \mathcal{T}_{0N}} E\ell(f_\tau - X_\tau^{\pi^*}),$$

so by the definition of $r_{\min}$ we obtain

$$r(\pi^*) = \sup_{\tau \in \mathcal{T}_{0N}} E\ell(f_\tau - X_\tau^{\pi^*}) = r_{\min}$$

completing the proof of the theorem. $\qquad\qquad\qquad\qquad\qquad \square$

### 5.1.3  *Komlos theorem*

For any subset $\Gamma$ of a linear space the convex hull conv$\Gamma$ of $\Gamma$ is defined as the set of all finite convex combinations $\sum_{i=1}^{n} p_i x_i : x_i \in \Gamma$, $p_i \geq 0$, $\sum_{i=1}^{n} p_i = 1$, $n = 1, 2, ...$ The following result which we used above and will be used later as well, is a version of the Komlos theorem (see [38], [23] and [2]). The proof below partially follows the arguments from [2] but we provide more details.

**Theorem 5.1.5.** *(i) Let $\xi_n$, $n \geq 1$ be a sequence of random $d$-dimensional vectors on a probability space $(\Omega, \mathcal{F}, P)$ such that*

$$\sup_n |\xi_n| < \infty \quad P\text{-a.s.} \tag{5.1.8}$$

*Then there exist $N_n \geq n$ and $p_j^{(n)} \geq 0$, $j = n, n+1, ..., N_n$, $n = 1, 2, ...$ with $\sum_{j=n}^{N_n} p_j^{(n)} = 1$ such that the convex combinations*

$$\eta_n = \sum_{j=n}^{N_n} p_j^{(n)} \xi_j \in conv(\xi_n, \xi_{n+1}, ...)$$

*converge P-a.s. to a random vector $\eta$.*

*(ii) In place of (5.1.8) assume that $\{\xi_n\}_{n\geq 1}$ is uniformly integrable, i.e.,*

$$\lim_{k\to\infty}\sup_{n\geq 1} E(|\xi_n|\mathbb{I}_{|\xi_n|>k}) = 0,$$

*then there exist $N_n \geq n$ and $p_j^{(n)}$, $j = n, ..., N_n$, $n = 1, 2, ...$ as in (i) such that $\eta_n = \sum_{j=n}^{N_n} p_j^{(n)}\xi_j$ converges in $L^1(\Omega, \mathcal{F}, P)$ to a random vector $\eta$.*

**Proof.** (i) Without loss of generality we can assume that $\sup_n |\xi_n| \leq 1$ $P$-a.s. since for otherwise we can consider $\tilde{\xi}_n = \frac{\xi_n}{\sup_n |\xi_n|}$ instead and if $\tilde{\eta}_n = \sum_{j=n}^{\infty} p_j^{(n)}\tilde{\xi}_j \to \tilde{\eta}$ $P$-a.s. as $n \to \infty$ then $\eta_n \to \tilde{\eta}\sup_n |\xi_n|$. Set $A = \sup_{n\geq 1}\inf\{\|\zeta\|_{L^2} : \zeta \in \text{conv}(\xi_n, \xi_{n+1}, ...)\} \leq 1$ where $L^2 = L^2(\Omega, \mathcal{F}, P)$. Then we can pick up a sequence $\zeta_n \in \text{conv}(\xi_n, \xi_{n+1}, ...)$, $n = 1, 2, ...$ such that $A - \frac{1}{n} \leq \|\zeta_n\|_{L^2} \leq A + \frac{1}{n}$. Clearly, for all $k, m \geq n$,

$$\left\|\frac{1}{2}(\zeta_k + \zeta_m)\right\|_{L^2} \geq \inf\{\|\zeta\|_{L^2} : \zeta \in \text{conv}(\xi_n, \xi_{n+1}, ...)\}.$$

Hence, for any $\varepsilon > 0$ there exists $n_\varepsilon$ such that $\|\frac{1}{2}(\zeta_k + \zeta_m)\|_{L^2} \geq A - \varepsilon$, whenever $k, m \geq n_\varepsilon$. Thus,

$$\|\zeta_k - \zeta_m\|_{L^2}^2 = 2\|\zeta_k\|_{L^2}^2 + 2\|\zeta_m\|_{L^2}^2 - \|\zeta_k + \zeta_m\|^2 \leq 4(A + \frac{1}{n})^2 - 4(A - \varepsilon)^2,$$

and so $\{\zeta_n \; n = 1, 2, ...\}$ is a Cauchy sequence in $L^2$. It follows by the completeness of $L^2$ that the $L^2$-limit $\lim_{n\to\infty} \zeta_n = \eta$ exists. Using the Chebyshev inequality and the Borel–Cantelli lemma we extract a subsequence $\eta_k = \zeta_{n_k}$ which converges to $\eta$ almost surely as $k \to \infty$, proving (i).

(ii) For each integer $k \geq 1$ set $\xi_n^{(k)} = \xi_n\mathbb{I}_{\{|\xi_n|\leq k\}}$. The sequence $\xi_n^{(k)}$, $n \geq 1$ is uniformly bounded by $k$, so in the same way as in the proof of (i) we see that there exist $N_n(k) \geq n$ and $p_j^{(n,k)} \geq 0$, $j = n, n+1, ..., N_n(k)$ with $\sum_{j=n}^{N_n(k)} p_j^{(n,k)} = 1$ such that

$$\eta_n^{(k)} = \sum_{j=n}^{N_n(k)} p_j^{(n,k)}\xi_j^{(k)}$$

converges in $L^2 = L^2(\Omega, \mathcal{F}, P)$ as $n \to \infty$ to a random vector $\eta^{(k)}$. Next, we will show that there exist $N_n \geq n$ and $q_j^{(n)} \geq 0$, $n = n, n+1, ..., N_n$ with $\sum_{j=n}^{N_n} q_j^{(n)} = 1$ such that

$$\eta_n^{(k)} = \sum_{j=n}^{N_n} q_j^{(n)}\xi_j^{(k)} \tag{5.1.9}$$

converges in $L^2$ as $n \to \infty$ for any $k = 1, 2, ...$

Indeed, set $q_j^{(n,1)} = p_j^{(n,1)}$, $M_n^{(1)} = N_n(1)$ and suppose that we found $M_n^{(i)} \geq n$ and $q_j^{(n,i)} \geq 0$, $j = n, n+1, ..., M_n^{(i)}$ such that

$$\sum_{j=n}^{M_n^{(i)}} q_j^{(n,i)} = 1 \text{ and } \zeta_n^{(i,k)} = \sum_{j=n}^{M_n^{(i)}} q_j^{(n,i)} \xi_j^{(k)} \to \zeta^{(i,k)} \text{ as } n \to \infty \text{ in } L^2$$

$$(5.1.10)$$

for all $k = 1, 2, ...i$ and some $\zeta^{(i,k)} \in L^2$. Next, for any $i \geq 1$ set $m_i(1) = M_1^{(i)}$ and inductively $m_i(k+1) = M_{m_i(k)}^{(i)} + 1$ defining also $m_0(n) = n$ for any $n$. The sequence

$$\tilde{\zeta}_n^{(i+1)} = \sum_{j=m_i(n)}^{m_i(n+1)-1} q_j^{(n,i)} \xi_j^{(i+1)}, \ n \geq 1$$

is uniformly bounded by $i + 1$, and so in the same way as in (i) we obtain the existence of $L_n^{(i+1)} \geq 0$ and $\tilde{q}_l^{(n,i+1)} \geq 0$, $l = n, ..., L_n^{(i+1)}$ such that

$$\sum_{l=n}^{L_n^{(i+1)}} \tilde{q}_l^{(n,i+1)} = 1 \text{ and } \sum_{l=n}^{L_n^{(i+1)}} \tilde{q}_l^{(n,i+1)} \tilde{\zeta}_l^{(i+1)} \to \tilde{\zeta}^{(i,k)} \text{ as } n \to \infty \text{ in } L^2$$

$$(5.1.11)$$

for some $\tilde{\zeta}^{(i,k)} \in L^2$.

Now set $M_n^{(i+1)} = m_i(L_n^{(i+1)}) - 1$ and $q_j^{(n,i+1)} = \tilde{q}_l^{(n,i+1)} q_j^{(n,i)}$ when $m_i(l) \leq j < m_i(l+1)$, $l = n, n+1, ..., L_n^{(i+1)} - 1$ and $q_j^{(n,i+1)} = 0$ when $j < m_i(n)$. Then

$$\sum_{j=n}^{M_n^{(i+1)}} q_j^{(n,i+1)} = \sum_{l=n}^{L_n^{(i+1)}-1} \tilde{q}_l^{(n,i+1)} \sum_{j=m_i(l)}^{m_i(l+1)-1} q_j^{(n,i)} = 1.$$

Furthermore,

$$\sum_{j=n}^{M_n^{(i+1)}} q_j^{(n,i+1)} \xi_j^{(i+1)} = \sum_{l=n}^{L_n^{(i+1)}-1} \tilde{q}_l^{(n,i+1)} \sum_{j=m_i(l)}^{m_i(l+1)-1} q_j^{(n,i)} \xi_j^{(i+1)}$$

$$= \sum_{l=n}^{L_n^{(i+1)}-1} \tilde{q}_l^{(n,i+1)} \tilde{\zeta}_l^{(i+1)}$$

converges in $L^2$ as $n \to \infty$ by the choice of $\tilde{q}_l^{(n,i+1)}$, $l = n, n+1, ..., L_n^{(i+1)}$ above. Next, for each $k = 1, 2, ..., i$,

$$\zeta_n^{(i+1,k)} = \sum_{j=n}^{M_n^{(i+1)}} q_j^{(n,i+1)} \xi_j^{(k)} \qquad (5.1.12)$$

$$= \sum_{l=n}^{L_n^{(i+1)}-1} \tilde{q}_l^{(n,i+1)} \sum_{j=m_i(l)}^{m_i(l+1)-1} q_j^{(n,i)} \xi_j^k = \sum_{l=n}^{L_n^{(i+1)}-1} \tilde{q}_l^{(n,i+1)} \zeta_{m_i(l)}^{(i,k)}$$

converges in $L^2$ as $n \to \infty$ since $\zeta_{m_i(l)}^{(i,k)}$ converges in $L^2$ as $l \to \infty$ by our assumption. Thus, we constructed by induction $M_n^{(i)} \geq n$ and $q_j^{(n,i)} \geq 0$, $j = n, n+1, ..., M_n^{(i)}$ such that (5.1.10) holds true for all $i = 1, 2, ...$

Observe that if for some $\varepsilon > 0$ and all $n \geq \tilde{n}$,

$$\|\zeta_n^{(i,k)} - \zeta\|_{L^2} < \varepsilon$$

for some $\tilde{n}$ and $\zeta \in L^2$, then taking into account that $m_i(l) \geq l$ we obtain from (5.1.11) and (5.1.12) that

$$\|\zeta_n^{(i+1,k)} - \zeta\|_{L^2} < \varepsilon \text{ for all } n \geq \tilde{n}.$$

Set $\zeta^{(k)} = \zeta^{(k,k)}$. Since $\zeta_n^{(k,k)} \to \zeta^{(k)}$ in $L^2$ as $n \to \infty$, it follows that for any $\varepsilon > 0$ there exists $n_{\varepsilon,k}$ such that

$$\|\zeta_n^{(k,k)} - \zeta^k\|_{L^2} < \varepsilon \text{ for all } n \geq n_{\varepsilon,k}.$$

Therefore, we conclude from above that for all $i \geq k$,

$$\|\zeta_n^{(i,k)} - \zeta^k\|_{L^2} < \varepsilon \text{ for all } n \geq n_{\varepsilon,k}.$$

It follows also that

$$\zeta_n^{(n,k)} \to \zeta^{(k)} \text{ in } L^2 \text{ as } n \to \infty,$$

so setting $q_j^{(n)} = q_j^{(n,n)}$ and $N_n = M_n^{(n)}$ we obtain that $\eta_n^{(k)}$ given by (5.1.9) converges in $L^2$ to $\zeta^{(k)}$ when $n \to \infty$ as required.

Now, set

$$\eta_n = \sum_{j=n}^{N_n} q_j^{(n)} \xi_j.$$

In view of the uniform integrability of the sequence $\{\xi_n\}_{n\geq 1}$, for any $\varepsilon > 0$ we can choose $k_\varepsilon \geq 1$ such that for all $k \geq k_\varepsilon$ and any $n \geq 1$,

$$\|\xi_n - \xi_n^{(k)}\|_{L^1} < \varepsilon/3,$$

where $L^1 = L^1(\Omega, \mathcal{F}, P)$, and so for such $k$ and $n$ we also have

$$\|\eta_n - \eta_n^{(k)}\|_{L^1} < \varepsilon/3.$$

Fix $k \geq k_\varepsilon$. Since $\eta_n^{(k)}$ converges in $L^2$ as $n \to \infty$, we can choose $n_\varepsilon$ such that for any $n, m \geq n_\varepsilon$,

$$\|\eta_n^{(k)} - \eta_m^{(k)}\|_{L^1} < \varepsilon/3.$$

By the triangle inequality it follows that

$$\|\eta_n - \eta_m\|_{L^1} < \varepsilon.$$

Hence, $\{\eta_n\}_{n\geq 1}$ is a Cauchy sequence in $L^1$. The latter is a complete space, so $\eta_n$ converges in $L^1$ to some $\eta \in L^1$ as $n \to \infty$, completing the proof of the theorem. $\square$

## 5.2 Lecture 10: Shortfall risk minimization for game options

Now, we consider the case of Israeli (game) contingent claims.

**Definition 5.2.1.** Let $\ell$ be a convex increasing function equal to zero on $\{x \leq 0\}$, $N$ be a positive integer and $R(m,n) = U_m \mathbb{I}_{m<n} + W_n \mathbb{I}_{n \leq m}$, $U_k \geq W_k \geq 0 \ \forall k \geq 0$ be payoff functions satisfying $E\ell(U_n) < \infty$ for $n = 0, 1, ..., N$. The shortfall risk $r(\pi)$ for a game contingent claim given a self-financing strategy $\pi$ is defined by

$$r(\pi) = \inf_{\sigma \in \mathcal{T}_{0N}} \sup_{\tau \in \mathcal{T}_{0N}} E\ell(R(\sigma,\tau) - X^\pi_{\sigma \wedge \tau})$$
$$= \inf_{\sigma \in \mathcal{T}_{0N}} \sup_{\tau \in \mathcal{T}_{0N}} E\ell((R(\sigma,\tau) - X^\pi_{\sigma \wedge \tau})^+)$$

where $\mathcal{T}_{0N}$ is the set of all stopping times $\tau$ taking values between 0 and $N$.

In the game contingent claim case a self-financing portfolio strategy is called admissible if the corresponding portfolio value $X^\pi_n$ is nonnegative for all $n = 0, 1, ..., N$. The set of admissible strategies with $X^\pi_0 \leq x$ will be denoted by $\Pi_x$. The existence of an admissible portfolio strategy $\pi^*$ such that $r(\pi^*) = inf_{\pi \in \Pi_x} r(\pi)$ for $x > 0$, $x < \inf_{\sigma \in \mathcal{T}_{0N}} \sup_{\tilde{P} \in \mathcal{P}(P), \tau \in \mathcal{T}_{0N}} E_{\tilde{P}}(R(\sigma,\tau))$ for markets defined on a finite probability space is not difficult to obtain. Indeed, for any $\sigma \in \mathcal{T}_{0N}$ and $\pi \in \Pi_x$ set

$$r(\sigma,\pi) = \sup_{\tau \in \mathcal{T}_{0N}} E\ell(R(\sigma,\tau) - X^\pi_{\sigma \wedge \tau}).$$

Then, in the same way as in the American contingent claims case above we can find $\pi_\sigma \in \Pi_x$ such that

$$r(\sigma,\pi_\sigma) = \inf_{\pi \in \Pi_x} r(\sigma,\pi).$$

Taking into account that there exist only finitely many stopping times in $\mathcal{T}_{0N}$ when the sample space $\Omega$ is finite (or when the filtration consists of finite $\sigma$-algebras) we obtain that there exists $\pi^* \in \Pi_x$ such that

$$r(\pi^*) = \min_{\sigma \in \mathcal{T}_{0N}} r(\sigma,\pi_\sigma) = \inf_{\pi \in \Pi_x} r(\pi).$$

For general markets the existence of such $\pi^*$ is a more difficult question due, in particular, to the fact that $r(\pi)$ defined above is not a convex function of $\pi$ in this case. Still, it turns out that the question can be answered affirmatively also for game contingent claims which will extend the results of Section 6 in [11]. Moreover, the method described below

provides a recursive way to compute numerically both the shortfall risk and
the corresponding minimizing portfolio strategies, unlike methods based
on the Komlos theorem which are purely of an existence theorem type.
By simplification the method here can be applied also to European and
American options studied in the previous lecture relying on the Komlos
theorem.

We consider a discrete time market on a general probability space
$(\Omega, \mathcal{F}, P)$ with discounted quantities, so that the bond price $B_n \equiv 1$, $\forall n \geq 0$
stays constant while the stock price at a time $n$ has the form

$$S_n = S_0 \prod_{k=1}^{n} (1 + \rho_k),$$

where $S_0$ is a positive constant and $\rho_1, \rho_2, ...$ is a sequence of random variables with values in $(-1, \infty)$. We assume also that the market admits no
arbitrage.

Next, consider a game contingent claim with the horizon $N < \infty$ and
the payoff function

$$R(m, n) = U_m \mathbb{I}_{m < n} + W_n \mathbb{I}_{n \leq m}$$

where $U_k = f_k(\rho_1, ..., \rho_k) \geq W_k = g_k(\rho_1, ..., \rho_k) \geq 0$ for some measurable
functions $f_k = f_k(x_1, ..., x_k) \geq g_k = g_k(x_1, ..., x_k)$ on $\mathbb{R}^k$, $k = 1, ..., N$. We
assume that $E\ell(U_k) < \infty$ for all $k = 1, 2, ...N$.

**Theorem 5.2.1.** *Under the above conditions for any $x > 0$ there exists a
stopping time $\sigma^* \leq N$ and an admissible self-financing portfolio strategy $\pi^*$
such that*

$$r(\sigma^*, \pi^*) = \inf_{\sigma \in \mathcal{T}_{0N}, \pi \in \Pi_x} r(\sigma, \pi).$$

**Proof.** Set $\rho^{(k)} = (\rho_1, ..., \rho_k)$ and denote by $\mu^{(k)}$ the distribution of the
random vector $\rho^{(k)}$ on $(-1, \infty)^k \subset \mathbb{R}^k$. Let $\mathcal{F}_0 = \{\emptyset, \Omega\}$, $\mathcal{F}_k = \sigma\{\rho_1, ..., \rho_k\}$,
$k = 1, 2, ...$ be $\sigma$-algebras generated by $\rho_1, \rho_2, ...$ It is well known (see, for
instance, Theorems 10.2.1 and 10.2.2 in [7] or Theorems 5.3 and 5.4 in [32])
that in this circumstances for each $x^{(k)} = (x_1, ..., x_k) \in (-1, \infty)^k$ there
exists a probability measure $\mu_{x^{(k)}}^{(k+1)}$ on $(-1, \infty)$ such that for any Borel set
$\Gamma \subset \mathbb{R}$,

$$\mu_{\rho^{(k)}}^{(k+1)}(\Gamma) = P\{\rho_{k+1} \in \Gamma | \mathcal{F}_k\} \quad P\text{-a.s.}$$

and for any Borel set $Q \subset \mathbb{R}^{k+1}$,

$$\mu^{(k+1)}(Q) = \int_{\mathbb{R}^k} \mu_{x^{(k)}}^{(k+1)}(Q_{x^{(k)}}) d\mu^{(k)}(x^{(k)})$$

where $Q_{x^{(k)}} = \{x \in \mathbb{R} : (x^{(k)}, x) \in Q\}$. Such measures $\mu_{x^{(k)}}^{(k+1)}$ are called regular conditional probabilities or disintegrations of the measure $\mu^{(k+1)}$. Observe that if $\rho_1, \rho_2, ...$ are independent then $\mu_{x^{(k)}}^{(k+1)}$ does not depend on $x^{(k)}$ and it is equal to the distribution $\nu_{k+1}$ of $\rho_{k+1}$, so that in this case $\mu^{(k+1)} = \nu_1 \times \nu_2 \times \cdots \times \nu_{k+1}$.

Recall, that the discounted portfolio value at time $n$ corresponding to a self-financing portfolio strategy $\pi = (\beta_n, \gamma_n)$, $n \geq 0$ can be written in the form

$$X_n^\pi = \beta_n + \gamma_n S_n = X_0^\pi + \sum_{k=1}^n \gamma_k \Delta S_k$$

where $\Delta S_k = S_k - S_{k-1}$. Since the market admits no arbitrage, then

$$\mu_{x^{(n)}}^{(n+1)}(0, \infty) < 1 \quad \text{for } \mu^{(n)}\text{-almost all } x^{(n)}$$

as for otherwise we could make a riskless profit on $(n+1)$-stage by buying stocks on the $n$-th stage when $\rho^{(n)}$ equals $x^{(n)}$ satisfying $\mu_{x^{(n)}}^{(n+1)}(0, \infty) = 1$ and selling them on the next stage. By the same no arbitrage argument, for $\mu^{(n)}$-almost all $x^{(n)}$ either

$$\text{supp}\mu_{x^{(n)}}^{(n+1)} \cap (-\infty, 0) \neq 0 \text{ or } \text{supp}\mu_{x^{(n)}}^{(n+1)} = \{0\} \qquad (5.2.1)$$

where supp denotes the support of a measure, i.e., the complement of the union of all open sets of zero measure. Let

$$a_{n+1}(x^{(n)}) = \inf(\text{supp}\mu_{x^{(n)}}^{(n+1)}).$$

If $\pi$ is admissible then with probability one

$$X_{n+1}^\pi = X_n^\pi + \gamma_{n+1}\Delta S_{n+1} = X_n^\pi + \gamma_{n+1}S_n\rho_{n+1} \geq 0, \qquad (5.2.2)$$

and so

$$X_n^\pi + \gamma_{n+1}S_n a_{n+1}(\rho^{(n)}) \geq 0. \qquad (5.2.3)$$

By (5.2.1) and (5.2.3) with probability one either $-\infty \leq a_{n+1}(\rho^{(n)}) < 0$ and

$$\gamma_{n+1} \leq \frac{X_n^\pi}{S_n(-a_{n+1}(\rho^{(n)}))} \quad \text{where} \quad \frac{C}{-\infty} = 0,$$

or $a_{n+1}(\rho^{(n)}) = 0$ and then $\text{supp}\mu_{\rho^{(n)}}^{(n+1)} = \{0\}$ when any choice of $\gamma_{n+1} \geq 0$ will preserve admissibility.

Next, set

$$b_{n+1}(x^{(n)}) = \sup(\text{supp}\mu_{x^{(n)}}^{(n+1)}).$$

and observe that for $\mu^{(n)}$-almost all $x^{(n)}$ either

$$\mathrm{supp}\mu_{x^{(n)}}^{(n+1)} \cap (0,\infty) \neq 0 \quad \text{or} \quad \mathrm{supp}\mu_{x^{(n)}}^{(n+1)} = \{0\}$$

since for otherwise we could make a riskless profit on the $(n+1)$-th stage by short selling the stocks on the $n$-th stage when $\mu_{x^{(n)}}^{(n+1)}(-\infty,0) = 1 - \mu_{x^{(n)}}^{(n+1)}(\{0\}) > 0$ and closing these positions on the next stage. By (5.2.2),

$$X_n^\pi + \gamma_{n+1} S_n b_{n+1}(\rho^{(n)}) \geq 0.$$

Hence, with probability one either $0 < b_{n+1}(\rho^{(n)}) \leq \infty$ and then

$$\gamma_{n+1} \geq -\frac{X_n^\pi}{S_n b_{n+1}(\rho^{(n)})} \quad \text{where} \quad \frac{C}{\infty} = 0,$$

or $b_{n+1}(\rho^{(n)}) = 0$ and then $\mathrm{supp}\mu_{\rho^{(n)}}^{(n+1)} = \{0\}$ when any choice of $\gamma_{n+1} \leq 0$ will preserve admissibility. The above conditions on $\gamma_{n+1}$ are the only constraints which keep the portfolio value nonnegative on the next $(n+1)$-th stage. This discussion motivates to consider the set $\mathcal{A}_n(X, \rho^{(n)})$ of all possible portfolio values at the time $n+1$, provided that the portfolio value at the time $n$ was $X$ which is a nonnegative $\mathcal{F}_n$-measurable random variable. Hence,

$$\mathcal{A}_n(X, \rho^{(n)}) = \big\{Y : Y = X + \alpha\Delta S_{n+1} \text{ for } -\frac{X}{S_n b_{n+1}(\rho^{(n)})} \leq \alpha$$
$$\leq -\frac{X}{S_n a_{n+1}(\rho^{(n)})} \text{ if } 0 > a_{n+1}(\rho^{(n)}) \geq -\infty \text{ and } 0 < b_{n+1}(\rho^{(n)}) \leq \infty$$
$$\text{while } Y = X \text{ if } a_{n+1}(\rho^{(n)}) = 0 \text{ or } b_{n+1}(\rho^{(n)}) = 0\big\}.$$

Next, we introduce the following optimal stopping (Dynkin's) game. For each admissible portfolio strategy $\pi \in \Pi_x$, $x > 0$ set

$$Q^\pi(m, n) = \ell(U_m - X_m^\pi)\mathbb{I}_{m<n} + \ell(W_n - X_n^\pi)\mathbb{I}_{m\geq n}$$

where, as before, $\ell$ is a convex increasing loss function equal to zero on $\{x \leq 0\}$ and such that $E\ell(U_n) < \infty$ for all $n \geq 0$. Define by the backward induction $\Psi_N^\pi = \ell(U_N - X_N^\pi)$ and for $n = N-1, N-2, ..., 0$,

$$\Psi_n^\pi = \min\big(\ell(U_n - X_n^\pi), \max(\ell(W_n - X_n^\pi), E(\Psi_{n+1}^\pi|\mathcal{F}_n))\big),$$

where, recall, $U_n = f_n(\rho^{(n)})$ and $W_n = g_n(\rho^{(n)})$. Then by the results on Dynkin games discussed in Lecture 3 we obtain that

$$\Psi_0^\pi = \inf_{\sigma\in\mathcal{T}_{0N}} \sup_{\tau\in\mathcal{T}_{0N}} EQ^\pi(\sigma, \tau) = r(\pi).$$

Moreover, there exists $\sigma = \sigma(\pi)$ such that

$$r(\pi, \sigma) = \sup_{\tau\in\mathcal{T}_{0N}} EQ^\pi(\sigma, \tau) = r(\pi).$$

In what follows we are going to construct $\pi^* \in \Pi_x$ and $\sigma^* \in \mathcal{T}_{0N}$ such that

$$r(\pi^*, \sigma^*) = \inf_{\pi \in \Pi_x} r(\pi). \qquad (5.2.4)$$

Recall that if $f : [0, \infty) \to [0, \infty)$ is a lower semi-continuous function, i.e., $\liminf_{z \to z_0} f(z) \geq f(z)$ for any $z_0$, then

$$\arg \min_{0 \leq z \leq a} f(z) = \min\{0 \leq \tilde{z} \leq a : f(\tilde{z}) = \min_{0 \leq z \leq a} f(z)\}$$

is well defined. We will need the following functions defined by the backward induction. First, we put $I_N(x^{(N)}, y, z) = J_N(x^{(N)}, y) = \ell(f_N(x^{(N)}) - y)$, $x^{(N)} = (x_1, ..., x_N) \in (-1, \infty)^N$, $y \geq 0$ and then for any $n < N$, $z \in (-\infty, \infty)$, $y \geq 0$ and $x^{(n)} = (x_1, ..., x_n) \in (-1, \infty)^n$ we set

$$I_n(x^{(n)}, y, z) = \min\big(\ell(f_n(x^{(n)}) - y),$$
$$\max(\ell(g_n(x^{(n)}) - y), \textstyle\int_{-1}^{\infty} J_{n+1}((x^{(n)}, u), y + zu\kappa_n(x^{(n)}))d\mu_{x^{(n)}}^{(n+1)}(u)))$$

and

$$J_n(x^{(n)}, y) = \inf_{z \in G_n(x^{(n)}, y)} I_n(x^{(n)}, y, z)$$

where $\kappa_n(x^{(n)}) = S_0 \prod_{k=1}^{n}(1 + x_k)$ and

$$G_n(x^{(n)}, y) = \{z \in \mathbb{R} : -y(\kappa_n(x^{(n)})b_{n+1}(x^{(n)}))^{-1} \leq z$$
$$\leq -y(\kappa_n(x^{(n)})a_{n+1}(x^{(n)}))^{-1} \text{ if } 0 > a_{n+1}(x^{(n)}) \geq -\infty \text{ and } 0 < b_{n+1}(x^{(n)})$$
$$\leq \infty \text{ while } z \text{ is arbitrary if } a_{n+1}(x^{(n)}) = 0 \text{ or } b_{n+1}(x^{(n)}) = 0\}.$$

In order to use the above argmin notion we will need to show that $I_n(x^{(n)}, \cdot, \cdot)$ and $J_n(x^{(n)}, \cdot)$ are lower semi-continuous in the arguments denoted by dots.

Since $\ell$ is a continuous function, $I_N(x^{(N)}, \cdot, \cdot) = J_N(x^{(N)}, \cdot)$ are continuous, thus they are lower semi-continuous. Suppose that the lower semi-continuity is established for $I_n$ and $J_n$ with $n = N, N-1, ..., m+1$ and we prove it for $n = m$. Let $\lim_{k \to \infty} y_k = y$ and $\lim_{k \to \infty} z_k = z$. Since $J_{m+1}((x^{(m)}, u), \cdot, \cdot)$ is lower semi-continuous we obtain by the Fatou lemma that

$$F_m(x^{(m)}, y, z) = \textstyle\int_{-1}^{\infty} J_{m+1}((x^{(m)}, u), y + zu\kappa_m(x^{(m)}))d\mu_{x^{(m)}}^{(m+1)}(u)$$
$$\leq \textstyle\int_{-1}^{\infty} \liminf_{k \to \infty} J_{m+1}((x^{(m)}, u), y_k + z_k u\kappa_m(x^{(m)}))d\mu_{x^{(m)}}^{(m+1)}(u)$$
$$\leq \liminf_{k \to \infty} \textstyle\int_{-1}^{\infty} J_{m+1}((x^{(m)}, u), y_k + z_k u\kappa_m(x^{(m)}))d\mu_{x^{(m)}}^{(m+1)}(u)$$
$$= \liminf_{k \to \infty} F_m(x^{(m)}, y_k, z_k),$$

hence $F_m(x, y, z)$ is lower semi-continuous in $y$ and $z$. As $\ell$ is a continuous function we obtain that $I_m(x^{(m)}, y, z)$ is lower semi-continuous in $y$ and $z$ as well.

Next, let $\lim_{k \to \infty} y_k = y$ be such that $\lim_{k \to \infty} J_m(x^{(m)}, y_k)$ exists. If $a_{m+1}(x^{(m)}) \neq 0$ and $b_{m+1}(x^{(m)}) \neq 0$, and since $I_m(x^{(m)}, y, z)$ is lower semi-continuous in $z$, for each $y$ and $x^{(m)} \in (-1, \infty)$ there exists $z_k \in G_m(x^{(m)}, y_k)$ such that $J_m(x^{(m)}, y_k) = I_m(x^{(m)}, y_k, z_k)$. When $a_{m+1}(x^{(m)}) \neq 0$ and $b_{m+1}(x^{(m)}) \neq 0$, the sequence $\{z_k\}_{k \geq 1}$ stays in a compact region, and so we can choose a convergent subsequence $z_{k_i} \to z$ as $i \to \infty$ with $z \in G(x^{(m)}, y)$. Then

$$J_m(x^{(m)}, y) \leq I_m(x^{(m)}, y, z) \leq \liminf_{i \to \infty} I_m(x^{(m)}, y_{k_i}, z_{k_i})$$
$$= \liminf_{i \to \infty} J_m(x^{(m)}, y_{k_i}) = \lim_{k \to \infty} J_m(x^{(m)}, y_k).$$

Since for any sequence $y_k \to y$ we can choose a subsequence $y_{k_i}$ such that

$$\liminf_{k \to \infty} J_m(x^{(m)}, y_k) = \lim_{i \to \infty} J_m(x^{(m)}, y_{k_i}),$$

we obtain that

$$J_m(x^{(m)}, y) \leq \liminf_{k \to \infty} J_m(x^{(m)}, y_k),$$

completing the induction step when $a_{m+1}(x^{(m)}) \neq 0$ and $b_{m+1}(x^{(m)}) \neq 0$. If $a_{m+1}(x^{(m)}) = 0$ or $b_{m+1}(x^{(m)}) = 0$ then $\mu_{x^{(m)}}^{(m+1)}\{0\} = 1$, thus

$$F_m(x^{(m)}, y, z) = J_{m+1}((x^{(m)}, 0), y)$$

does not depend on $z$, whence $I_m(x^{(m)}, y, z)$ does not depend on $z$ in this case as well. Then $J_m(x^{(m)}, y) = I_m(x^{(m)}, y, 0)$ and since $I_m(x^{(m)}, y, 0)$, is lower semi-continuous in $y$ we obtain that $J_m(x^{(m)}, y)$ is also lower semi-continuous in $y$, completing the induction step in this case.

Now we can construct $\pi^* \in \Pi_x$ and $\sigma^* \in \mathcal{T}_{0N}$ so that (5.2.4) holds true with such $\pi^*$ and $\sigma^*$. Set $X_0^{\pi^*} = x$ and inductively

$$X_{n+1}^{\pi^*} = X_n^{\pi^*} + \lambda_n(\rho^{(n)}, X_n^{\pi^*}) \Delta S_{n+1},$$

where

$$\lambda_n(x^{(n)}, y) = \mathrm{argmin}_{\gamma \in G_n(x^{(n)}, y)} I_n(x^{(n)}, y, \gamma)$$

if $a_{n+1}(x^{(n)}) \neq 0$ and $b_{n+1}(x^{(n)}) \neq 0$, while

$$\lambda_n(x^{(n)}, y) = I_n(x^{(n)}, y, 0)$$

if $a_{n+1}(x^{(n)}) = 0$ or $b_{n+1}(x^{(n)}) = 0$, recalling that in this case $I_n(x^{(n)}, y, z)$ does not depend on $z$. Thus, we set $\gamma_{n+1} = \lambda(\rho^{(n)}, X_n^{\pi^*})$ and $\beta_{n+1} = X_{n+1}^{\pi^*} - \gamma_{n+1} S_{n+1}$. Define also

$$\sigma^* = \min\{0 \leq n \leq N : \ell(U_n - X_n^{\pi^*}) = \Psi_n^{\pi^*}\}.$$

Verifying (5.2.4) we prove by the backward induction that

$$J_n(\rho^{(n)}, X_n^{\pi^*}) = \Psi_n^{\pi^*} \tag{5.2.5}$$

and that for any $\pi \in \Pi_x$,

$$J_n(\rho^{(n)}, X_n^{\pi}) \leq \Psi_n^{\pi}. \tag{5.2.6}$$

By definition $J_N(\rho^{(N)}, X_N^{\pi}) = \Psi_N^{\pi}$ for any admissible self-financing strategy $\pi$, so (5.2.5) and (5.2.6) are trivially satisfied for $n = N$. Suppose that (5.2.5) and (5.2.6) hold true for all $N \geq n \geq m+1$ and prove them for $n = m$. Relying on the properties of regular conditional probabilities discussed above we can write that with probability one,

$$E\big(J_{m+1}(\rho^{(m+1)}, X_{m+1}^{\pi})|\mathcal{F}_m\big)$$
$$= E\big(J_{m+1}((\rho^{(m)}, \rho_{m+1}), X_m^{\pi} + \gamma_{m+1}\rho_{m+1}\kappa_m(\rho^{(m)}))|\mathcal{F}_m\big)$$
$$= \int_{-1}^{\infty} J_{m+1}((\rho^{(m)}, u), X_m^{\pi} + \gamma_{m+1}u\kappa_m(\rho^{(m)}))d\mu_{\rho^{(m)}}^{(m+1)}(u).$$

It follows from the definition of $\Psi_m^{\pi}$, $I_m(\rho^{(m)}, X_m^{\pi}, \gamma_{m+1})$ and $J_m(\rho^{(m)}, X_m^{\pi})$ that with probability one,

$$\Psi_m^{\pi} \geq I_m(\rho^{(m)}, X_m^{\pi}, \gamma_{m+1}) \geq J_m(\rho^{(m)}, X_m^{\pi})$$

for any admissible self-financing strategy $\pi$, completing the induction step for (5.2.6).

On the other hand, if we choose $\gamma_{m+1} = \gamma_{m+1}^* = \lambda(\rho^{(m)}, X_m^{\pi^*}) \in G(\rho^{(m)}, X_m^{\pi^*})$ then by the construction of $\pi^*$,

$$I_m(\rho^{(m)}, X_m^{\pi^*}, \gamma_{m+1}^*) = J_m(\rho^{(m)}, X_m^{\pi^*}).$$

By the induction hypothesis with probability one,

$$I_m(\rho^{(m)}, X_m^{\pi^*}, \gamma_{m+1}^*) = \min\big(\ell(f_m(\rho^{(m)}) - X_m^{\pi^*}), \max(\ell(g_m(\rho^{(m)}) - X_m^{\pi^*}),$$
$$E(J_{m+1}(\rho^{(m+1)}, X_{m+1}^{\pi^*})|\mathcal{F}_m))) = \min\big(\ell(f_m(\rho^{(m)}) - X_m^{\pi}),$$
$$\max(\ell(g_m(\rho^{(m)}) - X_m^{\pi}), E(\Psi_{m+1}^{\pi^*}|\mathcal{F}_m))) = \Psi_m^{\pi^*}.$$

Hence, $J_m(\rho^{(m)}, X_m^{\pi^*}) = \Psi_m^{\pi^*}$ completing the induction step for (5.2.5).

Observe that $\pi^* = (\beta_n^*, \gamma_n^*)_{n=1}^N$ with $\beta_n^* = X_n^{\pi^*} - \gamma_n^* S_n$. The formula for the optimal stopping time $\sigma^*$ follows from the standard results about Dynkin's games which were discussed in Lecture 3. Finally,

$$r(\pi^*, \sigma^*) = r(\sigma^*) = \Psi_0^{\pi^*} = J_0(S_0, x) \leq \Psi_0^{\pi} = r(\pi)$$

for any admissible self-financing portfolio strategy $\pi \in \Pi_x$, and so (5.2.4) holds true, completing the proof of the theorem. $\square$

## 5.3   Exercises

1) Prove the existence of shortfall minimizing portfolio strategies for European and American contingent claims, adapting to these cases the method of Lecture 10.

2) Describe the minimization of the shortfall risk relying on recursive formulas obtained in the previous exercise for the specific case of the examples in Exercise 4.3 of Chapter 4, and explain the corresponding computations.

# PART 2
# Continuous Time

# Chapter 6

# Martingales in Continuous Time and Optimal Stopping

## 6.1 Lecture 11: Optional stopping and submartingale decomposition

### 6.1.1 *Martingales in continuous time*

In Lecture 1 we defined martingales, submartingales and supermartingales with respect to a filtration $\{\mathcal{F}_t\}_{t\geq 0}$ in both discrete and continuous time. In the continuous time case the time parameter runs over an uncountable set which causes additional technical difficulties in comparison with the discrete time case. Most of the results related to continuous time martingales are obtained by limiting procedures from a sequence of discretizations relying on the corresponding results in the discrete time case. This usually requires certain regularity in time of stochastic processes under consideration.

**Definition 6.1.1.** A stochastic process $X_t$, $t \in [0,T]$, $T \leq \infty$ on a probability space $(\Omega, P, \mathcal{F})$ is a modification of a stochastic process $Y_t$, $t \in [0,T]$ if $P\{X_t = Y_t\} = 1$ for any $t \in [0,T]$.

**Definition 6.1.2.** A filtration $\{\mathcal{F}_t\}$ in a complete probability space $(\Omega, P, \mathcal{F})$ is said to satisfy the usual conditions if it is right-continuous, i.e., $\mathcal{F}_t = \cap_{s>t}\mathcal{F}_s$, and $\mathcal{F}_0$ contains all events $\Gamma$ with $P(\Gamma) = 0$, and so all $\sigma$-algebras in question have the same $P$-null sets and all probability spaces $(\Omega, \mathcal{F}_t, P)$, $t \in [0,T]$ are complete.

**Theorem 6.1.1.** *(i) Let $X = \{X_t\}_{t\in[0,T]}$ be a submartingale with respect to a filtration $\{\mathcal{F}_t\}_{t\in[0,T]}$. Then $\hat{X}_t = \lim_{r\downarrow t, r\in\mathbb{Q}} X_r$ exists a.s., where $\mathbb{Q}$ is the set of rational numbers. The process $\hat{X} = \{\hat{X}_t\}_{t\in[0,T]}$ is a submartingale with respect to the filtration $\{\mathcal{F}_{t+}\}_{t\in[0,T]}$, $\mathcal{F}_{t+} = \cap_{s>t}\mathcal{F}_s$ with right-continuous paths $\hat{X}_t(\omega)$ and left hand limits in $t$ for $P$ almost all $\omega$.*

*If $X$ is a martingale then $\hat{X}$ is a martingale with respect to the filtration $\{\mathcal{F}_{t+}\}$.*

*(ii) If the filtration $\{\mathcal{F}_t\}_{t\in[0,T]}$ satisfies the usual conditions then $X_t \le \hat{X}_t$ a.s. for every $t \in [0,T]$ and*

$$P\{X_t = \hat{X}_t\} = 1 \quad \text{for each} \quad t \in [0,T]$$

*if and only if the expectation $EX_t$ is right-continuous in $t$, i.e., in the latter case $\hat{X}$ is the right-continuous with left limits modification of $X$.*

**Proof.** (i) Clearly, it suffices to prove the theorem for $T < \infty$. We will prove first that with probability one $X_t$ is bounded when $t \in [0,T]\cap\mathbb{Q}$ and that for any $t \in [0,T]$,

$$\lim_{r\downarrow t, r\in\mathbb{Q}} X_t \quad \text{and} \quad \lim_{r\uparrow t, r\in\mathbb{Q}} X_t$$

exist. Indeed, let $\{r_1, r_2, ...\}$ be an enumeration of the set $[0,T]\cap\mathbb{Q}$ and for any $n$ let $\{s_1, s_2, ..., s_n\}$ be the set $\{r_1, r_2, ..., r_n\}$ arranged according to the natural order of reals. Setting also $s_{n+1} = T$ we obtain a submartingale $Y_i = X_{s_i}$, $i = 1, 2, ..., n+1$. By the martingale inequalities and the upcrossings bound from Lecture 2 we have

$$P\{\max_{1\le i\le n+1} |Y_i| > a\} \le a^{-1}(E|X_0| + E|X_T|), \quad \forall a > 0$$

and

$$EN_{n+1}(a,b) \le (b-a)E(X_T - a)^+, \quad \forall b > a$$

where $N_{n+1}(a,b)$ is the number of upcrossings of the interval $(a,b)$ by the sequence $Y_1, Y_2, ..., Y_{n+1}$. This being true for all $n$ implies that

$$P\{\sup_{t\in[0,T]\cap\mathbb{Q}} |X_t| > a\} \le a^{-1}(E|X_0| + E|X_T|)$$

and

$$EN_\infty(a,b) \le (b-a)E(X_T - a)^+,$$

where $N_n(a,b) \uparrow N_\infty(a,b)$ as $n \uparrow \infty$. It follows that with probability one $N_\infty(a,b) < \infty$ simultaneously for all pairs $a < b$, $a, b \in \mathbb{Q}$. Hence, we conclude in the same way as in the martingale convergence theorem that for each $t \in [0,T]$ both one sided limits along rationals

$$\hat{X}_t = \lim_{r\downarrow t, r\in\mathbb{Q}} X_r \quad \text{and} \quad \bar{X}_t = \lim_{r\uparrow t, r\in\mathbb{Q}} X_r$$

exist with probability one.

It is easy to see that the process $\hat{X}_t$, $t \in [0,T]$ is right-continuous. Indeed, let

$$\hat{X}_t^+ = \limsup_{s \downarrow t} \hat{X}_s \quad \text{and} \quad \hat{X}_t^- = \liminf_{s \downarrow t} \hat{X}_s.$$

Choose $s_i^+ \downarrow t$ and $s_i^- \downarrow t$ such that

$$\hat{X}_t^+ = \limsup_{s_i^+ \downarrow t} \hat{X}_{s_i^+} \quad \text{and} \quad \hat{X}_t^- = \liminf_{s_i^- \downarrow t} \hat{X}_{s_i^-}.$$

Observe that the sequences $s_i^+$ and $s_i^-$, $i \geq 1$ depend on $\omega$, i.e., on the corresponding path $\hat{X}_s(\omega)$, $s \geq 0$ of the process $\hat{X}_s$, $s \geq 0$. Fix an arbitrary $\varepsilon > 0$ and choose $r_i^+, r_i^- \in \mathbb{Q}$ such that $0 \leq r_i^\pm - s_i^\pm \leq 1/i$, $i \geq 1$ and $|X_{r_i^\pm} - \hat{X}_{s_i^\pm}| < \varepsilon$. Then with probability one,

$$\hat{X}_t = \lim_{i \to \infty} X_{r_i^+} = \lim_{i \to \infty} X_{r_i^-}$$

and

$$\hat{X}_t + \varepsilon \geq \limsup_{s_i^\pm \downarrow t} \hat{X}_{s_i^\pm} \geq \liminf_{s_i^\pm \downarrow t} \hat{X}_{s_i^\pm} \geq \hat{X}_t - \varepsilon,$$

whence

$$\hat{X}_t + \varepsilon \geq \limsup_{s \downarrow t} \hat{X}_s \geq \liminf_{s \downarrow t} \hat{X}_s \geq \hat{X}_t - \varepsilon.$$

Since $\varepsilon > 0$ is arbitrary we obtain that with probability one

$$\hat{X}_t = \lim_{s \downarrow t} \hat{X}_s,$$

i.e., $\{\hat{X}_t\}$ is right-continuous.

In the same way we derive that $\hat{X}_t$, $t \in [0,T]$ has left hand limits. Namely, we define

$$\bar{X}_t^+ = \limsup_{s \uparrow t} \hat{X}_s \quad \text{and} \quad \bar{X}_t^- = \liminf_{s \uparrow t} \hat{X}_s.$$

Choose $s_i^+ \uparrow t$ and $s_i^- \uparrow t$ such that

$$\bar{X}_t^+ = \limsup_{s_i^+ \uparrow t} \hat{X}_{s_i^+} \quad \text{and} \quad \bar{X}_t^- = \liminf_{s_i^- \uparrow t} \hat{X}_{s_i^-}.$$

Fix an $\varepsilon > 0$ and choose $r_i^+, r_i^- \in \mathbb{Q}$ such that $0 \leq r_i^\pm - s_i^\pm \leq 1/i$, $i \geq 1$, $r_i^\pm \uparrow t$ as $i \uparrow \infty$ and $|X_{r_i^\pm} - \hat{X}_{s_i^\pm}| < \varepsilon$. Since, as explained above, the limit

$$\bar{X}_t = \lim_{r \uparrow t, r \in [0,T] \cap \mathbb{Q}} X_r = \lim_{i \uparrow \infty} X_{r_i^+}$$

exists with probability one, we obtain in the same way as above that with probability one, $\bar{X}_t = \lim_{s \uparrow t} \hat{X}_s$.

Next, $\hat{X}_t$ is measurable with respect to $\mathcal{F}_{t+} = \cap_{u>t}\mathcal{F}_u$. Let $r_n \downarrow t$, $r_n \in \mathbb{Q}$ as $n \to \infty$. Then $EX_{r_n} \downarrow A = \lim_{n \uparrow \infty} EX_{r_n} \geq EX_t$ since $\{X_t\}$ is a submartingale. For an arbitrary $\varepsilon > 0$ choose an integer $n(\varepsilon)$ such that $EX_{r_n} - A < \varepsilon$ for any $n \geq n(\varepsilon)$. By the submartingale property of $\{X_t\}_{t\in[0,T]}$ for each $n \geq n(\varepsilon)$ and $a > 0$,

$$E(|X_{r_n}|\mathbb{I}_{|X_{r_n}|>a}) = E(X_{r_n}\mathbb{I}_{X_{r_n}>a}) + E(X_{r_n}\mathbb{I}_{X_{r_n}\geq-a}) - EX_{r_n}$$
$$\leq E(X_{r_{n(\varepsilon)}}\mathbb{I}_{X_{r_n}>a}) + E(X_{r_{n(\varepsilon)}}\mathbb{I}_{X_{r_n}\geq-a}) - EX_{r_{n(\varepsilon)}} + \varepsilon$$
$$\leq E(|X_{r_{n(\varepsilon)}}|\mathbb{I}_{|X_{r_n}|>a}) + \varepsilon.$$

In addition,

$$P\{|X_{r_n}| > a\} \leq a^{-1}E|X_{r_n}| = a^{-1}(2EX_{r_n}^+ - EX_{r_n}) \leq a^{-1}(2EX_{r_1}^+ - A).$$

It follows from here that

$$\lim_{a\to\infty} \sup_{n\geq 1} E(|X_{r_n}|\mathbb{I}_{|X_{r_n}|>a}) = 0,$$

i.e., the sequence $\{X_{r_n}\}_{n\geq 1}$ is uniformly integrable.

Let now $t < s$, $r_n \downarrow t$, $q_n \downarrow s$, $r_n, q_n \in \mathbb{Q}$ and $r_1 < s$. We conclude as above that $\{X_{q_n}\}_{n\geq 1}$ is also uniformly integrable, and so for any $\Gamma \in \mathcal{F}_{t+}$ by the submartingale property of $\{X_t\}$,

$$E(\hat{X}_t\mathbb{I}_\Gamma) = \lim_{n\to\infty} E(X_{r_n}\mathbb{I}_\Gamma) \leq \lim_{n\to\infty} E(X_{q_n}\mathbb{I}_\Gamma) = E\hat{X}_s\mathbb{I}_\Gamma,$$

proving that $\{\hat{X}_t\}_{t\in[0,T]}$ is a submartingale. If $X$ is a martingale then we have equalities here, and so $\hat{X}$ is a martingale with respect to the filtration $\{\mathcal{F}_{t+}\}$.

(ii) In the same way as above, for any $\Gamma \in \mathcal{F}_t$ and $r_n \downarrow t$, $r_n \in \mathbb{Q}$,

$$E\hat{X}_t\mathbb{I}_\Gamma = \lim_{n\to\infty} E(X_{r_n}\mathbb{I}_\Gamma) \geq EX_t\mathbb{I}_\Gamma.$$

If the filtration $\{\mathcal{F}\}_{t\in[0,T]}$ is right-continuous then $\hat{X}_t$ is $\mathcal{F}_t$-measurable and the above inequality valid for any $\Gamma \in \mathcal{F}_t$ which yields that $\hat{X}_t \geq X_t$ a.s. If $s > t$ and $r_1 < s$ then

$$E\hat{X}_t\mathbb{I}_\Gamma = \lim_{n\to\infty} E(X_{r_n}\mathbb{I}_\Gamma) \leq E(X_s\mathbb{I}_\Gamma),$$

so $\hat{X}_t \leq E(X_s|\mathcal{F}_t)$ a.s. Hence,

$$EX_s - EX_t \geq E(\hat{X}_t - X_t) \geq 0,$$

thus $\hat{X}_t = X_t$ a.s. if and only if $EX_s \downarrow EX_t$ as $s \downarrow t$, completing the proof of the theorem. $\qquad\square$

Because of this theorem we can usually assume in applications that the martingales (submartingales, supermartingales) under consideration are right-continuous and have left hand limits, abbreviated RCLL. Such processes are usually called cádlág (in French: continue á droite, limite á gauche). To avoid unnecessary complications, filtrations in question are often assumed to satisfy usual conditions.

**Remark 6.1.1.** Observe that if $X_t$, $t \in [0, T]$ is an adapted right-continuous process then $\sup_{t \in [0,T]} X_t$ and $\inf_{t \in [0,T]} X_t$ are measurable and there is no need in ess sup and ess inf since these sup and inf will remain the same if taken over rationals (which is a countable set).

### 6.1.2 Optional stopping (sampling) theorem and martingale inequalities

**Theorem 6.1.2.** *Let $X = \{X_t\}_{t \in [0,T]}$ be a right-continuous submartingale with respect to a filtration $\{\mathcal{F}_t\}_{t \in [0,T]}$ and $\sigma \leq \tau$ be two bounded stopping times. Then*

$$E(X_\tau | \mathcal{F}_\sigma) \geq X_\sigma \quad a.s.$$

*where, in the same way as in the discrete time case, $\mathcal{F}_\sigma = \{\Gamma : \Gamma \cap \{\sigma \leq t\} \in \mathcal{F}_t\}$. If $X$ is a martingale then we have the equality here. If $T = \infty$ and*

$$E(\sup_{t \geq 0} |X_t|) < \infty \tag{6.1.1}$$

*then the theorem remains true for any stopping times $\sigma < \tau < \infty$ a.s.*

**Proof.** Define $\sigma_n(\omega) = \frac{k}{2^n}$ if $\sigma(\omega) \in [(k-1)/2^n, k/2^n)$ and $\tau_n(\omega) = \frac{k}{2^n}$ if $\tau(\omega) \in [(k-1)/2^n, k/2^n)$. Then $\sigma_n \leq \tau_n$ are decreasing sequences of stopping times taking on values in the set $\mathcal{D}_n = \{\frac{k}{2^n}, k = 0, 1, ...\}$ such that $\sigma_n \leq \sigma + 2^{-n}$, $\tau_n \leq \tau + 2^{-n}$, $\sigma_n \downarrow \sigma$ and $\tau_n \downarrow \tau$ as $n \uparrow \infty$. By the discrete time optional sampling theorem from Lecture 2 we obtain

$$E(X_{\tau_n} | \mathcal{F}_{\sigma_n}) \geq X_{\sigma_n} \quad a.s.$$

In particular, for any $\Gamma \in \mathcal{F}_\sigma \subset \mathcal{F}_{\sigma_n}$,

$$E(X_{\tau_n} \mathbb{I}_\Gamma) \geq E(X_{\sigma_n} \mathbb{I}_\Gamma).$$

Similarly to the theorem in the previous subsection we observe that the sequences $\{X_{\sigma_n}\}_{n \geq 1}$ and $\{X_{\tau_n}\}_{n \geq 1}$ are uniformly integrable, so letting $n \to \infty$ it follows that

$$E(X_\tau \mathbb{I}_\Gamma) \geq E(X_\sigma \mathbb{I}_\Gamma) \quad \text{for any} \quad \Gamma \in \mathcal{F}_\sigma,$$

and if $X$ is a martingale then all above inequalities become equalities. Under (6.1.1) the sequences $\{X_{\sigma_n}\}_{n\geq 1}$ and $\{X_{\tau_n}\}_{n\geq 1}$ remain uniformly integrable without boundedness of stopping times assumption, and so we can complete the proof as above in this case as well. $\qquad\square$

The martingale inequalities in continuous time are easy corollaries of the corresponding results in the discrete time case.

**Theorem 6.1.3.** *Let* $X = \{X_t\}_{t\geq 0}$ *be a right-continuous submartingale.*
*(i) For any* $a, T > 0$,

$$aP\{\sup_{t\in[0,T]} X_t \geq a\} \leq E(X_T\mathbb{I}_{\{\sup_{t\in[0,T]} X_t\geq a\}}) \leq EX_T^+ \leq E|X_T|.$$

*In particular, if* $X_t$ *is a martingale and* $p \geq 1$ *then*

$$P\{\sup_{t\in[0,T]} X_t \geq a\} \leq \frac{E|X_T|^p}{a^p}.$$

*(ii) Suppose that* $X_t \geq 0$, $T > 0$ *and* $p > 1$. *Then*

$$E(\sup_{t\in[0,T]} X_t^p) \leq (\frac{p}{p-1})^p EX_T^p.$$

*In particular, if* $\{X_t\}_{t\in[0,T]}$ *is a martingale (not assuming nonnegativity) then*

$$E(\sup_{t\in[0,T]} |X_t|^p) \leq (\frac{p}{p-1})^p E|X_T|^p.$$

*(iii) More generally, let* $\tau < \infty$ *a.s. be a stopping time. Then (i)–(ii) remain true with* $\sup_{t\in[0,\tau]} X_t$ *and* $X_\tau$ *in place of* $\sup_{t\in[0,T]} X$ *and* $|X_T|$, *respectively.*

**Proof.** Let $\mathcal{D}_{n,T} = \{\frac{kT}{2^n}, k = 0, 1, ..., 2^n\}$. Then by the discrete time martingale inequalities from Lecture 2 all above inequalities hold true if we replace $\sup_{t\in[0,T]} X_t$ by $\max_{t\in\mathcal{D}_{n,T}} X_t$. Since $\max_{t\in\mathcal{D}_{n,T}} X_t \uparrow \sup_{t\in[0,T]} X_t$ as $n \uparrow \infty$ we obtain (i) by the $\sigma$-additivity of the probability $P$ and (ii) follows by the integral monotone convergence theorem. In order to prove (iii) consider the stopped submartingale $X_{\tau\wedge t}$, $t \in [0,T]$, use (i)–(ii) and pass to the limit $T \to \infty$. $\qquad\square$

### 6.1.3 The Doob–Meyer decomposition

Let $(\Omega, \mathcal{F}, P)$ be a complete probability space and $\{\mathcal{F}_t\}_{t \in [0,T]}$ be a filtration satisfying usual conditions.

**Definition 6.1.3.** An adapted to the filtration $\{\mathcal{F}_t\}_{t \in [0,T]}$ process $X_t$, $t \in [0,T]$ is said to be of class $D$ if the family of random variables $\{X_\tau\}$, where $\tau$ runs over all stopping times with $0 \leq \tau \leq T$, is uniformly integrable.

**Definition 6.1.4.** An adapted (one dimensional) process $\{A_t\}_{t \in [0,T]}$ is called increasing if $A_t(\omega)$ is a non-decreasing function of $t \in [0,T]$ for $P$ almost all $\omega \in \Omega$.

**Definition 6.1.5.** Let $\mathcal{J}$ be the smallest $\sigma$-field on $[0,T] \times \Omega$ such that all left-continuous $\{\mathcal{F}_t\}$-adapted processes $Y : [0,T] \times \Omega \to \mathbb{R}^d$, $(t, \omega) \to Y_t(\omega)$ are measurable. A process $X = \{X_t\}_{t \in [0,T]}$ is called predictable if the mapping $X : ([0,T] \times \Omega, \mathcal{J}) \to (\mathbb{R}^d, \mathcal{B})$, $(t, \omega) \to X_t(\omega)$ is measurable, where $\mathcal{B}$ is the Borel $\sigma$-algebra on $\mathbb{R}^d$.

It is customary (see [28], [32], [44] and [2]) to formulate the Doob–Meyer decomposition theorem for submartingales, though usually it is used in mathematical finance for supermartingales, of course, the latter follows from the former just by multiplying all terms of the equality by $-1$. The uniqueness part of the proof below is taken from Chapter 22 of [32] while the existence part follows mostly [2] and [29] with the use of the Komlos theorem from Lecture 9.

**Theorem 6.1.4.** *Any cádlág submartingale $X = \{X_t\}_{t \in [0,T]}$ of class $D$ can be written in the unique way in the form*

$$X = M + A$$

*where $M = \{M_t\}_{t \in [0,T]}$ is a cádlág martingale and $A$ is a predictable increasing cádlág process starting at 0.*

**Proof.** For more details related to the uniqueness proof we refer the reader to [32] providing here its main points. Suppose that there exists another decomposition $X = M' + A'$ with the same properties. Since $A$ and $A'$ are predictable processes, the cádlág martingale $\mathcal{M} = M - M' = A' - A$ is predictable as well. On the predictable $\sigma$-field $\mathcal{J}$ define the signed measure

$$\mu(\Gamma) = E \int_0^\infty \mathbb{I}_\Gamma(t) d\mathcal{M}_t, \ \Gamma \in \mathcal{J},$$

where $\int_0^\infty$ is an ordinary Lebesgue–Stiltjes integral. Using the martingale property it is easy to see that $\mu$ vanishes for sets $\Gamma$ of the form $G \times (t, \infty)$ with $G \in \mathcal{F}_t$. Applying monotone limits we conclude that these sets generate the whole $\sigma$-algebra $\mathcal{J}$, and so $\mu = 0$ on $\mathcal{J}$. Since $\mathcal{M}$ is predictable, $\Delta \mathcal{M}_t = \mathcal{M}_t - \mathcal{M}_{t-}$ is predictable as well, and $J_\pm = \{t > 0 : \pm \Delta \mathcal{M}_t > 0\} \in \mathcal{J}$. Thus $\mu(J_\pm) = 0$, so $\Delta \mathcal{M} = 0$ a.s., i.e., $\mathcal{M}$ is continuous. Consider the total variation $V_t \leq A_t' + A_t$ of $\mathcal{M}$ on each interval $[0, t]$, $t \leq T$. Then $V$ is continuous and adapted to the same filtration. Without loss of generality we can assume that $V_t$ is bounded by a constant. Indeed, introduce stopping times $\tau_n = \inf\{t \geq 0 : V_t = n\}$, then the continuous martingale $\mathcal{M}_s^{\tau_n} = \mathcal{M}_{s \wedge \tau_n}$, $s \geq 0$ has total variation bounded by $n$. If we show that $\mathcal{M}^{\tau_n} \equiv 0$ for each $n$ then, of course, $\mathcal{M} \equiv 0$, and we assume that $V_t$ is bounded by a constant. It follows from the continuity of $\mathcal{M}$ that with probability one, as $n \to \infty$,

$$Q_n = \sum_{k \leq n} (\mathcal{M}_{kt/n} - \mathcal{M}_{(k-1)t/n})^2 \leq V_t \max_{k \leq n} |\mathcal{M}_{kt/n} - \mathcal{M}_{(k-1)t/n}| \to 0.$$

Since $Q_n \leq V_t^2$, which is bounded by a constant, it follows by the martingale property and the dominated convergence theorem that

$$E\mathcal{M}_t^2 = E\Big( \sum_{k=1}^n (\mathcal{M}_{kt/n} - \mathcal{M}_{(k-1)t/n}) \Big)^2 = EQ_n \to 0 \quad \text{as} \quad n \to \infty,$$

and so $\mathcal{M}_t = 0$ a.s. for each $t > 0$.

Next, we will deal with the construction of the required decomposition. If $T < \infty$ then just by rescaling time we can assume for simplicity that $T = 1$. If $T = \infty$ then we can first prove the result on each interval $[n, n+1]$ separately which will give the decomposition for all $t \geq 0$. In the latter case it suffices to assume that the submartingale $X$ is of class $DL$ which means that the family of random variables $\{X_\tau\}$ is uniformly integrable where $\tau$ runs over all stopping times with ess $\sup \tau < \infty$.

Set $\mathcal{D}_n = \{\frac{k}{2^n} : k = 0, 1, ..., 2^n\}$ and $\mathcal{D} = \cup_{n \geq 1} \mathcal{D}_n$. For each $n$ let

$$X_t = X_t^{(n)} = M_t^{(n)} + A_t^{(n)}, \quad t \in \mathcal{D}_n$$

be the discrete time Doob submartingale decomposition discussed in Lecture 2 so that $A_0^{(n)} = 0$,

$$A_t^{(n)} - A_{t-2^{-n}}^{(n)} = E(X_t - X_{t-2^{-n}} | \mathcal{F}_{t-2^{-n}}), \quad t \in \mathcal{D}_n,$$

$\{M_t^{(n)}\}_{t \in \mathcal{D}_n}$ is a martingale and $\{A_t^{(n)}\}_{t \in \mathcal{D}_n}$ is increasing and predictable with respect to the filtration $\{\mathcal{F}_t\}_{t \in \mathcal{D}_n}$.

Next, we prove that the sequence $\{M_1^{(n)}\}_{n \geq 1}$ is uniformly integrable. First, observe that subtracting the martingale $E(X_1|\mathcal{F}_t)$ from the submartingale $X_t$ we can assume that $X_1 = 0$ and $X_t \leq 0$ for $0 \leq t \leq 1$. Then $M_1^{(n)} = -A_1^{(n)}$ and by the optional sampling theorem for any stopping time $\tau$ with respect to the filtration $\{\mathcal{F}_t\}_{t \in \mathcal{D}_n}$,

$$X_\tau^{(n)} = -E(A_1^{(n)}|\mathcal{F}_\tau) + A_\tau^{(n)}. \tag{6.1.2}$$

Since $M_1^{(n)} = X_1 - A_1^{(n)}$ it suffices to prove that $\{A_1^{(n)}\}_{n \geq 1}$ is a uniformly integrable sequence.

For $c > 0$, $n \geq 1$ define the stopping time

$$\tau_n(c) = \min\{(j-1)2^{-n} : A_{j2^{-n}}^{(n)} > c\} \wedge 1.$$

Since $A_{\tau_n(c)}^{(n)} \leq c$ we obtain from (6.1.2) that

$$X_{\tau_n(c)} \leq -E(A_1^{(n)}|\mathcal{F}_{\tau_n(c)}) + c. \tag{6.1.3}$$

By (6.1.3) and since $A_t^{(n)}$ is increasing,

$$\begin{aligned}\int_{\{A_1^{(n)}>c\}} A_1^{(n)}dP &= \int_{\{\tau_n(c)<1\}} E(A_1^{(n)}|\mathcal{F}_{\tau_n(c)})dP \\ &\leq cP\{\tau_n(c)<1\} - \int_{\{\tau_n(c)<1\}} X_{\tau_n(c)}dP.\end{aligned} \tag{6.1.4}$$

Clearly, $\{\tau_n(c) < 1\} \subset \{\tau_n(c/2) < 1\}$, and so by (6.1.2),

$$\begin{aligned}\int_{\{\tau_n(c/2)<1\}}(-X_{\tau_n(c/2)})dP &= \int_{\{\tau_n(c/2)<1\}}(A_1^{(n)} - A_{\tau_n(c/2)}^{(n)})dP \\ &\geq \int_{\{\tau_n(c)<1\}}(A_1^{(n)} - A_{\tau_n(c/2)}^{(n)})dP \geq \tfrac{c}{2}P\{\tau_n(c)<1\}.\end{aligned} \tag{6.1.5}$$

Combining (6.1.4) and (6.1.5) we obtain

$$\int_{\{A_1^{(n)}>c\}} A_1^{(n)}dP \leq -2\int_{\{\tau_n(c/2)<1\}} X_{\tau_n(c/2)}dP - \int_{\{\tau_n(c)<1\}} X_{\tau_n(c)}dP. \tag{6.1.6}$$

On the other hand, by Chebyshev's inequality and the martingale property for any $a > 0$,

$$\begin{aligned}P\{\tau_n(a)<1\} = P\{A_1^{(n)} > a\} &\leq a^{-1}EA_1^{(n)} = -a^{-1}EM_1^{(n)} \\ &= -a^{-1}EM_0^{(n)} = -a^{-1}EX_0.\end{aligned}$$

Hence, for any $K > 0$,

$$\begin{aligned}\int_{\{\tau_n(a)<1\}}(-X_{\tau_n(a)})dP &\leq \int_{\{\tau_n(a)<1\}\cap\{-X_{\tau_n(a)}>K\}}(-X_{\tau_n(a)})dP \\ +\int_{\{\tau_n(a)<1\}\cap\{-X_{\tau_n(a)}\leq K\}}(-X_{\tau_n(a)})dP &\leq \int_{\{-X_{\tau_n(a)}>K\}}(-X_{\tau_n(a)})dP \\ +KP\{\tau_n(a)<1\} &\leq \sup_\tau \int_{\{-X_\tau>K\}}(-X_\tau)dP - \tfrac{K}{a}EX_0\end{aligned}$$

where the supremum is taken over all stopping times $\tau$ with $0 \leq \tau \leq 1$. Thus, by (6.1.6),

$$\sup_{n \geq 1} \int_{\{A_1^{(n)} > c\}} A_1^{(n)} dP \leq 3 \sup_{\tau} \int_{\{-X_\tau > K\}} (-X_\tau) dP - \frac{3K}{c} E X_0. \qquad (6.1.7)$$

Since $\{X_t\}_{t \in [0,T]}$ is of class $D$ we obtain by letting $c \to \infty$ and then $K \to \infty$ that the limit as $c \to \infty$ of the left hand side in (6.1.7) is zero, i.e., that the sequence $\{A_1^{(n)}\}_{n \geq 1}$ is uniformly integrable, as required.

Next, we extend $M^{(n)}$ to a cádlág martingale on $[0,1]$ setting $M_1^{(n)} = E(M_1^{(n)}|\mathcal{F}_t)$ and using an appropriate modification available by the theorem at the beginning of this lecture. By the Komlos theorem proved in Lecture 9 there exists $M_1 \in L^1(\Omega, P)$, an integer $N_n \geq n$ and numbers $p_j^{(n)} \geq 0$, $j = n, ..., N_n$ with $\sum_{j=n}^{N_n} p_j^{(n)} = 1$ such that if

$$\mathcal{M}^{(n)} = p_n^{(n)} M_t^{(n)} + \cdots + p_{N_n}^{(n)} M_t^{(N_n)},$$

then $\mathcal{M}_1^{(n)} \to M_1$ in $L^1(\Omega, P)$ as $n \to \infty$. Then by Jensen's inequality, $\mathcal{M}_t^{(n)} = E(\mathcal{M}_1^{(n)}|\mathcal{F}_t)$ converges in $L^1(\Omega, P)$ to $M_t = E(M_1|\mathcal{F}_t)$ as $n \to \infty$ where we take cádlág versions of these conditional expectations which is possible since the filtration is right-continuous.

For each $n \geq 1$ extend $A$ to $[0,1]$ by

$$A^{(n)} = \sum_{t \in \mathcal{D}_n} A_t^{(n)} \mathbb{I}_{(t-2^{-n},t]}$$

and set

$$\mathcal{A}_t^{(n)} = p_n^{(n)} A_t^{(n)} + \cdots + p_{N_n}^{(n)} A_t^{(N_n)}, \ t \in [0,1].$$

Then for every $t \in \mathcal{D}$, the cádlág process

$$A_t = X_t - M_t, \ 0 \leq t \leq 1$$

satisfies

$$\mathcal{A}_t^{(n)} = (X_t - \mathcal{M}_t^{(n)}) \to (X_t - M_t) = A_t \text{ in } L^1(\Omega, P) \text{ as } n \to \infty.$$

Passing to a subsequence, denoted again by $n$, we ensure that the convergence holds true also almost surely. Then, $A$ is almost surely increasing on $\mathcal{D}$ and, by right continuity, also on the whole $[0,1]$. Since the processes $A^{(n)}$ and $\mathcal{A}^{(n)}$ are left-continuous and adapted, they are predictable. In order to show that $A$ is predictable it suffices to prove that for almost all $\omega$ and each $t \in [0,1]$,

$$\limsup_{n \to \infty} \mathcal{A}_t^{(n)}(\omega) = A_t(\omega). \qquad (6.1.8)$$

Let $f_n, f : [0,1] \to \mathbb{R}$ be increasing functions such that $f$ is right-continuous and $\lim_{n\to\infty} f_n(t) = f(t)$ for $t \in \mathcal{D}$. Then it is easy to see that

$$\limsup_{n\to\infty} f_n(t) \leq f(t) \text{ for all } t \in [0,1] \text{ and} \qquad (6.1.9)$$

$$\lim_{n\to\infty} f_n(t) = f(t) \text{ if } f \text{ is continuous at } t.$$

Thus, (6.1.8) may not hold true only at discontinuity points of $A$. Since $A$ is a cádlág, every path of $A$ can have only finitely many jumps larger than $1/k$. Set $B_t = A_t - \lim_{s \uparrow t} A_t$ and introduce stopping times $\tau_1^{(k)} = \inf\{t > 0 : B_t > 1/k\} \wedge 1$ and inductively $\tau_{i+1}^{(k)} = \inf\{t > \tau_i^{(k)} : B_t > 1/k\} \wedge 1$. It follows that the points of discontinuity of $A$ can be exhausted by a countable sequence of stopping times $\tau_i^{(k)}$, $i, k = 1, 2, ...$, and so it suffices to prove that

$$\limsup_{n\to\infty} A_\tau^{(n)} = A_\tau \qquad (6.1.10)$$

for every stopping time $\tau \leq 1$.

Recall, that

$$\mathcal{A}_1^{(n)} \to A_1 \quad \text{in} \quad L^1(\Omega, P) \quad \text{as} \quad n \to \infty. \qquad (6.1.11)$$

It follows that the sequence $\{\mathcal{A}_1^{(n)}\}_{n\geq 1}$ is uniformly integrable. Indeed, $\mathcal{A}_1^{(n)} \geq 0$, $A_1 \geq 0$ and

$$\int \mathcal{A}_1^{(n)} \mathbb{I}_{\{\mathcal{A}_1^{(n)}>K\}} dP \leq \int A_1 \mathbb{I}_{\{A_1>K/2\}} dP$$

$$+ \int A_1 \mathbb{I}_{\{|\mathcal{A}_1^{(n)}-A_1|>K/2\}} dP + \|\mathcal{A}_1^{(n)} - A_1\|_{L^1(\Omega,P)}$$

for all $n, K \geq 1$. By the absolute continuity of the Lebesgue integral, for any $\varepsilon > 0$ there exists $\delta(\varepsilon) > 0$ such that if $P(\Gamma) \leq \delta(\varepsilon)$ then $\int A_1 \mathbb{I}_\Gamma dP \leq \varepsilon$. Observe that

$$P\{|\mathcal{A}_1^{(n)} - A_1| > K/2\} \leq 2K^{-1}\|\mathcal{A}_1^{(n)} - A_1\|_{L^1(\Omega,P)}.$$

By (6.1.11) there exists $n(\varepsilon)$ such that for all $n \geq n(\varepsilon)$,

$$\|\mathcal{A}_1^{(n)} - A_1\|_{L^1(\Omega,P)} \leq \min(\varepsilon, \frac{1}{2}\delta(\varepsilon)).$$

Since by Chebyshev's inequality,

$$P\{A_1 \geq K/2\} \leq 2K^{-1}\int A_1 dP,$$

it follows that for any $K > K(\varepsilon) \geq 2(\delta(\varepsilon))^{-1} \int A_1 dP$,

$$\sup_{n\geq n(\varepsilon)} \int \mathcal{A}_1^{(n)} \mathbb{I}_{\{\mathcal{A}_1^{(n)}>K\}} dP \leq 3\varepsilon.$$

Applying again the absolute continuity of the Lebesgue integral theorem we conclude that there exists $\tilde{K}(\varepsilon)$ such that for all $K > \tilde{K}(\varepsilon)$,

$$\max_{1 \le n < n(\varepsilon)} \int \mathcal{A}_1^{(n)} \mathbb{I}_{\{\mathcal{A}_1^{(n)} > K\}} dP \le \varepsilon,$$

and so for such $K$,

$$\sup_{n \ge 1} \int \mathcal{A}_1^{(n)} \mathbb{I}_{\{\mathcal{A}_1^{(n)} > K\}} dP \le 4\varepsilon.$$

Letting $K \to \infty$ and taking into account that $\varepsilon > 0$ is arbitrary, we obtain uniform integrability of the sequence $\{\mathcal{A}_1^{(n)}\}_{n \ge 1}$.

Since $0 \le \mathcal{A}_\tau^{(n)} \le \mathcal{A}_1^{(n)}$, it follows that the sequence $\{\mathcal{A}_\tau^{(n)}\}_{n \ge 1}$ is also uniformly integrable. Hence, by the slightly extended Fatou lemma

$$\limsup_{n \to \infty} E\mathcal{A}_\tau^{(n)} \le E \limsup_{n \to \infty} \mathcal{A}_\tau^{(n)}. \tag{6.1.12}$$

Indeed, since $\{\mathcal{A}_\tau^{(n)}\}_{n \ge 1}$ is uniformly integrable, for any $\varepsilon > 0$, we can choose $\hat{K}(\varepsilon)$ such that

$$\int \mathcal{A}_\tau^{(n)} \mathbb{I}_{\{\mathcal{A}_\tau^{(n)} > K\}} dP < \varepsilon \quad \text{for all} \quad K \ge \hat{K}(\varepsilon).$$

Hence,

$$\limsup_{n \to \infty} \int \mathcal{A}_\tau^{(n)} dP = \limsup_{n \to \infty} (\int \mathcal{A}_\tau^{(n)} \mathbb{I}_{\{\mathcal{A}_\tau^{(n)} \le K\}} dP$$
$$+ \int \mathcal{A}_\tau^{(n)} \mathbb{I}_{\{\mathcal{A}_\tau^{(n)} > K\}} dP) \le \int \limsup_{n \to \infty} \mathcal{A}_\tau^{(n)} \mathbb{I}_{\{\mathcal{A}_\tau^{(n)} \le K\}} dP + \varepsilon$$

where we employed the standard Fatou lemma for a bounded sequence. Letting $K \uparrow \infty$ in this inequality and taking into account that $\varepsilon > 0$ is arbitrary we derive (6.1.12).

Next, observe that

$$\inf_{n \ge k} E\mathcal{A}_\tau^{(n)} \le \inf_{n \ge m} E\mathcal{A}_\tau^{(n)} \le E\mathcal{A}_\tau^{(m)} = \sum_{j=m}^{N_m} p_j^{(m)} E\mathcal{A}_\tau^{(j)}$$

for any $m \ge k$. Hence,

$$\liminf_{n \to \infty} E\mathcal{A}_\tau^{(n)} \le \limsup_{n \to \infty} E\mathcal{A}_\tau^{(n)}. \tag{6.1.13}$$

By the first inequality in (6.1.9),

$$\limsup_{n \to \infty} \mathcal{A}_\tau^{(n)} \le \mathcal{A}_\tau, \tag{6.1.14}$$

and so combining this with (6.1.12) and (6.1.13) we obtain that

$$\liminf_{n \to \infty} E\mathcal{A}_\tau^{(n)} \le E \limsup_{n \to \infty} \mathcal{A}_\tau^{(n)} \le E\mathcal{A}_\tau.$$

If we show that

$$\lim_{n\to\infty} EA_\tau^{(n)} = EA_\tau \qquad (6.1.15)$$

then we obtain

$$E \limsup_{n\to\infty} \mathcal{A}_\tau^{(n)} = EA_\tau,$$

which together with (6.1.14) yields (6.1.10), as required.

In order to have (6.1.15), for each $n \geq 1$ set $\sigma_n = \inf\{t \in \mathcal{D}_n : t \geq \tau\}$. Then $A_\tau^{(n)} = A_{\sigma_n}^{(n)}$ and $\sigma_n \downarrow \tau$. Since $X$ is of class $D$ we obtain as $n \to \infty$,

$$EA_\tau^{(n)} = EA_{\sigma_n}^{(n)} = EX_{\sigma_n} - EM_0 \to EX_\tau - EM_0 = EA_\tau,$$

where we use that by the optional sampling theorem $EM_{\sigma_n}^{(n)} = EM_0^{(n)} = EX_0 = EM_0$, completing the proof of the theorem. $\qquad\square$

## 6.2 Lecture 12: Optimal stopping in continuous time

### 6.2.1 *Single player*

Let $(\Omega, \mathcal{F}, P)$ be a complete probability space and $\{\mathcal{F}_t\}_{t\geq 0}$ be a filtration satisfying the usual conditions. For $T \leq \infty$ and $t \leq T$, $t < \infty$ denote by $\Lambda_{tT}$ the set of all stopping times $\tau < \infty$ a.s., such that $t \leq \tau \leq T$. All processes below will be considered only at finite time so that when we write $s \leq t \leq T$, $t \in [s,T]$, $s \leq \tau \leq T$, etc., and $T = \infty$ we mean also that $t < \infty$, $\tau < \infty$ a.s., etc. Again, an observer (player) can stop the reward process at any time $t \leq T$ receiving then a reward $R_t$, and the goal is to maximize the average reward using stopping times as allowed strategies.

**Theorem 6.2.1.** *Let* $R = \{R_t\}_{0\leq t\leq T}$ *be an adapted to the filtration* $\{\mathcal{F}_t\}_{t\geq 0}$ *stochastic process satisfying*

$$E(\sup_{t\in[0,T]} |R_t|) < \infty. \qquad (6.2.1)$$

*(i) Then the process* $L = \{L_t\}_{0\leq t\leq T}$ *defined by*

$$L_t = \operatorname*{ess\,sup}_{\tau\in\Lambda_{tT}} E(R_\tau|\mathcal{F}_t)$$

*is a supermartingale and*

$$EL_t = \sup_{\tau\in\Lambda_{tT}} ER_\tau. \qquad (6.2.2)$$

*(ii) If $R$ is right-continuous then $L$ admits a right-continuous modification which is called the Snell envelope and it is the smallest right-continuous supermartingale which dominates $R$ (i.e., $L_t \geq R_t$ for each $t \in [0,T]$).*

**Proof.** The proof of (i) is similar to the discrete time case. First, observe that by (6.2.1) the process $R$ is of class $D$ since for any $\tau \in \Lambda_{0T}$,

$$|R_\tau| \leq \sup_{t\in[0,T]} |R_t| \quad \text{a.s.} \tag{6.2.3}$$

Next, for any $\tau_1, \tau_2 \in \Lambda_{tT}$, $t \leq T$ set $\tau = \tau_1 \mathbb{I}_{\Omega\backslash\Gamma} + \tau_2 \mathbb{I}_\Gamma$ where $\Gamma = \{\omega : E(R_{\tau_1}|\mathcal{F}_t) < E(R_{\tau_2}|\mathcal{F}_t)\}$. Then $\tau \in \Lambda_{tT}$ and

$$E(R_\tau|\mathcal{F}_t) = \mathbb{I}_{\Omega\backslash\Gamma}E(R_{\tau_1}|\mathcal{F}_t) + \mathbb{I}_\Gamma E(R_{\tau_2}|\mathcal{F}_t) = \max((R_{\tau_1}|\mathcal{F}_t),(R_{\tau_2}|\mathcal{F}_t))$$

(meaning that the set $\{(R_\tau|\mathcal{F}_t), \tau \in \Lambda_{tT}\}$ is a lattice as it contains maximum of each pair of its elements).

By the property of ess sup (see, for instance, [58]) there exists a sequence $\tau_k \in \Lambda_{tT}$, $k = 0,1,2,...$ with $\tau_0 \equiv t$ such that $\sup_{k\geq1} E(R_{\tau_k}|\mathcal{F}_t) = L_t$ a.s., and by the above argument we can choose this sequence so that

$$E(R_{\tau_k}|\mathcal{F}_t) \uparrow L_t \quad \text{as} \quad k \uparrow \infty \quad a.s.$$

Now,

$$L_t \geq E(R_{\tau_k}|\mathcal{F}_t) \geq E(R_{\tau_0}|\mathcal{F}_t) = R_t.$$

In view of (6.2.1) and (6.2.3) we can apply the dominated convergence theorem to obtain that for $s \leq t$,

$$E(L_t|\mathcal{F}_s) = E(\lim_{k\uparrow\infty} \uparrow E(R_{\tau_k}|\mathcal{F}_t)|\mathcal{F}_s) = \lim_{k\uparrow\infty} \uparrow E(R_{\tau_k}|\mathcal{F}_s) \leq L_s$$

since $\tau_k \in \Lambda_{tT} \subset \Lambda_{sT}$, and so the process $L$ is a supermartingale.

Next,

$$EL_t = E(\text{ess}\sup_{\tau\in\Lambda_{tT}} E(R_\tau|\mathcal{F}_t)) \geq \sup_{\tau\in\Lambda_{tT}} E(E(R_\tau|\mathcal{F}_t)) = \sup_{\tau\in\Lambda_{tT}} ER_\tau.$$

On the other hand, by the dominated convergence theorem

$$EL_t = E(\text{ess}\sup_{\tau\in\Lambda_{tT}} E(R_\tau|\mathcal{F}_t)) = E(\lim_{k\uparrow\infty} \uparrow E(R_{\tau_k}|\mathcal{F}_t))$$
$$= \lim_{k\uparrow\infty} \uparrow E(E(R_{\tau_k}|\mathcal{F}_t)) = \lim_{k\uparrow\infty} \uparrow ER_{\tau_k} \leq \sup_{t\in\Lambda_{tT}} ER_t.$$

Hence, (6.2.2) holds true completing the proof of (i).

(ii) In order to prove the existence of a right-continuous modification of $L_t$, $t \geq 0$ it suffices in view of Theorem 6.1.1 to show that $EL_t$ is right-continuous as a function of $t \in [0,T]$. Let $t_n \downarrow t$ as $n \uparrow \infty$. Then by (6.2.2),

$$EL_{t_n} = \sup_{\tau\in\Lambda_{t_nT}} ER_\tau \leq \sup_{\tau\in\Lambda_{tT}} ER_\tau = EL_t$$

since $\Lambda_{t_n T} \subset \Lambda_{tT}$. On the other hand, if $\tau \in \Lambda_{tT}$ then $\tau_n = \tau \vee t_n \in \Lambda_{t_n T}$ and $\lim_{n \uparrow \infty} R_{\tau_n} = R_\tau$ by the right-continuity of $R$. By (6.2.1) and (6.2.3) we can apply the dominated convergence theorem to obtain that

$$ER_\tau = E(\lim_{n \uparrow \infty} R_{\tau_n}) = \lim_{n \uparrow \infty} ER_{\tau_n} \leq \lim_{n \uparrow \infty} EL_{t_n}.$$

Hence,

$$EL_t = \sup_{\tau \in \Lambda_{tT}} ER_\tau \leq \lim_{n \uparrow \infty} EL_{t_n}.$$

This together with the inequality in the other direction above yields that

$$\lim_{n \uparrow \infty} EL_{t_n} = EL_t,$$

and so $EL_t$, $t \geq 0$ is right-continuous.

Finally, let $\tilde{L}_t$, $t \geq 0$ be another right-continuous supermartingale such that $\tilde{L}_t \geq R_t$ a.s. for any $t \in [0, T]$. Then for any $\tau \in \Lambda_{tT}$ by the optional stopping theorem

$$\tilde{L}_t \geq E(\tilde{L}_\tau | \mathcal{F}_t) \geq E(R_\tau | \mathcal{F}_t) \tag{6.2.4}$$

where we use that $\tilde{L}_\tau \geq R_\tau$ in view of the right-continuity of both $\tilde{L}$ and $R$. Indeed, since $\tilde{R}_t \geq R_t$ a.s. for each $t \in [0, T]$, we have

$$P\{\tilde{L}_t \geq R_t \quad \text{for all rational} \quad t \in [0, T]\} = 1,$$

and by the right-continuity of these processes,

$$P\{\tilde{L}_t \geq R_t \quad \text{for all} \quad t \in [0, T]\} = 1.$$

Hence, $\tilde{L}_\tau \geq R_\tau$ a.s. for any $\tau \in \Lambda_{0T}$. Taking in (6.2.4) the supremum in $\tau \in \Lambda_{tT}$ we obtain $\tilde{L}_t \geq L_t$ a.s. for each $t \in [0, T]$, completing the proof of the theorem. $\square$

For each $t \in [0, T]$, $T < \infty$ and $\varepsilon > 0$, introduce the following random variables

$$\tau(t, \varepsilon) = \inf\{s \geq t : L_s \leq R_s + \varepsilon\}, \quad \sigma(t, \varepsilon) = \inf\{s > t : L_s \leq R_s + \varepsilon\}$$
$$\text{and} \quad \tau(t) = \inf\{s \geq t : L_s = R_s\}$$

where $L$ is the Snell envelope of $R$ satisfying the conditions of Theorem 6.2.1. These random variables are well defined since $L_T = R_T$ and they satisfy $\tau(t, \varepsilon) \leq \sigma(t, \varepsilon) \leq \tau(t)$. Furthermore, $\tau(t, \varepsilon)$, $\sigma(t, \varepsilon)$ and $\tau(t)$ are stopping times. Indeed, since $R$ and $L$ are right-continuous we have that

$$\{\tau(t, \varepsilon) \geq u\} = \{L_s > R_s + \varepsilon \text{ for all } t \leq s < u\}$$
$$= \cap_{r \in Q \cap [t,u)} \{L_r > R_r + \varepsilon\} \in \mathcal{F}_u,$$

where $Q$ is the set of rational numbers, and, similarly,

$$\{\sigma(t,\varepsilon) \geq u\} = \cap_{r \in Q \cap (t,u)}\{L_r > R_r + \varepsilon\} \in \mathcal{F}_u.$$

Hence, $\{\tau(t,\varepsilon) < u\} \in \mathcal{F}_u$ and $\{\sigma(t,\varepsilon) < u\} \in \mathcal{F}_u$. Then

$$\{\tau(t,\varepsilon) \leq u\} = \cap_{n=1}^{\infty}\{\tau(t,\varepsilon) < u + \frac{1}{n}\} \in \mathcal{F}_{u+} = \mathcal{F}_u$$

since the filtration is right-continuous. Similarly, $\{\sigma(t,\varepsilon) \leq u\} \in \mathcal{F}_u$. These arguments remain valid for all $\varepsilon \geq 0$, in particular, for $\varepsilon = 0$, so $\tau(t)$ is a stopping time as well.

It is also easy to see that $\sigma(s,\varepsilon) \downarrow \sigma(t,\varepsilon)$ as $s \downarrow t$ while this may not be true for the family of stopping times $\{\tau(t,\varepsilon)\}_{t \geq 0}$. Indeed, by the definition of $\sigma$ for each $\omega \in \Omega$ we must have $s_n = s_n(\omega) > t$ and $s_\infty = s_\infty(\omega)$, such that $s_n \downarrow s_\infty$ and $L_{s_n} \leq R_{s_n} + \varepsilon$. Hence, $t < s_n \leq \sigma(s_n,\varepsilon) \leq s_{n-1}$, and so $\sigma(s_n,\varepsilon) \downarrow \sigma(t,\varepsilon)$ as $n \uparrow \infty$. On the other hand, we may have $L_t \leq R_t + \varepsilon$ and then $\tau(t,\varepsilon) = t$ while $L_{t+\delta} > R_{t+\delta} + \varepsilon$ for all $\delta > 0$ small enough.

As in [62] the process $R_t$, $t \in [0,T]$ will be called left-continuous over stopping times, if $R_{\tau_n} \to R_\tau$ $P$-a.s. as $n \to \infty$ whenever $\tau_n \uparrow \tau$ as $n \uparrow \infty$, $\tau, \tau_n \in \Lambda_{0T}$, $n = 1,2,\ldots$ This property is weaker than left continuity since the $P$-null set for which the convergence above fails may depend on the sequence of stopping times $\tau_n$, $n \geq 1$. The following result was obtained in [17].

**Theorem 6.2.2.** *Suppose that conditions of Theorem 6.2.1 hold true.*
*(i) The stopping times $\tau(t,\varepsilon)$ and $\sigma(t,\varepsilon)$ are $\varepsilon$-optimal in the sense that*

$$\sup_{\tau \in \Lambda_{tT}} ER_\tau = EL_t \geq ER_{\tau(t,\varepsilon)} \geq EL_t - \varepsilon \ \ \text{and} \ \ EL_t \geq ER_{\sigma(t,\varepsilon)} \geq EL_t - \varepsilon.$$

(6.2.5)

*(ii) If the process $R_t$, $t \in [0,T]$ is left-continuous over stopping times, then $\tau(t)$ is the optimal stopping time in the sense that*

$$E(R_{\tau(t)}|\mathcal{F}_t) = EL_t = \text{ess} \sup_{\tau \in \Lambda_{tT}} E(R_\tau|\mathcal{F}_t) \quad a.s. \tag{6.2.6}$$

*(meaning that $\tau(t)$ maximizes the average "gain" $E(R_\tau|\mathcal{F}_t)$ among all stopping times $\tau \geq t$, i.e., in the "game" starting at $t$). In particular, if $\mathcal{F}_0 = \{\emptyset,\Omega\}$ is the trivial $\sigma$-algebra then*

$$\sup_{\tau \in \Lambda_{0T}} ER_\tau = L_0 = ER_{\tau(0)}.$$

**Proof.** First, consider the process

$$Q_t = Q_t^{(\varepsilon)} = E(L_{\sigma(t,\varepsilon)}|\mathcal{F}_t), \ \ t \in [0,T]. \tag{6.2.7}$$

Since $\sigma(u, \varepsilon) \geq \sigma(t, \varepsilon)$ for $u \geq t$ and $L_s$, $s \geq 0$ is a supermartingale,

$$E(Q_u | \mathcal{F}_t) = E(L_{\sigma(u,\varepsilon)} | \mathcal{F}_t) \leq E(L_{\sigma(t,\varepsilon)} | \mathcal{F}_t) = Q_t,$$

and so $Q_t$, $t \geq 0$ is a supermartingale. Next, $EQ_t = EL_{\sigma(t,\varepsilon)}$ and since $\sigma(s, \varepsilon) \downarrow \sigma(t, \varepsilon)$ as $s \downarrow t$ and $L_u$, $u \geq 0$ is a right-continuous supermartingale we obtain that $EQ_t$ is right-continuous in $t$ as well. Hence, by Theorem 6.1.1 in Lecture 11 it follows that $Q_t$, $t \geq 0$ has a right-continuous modification which we denote by the same letter $Q$ and work with in what follows.

Next, we show that (6.2.7) can be extended to stopping times, i.e., for any $\zeta \in \Lambda_{0T}$,

$$Q_\zeta = E(L_{\sigma(\zeta,\varepsilon)} | \mathcal{F}_\zeta) \quad \text{a.s.} \tag{6.2.8}$$

Indeed, if $\zeta$ takes on only finite or countable number of values then (6.2.8) follows directly from (6.2.7) considering $\zeta$ separately on each its constant value set. Now, let $\zeta_n \downarrow \zeta$ as $n \uparrow \infty$, $\zeta_n \in \Lambda_{0T}$ and $\zeta_n$, $n \geq 1$ take on finite or countable number of values. Since $\zeta_n \geq \zeta$,

$$E(Q_{\zeta_n} | \mathcal{F}_\zeta) = E(L_{\sigma(\zeta_n,\varepsilon)} | \mathcal{F}_\zeta) \quad \text{a.s.}$$

and letting $n \uparrow \infty$ we obtain (6.2.8) by the right-continuity of $Q$ and $L$ taking into account (6.2.1), which makes the sequences in the above expectations uniformly integrable.

Next, we proceed with the proof of the assertion (i) of the theorem. By the right-continuity of $R$ and $L$ it follows that

$$R_{\tau(t,\varepsilon)} \geq L_{\tau(t,\varepsilon)} - \varepsilon \text{ and } R_{\sigma(t,\varepsilon)} \geq L_{\sigma(t,\varepsilon)} - \varepsilon.$$

This together with (6.2.2) gives

$$EL_t \geq ER_{\tau(t,\varepsilon)} \geq EL_{\tau(t,\varepsilon)} - \varepsilon \text{ and } EL_t \geq ER_{\sigma(t,\varepsilon)} \geq EL_{\sigma(t,\varepsilon)} - \varepsilon. \tag{6.2.9}$$

Since $t \leq \tau(t, \varepsilon) \leq \sigma(t, \varepsilon)$ then by the optional stopping Theorem 6.1.2,

$$EL_t \geq EL_{\tau(t,\varepsilon)} \geq EL_{\sigma(t,\varepsilon)}. \tag{6.2.10}$$

Next, we show that (6.2.10) is, in fact, an equality, which together with (6.2.9) will yield (6.2.5). The equality in (6.2.10) will follow if we show that

$$Q_t = E(L_{\sigma(t,\varepsilon)} | \mathcal{F}_t) = L_t \quad \text{a.s.} \tag{6.2.11}$$

Since $L_t$, $t \geq 0$ is a right-continuous supermartingale then by the optional stopping theorem $L_t \geq E(L_{\sigma(t,\varepsilon)} | \mathcal{F}_t) = Q_t$. We will show next that $Q_t$, $t \in [0, T]$ dominates $R_t$, $t \in [0, T]$, i.e., $Q_t \geq R_t$ a.s. for each $t \in [0, T]$, and since $Q_t$, $t \in [0, T]$ is a right-continuous supermartingale while $L_t$, $t \geq 0$ is the Snell envelope, it will follow that $Q_t \geq L_t$ a.s. for each $t \in [0, T]$, which will give (6.2.11).

Set $a(\omega) = \sup_{t \in [0,T]} (R_t(\omega) - Q_t(\omega))^+$ which is dominated by the integrable random variable $2 \sup_{t \in [0,T]} |R_t|$. Consider the martingale $a_t = E(a|\mathcal{F}_t)$, $t \geq 0$ and take its right-continuous modification denoted by the same letter $a_t$, $t \geq 0$. If $Ea = 0$ then $Q_t \geq R_t$ a.s. for each $t \in [0, T]$ and we are done. Suppose that $Ea_t = Ea = \lambda > 0$. The supermartingale $G_t = Q_t + a_t$ is right-continuous and it dominates $R_t$, $t \geq 0$, whence $G_t \geq L_t$ a.s. for each $t \in [0, T]$ since $L$ is Snell's envelope of $R$.

Choose $\alpha < \min(\varepsilon, \lambda)$ and set $\zeta(t) = \inf\{s \in (t, T] : R_s + \alpha \geq Q_s + a_s\}$ where we put $\zeta(t) = T$ if there exists no $s$ satisfying the conditions in braces. Then by the above $R_\zeta + \alpha \geq Q_\zeta + a_\zeta \geq L_\zeta$ if $\zeta < T$, and so $R_\zeta + \varepsilon \geq L_\zeta$. If $\zeta = T$ then this also holds true since $R_T = L_T$. Since $\zeta(t)$, $t \geq 0$ is right-continuous we obtain from here and the definition of $\sigma(t, \varepsilon)$ that $\zeta = \sigma(\zeta, \varepsilon)$. Then

$$Q_\zeta = E(L_{\sigma(\zeta,\varepsilon)}|\mathcal{F}_\zeta) = L_\zeta.$$

Since $L_t \geq R_t$ a.s. for each $t \in [0, T]$ and both processes are right-continuous, it follows by the definition of $\zeta$ that

$$0 \geq E(R_\zeta - L_\zeta) = E(R_\zeta - Q_\zeta) \geq E(a_\zeta - \alpha) = \lambda - \alpha,$$

contradicting the choice of $\alpha$. Hence, $Ea = 0$, $Q_t \geq R_t$ a.s. for each $t \in [0, T]$ and (6.2.11) is proved.

(ii) Clearly, $\tau(t, \varepsilon) \uparrow \tau^*(t)$ as $\varepsilon \downarrow 0$ where $\tau^*(t) \leq \tau(t) \leq T$ is a stopping time as

$$\{\tau^*(t) > s\} = \cup_{k=1}^{\infty} \cap_{n \geq k} \left\{\tau\left(t, \frac{1}{n}\right) > s\right\} \in \mathcal{F}_s.$$

Since the process $R$ is left-continuous over stopping times,

$$\lim_{n \uparrow \infty} R_{\tau(t, \frac{1}{n})} = R_{\tau^*(t)} \quad \text{a.s.}.$$

This together with (6.2.1) and (6.2.5) yields

$$EL_t = \lim_{n \uparrow \infty} ER_{\tau(t, \frac{1}{n})} = ER_{\tau^*(t)}.$$

Having that $L_t \geq E(R_{\tau^*(t)}|\mathcal{F}_t)$ a.s. by the definition of $L$ we obtain that

$$L_t = E(R_{\tau^*(t)}|\mathcal{F}_t) \quad \text{a.s.} \tag{6.2.12}$$

Taking into account that $L$ is a right-continuous supermartingale dominating the right-continuous process $R$ and that $\tau^*(t) \geq t$ we conclude by the optional stopping theorem that

$$L_t \geq E(L_{\tau^*(t)}|\mathcal{F}_t) \quad \text{and} \quad L_{\tau^*(t)} \geq R_{\tau^*(t)} \quad \text{a.s.}$$

It follows that $L_{\tau^*(t)} = R_{\tau^*(t)}$ a.s. But $\tau(t)$ is the minimal stopping time satisfying such an equality, then $\tau(t) \leq \tau^*(t)$. Hence, $\tau(t) = \tau^*(t)$ a.s., (6.2.12) yields (6.2.6), completing the proof of the theorem. $\square$

**Remark 6.2.1.** The assertion (i) of Theorem 6.2.2 can be extended to the case $T = \infty$ by considering first $T < \infty$ and then letting $T \to \infty$, which is possible under (6.2.1) taken with $T = \infty$. In this case we can consider all finite stopping times $\tau$ with $P\{\tau < \infty\} = 1$. On the other hand, when $T = \infty$, $\tau(t)$ may become infinite with positive probability, but if we define $R_\infty = \limsup_{t \to \infty} R_t$ then, still, $\tau(t)$ will remain the optimal stopping time in the same sense as above.

### 6.2.2 Dynkin games in continuous time

We start with a complete probability space $(\Omega, \mathcal{F}, P)$ and a filtration $\{\mathcal{F}_t\}_{t \geq 0}$ satisfying the usual conditions. The modern Dynkin game setup consists of two players and two $\{\mathcal{F}_t\}_{t \geq 0}$-adapted stochastic processes $Y_t$ and $Z_t$, $t \geq 0$, so that when the first player stops the game at time $s$ and the second one stops at time $t$, the former pays to the latter the amount

$$R(s,t) = Y_s \mathbb{I}_{\{s < t\}} + Z_t \mathbb{I}_{\{s \geq t\}}.$$

The time runs up to $T < \infty$ when the game stops automatically if it was not stopped before, then the first player pays to the second one the amount

$$Y_T = Z_T. \tag{6.2.13}$$

We assume that

$$Z_t \leq Y_t \quad \text{for all} \quad t \in [0, T] \quad \text{and} \tag{6.2.14}$$

$$E \sup_{0 \leq t \leq T} (|Y_t| + |Z_t|) < \infty. \tag{6.2.15}$$

Most of the results can be easily extended to the perpetual case $T = \infty$ if, say, $\lim_{T \to \infty} Y_T = \lim_{T \to \infty} Z_T = 0$.

For each stopping time $\zeta \geq 0$, $\zeta \leq T$ denote by $\Lambda_{\zeta T}$ the set of all stopping times $\tau$ such that $\zeta \leq \tau \leq T$. Introduce the upper and the lower values of the game starting at $\zeta \in \Lambda_{0T}$ by

$$\overline{V}_\zeta = ess\inf_{\sigma \in \Lambda_{\zeta T}} ess\sup_{\tau \in \Lambda_{\zeta T}} E(R(\sigma, \tau) | \mathcal{F}_\zeta) \quad \text{and}$$

$$\underline{V}_\zeta = ess\sup_{\tau \in \Lambda_{\zeta T}} ess\inf_{\sigma \in \Lambda_{\zeta T}} E(R(\sigma, \tau) | \mathcal{F}_\zeta).$$

It is clear that $\overline{V}_\zeta \geq \underline{V}_\zeta$ a.s. for each $\zeta \in \Lambda_{0T}$.

**Theorem 6.2.3.** *Suppose that the processes $Y_t$ and $Z_t$, $t \in [0,T]$ are right-continuous and that (6.2.13)–(6.2.15) hold true. Then there exist right-continuous modifications of $\overline{V}_t$ and $\underline{V}_t$, $t \in [0,T]$ denoted by the same letters and for them and each $\zeta \in \Lambda_{0T}$,*

$$\overline{V}_\zeta = \underline{V}_\zeta \quad a.s. \tag{6.2.16}$$

*and, in particular, the Dynkin game has a value $V = V_0 = \overline{V}_0 = \underline{V}_0$ a.s. Furthermore, for each $\varepsilon > 0$ the stopping times*

$$\sigma_0^\varepsilon = \inf\{u \geq 0 : Y_t \leq V_t + \varepsilon\} \quad and \quad \tau_0^\varepsilon = \inf\{u \geq 0 : Z_t \geq V_t - \varepsilon\}, \tag{6.2.17}$$

*where $V_t = \overline{V}_t = \underline{V}_t$, $t \in [0,T]$, are $\varepsilon$-optimal in the sense that*

$$ER(\sigma_0^\varepsilon, \tau) - \varepsilon \leq V \leq ER(\sigma, \tau_0^\varepsilon) + \varepsilon \tag{6.2.18}$$

*for any $\sigma, \tau \in \Lambda_{0T}$. If, in addition, the processes $Y_t$ and $Z_t$, $t \in [0,T]$ are left-continuous then there exist optimal stopping times $\sigma^*, \tau^* \in \Lambda_{0T}$ (called a saddle point of the game) so that*

$$ER(\sigma^*, \tau) \leq V \leq ER(\sigma, \tau^*) \tag{6.2.19}$$

*for any $\sigma, \tau \in \Lambda_{0T}$.*

The proof will be similar to [51] but our exposition is self-contained while in [51] it relies on some general theory of stochastic processes. The result will follow from the series of lemmas.

**Lemma 6.2.1.** *Consider another payoff function $R'(s,t) = Y_s \mathbb{I}_{\{s \leq t\}} + Z_t \mathbb{I}_{\{s > t\}}$ and for each $t \in [0,T]$ define new upper and lower game values for $\zeta \in \Lambda_{0T}$ by*

$$\overline{V}'_\zeta = ess\inf_{\sigma \in \Lambda_{\zeta T}} ess\sup_{\tau \in \Lambda_{\zeta T}} E(R'(\sigma, \tau)|\mathcal{F}_\zeta) \quad and$$
$$\underline{V}'_\zeta = ess\sup_{\tau \in \Lambda_{\zeta T}} ess\inf_{\sigma \in \Lambda_{\zeta T}} E(R'(\sigma, \tau)|\mathcal{F}_\zeta).$$

*Then*

$$\overline{V}_\zeta = \overline{V}'_\zeta \quad a.s. \quad and \quad \underline{V}_\zeta = \underline{V}'_\zeta \quad a.s. \tag{6.2.20}$$

**Proof.** For any $\sigma, \tau \in \Lambda_{\zeta T}$ set

$$K(\sigma, \tau) = E(R(\sigma, \tau)|\mathcal{F}_\zeta) \quad and \quad K'(\sigma, \tau) = E(R'(\sigma, \tau)|\mathcal{F}_\zeta).$$

We will show first that $\overline{V}_\zeta = \overline{V}'_\zeta$ a.s. for which it suffices to prove that, for any $\sigma \in \Lambda_{\zeta T}$,

$$ess\sup_{\tau \in \Lambda_{\zeta T}} K(\sigma, \tau) = ess\sup_{\tau \in \Lambda_{\zeta T}} K'(\sigma, \tau) \quad a.s. \tag{6.2.21}$$

Since $Y_t \geq Z_t$ for each $t \in [0, T]$, we have $R(\sigma, \tau) \leq R'(\sigma, \tau)$, and so always

$$K(\sigma, \tau) \leq K'(\sigma, \tau) \quad \text{a.s.,} \qquad (6.2.22)$$

implying that

$$\text{ess} \sup_{\tau \in \Lambda_{\zeta T}} K(\sigma, \tau) \leq \text{ess} \sup_{\tau \in \Lambda_{\zeta T}} K'(\sigma, \tau) \quad \text{a.s.} \qquad (6.2.23)$$

In order to prove the inequality in the other direction set $\tau_n = \tau$ if $\tau < \sigma$ and $\tau_n = (\tau + \frac{1}{n}) \wedge T$ if $\tau \geq \sigma$. Observe that

$$\{\tau_n \leq t\} = (\{\tau < \sigma\} \cap \{\tau \leq t\}) \cup (\{\tau \geq \sigma\} \cap \{\tau \leq t - \frac{1}{n}\}),$$
$$\{\sigma > t\} \cap \{\tau \leq t\}$$
$$= (\{\sigma > t\} \cap \{\tau \leq t\}) \cup (\{\sigma \leq t\} \cap \{\tau \leq t\} \cap \{\tau \wedge t < \sigma \wedge t\}) \in \mathcal{F}_t,$$
$$\{\tau \geq \sigma\} \cap \{\tau \leq t - \frac{1}{n}\}$$
$$= \{\sigma \leq t - \frac{1}{n}\} \cap \{\tau \leq t - \frac{1}{n}\} \cap \{\sigma \wedge (t - \frac{1}{n}) \leq \tau \wedge (t - \frac{1}{n})\}$$
$$\in \mathcal{F}_{t - \frac{1}{n}} \subset \mathcal{F}_t,$$

and so $\tau_n$ is a stopping time. Clearly, $\tau_n < \sigma$ if $\tau < \sigma$ and $\tau_n > \sigma$ if $\tau \geq \sigma$ unless $\sigma = \tau = \tau_n = T$. Hence, by (6.2.13),

$$R(\sigma, \tau_n) = Y_\sigma \mathbb{I}_{\{\sigma < \tau_n\}} + Z_{\tau_n} \mathbb{I}_{\{\tau_n \leq \sigma\}} = Y_\sigma \mathbb{I}_{\{\sigma \leq \tau_n\}} + Z_{\tau_n} \mathbb{I}_{\{\tau_n < \sigma\}} = R'(\sigma, \tau_n).$$

Observe that the process $G'(t) = R'(\sigma, t)$, $t \geq 0$ is right-continuous and $\tau_n \downarrow \tau$ as $n \uparrow \infty$. Hence,

$$\lim_{n \to \infty} (Y_\sigma \mathbb{I}_{\{\sigma < \tau\}} + Z_{\{\tau_n \leq \sigma\}}) = Y_\sigma \mathbb{I}_{\{\sigma \leq \tau\}} + Z_\tau \mathbb{I}_{\{\tau < \sigma\}} = R'(\sigma, \tau).$$

In view of (6.2.15) we can apply the dominated convergence theorem for conditional expectations in order to obtain that

$$K'(\sigma, \tau) = \lim_{n \to \infty} K(\sigma, \tau_n) \leq \text{ess} \sup_{\eta \in \Lambda_{\zeta T}} K(\sigma, \eta).$$

Taking now ess sup in $\tau \in \Lambda_{\zeta T}$ we obtain the inequality opposite to (6.2.23) proving (6.2.21) which completes the proof of the first equality in (6.2.20). The proof of the second equality in (6.2.20) proceeds in the same way taking into account that

$$-\underline{V}_\zeta = \text{ess} \inf_{\tau \in \Lambda_{\zeta T}} \text{ess} \sup_{\sigma \in \Lambda_{\zeta T}} E(-R(\sigma, \tau)|\mathcal{F}_\zeta) \quad \text{and} \quad (6.2.24)$$
$$-\underline{V}'_\zeta = \text{ess} \inf_{\tau \in \Lambda_{\zeta T}} \text{ess} \sup_{\sigma \in \Lambda_{\zeta T}} E(-R'(\sigma, \tau)|\mathcal{F}_\zeta)$$

and replacing in the above proof $Y$ by $-Y$ and $Z$ by $-Z$. $\qquad \square$

**Lemma 6.2.2.** *For any* $\zeta \in \Lambda_{0T}$,

$$Z_\zeta \leq \underline{V}_\zeta \leq \overline{V}_\zeta \leq Y_\zeta \quad \text{a.s.} \qquad (6.2.25)$$

**Proof.** By (6.2.14) and (6.2.20) with probability one,

$$\overline{V}_\zeta = ess \inf_{\sigma \in \Lambda_{\zeta T}} ess \sup_{\tau \in \Lambda_{\zeta T}} E(R'(\sigma, \tau)|\mathcal{F}_\zeta) \leq ess \sup_{\tau \in \Lambda_{\zeta T}} E(R'(\zeta, \tau)|\mathcal{F}_\zeta) = Y_\zeta$$

and

$$\overline{V}_\zeta = ess \inf_{\sigma \in \Lambda_{\zeta T}} ess \sup_{\tau \in \Lambda_{\zeta T}} E(R'(\sigma, \tau)|\mathcal{F}_\zeta) \geq ess \inf_{\sigma \in \Lambda_{\zeta T}} E(R'(\sigma, \zeta)|\mathcal{F}_\zeta) = Z_\zeta.$$

Hence, $Z_\zeta \leq \overline{V}_\zeta \leq Y_\zeta$ a.s. and, similarly, $Z_\zeta \leq \underline{V}_\zeta \leq Y_\zeta$ a.s. $\qquad \square$

**Lemma 6.2.3.** *For any* $\eta, \zeta \in \Lambda_{0T}$ *with* $\eta \geq \zeta$,

$$E(\overline{V}_\eta|\mathcal{F}_\zeta) = ess \inf_{\sigma \in \Lambda_{\eta T}} ess \sup_{\tau \in \Lambda_{\eta T}} E(R(\sigma, \tau)|\mathcal{F}_\zeta) \quad a.s. \qquad (6.2.26)$$

*and*

$$E(\underline{V}_\eta|\mathcal{F}_\zeta) = ess \sup_{\tau \in \Lambda_{\eta T}} ess \inf_{\sigma \in \Lambda_{\eta T}} E(R(\sigma, \tau)|\mathcal{F}_\zeta) \quad a.s. \qquad (6.2.27)$$

**Proof.** By the basic property of *ess* sup (see, for instance, [58] or [62]) it can be obtained by taking the supremum over a sequence, and so for each $\sigma \in \Lambda_{\eta T}$ there exists a sequence $\tau_n \in \Lambda_{\eta T}$ such that

$$ess \sup_{\tau \in \Lambda_{\eta T}} E(R(\sigma, \tau)|\mathcal{F}_\eta) = \sup_{n \geq 1} E(R(\sigma, \tau_n)|\mathcal{F}_\eta) \quad a.s.$$

For any $\xi_1, \xi_2 \in \Lambda_{\eta T}$ define the stopping time $\tau(\xi_1, \xi_2)$ by $\tau(\xi_1, \xi_2)(\omega) = \xi_1(\omega)$ if $E(R(\sigma, \xi_1)|\mathcal{F}_\eta)(\omega) \geq E(R(\sigma, \xi_2)|\mathcal{F}_\eta)(\omega)$ and $\tau(\xi_1, \xi_2)(\omega) = \xi_2(\omega)$ if $E(R(\sigma, \xi_1)|\mathcal{F}_\eta)(\omega) < E(R(\sigma, \xi_2)|\mathcal{F}_\eta)(\omega)$. Then $\tau(\xi_1, \xi_2) \in \Lambda_{\eta T}$ and

$$E(R(\sigma, \tau(\xi_1, \xi_2))|\mathcal{F}_\eta) = \max(E(R(\sigma, \xi_1)|\mathcal{F}_\eta), E(R(\sigma, \xi_2)|\mathcal{F}_\eta)).$$

Set $\tau^{(1)} = \tau_1$ and inductively $\tau^{(n)} = \tau(\tau_n, \tau^{(n-1)})$, $n = 2, 3, ...$ Then

$$ess \sup_{\tau \in \Lambda_{\eta T}} E(R(\sigma, \tau)|\mathcal{F}_\eta) = \lim_{n \uparrow \infty} \uparrow E(R(\sigma, \tau^{(n)})|\mathcal{F}_\eta) \quad a.s.,$$

where the right hand side is a monotone non-decreasing limit. In view of (6.2.15) we can apply the monotone convergence theorem for conditional expectations which implies that with probability one,

$$E\big(ess \sup_{\tau \in \Lambda_{\eta T}} E(R(\sigma, \tau)|\mathcal{F}_\eta)|\mathcal{F}_\zeta\big) = E\big(\lim_{n \uparrow \infty} \uparrow E(R(\sigma, \tau^{(n)})|\mathcal{F}_\eta)|\mathcal{F}_\zeta\big)$$

$$= \lim_{n \uparrow \infty} \uparrow E(R(\sigma, \tau^{(n)})|\mathcal{F}_\zeta) \leq ess \sup_{\tau \in \Lambda_{\eta T}} E(R(\sigma, \tau)|\mathcal{F}_\zeta).$$

On the other hand, always,

$$E\big(ess \sup_{\tau \in \Lambda_{\eta T}} E(R(\sigma, \tau)|\mathcal{F}_\eta)|\mathcal{F}_\zeta\big) \geq ess \sup_{\tau \in \Lambda_{\eta T}} E(R(\sigma, \tau)|\mathcal{F}_\zeta) \quad a.s.,$$

and so,

$$E\big(ess \sup_{\tau \in \Lambda_{\eta T}} E(R(\sigma,\tau)|\mathcal{F}_\eta)|\mathcal{F}_\zeta\big) = ess \sup_{\tau \in \Lambda_{\eta T}} E(R(\sigma,\tau)|\mathcal{F}_\zeta) \quad \text{a.s.} \quad (6.2.28)$$

Next, set $Q_\eta(\sigma) = ess\sup_{\tau \in \Lambda_{\eta T}} E(R(\sigma,\tau)|\mathcal{F}_\eta)$ and observe again that by the basic property of $ess\inf$ there exists a sequence $\sigma_n \in \Lambda_{\eta T}$ such that

$$ess \inf_{\sigma \in \Lambda_{\eta T}} Q_\eta(\sigma) = \inf_{n \geq 1} Q_\eta(\sigma_n).$$

For any $\xi_1, \xi_2 \in \Lambda_{\eta T}$ define the stopping time $\sigma(\xi_1, \xi_2)$ by $\sigma(\xi_1, \xi_2)(\omega) = \xi_1(\omega)$ if $Q_\eta(\xi_1)(\omega) \leq Q_\eta(\xi_2)(\omega)$ and $\sigma(\xi_1, \xi_2)(\omega) = \xi_2(\omega)$ if $Q_\eta(\xi_1)(\omega) > Q_\eta(\xi_2)(\omega)$. Since $Q_\eta$ is $\mathcal{F}_\eta$-measurable we have that $\sigma(\xi_1, \xi_2) \in \Lambda_{\eta T}$ and

$$Q_\eta(\sigma(\xi_1, \xi_2)) = \min\big(Q_\eta(\sigma(\xi_1)), Q_\eta(\sigma(\xi_2))\big).$$

Set $\sigma^{(1)} = \sigma_1$ and inductively $\sigma^{(n)} = \sigma(\sigma_n, \sigma^{(n-1)})$, $n = 2, 3, \ldots$ Then

$$ess \inf_{\sigma \in \Lambda_{\eta T}} Q_\eta(\sigma) = \lim_{n \uparrow \infty} \downarrow Q_\eta(\sigma^{(n)}) \quad \text{a.s.},$$

where the right hand side is a monotone nonincreasing limit. Employing again the monotone convergence theorem for conditional expectations we obtain that

$$E(ess\inf_{\sigma \in \Lambda_{\eta T}} Q_\eta(\sigma)|\mathcal{F}_\zeta) = E(\lim_{n \uparrow \infty} \downarrow Q_\eta(\sigma^{(n)})|\mathcal{F}_\zeta)$$
$$= \lim_{n \uparrow \infty} \downarrow E(Q_\eta(\sigma^{(n)})|\mathcal{F}_\zeta) \geq ess\inf_{\sigma \in \Lambda_{\eta T}} E(Q_\eta(\sigma)|\mathcal{F}_\zeta) \quad \text{a.s.}$$

On the other hand, always,

$$E(ess \inf_{\sigma \in \Lambda_{\eta T}} Q_\eta(\sigma)|\mathcal{F}_\zeta) \leq ess \inf_{\sigma \in \Lambda_{\eta T}} E(Q_\eta(\sigma)|\mathcal{F}_\zeta) \quad \text{a.s.}$$

Hence,

$$E(ess \inf_{\sigma \in \Lambda_{\eta T}} Q_\eta(\sigma)|\mathcal{F}_\zeta) = ess \inf_{\sigma \in \Lambda_{\eta T}} E(Q_\eta(\sigma)|\mathcal{F}_\zeta) \quad \text{a.s.}$$

This together with (6.2.28) yields (6.2.26) and (6.2.27) follows in the same way using (6.2.24). $\square$

**Lemma 6.2.4.** *There exist right-continuous modifications $\overline{W}_t$ and $\underline{W}_t$, $t \in [0, T]$ of the processes $\overline{V}_t$ and $\underline{V}_t$, $t \in [0, T]$, respectively. Moreover, $\overline{W}_\zeta = \overline{V}_\zeta$ a.s. and $\underline{W}_\zeta = \underline{V}_\zeta$ a.s., for any $\zeta \in \Lambda_{0T}$.*

**Proof.** Observe that for any $\zeta \in \Lambda_{0T}$ and $\zeta_n \downarrow \zeta$, $\zeta_n \in \Lambda_{\zeta T}$,

$$\limsup_{\zeta_n \downarrow \zeta} \overline{V}_{\zeta_n} = \limsup_{\zeta_n \downarrow \zeta} ess\inf_{\sigma \in \Lambda_{\zeta_n T}} ess\sup_{\tau \in \Lambda_{\zeta_n T}} E(R'(\sigma,\tau)|\mathcal{F}_{\zeta_n})$$
$$\leq \limsup_{\zeta_n \downarrow \zeta} ess\sup_{\tau \in \Lambda_{\zeta_n T}} E(R'(\sigma,\tau)|\mathcal{F}_{\zeta_n}) \quad \text{a.s.},$$

where $\sigma \in \Lambda_{\zeta_m T}$ for some $m$. Since $G'(t) = R'(\sigma, t)$, $t \in [0, T]$ is right-continuous we can consider its Snell envelope $Q_\sigma = Q_\sigma(t)$ which is right-continuous as well, and so $\lim_{\zeta_n \downarrow \zeta} Q_\sigma(\zeta_n) = Q_\sigma(\zeta)$. Hence, with probability one

$$\limsup_{\zeta_n \downarrow \zeta} ess \sup_{\tau \in \Lambda_{\zeta_n T}} E(R'(\sigma, \tau)|\mathcal{F}_{\zeta_n}) = ess \sup_{\tau \in \Lambda_{\zeta T}} E(R'(\sigma, \tau)|\mathcal{F}_\zeta) \text{ a.s.}$$

for each $\sigma \in \hat{\Lambda}_{\zeta T} = \cup_{m \geq 1} \Lambda_{\zeta_m T}$. Thus,

$$\limsup_{\zeta_n \downarrow \zeta} \overline{V}_{\zeta_n} \leq ess \inf_{\sigma \in \hat{\Lambda}_{\zeta T}} ess \sup_{\tau \in \Lambda_{\zeta T}} E(R'(\sigma, \tau)|\mathcal{F}_\zeta) \text{ a.s.} \qquad (6.2.29)$$

For each $\sigma \in \Lambda_{\zeta T}$ define $\sigma_n = \sigma \vee \zeta_n \in \Lambda_{\zeta_n T}$. Then

$$R'(\sigma_n, \tau) \leq R'(\sigma, \tau) + 2 \sup_{\zeta \leq s \leq \zeta_n} \max(|Y_s - Y_\zeta|, |Z_s - Z_\zeta|).$$

Hence, for any $n \geq 1$,

$$\overline{V}_\zeta \leq ess\inf_{\sigma \in \hat{\Lambda}_{\zeta T}} ess\sup_{\tau \in \Lambda_{\zeta T}} E(R'(\sigma, \tau)|\mathcal{F}_\zeta)$$
$$\leq \overline{V}_\zeta + 2E(\sup_{\zeta \leq s \leq \zeta_n} \max(|Y_s - Y_\zeta|, |Z_s - Z_\zeta|)|\mathcal{F}_\zeta)$$

Letting $n \uparrow \infty$ we obtain by the dominated convergence theorem for conditional expectations that the last conditional expectation tends to zero a.s. It follows that the first inequality in the above formula is, in fact, an equality which together with (6.2.29) yields that

$$\limsup_{\zeta_n \downarrow \zeta} \overline{V}_{\zeta_n} \leq \overline{V}_\zeta \quad \text{a.s.} \qquad (6.2.30)$$

Next, we claim that

$$\liminf_{\zeta_n \downarrow \zeta} E\overline{V}_{\zeta_n} \geq E\overline{V}_\zeta, \qquad (6.2.31)$$

in particular, that the process $\overline{V}_t$, $t \in [0, T]$ is right lower semi-continuous in expectation. Indeed, by Lemma 6.2.3 for each $n \geq 1$,

$$E\overline{V}_\zeta = \inf_{\sigma \geq \zeta} \sup_{\tau \geq \zeta} ER(\sigma, \tau) \leq \inf_{\sigma \geq \zeta_n} \sup_{\tau \geq \zeta} ER(\sigma, \tau).$$

Now, for any $\sigma \geq \zeta_n$,

$$ER(\sigma, \tau) = E(\mathbb{I}_{\tau < \zeta_n} R(\sigma, \tau)) + E(\mathbb{I}_{\tau \geq \zeta_n} R(\sigma, \tau)) = E(\mathbb{I}_{\tau < \zeta_n} Z_\tau)$$
$$+ E(\mathbb{I}_{\tau \geq \zeta_n} (Z_{\tau \vee \zeta_n} \mathbb{I}_{\tau \vee \zeta_n \leq \sigma} + Y_\sigma \mathbb{I}_{\sigma < \tau \vee \zeta_n}))$$
$$= E(\mathbb{I}_{\tau < \zeta_n} (Z_\tau - Z_{\zeta_n})) + ER(\sigma, \tau \vee \zeta_n).$$

Hence,

$$\sup_{\tau \geq \zeta} ER(\sigma, \tau) \leq \sup_{\tau \geq \zeta} E(Z_{\tau \wedge \zeta_n} - Z_{\zeta_n}) + \sup_{\tau \geq \zeta_n} ER(\sigma, \tau).$$

Now, taking the infimum over $\sigma \geq t_n$ and relying again on Lemma 6.2.3 we obtain

$$E\overline{V}_\zeta \leq \sup_{\tau \geq \zeta} E Z_{\tau \wedge \zeta_n} - E Z_{\zeta_n} + E\overline{V}_{\zeta_n} \leq E \sup_{\zeta_n \geq s \geq \zeta} |Z_s - Z_{\zeta_n}| + E\overline{V}_{\zeta_n}.$$

Since the process $Z_s$, $s \in [0, T]$ is right-continuous and (6.2.15) holds true we can apply the Lebesgue dominated convergence theorem taking here $\liminf_{\zeta_n \downarrow \zeta}$ and deriving (6.2.31).

Now, we are ready to construct a right-continuous modification of the process $\overline{V}_t$, $t \in [0, T]$. Observe that by (6.2.30) and (6.2.31),

$$E(\limsup_{\zeta_n \downarrow \zeta} \overline{V}_{\zeta_n} - \liminf_{\zeta_n \downarrow \zeta} \overline{V}_{\zeta_n}) \leq E\overline{V}_\zeta - E\overline{V}_\zeta = 0,$$

and so with probability one,

$$\overline{V}_\zeta = \limsup_{\zeta_n \downarrow \zeta} \overline{V}_{\zeta_n} = \liminf_{\zeta_n \downarrow \zeta} \overline{V}_{\zeta_n} = \lim_{\zeta_n \downarrow \zeta} \overline{V}_{\zeta_n}. \qquad (6.2.32)$$

Let $\mathbb{Q}$ be the set of rational numbers and define

$$\tilde{\Omega} = \{\omega \in \Omega : \overline{V}_t(\omega) = \lim_{s \downarrow t, s \in \mathbb{Q}} \overline{V}_s \text{ for all } t \in \mathbb{Q} \cap [0, T]\}.$$

Then $P(\tilde{\Omega}) = 1$ by (6.2.32). Now, for each $t \in [0, T]$ and $\omega \in \tilde{\Omega}$ we set $\overline{W}_t(\omega) = \limsup_{s \downarrow t, s \in \mathbb{Q}} \overline{V}_s(\omega)$ and $\overline{W}_t(\omega) = 0$ if $\omega \notin \tilde{\Omega}$. It follows from (6.2.32) that $\overline{W}_t$, $t \in [0, T]$ is the right-continuous modification of $\overline{V}_t$, $t \in [0, T]$ and the right-continuous modification $\underline{W}_t$, $t \in [0, T]$ of $\underline{V}_t$, $t \in [0, T]$ is constructed in a similar way.

In order to show that for any $\zeta \in \mathcal{T}_{0T}$,

$$\overline{W}_\zeta = \overline{V}_\zeta \text{ a.s. and } \underline{W}_\zeta = \underline{V}_\zeta \text{ a.s.}, \qquad (6.2.33)$$

first observe that this holds true when $\zeta \equiv t \in [0, T]$ is a constant by the definition of a modification. It is easy to see that (6.2.33) remains true for any $\zeta \in \mathcal{T}_{0T}$ taking on values in the set of rational numbers (or in any countable set). Next, let $\zeta \in \mathcal{T}_{0T}$ be arbitrary. Define the sequence of stopping times $\zeta_n$ so that $\zeta_n(\omega) = \frac{k+1}{n}$ if $\frac{k}{n} \leq \zeta(\omega) < \frac{k+1}{n}$. Then $\zeta_n \downarrow \zeta$ as $n \uparrow \infty$ and $\zeta_n$ takes on values in $\mathbb{Q}$. By (6.2.32) and the right-continuity of $\overline{W}_t$, $t \in [0, T]$ we obtain

$$\overline{W}_\zeta = \lim_{\zeta_n \downarrow \zeta} \overline{W}_{\zeta_n} = \lim_{\zeta_n \downarrow \zeta} \overline{V}_{\zeta_n} = \overline{V}_\zeta \text{ a.s.},$$

proving the first equality in (6.2.33) and the second one follows in a similar way. $\qquad \square$

From now on, in view of Lemma 6.2.4, we can and will assume without loss of generality that the processes $\overline{V}_t$ and $\underline{V}_t$, $t \in [0, T]$ themselves are right-continuous since, anyway, they are defined for each $t$ only almost surely, and so we can take from the beginning their right-continuous modifications. We will also need a slight generalization of the result appeared in the single player case above which is formulated in the following lemma.

**Lemma 6.2.5.** *Let $U_t$, $t \in [0, T]$ be a right-continuous stochastic process adapted to a filtration $\{\mathcal{F}_t, t \in [0, T]\}$ satisfying the usual conditions and assume that*

$$E(\sup_{t \in [0,T]} |U_t|) < \infty. \tag{6.2.34}$$

*Denote by $L_t$, $t \in [0, T]$ the Snell envelope of the process $U_t$, $t \in [0, T]$ and set*

$$\sigma(t, \varepsilon) = \inf\{s > t : L_s \leq U_s + \varepsilon\} \quad and \quad \tau(t, \varepsilon) = \inf\{s \geq t : L_s \leq U_s + \varepsilon\}.$$

*Then for any $\zeta \in \mathcal{T}_{0T}$ and $\tau \in \mathcal{T}_{\zeta T}$,*

$$L_\zeta = E(L_{\tau \wedge \tau(\zeta, \varepsilon)} | \mathcal{F}_\zeta) = E(L_{\tau \wedge \sigma(\zeta, \varepsilon)} | \mathcal{F}_\zeta) \quad a.s. \tag{6.2.35}$$

**Proof.** Since $L_t$, $t \in [0, T]$ is a supermartingale and $\sigma(\zeta, \varepsilon) \geq \tau(\zeta, \varepsilon)$ we have by the optional stopping theorem that for any $\tau \in \mathcal{T}_{\zeta T}$,

$$L_\zeta \geq E(L_{\tau \wedge \tau(\zeta, \varepsilon)} | \mathcal{F}_\zeta) \geq E(L_{\tau \wedge \sigma(\zeta, \varepsilon)} | \mathcal{F}_\zeta). \tag{6.2.36}$$

For each $\tau \in \mathcal{T}_{0T}$ set $\tau_t = \tau \vee t$ and consider the process

$$Q_t = Q_t^{(\varepsilon)} = E(L_{\tau_t \wedge \sigma(t, \varepsilon)} | \mathcal{F}_t), \, t \in [0, T]. \tag{6.2.37}$$

Since $\sigma(u, \varepsilon) \geq \sigma(t, \varepsilon)$ and $\tau_u \geq \tau_t$ for $u \geq t$ and $L_s$, $s \in [0, T]$ is a supermartingale,

$$E(Q_u | \mathcal{F}_t) = E(L_{\tau_t \wedge \sigma(u, \varepsilon)} | \mathcal{F}_t) \leq E(L_{\tau_t \wedge \sigma(t, \varepsilon)} | \mathcal{F}_t) = Q_t,$$

so $Q_t$, $t \in [0, T]$ is a supermartingale as well. Next, $EQ_t = EL_{\tau_t \wedge \sigma(t, \varepsilon)}$, and since $\sigma(s, \varepsilon) \downarrow \sigma(t, \varepsilon)$, $\tau_s \downarrow \tau_t$ as $s \downarrow t$ and $L_u$, $u \in [0, T]$ is a right-continuous supermartingale, we obtain that $EQ_t$ is right-continuous, thus $Q_t$, $t \in [0, T]$ has a right-continuous modification which we denote by the same letter.

Next, we show that (6.2.37) can be extended to stopping times, i.e., for any $\zeta \in \mathcal{T}_{0T}$,

$$Q_\zeta = E(L_{\tau_\zeta \wedge \sigma(\zeta, \varepsilon)} | \mathcal{F}_\zeta) \quad a.s. \tag{6.2.38}$$

Indeed, if $\zeta$ takes on only countable number of values then (6.2.38) follows from (6.2.37) considering $\zeta$ separately on each its constant value set. Now, let $\zeta_n \downarrow \zeta$ as $n \uparrow \zeta$ as $n \uparrow \infty$, $\zeta_n \in \mathcal{T}_{0T}$ and $\zeta_n$, $n \geq 1$ take on only rational values as constructed in Lemma 6.2.4. Then

$$E(Q_{\zeta_n}|\mathcal{F}_\zeta) = E(L_{\tau_{\zeta_n} \wedge \sigma(\zeta_n, \varepsilon)}|\mathcal{F}_\zeta) \quad \text{a.s.}$$

Since $\zeta_n \geq \zeta$, we let $n \uparrow \infty$ and obtain (6.2.37) by the right-continuity of $Q_t$ and $L_t$ taking into account (6.2.34), which makes the sequences of random variables in the above expectations uniformly integrable.

Next, by the optional stopping theorem,

$$L_t \geq E(L_{\tau_t \wedge \sigma(t,\varepsilon)}|\mathcal{F}_t) = Q_t \quad \text{a.s.}$$

We will show that $Q_t$, $t \in [0,T]$ dominates $U_t$, $t \in [0,T]$, i.e., $Q_t \geq U_t$ a.s. for each $t \in [0,T]$. This will imply that $Q_t \geq L_t$ a.s. for each $t \in [0,T]$ since $Q_t$, $t \in [0,T]$ is a right-continuous supermartingale and $L_t$, $t \in [0,T]$ is the Snell envelope. It will then follow that $Q_t = L_t$ a.s. for each $t \in [0,T]$ implying that $Q_\zeta = L_\zeta$ a.s. for any $\zeta \in \mathcal{T}_{0,T}$, since both $Q$ and $L$ are right-continuous and having this equality for $\zeta = \zeta_n$, $n \geq 1$ which takes on only rational values, and choosing $\zeta_n \downarrow \zeta$ as $n \uparrow \infty$ we obtain the equality for any $\zeta \in \mathcal{T}_{0T}$ which will yield the equality in (6.2.36), i.e., (6.2.35) will hold true.

It remains to establish the domination $Q_t \geq U_t$ a.s. Set $a(\omega) = \sup_{t \in [0,T]}(U_t(\omega) - Q_t(\omega))^+$, consider the martingale $a_t = E(a|\mathcal{F}_t)$, $t \in [0,T]$ and take its right-continuous modification denoted by the same letter. If $Ea = 0$ then $Q_t \geq U_t$ a.s. for each $t \in [0,T]$ and we are done. If $Ea_t = Ea = \lambda > 0$ then consider the right-continuous supermartingale $G_t = Q_t + a_t$ which dominates $U_t$, $t \in [0,T]$, and so $G_t \geq L_t$ a.s. for each $t \in [0,T]$ since $L$ is the Snell envelope of $U$. Choose $\alpha < \min(\varepsilon, \lambda)$ and set $\zeta = \zeta(t) = \inf\{s > t : U_s + \alpha \geq Q_s + a_s\}$ where $\zeta = \zeta(t) = T$ if $U_s + \alpha < Q_s + a_s$ for all $s \in (0,T]$. Then $U_\zeta + \alpha \geq Q_\zeta + a_\zeta \geq L_\zeta$ if $\zeta < T$, and so $U_\zeta + \varepsilon \geq L_\zeta$. If $\zeta = T$ then the latter inequality remains true as $U_T = L_T$. This together with the definition of $\sigma(t, \varepsilon)$ and the right-continuity of $\zeta(t)$, $t \in [0,T]$ yields that $\zeta = \sigma(\zeta, \varepsilon)$. Hence,

$$Q_\zeta = E(L_{\tau_\zeta \wedge \sigma(\zeta, \varepsilon)}|\mathcal{F}_\zeta) = E(L_{\tau_\zeta \wedge \zeta}|\mathcal{F}_\zeta) = L_\zeta.$$

Since $L_t \geq U_t$ a.s. for each $t \in [0,T]$ and both processes are right-continuous it follows by the definition of $\zeta$ that

$$0 \geq E(U_\zeta - L_\zeta) \geq E(U_\zeta - Q_\zeta) \geq E(a_\zeta - \alpha) = \lambda - \alpha$$

contradicting the choice of $\alpha$. Hence, $Ea = 0$ and $Q_t \geq U_t$ a.s. for each $t \in [0,T]$, as required. $\qquad \square$

**Lemma 6.2.6.** *For $\zeta \in \mathcal{T}_{0T}$ and $\varepsilon > 0$ let*

$$\sigma_\zeta^\varepsilon = \inf\{s \geq \zeta : \overline{V}_s \geq Z_s + \varepsilon\} \quad \text{and} \quad \tau_\zeta^\varepsilon = \inf\{s \geq \zeta : \underline{V}_s \geq Y_s - \varepsilon\}.$$

*Then for any $\eta \in \mathcal{T}_{\zeta T}$ with probability one,*

$$\overline{V}_\zeta \leq E(\overline{V}_{\tau_\zeta^\varepsilon \wedge \eta}|\mathcal{F}_\zeta) \quad \text{and} \quad \underline{V}_\zeta \geq E(\underline{V}_{\sigma_\zeta^\varepsilon \wedge \eta}|\mathcal{F}_\zeta). \tag{6.2.39}$$

**Proof.** For each $\sigma \in \mathcal{T}_{\zeta T}$ consider the process $G_\sigma(t) = R'(\sigma, t)$, $t \in [0, T]$ which is right-continuous. Denote by $Q_\sigma$ the Snell envelope of $G_\sigma$ and for each $\varepsilon > 0$ set

$$\nu_{\sigma,\zeta}^\varepsilon = \inf\{t \geq \zeta : Q_\sigma(t) \leq G_\sigma + \varepsilon\} \in \mathcal{T}_{\zeta T}.$$

Then for any $\eta \in \mathcal{T}_{0T}$ and $\tilde{\sigma} \in \mathcal{T}_{\eta T}$ with probability one,

$$\overline{V}_\eta = ess\inf_{\sigma \in \mathcal{T}_{\eta T}} ess\sup_{\tau \in \mathcal{T}_{\eta T}} ER'(\sigma, \tau) = ess\inf_{\sigma \in \mathcal{T}_{\eta T}} Q_\sigma(\eta) \leq Q_{\tilde{\sigma}}(\eta).$$

In particular, $Q_\sigma(t) \geq \overline{V}_t$ a.s. when $t \in [0, T]$ and $\sigma \in \mathcal{T}_{tT}$.

Observe that $\overline{V}_t > Z_t + \varepsilon$ and $G_\sigma(t) = Z_t$ when $\zeta \leq t < \tau_\zeta^\varepsilon \wedge \eta$, $\eta \in \mathcal{T}_{\zeta T}$ and $\sigma \in \mathcal{T}_{\tau_\zeta^\varepsilon \wedge \eta, T}$. Hence, for such $t$ and $\sigma$,

$$Q_\sigma(t) \geq \overline{V}_t > Z_t + \varepsilon = G_\sigma(t) + \varepsilon \quad \text{a.s.},$$

and so $\nu_{\sigma,\zeta}^\varepsilon \geq \tau_\zeta^\varepsilon \wedge \eta$ for any $\sigma \in \mathcal{T}_{\tau_\zeta^\varepsilon \wedge \eta, T}$. By Lemma 6.2.5 for any $\eta \in \mathcal{T}_{\zeta T}$ and $\sigma \in \mathcal{T}_{\tau_\zeta^\varepsilon \wedge \eta, T}$,

$$Q_\sigma(\zeta) = E(Q_\sigma(\tau_\zeta^\varepsilon \wedge \eta)|\mathcal{F}_\zeta) \quad \text{a.s.}$$

since by the above $\tau_\zeta^\varepsilon \wedge \eta \leq \nu_{\sigma,\varepsilon}^\varepsilon$. Similarly to (6.2.28) we obtain that

$$E(Q_\sigma(\tau_\zeta^\varepsilon \wedge \eta)|\mathcal{F}_\zeta) = E\big(ess\sup_{\tau \in \mathcal{T}_{\tau_\zeta^\varepsilon \wedge \eta, T}} E(R'(\sigma, \tau)|\mathcal{F}_{\tau_\zeta^\varepsilon \wedge \eta, T})|\mathcal{F}_\zeta\big)$$

$$= ess\sup_{\tau \in \mathcal{T}_{\tau_\zeta^\varepsilon \wedge \eta, T}} E(R'(\sigma, \tau)|\mathcal{F}_\zeta) \quad \text{a.s.}$$

By Lemmas 6.2.1 and 6.2.3 it follows from here that

$$\overline{V}_\zeta = ess\inf_{\sigma \in \mathcal{T}_{\zeta T}} Q_\sigma(\zeta) \leq ess\inf_{\sigma \in \mathcal{T}_{\tau_\zeta^\varepsilon \wedge \eta, T}} Q_\sigma(\zeta)$$

$$= ess\inf_{\sigma \in \mathcal{T}_{\tau_\zeta^\varepsilon \wedge \eta, T}} ess\sup_{\tau \in \mathcal{T}_{\tau_\zeta^\varepsilon \wedge \eta, T}} E(R'(\sigma, \tau)|\mathcal{F}_\zeta) = E(\overline{V}_{\tau_\zeta^\varepsilon \wedge \eta}|\mathcal{F}_\zeta),$$

proving the first inequality in (6.2.39), while the second inequality there follows in the same way relying on (6.2.24). □

Now we can prove that the game has a value.

**Corollary 6.2.1.** *For any $\zeta \in \mathcal{T}_{0T}$, $\overline{V}_\zeta = \underline{V}_\zeta$ a.s. In particular, for $\zeta \equiv 0$,*

$$\overline{V}_0 = \inf_{\sigma \in \mathcal{T}_{0T}} \sup_{\tau \in \mathcal{T}_{0T}} ER(\sigma, \tau) = \underline{V}_0 = \sup_{\tau \in \mathcal{T}_{0T}} \inf_{\sigma \in \mathcal{T}_{0T}} ER(\sigma, \tau),$$

*i.e., there exists a value for the game.*

**Proof.** By the first inequality in (6.2.39) for any $\eta \in \mathcal{T}_{\zeta T}$,

$$\overline{V}_\zeta \le E(\overline{V}_{\tau_\zeta^\varepsilon} \mathbb{I}_{\tau_\zeta^\varepsilon \le \eta} + \overline{V}_\eta \mathbb{I}_{\eta < \tau_\zeta^\varepsilon} | \mathcal{F}_\zeta) \quad \text{a.s.} \qquad (6.2.40)$$

Recall, that relying on Lemma 6.2.4 we chose $\overline{V}_t$, $t \in [0, T]$ to be right-continuous, and so $\overline{V}_t - Y_t$, $t \in [0, T]$ is right-continuous as well. This together with the definition of $\tau_\zeta^\varepsilon$ yields that

$$\overline{V}_{\tau_\zeta^\varepsilon} \le Z_{\tau_\zeta^\varepsilon} + \varepsilon. \qquad (6.2.41)$$

Since $\overline{V}_\eta \le Y_\eta$ a.s. by (6.2.25) for any $\eta \in \mathcal{T}_{\zeta T}$, we obtain from (6.2.40) and (6.2.41) that

$$\overline{V}_\zeta \le E(Z_{\tau_\zeta^\varepsilon} \mathbb{I}_{\tau_\zeta^\varepsilon \le \eta} + Y_\eta \mathbb{I}_{\eta < \tau_\zeta^\varepsilon} | \mathcal{F}_\zeta) + \varepsilon = E(R(\eta, \tau_\zeta^\varepsilon) | \mathcal{F}_\zeta) + \varepsilon \quad \text{a.s.}$$

In the same way, using the second inequality in (6.2.39) we obtain that for any $\eta \in \mathcal{T}_{\zeta T}$,

$$\underline{V}_\zeta \le E(Z_\eta \mathbb{I}_{\eta \le \sigma_\zeta^\varepsilon} + Y_{\sigma_\zeta^\varepsilon} \mathbb{I}_{\sigma_\zeta^\varepsilon < \eta} | \mathcal{F}_\zeta) - \varepsilon = E(R(\sigma_\zeta^\varepsilon, \eta) | \mathcal{F}_\zeta) - \varepsilon \quad \text{a.s.}$$

Hence, for any $\eta \in \mathcal{T}_{\zeta T}$ with probability one,

$$E(R(\eta, \tau_\zeta^\varepsilon) | \mathcal{F}_\zeta) + \varepsilon \ge \overline{V}_\zeta \ge \underline{V}_\zeta \ge E(R(\sigma_\zeta^\varepsilon, \eta) | \mathcal{F}_\zeta) - \varepsilon, \qquad (6.2.42)$$

and so

$$\underline{V}_\zeta + \varepsilon = ess\sup_{\tau \in \mathcal{T}_{\zeta T}} ess\inf_{\sigma \in \mathcal{T}_{\zeta T}} E(R(\sigma, \tau) | \mathcal{F}_\zeta) + \varepsilon \ge \overline{V}_\zeta \ge \underline{V}_\zeta$$
$$\ge ess\inf_{\sigma \in \mathcal{T}_{\zeta T}} ess\sup_{\tau \in \mathcal{T}_{\zeta T}} E(R(\sigma, \tau) | \mathcal{F}_\zeta) - \varepsilon = \overline{V}_\zeta - \varepsilon \quad \text{a.s.}$$

Since $\varepsilon > 0$ is arbitrary we obtain $\overline{V}_\zeta = \underline{V}_\zeta$ a.s, proving the corollary as well, as (6.2.16) and (6.2.18) of Theorem 6.2.3. $\qquad \square$

Finally, we will show that under the additional left-continuity condition optimal stopping times exist.

**Corollary 6.2.2.** *Assume, in addition, that $Y_t$ and $Z_t$, $t \in [0, T]$ are left-continuous (i.e., in fact, they are continuous processes). Then*

$$\sigma^* = \lim_{\varepsilon \downarrow 0} \sigma_0^\varepsilon \quad and \quad \tau^* = \lim_{\varepsilon \downarrow 0} \tau_0^\varepsilon$$

*satisfy (6.2.19) where $\sigma_0^\varepsilon$ and $\tau_0^\varepsilon$ are defined by (6.2.17) with $V_t = \overline{V}_t = \underline{V}_t$, $t \in [0, T]$.*

**Proof.** Let $\zeta \in \mathcal{T}_{0T}$ and $V_\zeta = \overline{V}_\zeta = \underline{V}_\zeta$. Then by (6.2.42) for any $\eta \in \mathcal{T}_{\zeta T}$ with probability one,

$$\liminf_{\varepsilon \downarrow 0} E(R(\sigma_\zeta^\varepsilon, \eta) | \mathcal{F}_\zeta) \le V_\zeta \le \limsup_{\varepsilon \downarrow 0} E(R(\eta, \tau_\zeta^\varepsilon) | \mathcal{F}_\zeta).$$

As $\varepsilon \downarrow 0$ the stopping times $\sigma_\zeta^\varepsilon$ and $\tau_\zeta^\varepsilon$ form non-decreasing families which converge to some stopping times $\sigma_\zeta^*$ and $\tau_\zeta^*$, respectively. Applying the dominated convergence theorem in view of (6.2.15) and taking into account that $R(\eta, t)$ is left-continuous in $t$ we obtain

$$\limsup_{\varepsilon \downarrow 0} E(R(\eta, \tau_\zeta^\varepsilon)|\mathcal{F}_\zeta) = E(R(\eta, \tau_\zeta^*)|\mathcal{F}_\zeta) \quad \text{a.s.}$$

Since $Y_t \geq Z_t, t \in [0, T]$, the function $R(t, \eta)$ in $t$ is left lower semi-continuous, i.e.,

$$\lim_{s \uparrow t} R(s, \eta) \geq R(t, \eta),$$

and so

$$\liminf_{\varepsilon \downarrow 0} E(R(\sigma_\zeta^\varepsilon, \eta)|\mathcal{F}_\zeta) \geq E(R(\sigma_\zeta^*, \eta)|\mathcal{F}_\zeta) \quad \text{a.s.}$$

Hence, for any $\eta \in \mathcal{T}_{\zeta T}$,

$$E(R(\sigma_\zeta^*, \eta)|\mathcal{F}_\zeta) \leq V_\zeta \leq E(R(\eta, \tau_\zeta^*)|\mathcal{F}_\zeta) \quad \text{a.s.}$$

and taking $\zeta \equiv 0$ we obtain (6.2.19), completing the proof of both Corollary 6.2.2 and of Theorem 6.2.3 itself. $\qquad\qquad\qquad\qquad\qquad\qquad\qquad\qquad\square$

## 6.3   Exercises

1) Let $\xi(t)$, $t \geq 0$ be a stochastic process with continuous time (i.e., just a family of random variables indexed by nonnegative reals playing the role of time) with independent increments, i.e., any increments $\xi(t_i) - \xi(t_{i-1})$, $i = 1, 2, ..., n$ are independent provided $t_i \geq t_{i-1}$, $i = 1, 2, ..., n$. Suppose that $E\xi^2(t) < \infty$ for all $t \geq 0$ and $\xi(0) \equiv E\xi(t) \equiv a$ is a constant. Prove that $M(t) = \xi(t) - a$, $t \geq 0$ is a martingale, and so $M^2(t)$, $t \geq 0$ is a submartingale. Find the submartingale decomposition of $M^2(t)$, $t \geq 0$.

2) The following random variables appeared in proofs and it is important to be sure that they are indeed stopping times.

(a) Let $\tau_1, \tau_2$ and $\sigma$ be stopping times such that $\tau_1 \wedge \tau_2 \geq \sigma$ and let $\Gamma \in \mathcal{F}_\sigma$. Set $\tau(\omega) = \tau_1(\omega)$ if $\omega \in \Gamma$ and $\tau(\omega) = \tau_2(\omega)$ if $\omega \in \Omega \setminus \Gamma$. Then $\tau$ is a stopping time.

(b) Let $Z_t$, $t \geq 0$ be a right-continuous process adapted to a right-continuous filtration, $\zeta$ be a stopping time and $c$ be a constant. Then

$$\sigma_c = \inf\{t \geq \zeta : Z_t < c\} \quad \text{and} \quad \tilde{\sigma}_c = \inf\{t \geq \zeta : Z_t \leq c\}$$

are stopping times.

# Chapter 7

# Introduction to Stochastic Analysis

## 7.1 Lecture 13: Brownian motion

### 7.1.1 *Definition and a direct construction*

Recall, that a stochastic process in the Euclidean space $\mathbb{R}^d$, $d \geq 1$ defined on a probability space $(\Omega, \mathcal{F}, P)$ is a collection $X(t) = X(t, \omega)$, $t \in \mathbf{T}$, $\omega \in \Omega$ of $d$-dimensional random vectors (i.e., vectors with components being random variables). A parameter $t$ plays the role of time, and the parameter space $\mathbf{T}$ is usually the halfline $[0, \infty)$, an interval $[a, b]$, $0 \leq a \leq b$ or the non-negative integers (the discrete time case). Sometimes, the negative time is considered, as well. In this lecture we will introduce and study the very important stochastic process called the Brownian motion or the Wiener process which we will denote by the letter $W$ since $B$ is reserved in this book (as it is usual in mathematical finance) for bonds.

**Definition 7.1.1.** A $d$-dimensional Brownian motion or the Wiener process is an $\mathbb{R}^d$-valued stochastic process $W(t) = W(t, \omega) = (W_1(t, \omega), ..., W_d(t, \omega))$, $t \geq 0$, $\omega \in \Omega$ on a probability space $(\Omega, \mathcal{F}, P)$ satisfying the following conditions:
    (i) The process starts at 0, i.e.,

$$P\{W(0) = 0\} = 1.$$

    (ii) The increments are independent, i.e., if $0 \leq t_0 < t_1 < \cdots < t_k$ then

$$P\{W(t_i) - W(t_{i-1}) \in U_i, \, i = 1, 2, ..., k\} = \prod_{1 \leq i \leq k} P\{W(t_i) - W(t_{i-1}) \in U_i\}$$

for any Borel sets $U_i \subset \mathbb{R}^d$, $i = 1, 2, ..., k$.
    (iii) For any $0 \leq s < t$ the increment $W(t) - W(s)$ has the $d$-dimensional normal distribution with zero mean and the covariance matrix $(t - s)I_d$

where $I_d$ is the $d \times d$ identity matrix, i.e.,

$$P\{W(t) - W(s) \in U\} = (2\pi(t - s))^{-d/2} \int_U \exp\left(-\frac{|x|^2}{2(t - s)}\right) dx,$$

where $|x| = (x_1^2 + \cdots x_d^2)^{1/2}$ denotes the length of a vector $x = (x_1, ..., x_d)$.

Sometimes it is convenient to give up the property (i) and to consider the Brownian motion starting at another point (vector) $a \neq 0$ or, more generally, to assume that $W(0)$ is a random vector independent of all increments $W(t) - W(s)$, $t > s \geq 0$. The family of $\sigma$-algebras $\mathcal{F}_t = \sigma\{W_s, 0 \leq s \leq t\}$, $t \geq 0$ complemented by all events of zero probability is called the Brownian filtration . By (ii) it follows that each increment $W(t) - W(s)$, $t > s \geq 0$ is independent of any $\mathcal{F}_u$ with $s \geq u \geq 0$. Sometimes it is convenient to consider a richer filtration $\mathcal{G}_s$, $s \geq 0$, $\mathcal{G}_s \supset \mathcal{F}_s$ such that, still, each increment $W(t) - W(s)$, $t > s \geq 0$ is independent of any $\mathcal{G}_u$ with $s \geq u \geq 0$ and then the process $W(t)$, $t \geq 0$ together with the latter filtration is called again the Brownian motion. When we define a mathematical object by conditions it is supposed to satisfy, it is necessary to show that such an object exists. In fact, we will construct the Brownian motion explicitly following Lévy and Ciesielski. It suffices to produce a one-dimensional Brownian motion since a $d$-dimensional Brownian motion is just a vector valued process whose components are independent one-dimensional Brownian motions.

The construction uses the Haar functions $H_0(t)$, $H_1(t)$, $H_2(t)$, ... defined for $0 \leq t \leq 1$, $k = 2^n + l$, $l = 0, 1, ..., r^n - 1$, $n = 0, 1, 2, ...$ by

$$H_k(t) = \begin{cases} 2^{n/2} & if \quad l2^{-n} \leq t < l2^{-n} + 2^{-(n+1)}, \\ -2^{n/2} & if \quad l2^{-n} + 2^{-(n+1)} \leq t < (l+1)2^{-n}, \\ 0 & otherwise \end{cases}$$

and $H_0(t) \equiv 1$. Consider also

$$S_k(t) = \int_0^t H_k(u) du, \quad k = 0, 1, 2, ...$$

which are called Schauder's functions whose graphs have the form of little disjoint tents of the height $2^{-(n+2)/2}$ with the base centered at $l2^{-n} + 2^{-(n+1)}$ (with $k$ and $l$ as above). Let now $\xi_0, \xi_1, \xi_2, ...$ be a sequence of i.i.d. random variables with the standard normal distribution

$$P\{\xi_i < x\} = (2\pi)^{-1/2} \int_{-\infty}^x e^{-y^2/2} dy.$$

For each $t \in [0, 1]$ define formally

$$W(t) = \sum_{k=0}^{\infty} \xi_k S_k(t). \tag{7.1.1}$$

**Theorem 7.1.1.** *With probability one the series (7.1.1) converges uniformly in $t \in [0, 1]$ to a stochastic process continuous in $t$ and satisfying the conditions (i)–(iii), i.e., (7.1.1) determines a continuous Brownian motion defined for $t \in [0, 1]$.*

**Proof.** First, we will show the convergence. Since $S_k(t)S_{\tilde{k}}(t) = 0$ if $2^n \le k, \tilde{k} < 2^{n+1}$, $k \ne \tilde{k}$ then

$$q_n = \sup_{0 \le t \le 1} \Big| \sum_{2^n \le k < 2^{n+1}} \xi_k S_k(t) \Big| = 2^{-(n+2)/2} \max_{2^n \le k < 2^{n+1}} |\xi_k|.$$

Thus,

$$P\{q_n > (2^{-n} \ln 2^n)^{1/2}\} \le P\{\max_{2^n \le k < 2^{n+1}} |\xi_k| > (4n \ln 2)^{1/2}\}$$
$$\le \sum_{2^n \le k < 2^{n+1}} P\{|\xi_k| > (4n \ln 2)^{1/2}\}$$
$$= 2^{n+1} \int_{(4n \ln 2)^{1/2}}^{\infty} (2\pi)^{-1/2} e^{-z^2/2} dz < c_n = (8\pi \ln 2)^{-1/2} n^{-1/2} 2^{-n}$$

where we used the inequality $\int_a^{\infty} e^{z^2/2} dz < a^{-1} \int_a^{\infty} e^{-z^2/2} z dz = a^{-1} e^{-a^2/2}$ for $a > 0$.

Since $\sum_{n \ge 1} c_n < \infty$ then by the Borel–Cantelli lemma there exists a random variable $N = N(\omega) < \infty$ a.s. such that $q_n \le 2^{-n/2} n^{1/2} (\ln 2)^{1/2}$ provided $n \ge N$ which implies the absolute and uniform convergence in $t$ of the series (7.1.1). Since all members of the series (7.1.1) are continuous functions of $t$ and the convergence is uniform in $t$, it follows that the sum $W(t)$ is continuous in $t$, or, in other words, $W(t)$, $t \in [0, 1]$ is a continuous stochastic process. It remains to show that $W(t)$ satisfies the conditions (i)–(iii). The condition (i) is clear. To establish (ii) and (iii) we will need several lemmas about characteristic functions which can be found in any advanced probability textbook but for readers' convenience we provide them here as well.

The characteristic function $\varphi_Z(\lambda)$, $\lambda = (\lambda_1, ..., \lambda_d) \in \mathbb{R}^d$ of a random vector $Z = (Z_1, ..., Z_d)$ is defined by $\varphi_Z(\lambda) = E e^{i\langle \lambda, Z \rangle}$ where $\langle \lambda, Z \rangle = \sum_{k=1}^{d} \lambda_k Z_k$ denotes the inner product in $\mathbb{R}^d$. The following property of characteristic functions is the origin of their important applications.

**Lemma 7.1.1.** *Let $F_1$ and $F_2$ be probability distributions of random vectors $X^{(j)} = (X_1^{(j)}, ..., X_n^{(j)})$, $j = 1, 2$, i.e., $P\{X^{(j)} \in Q\} = F_j(Q)$ for any Borel* set $Q \subset \mathbb{R}^d$. Suppose that the characteristic functions

$$\varphi_j(\lambda) = Ee^{i\langle\lambda, X^{(j)}\rangle} = \int_{\mathbb{R}^d} e^{i\langle\lambda, x\rangle} dF_j(x), \ j = 1, 2$$

coincide, i.e., $\varphi_1 = \varphi_2$. Then $F_1 = F_2$.

**Proof.** Let $G_a$ be the $d$-dimensional normal distribution with the parameters $(0, a^{-2})$, i.e.,

$$G_a(Q) = a^d(2\pi)^{-d/2} \int_Q e^{-\frac{1}{2}a^2|z|^2} dz$$

for any Borel $Q \subset \mathbb{R}^d$. Then

$$e^{-i\langle\lambda, y\rangle} \varphi_1(\lambda) = \int_{\mathbb{R}^d} e^{i\langle\lambda, x-y\rangle} dF_1(x).$$

Integrating this with respect to $dG_a(\lambda)$ and changing the order of integration, which is possible by the Fubini theorem, we obtain

$$\int_{\mathbb{R}^d} e^{-i\langle\lambda, y\rangle} \varphi_1 dG_a(\lambda)$$

$$= a^d(2\pi)^{-d/2} \int_{\mathbb{R}^d} \int_{\mathbb{R}^d} e^{i\langle\lambda, x-y\rangle} e^{-\frac{a^2|\lambda|^2}{2}} d\lambda dF_1(x) = \int_{\mathbb{R}^d} e^{-\frac{|x-y|^2}{2a^2}} dF_1(x).$$

Let $h$ be a continuous function on $\mathbb{R}^d$ with a compact support then it is easy to see that

$$\lim_{a\to 0} (2\pi a^2)^{-d/2} \int_{\mathbb{R}^d} \int_{\mathbb{R}^d} e^{-\frac{|x-y|^2}{2a^2}} h(y) dy dF_1(x) = \int_{\mathbb{R}^d} h(x) dF_1(x).$$

Thus

$$\int_{\mathbb{R}^d} h(x) dF_1(x) = \lim_{a\to 0} \int_{\mathbb{R}^d} \int_{\mathbb{R}^d} e^{-i\langle\lambda, y\rangle} h(y) \varphi_1(\lambda) dy dG_a(\lambda).$$

Since $\varphi_1 \equiv \varphi_2$ we obtain that

$$\int_{\mathbb{R}^d} h(x) dF_1(x) = \int_{\mathbb{R}^d} h(x) dF_2(x)$$

for any continuous function with a compact support, and so $F_1 = F_2$. $\square$

Another important property of characteristic functions is the following

**Lemma 7.1.2.** *The components of a random vector $Z = (Z_1, ..., Z_d)$ are independent if and only if the characteristic function of $Z$ is the product of the characteristic functions of $Z_1, ..., Z_d$, i.e., for any $\lambda = (\lambda_1, ..., \lambda_d) \in \mathbb{R}^d$,*

$$Ee^{i\langle\lambda, Z\rangle} = \prod_{k=1}^{d} Ee^{i\lambda_k Z_k}. \tag{7.1.2}$$

**Proof.** If $Z_1, ..., Z_d$ are independent then the assertion is clear. Let now (7.1.2) hold true. Denote by $F$ the probability distribution of $Z$ and let $F_k$ be the one-dimensional probability distribution of $Z_k$, $k = 1, ..., d$. Denote by $G$ the product measure $F_1 \times F_2 \times \cdots \times F_d$. Then by (7.1.2) and the Fubini theorem

$$\int_{\mathbb{R}^d} e^{i\langle \lambda, x \rangle} dG(x) = \prod_{k=1}^{d} \int_{\mathbb{R}^1} e^{i\lambda_k x_x} dF_k(x_k)$$

$$= \prod_{k=1}^{d} E e^{i\lambda_k Z_k} = E e^{i\langle \lambda, Z \rangle} = \int_{\mathbb{R}^d} e^{i\langle \lambda, x \rangle} dF(x),$$

and so by the uniqueness Lemma 7.1.1, $F = G$ implying independence of the components of $Z$. $\qquad\square$

Now, to complete the proof of Theorem 7.1.1 it suffices to show that for any numbers $\lambda_1, ..., \lambda_k$ and $0 \le t_0 \le t_1 \le \cdots \le t_k$,

$$E \exp\left(i \sum_{l=1}^{k} \lambda_l (W(t_l) - W(t_{l-1}))\right) = \prod_{l=1}^{k} \exp(-\frac{1}{2}\lambda_l^2(t_l - t_{l-1})). \quad (7.1.3)$$

Indeed, the left hand side of (7.1.3) is the characteristic function of the random vector $(W(t_1) - W(t_0), ..., W(t_k) - W(t_{k-1}))$ and the right hand side of (7.1.3) is the characteristic function of a random vector $Z = (Z_1, ..., Z_k)$ with independent components $Z_l$, $l = 1, ..., k$ such that each $Z_l$ has normal distribution with the parameters $(0, t_l - t_{l-1})$. By Lemmas 7.1.1 and 7.1.2 this implies the properties (ii) and (iii) of the process $W$.

To prove (7.1.3) we substitute (7.1.1) into the left hand side of (7.1.3),

$$E \exp\left(i \sum_{l=1}^{k} \lambda_l (W(t_l) - W(t_{l-1}))\right) \qquad (7.1.4)$$

$$= E \exp\left(i \sum_{n=0}^{\infty} \xi_n (\sum_{l=1}^{k} \lambda_l (S_n(t_l) - S_n(t_{l-1})))\right)$$

$$= \prod_{n=0}^{\infty} E \exp\left(i\xi_n \sum_{l=1}^{k} \lambda_l (S_n(t_l) - S_n(t_{l-1}))\right)$$

$$= \prod_{n=0}^{\infty} \exp\left(-\frac{1}{2}(\sum_{l=1}^{k} \lambda_l (S_n(t_l) - S_n(t_{l-1})))^2\right)$$

$$= \exp\left(-\frac{1}{2}\sum_{1 \le l,m \le k} \lambda_l \lambda_m \sum_{n=0}^{\infty}(S_n(t_l)S_n(t_m)\right.$$

$$\left. -S_n(t_{l-1})S_n(t_m) - S_n(t_l)S_n(t_{m-1}) + S_n(t_{l-1})S(t_{m-1}))\right)$$

where we used the independence of $\xi_0, \xi_1, \xi_2, ...$ and the absolute convergence in (7.1.1). It is easy to see by direct computations that the right hand side of (7.1.4) transforms into the right hand side of (7.1.3) provided that for any $s, t \in [0, 1]$,

$$\sum_{n=0}^{\infty} S_n(s)S_n(t) = \min(s, t). \qquad (7.1.5)$$

This relation is a version of the Parseval equality and it is the last thing we have to prove to establish Theorem 7.1.1.

To prove (7.1.5) we will discuss further properties of Haar functions. Notice that the Haar functions form an orthonormal basis of the space $L^2([0,1])$ of real measurable functions $\varphi$ on $[0,1]$ satisfying $(\varphi, \varphi) < \infty$ where $(\varphi_1, \varphi_2) = \int_0^1 \varphi_1(t)\varphi_2(t)dt$ is the inner product. Indeed, it is clear from the definition that

$$\int_0^1 H_k(t)H_l(t)dt = \delta_{kl}$$

where $\delta_{kl} = 1$ if $k = l$ and $\delta_{kl} = 0$ for otherwise. If $\varphi \in L^2([0,1])$ is orthogonal to all Haar function then it is easy to see that the integral $\int_{(k-1)2^{-n}}^{k2^{-n}} \varphi(t)dt$ is the same for all $k = 0, 1, ..., 2^n$. Thus

$$\int_a^b \varphi(t)dt = (b-a)\int_0^1 \varphi(t)dt = (b-a)(\varphi, H_0) = 0$$

for any dyadic $a, b \in [0,1]$ (i.e., for $a$ and $b$ having the form $\frac{k}{2^n}$). Hence, $\int_U \varphi(t)dt = 0$ for any open set $U \subset [0,1]$, and so by the regularity of the Lebesgue measure and by the absolute continuity of the Lebesgue integral $\int_\Gamma \varphi(t)dt = 0$ for any measurable set $\Gamma \subset [0,1]$, which implies that $\varphi \equiv 0$ almost everywhere with respect to the Lebesgue measure on $[0,1]$. It follows that $\{H_k, \ k = 0, 1, ...\}$ is the complete basis of $L^2([0,1])$. Let $\mathbb{I}_{[0,t]}$ be the indicator function of $[0,t]$. Then clearly,

$$(\mathbb{I}_{[0,t]}, H_k) = S_k(t).$$

We will show next that

$$\mathbb{I}_{[0,t]}(u) = \sum_{k=0}^\infty S_k(t)H_k(u)$$

in the sense that

$$\left\| \mathbb{I}_{[0,t]} - \sum_{k=0}^n S_k(t)H_k \right\|^2 \to 0 \quad \text{as} \quad n \to \infty \qquad (7.1.6)$$

where $\|\varphi\|^2 = (\varphi, \varphi)$.

Indeed,

$$\left\| \mathbb{I}_{[0,t]} - \sum_{k=0}^n S_k(t)H_k \right\|^2 = \|\mathbb{I}_{[0,t]}\|^2 - \sum_{k=0}^n (S_k(t))^2 = t - \sum_{k=0}^n (S_k(t))^2.$$

Hence, $t \geq \sum_{k=0}^n (S_k)^2$, and so

$$\left\| \sum_{k=l}^n S_k(t)H_k \right\|^2 = \sum_{k=l}^n (S_k(t))^2 \to 0 \quad \text{as} \quad n, l \to \infty,$$

which means that $\sum_{k=0}^{n} S_k(t)H_k$ is a fundamental sequence. Hence, there exists $f = \sum_{k=0}^{\infty} S_k(t)H_k \in L^2([0,1])$. But,

$$(\mathbb{I}_{[0,t]} - f, H_m) = \lim_{n \to \infty} (\mathbb{I}_{[0,t]} - \sum_{k=0}^{n} S_k(t)H_k, H_m) = S_m(t) - S_m(t) = 0,$$

and so $f = \mathbb{I}_{[0,t]}$, proving (7.1.6). Finally,

$$0 = \lim_{n \to \infty} (\mathbb{I}_{[0,t]} - \sum_{k=0}^{n} S_k(t)H_k, \mathbb{I}_{[0,s]} - \sum_{k=0}^{n} S_k(s)H_k) = (\mathbb{I}_{[0,t]}, \mathbb{I}_{[0,s]})$$
$$- \lim_{n \to \infty} \sum_{k=0}^{n} S_k(t)S_k(s) = \min(t,s) - \sum_{k=0}^{\infty} S_k(t)S_k(s)$$

which proves (7.1.5) and completes the proof of Theorem 7.1.1. □

We just constructed the Brownian motion $W(t)$, but only for $t \in [0,1]$. To extend the construction to the whole half line $t \geq 0$ we proceed as follows. Recall, that the representation (7.1.1) is based on a sequence of independent normal random variables $\xi_k$, $k = 0, 1, \ldots$ We can renumber this sequence arriving at a two-parameter sequence $\eta_k^l$, $k, l = 0, 1, \ldots$ of independent normal random variables. Set $W_l(t) = \sum_{k=0}^{\infty} \eta_k^l S_k(t)$ then $W_l$, $l = 0, 1, \ldots$ is a sequence of independent Brownian motions defined for $t \in [0,1]$. Now we define the Brownian motion $W(t)$ for all $t \geq 0$ successively by setting $W(t) = W_0(t)$ for $t \in [0,1]$ and $W(t) = W(n) + W_n(t - n)$ for $t \in [n, n+1]$, $n = 1, 2, \ldots$ This completes the construction of the continuous one-dimensional Brownian motion and its $d$-dimensional counterpart $W(t) = (W^1(t), \ldots, W^d(t))$ is obtained by taking $W^1, \ldots, W^d$ to be independent one-dimensional Brownian motions.

## 7.1.2 *Gaussian random vectors and some properties of Brownian motion*

A random variable $Z$ is called Gaussian with parameters $(m, \sigma^2)$ if its characteristic function $\varphi_Z$ has the form

$$\varphi_Z(\zeta) = Ee^{i\zeta Z} = e^{i\zeta m - \frac{1}{2}\zeta^2 \sigma^2}.$$

It is easy to check that $\varphi_Z$ has this form if $Z$ is a normal random variable with mean $m = EZ$ and variance $\sigma^2 = \text{Var} Z = E(Z - m)^2$. A random vector $Z = (Z_1, \ldots, Z_d)$, where $Z_1, \ldots, Z_d$ are random variables, is called Gaussian if $\langle \lambda, Z \rangle = \sum_{j=1}^{d} \lambda_j Z_j$ is a Gaussian random variable for any nonrandom vector $\lambda = (\lambda_1, \ldots, \lambda_d) \in \mathbb{R}^d$. Hence, the characteristic function of such random vector $Z$ has the form

$$\varphi_Z(\lambda) = Ee^{i\langle \lambda, Z \rangle} = e^{iE\langle \lambda, Z \rangle - \frac{1}{2}\text{Var}\langle \lambda, Z \rangle}$$
$$= \exp\left(i \sum_{k=1}^{d} \lambda_k E Z_k - \frac{1}{2} \sum_{k,l=1}^{d} \lambda_k \lambda_l \text{Cov}(Z_k, Z_l)\right)$$

where $\mathrm{Cov}(X, Y) = E(X - EX)(Y - EY)$ denotes the covariance of random variables $X$ and $Y$.

The matrix $R = (R_{kl})_{k,l=1}^d$ with $R_{kl} = \mathrm{Cov}(Z_k, Z_l)$ is called the covariance matrix. It is symmetric and nonnegative definite since $\langle R\lambda, \lambda \rangle = E(\sum_{k=1}^d \lambda_k Z_k)^2 \geq 0$ for any vector $\lambda$. If $m = EZ$ we can rewrite $\varphi_Z$ in the form

$$\varphi_Z(\lambda) = \exp\left(i\langle \lambda, m \rangle - \frac{1}{2}\langle R\lambda, \lambda \rangle\right).$$

Suppose that $R$ is nonsingular and let $A = R^{-1}$ be its inverse. Set

$$f(x) = \frac{|A|^{1/2}}{(2\pi)^{d/2}} \exp\left(-\frac{1}{2}\langle A(x - m), (x - m)\rangle\right)$$

where $x = (x_1, ..., x_d) \in \mathbb{R}^d$ and $|A| = det A$. Diagonalizing the matrix $R$ we can see easily that

$$\int_{\mathbb{R}^d} e^{i\langle \lambda, x \rangle} f(x) dx = \exp\left(i\langle \lambda, m \rangle - \frac{1}{2}\langle R\lambda, \lambda \rangle\right).$$

By the uniqueness property of characteristic functions it follows that $f$ is the density of the probability distribution of the Gaussian random vector $Z$, i.e.,

$$P\{Z \in U\} = \int_U f(x) dx$$

for any Borel set $U \subset \mathbb{R}^d$.

In the following assertion we collect some useful properties of Gaussian random vectors.

**Proposition 7.1.1.** *a) The components of a Gaussian random vector $Z = (Z_1, ..., Z_d)$ are uncorrelated if and only if they are independent;*

*b) If components of the random vector $Z = (Z_1, ..., Z_d)$ are independent Gaussian random variables then $Z$ is a Gaussian random vector;*

*c) If $Z$ is a d-dimensional Gaussian random vector and $A$ is a $d \times k$ non-random matrix then $AZ$ is a k-dimensional Gaussian random vector;*

*d) The items b) and c) exhaust all Gaussian random vectors in the sense that any such vector can be represented as $AZ$ where $A$ is a non-random matrix and $Z$ is a random vector whose components are independent Gaussian random variables;*

*e) Let $Z^{(1)}, Z^{(2)}, ...$ be a sequence of Gaussian random vectors that converge in probability to $Z$. Then $Z$ is also a Gaussian random vector.*

**Proof.** a) The independence always implies lack of correlation. If the components $Z_1, ..., Z_d$ of a Gaussian vector $Z$ are uncorrelated, i.e., $\text{Cov}(Z_k, Z_l) = 0$, $k \neq l$ then

$$\varphi_Z(\lambda) = \exp\left(i\sum_{k=1}^d \lambda_k E Z_k - \frac{1}{2}\sum_{k=1}^d \lambda_k^2 \text{Var} Z_k\right)$$
$$= \prod_{k=1}^d E\exp\left(i\lambda_k E Z_k - \frac{1}{2}\lambda_k^2 \text{Var} Z_k\right),$$

and so by Lemma 7.1.2 the components $Z_1, ..., Z_d$ are independent.

b) Since $Z_1, ..., Z_d$ are independent,

$$Ee^{i\zeta\langle\lambda, Z\rangle} = \prod_{k=1}^d Ee^{i\zeta\lambda_k Z_k} = \prod_{k=1}^d E\exp\left(i\zeta\lambda_k E Z_k - \frac{1}{2}\zeta^2\lambda_k^2 \text{Var} Z_k\right)$$
$$= E\exp\left(i\zeta(\sum_{k=1}^d \lambda_k E Z_k) - \frac{1}{2}\zeta^2(\sum_{k=1}^d \lambda_k^2 \text{Var} Z_k)\right)$$

which means that $\langle\lambda, Z\rangle$ is a Gaussian random variable, proving b).

c) Follows that, since $\langle\lambda, AZ\rangle = \langle A^*\lambda, Z\rangle$, where $A^*$ is conjugate to $A$, and by definition the right hand side in this equality must be a Gaussian random variable.

d) Let $Y = (Y_1, ..., Y_d)$ be a Gaussian random vector and let $R$ be its covariance matrix. Without loss of generality we can assume that $\text{rank} R = d$ and $EY = 0$. Indeed, if $\text{rank} R = r < d$ then it is easy to show that there are $n - r$ linear relations between $Y_1, ..., Y_d$ (since Gaussian random vectors are characterized by expectations and covariances of their components). Thus, in this case we can assume that, say, $Y_1, ..., Y_r$ are linearly independent while other $Y_j$'s are linear combinations of them. Then we have to consider only the $r$-vector $(Y_1, ..., Y_r)$ whose covariance matrix is already nonsingular. So, suppose that $\det R > 0$ and since $R$ is symmetric there exists an orthogonal matrix $K$ diagonalizing $R$, i.e., $K^*RK = D$ where $D$ is a diagonal matrix with positive diagonal elements. Put $B^2 = D$ and $Z = B^{-1}K^*Y$. By c) the random vector $Z$ is Gaussian. The covariance matrix of $Z$ is diagonal, i.e., the components of $Z$ are uncorrelated, and so by a) they are independent. Hence, $Y = KBZ$ is the required representation.

e) Let $E\langle\lambda, Z^{(k)}\rangle = m_k(\lambda)$ and $\text{Var}\langle\lambda, Z^{(k)}\rangle = E(\langle\lambda, Z^{(k)}\rangle - m_k(\lambda))^2 = \sigma_k^2(\lambda)$, $k = 1, 2, ...$ Then for any $\zeta$,

$$\lim_{n\to\infty} e^{i\zeta m_n(\lambda) - \frac{1}{2}\zeta^2\sigma_n^2(\lambda)} = \lim_{n\to\infty} Ee^{i\zeta\langle\lambda, Z^{(n)}\rangle} = Ee^{i\zeta\langle\lambda, Z\rangle}.$$

The existence of the limit in the left hand side above implies that there are numbers $m(\lambda) = \lim_{n\to\infty} m_n(\lambda)$ and $\sigma^2(\lambda) = \lim_{n\to\infty} \sigma_n^2(\lambda)$, and so

$$Ee^{i\zeta\langle\lambda, Z\rangle} = e^{i\zeta m(\lambda) - \frac{1}{2}\zeta^2\sigma^2(\lambda)}.$$

Since this holds true for any non-random vector $\lambda$, it follows by definition that $Z$ is a Gaussian vector. $\qquad\square$

Next, we derive additional properties of the Brownian motion $W(t), t \geq 0$ where we will not refer to the explicit construction above but will rely only on the conditions (i)–(iii) in the definition of $W$.

**Proposition 7.1.2.** *a)* $W_1(t) \equiv W(t+s) - W(s)$, $t \geq 0$ *is a Brownian motion independent of* $\mathcal{F}_s = \sigma\{W(u), u \leq s\}$ *and* $W(t)$, $t \geq 0$ *is a martingale with respect to the filtration* $\mathcal{F}_t$, $t \geq 0$.

*b) Let* $W(t)$, $t \geq 0$ *be the Brownian motion with continuous paths and* $\tau$ *be a stopping time with respect to the filtration* $\mathcal{F}_t = \sigma\{W(s), s \leq t\}$, $T \geq 0$ *supplemented by sets of zero probability. Then* $W_2(t) = W(\tau + t) - W(\tau)$ *is the Brownian motion independent of* $\mathcal{F}_\tau$;

*c)(Brownian scaling)* $W_3(t) = cW(t/c^2)$, $t \geq 0$ *is a Brownian motion for each* $c > 0$;

*d) (Symmetry)* $W_4(t) = -W(t)$ *is a Brownian motion;*

*e)* $W_5(t) = tW(1/t)$ *for* $t > 0$ *and* $W_5(0) = 0$ *is a Brownian motion.*

**Proof.** a) The properties (i)–(iii) are verified for $W_1$ directly. The independence of $W_1$ of $\mathcal{F}_s$ follows from (ii). Since $E(W(t+s)|\mathcal{F}_s) = W(s) + E((W(t+s) - W(s)|\mathcal{F}_s) = W(s) + E(W(t+s) - W(s)) = W(s)$ for $s, t \geq 0$, it follows that $W(t)$, $t \geq 0$ is a martingale.

b) We assume that $W$ is one dimensional as the proof in the multidimensional case is the same. Set

$$
\tau_n = \begin{cases} k2^{-n} & \text{if} \quad (k-1)2^{-n} < \tau \leq k2^{-n}, \quad k = 1, 2, ..., \\ 0 & \text{if} \quad \tau = 0 \end{cases}
$$

and let $W^{(n)}(t) = W(\tau_n + t) - W(\tau_n)$. Since $\Gamma_{k,n} = \Gamma \cap \{\tau_n = k2^{-n}\} \in \mathcal{F}_{k2^{-n}}$ for any $\Gamma \in \mathcal{F}_{\tau_n}$, it follows using a) that for every Borel $U \subset \mathbb{R}^m$ and all $0 \leq t_1 < t_2 < ... < t_m$,

$$
P\{\{(W^{(n)}(t_1), ..., W^{(n)}(t_m)) \in U\} \cap \Gamma\} = \sum_{k=0}^{\infty} \qquad (7.1.7)
$$
$$
P\{\{(W(k2^{-n} + t_1) - W(k2^{-n}), ..., W(k2^{-n} + t_m)
$$
$$
-W(k2^{-n})) \in U\} \cap \Gamma_{k,n}\} = P\{(W(t_1), ..., W(t_m)) \in U\}P(\Gamma).
$$

This proves the claim for each $\tau_n$. Observe that $\tau_n \downarrow \tau$ as $n \uparrow \infty$, and so $W^{(n)}(t) \to W_2(t)$ as $n \to \infty$ by the continuity of the sample paths of $W$. Hence, for any bounded continuous function $f$ on $\mathbb{R}^m$,

$$
\lim_{n \to \infty} Ef(W^{(n)}(t_1), ..., W^{(n)}(t_m)) = Ef(W_2(t_1), ..., W_2(t_m)).
$$

It follows that $(W^{(n)}(t_1), ..., W^{(n)}(t_m))$ converges in distribution to $(W_2(t_1), ..., W_2(t_m))$ as $n \to \infty$. If $\Gamma \in \mathcal{F}_\tau$ then $\Gamma \in \mathcal{F}_{\tau_n}$ for all $n$ since

$\mathcal{F}_\tau \subset \mathcal{F}_{\tau_n}$, and so for such $\Gamma$ the right hand side of (7.1.7) does not depend on $n$. Thus, the limiting in distribution random vector $(W_2(t_1), ..., W_2(t_m))$ is independent of $\mathcal{F}_\tau$ and it has the same distribution as $(W(t_1), ..., W(t_m))$, completing the proof of (b).

c) The properties (i) and (ii) for $W_3$ follow immediately from the corresponding properties for $W$ itself. By (iii), $W(t/c^2) - W(s/c^2)$, $t > s$ has the normal distribution with zero expectation and the variance $(t/c^2 - s/c^2)$. Hence,

$$E(W_3(t) - W_3(s))^2 = c^2(t/c^2 - s/c^2) = t - s,$$

and so $W_3$ satisfies (iii) as well.

d) Properties (i)–(iii) for $W_4 = -W$ are clear.

e) It suffices to consider a one-dimensional Brownian motion since the components of $W_5(t) = tW(1/t)$ are independent, and so if each component will be a one-dimensional Brownian motion then $W_5$ will be a $d$-dimensional Brownian motion. Take $0 < t_1 < t_2 < ... < t_m$ and consider the vector

$$\begin{aligned} X &= (W_5(t_1), W_5(t_2) - W_5(t_1), ..., W_5(t_m) - W_5(t_{m-1})) \\ &= (t_1 W(1/t_1), t_2 W(1/t_2) - t_1 W(1/t_1), \\ & \qquad ..., t_m W(1/t_m) - t_{m-1} W(1/t_{m-1})) \end{aligned}$$

where $W(t)$ is a one-dimensional Brownian motion. Introduce also the vector $Y = (W(1/t_1) - W(1/t_2), ..., W(1/t_{m-1}) - W(1/t_m), W(1/t_m))$ which by the definition of the Brownian motion and Proposition 7.1.1(b) is a Gaussian random vector. Then $X = AY$ where $A = (a_{kl})$ is the $m \times m$ non-random matrix with $a_{1l} = t_1$ for $l = 1, ..., m$, $a_{k,k-1} = t_{k-1}$ for $k = 2, ..., m - 1$, $a_{kl} = t_k - t_{k-1}$ for $l \geq k = 2, ..., m$ and $a_{kl} = 0$ when $l < k - 1$. Hence, by Proposition 7.1.1(c) the random vector $X$ is also Gaussian. It remains to check only the covariances where we use that $EW(t)W(s) = \min(t, s)$ and obtain

$$E(W_5(t_{k+1}) - W_5(t_k))(W_5(t_{l+1}) - W_5(t_l)) = t_{k+1}t_{l+1}EW(t_{k+1}^{-1})W(t_{l+1}^{-1})$$
$$-t_k t_{l+1}EW(t_k^{-1})W(t_{l+1}^{-1}) - t_{k+1}t_l EW(t_{k+1}^{-1})W(t_l^{-1})$$
$$+t_k t_l EW(t_k^{-1})W(t_l^{-1})$$
$$= \min(t_{k+1}, t_{l+1}) - \min(t_k, t_{l+1}) - \min(t_{k+1}, t_l) + \min(t_k, t_l) = t_{k+1} - t_k$$

if $k = l$ and the right hand side equals 0 if $k \neq l$, which by Proposition 7.1.1(a) proves (e). $\qquad\square$

**Remark 7.1.1.** The assertions a) and b) of Proposition 7.1.2 exhibit the Markov and the strong Markov properties of the Brownian motion, respectively. The former says that the evolution of the Brownian motion $W$ after time $s$ depends only on $W(s)$ and is independent of its behavior before $s$. The latter extends this to any stopping time $\tau$, i.e., given $W(\tau)$ the behaviors of $W$ after time $\tau$ and before it are independent.

### 7.1.3 *Kolmogorov's theorems*

We will exhibit here two theorems of Kolmogorov which also yield the existence of the Brownian motion with continuous, even Hölder continuous, paths without providing its explicit construction as demonstrated above. The importance of these theorems is in their generality as they yield the existence of many classes of stochastic processes on quite general topological spaces and not only of the Brownian motion in $\mathbb{R}^d$. In general, we can define a stochastic process $\xi = \xi(t, \omega)$ on a probability space $(\Omega, \mathcal{F}, P)$ with $t$ belonging to some index set $\mathbf{T}$ as a family of measurable maps $\xi(t, \cdot)$ from $(\Omega, \mathcal{F})$ to a measure space $(M, \mathcal{B})$, where $\mathcal{B}$ is usually the Borel $\sigma$-algebra if $M$ is a topological space. Here we will consider a real valued stochastic process $\xi = \xi(t, \omega)$ with $\omega \in \Omega$ and the time parameter $t$ running over a subset $\mathbf{T}$ of the reals $\mathbb{R}$ and we can view a $\mathbb{R}^d$-valued stochastic process as a collection of $d$ one-dimensional processes. Most frequent choices for $\mathbf{T}$ are nonnegative reals $\{t \geq 0\}$ and nonnegative integers $\{0, 1, ...\}$ or intervals $[a, b]$, $0 \leq a < b$ and $m, m + 1, ..., n$ with $0 \leq m < n$. For a fixed $\omega \in \Omega$ the set $\{\xi(t, \omega),\, t \in \mathbf{T}\}$ will be called a (sample) path (or a trajectory, or a realization) of the process. This is also a real valued function with the domain $\mathbf{T}$, and so the set of paths coincides with the set of real valued functions $x(t)$, $t \in \mathbf{T}$ defined on $\mathbf{T}$ which is denoted by $\mathbb{R}^{\mathbf{T}}$.

Since each $\xi(t) = \xi(t, \cdot)$, $t \in \mathbf{T}$ is a random variable then $\{\omega \in \Omega : \xi(t, \omega) \in B\} \in \mathcal{F}$ for any $B$ from the Borel $\sigma$-algebra $\mathcal{B} = \mathcal{B}(\mathbb{R})$ on the real line $\mathbb{R}$. Hence,

$$\{\omega \in \Omega : \xi(t_j, \omega) \in B_j,\, j = 1, ..., k\} = \cap_{j=1}^{k} \{\omega \in \Omega : \xi(t_j, \omega) \in B_j\}$$
$$= \{\omega \in \Omega : (\xi(t_1, \omega), ..., \xi(t_k, \omega)) \in B_1 \times \cdots \times B_k\} \in \mathcal{F}$$

for any $t_j \in \mathbf{T}$ and $B_j \in \mathcal{B}$, $i = 1, ..., k$. Since product sets $B_1 \times \cdots \times B_k$ form a semi-algebra in the Borel $\sigma$-algebra $\mathcal{B}(\mathbb{R}^k)$ on $\mathbb{R}^k$ we conclude that for any Borel set $B$ in $\mathbb{R}^k$,

$$\{\omega \in \Omega : (\xi(t_1, \omega), ..., \xi(t_k, \omega)) \in B\} \in \mathcal{F}.$$

Define cylinder sets in $\mathbb{R}^{\mathbf{T}}$ by $C_{t_1,...,t_k;B} = \{x \in \mathbb{R}^{\mathbf{T}} : (x(t_1), ..., x(t_k)) \in B\}$ where $B \in \mathcal{B}(\mathbb{R}^k)$, $t_j \in \mathbf{T}$, $j = 1, ..., k$ are distinct and $k$ is a positive integer. Then we can rewrite the above formula as

$$\Gamma_{t_1,...,t_k;B} = \{\omega \in \Omega : \xi(\cdot, \omega) \in C_{t_1,...,t_k;B}\} \in \mathcal{F}.$$

Let $\mathcal{B}(\mathbb{R}^{\mathbf{T}})$ denote the minimal $\sigma$-algebra in $\mathbb{R}^{\mathbf{T}}$ which contains all cylinder sets $C_{t_1,\ldots,t_k;B}$ as above. Since unions and intersections of sets $\Gamma_{t_1,\ldots,t_k;B}$ are determined by the corresponding unions and intersections of sets $C_{t_1,\ldots,t_k;B}$ we obtain that $\Gamma = \{\omega \in \Omega : \xi(\cdot,\omega) \in D\} \in \mathcal{F}$ for any $D \in \mathcal{B}(\mathbb{R}^{\mathbf{T}})$. Thus, $\xi$ becomes a measurable map from $(\Omega, \mathcal{F})$ to $(\mathbb{R}^{\mathbf{T}}, \mathcal{B}(\mathbb{R}^{\mathbf{T}}))$ sending $\omega \in \Omega$ to $\xi(\cdot,\omega) \in \mathbb{R}^{\mathbf{T}}$. This map produces also a probability measure $Q = \xi P$ on $\mathbb{R}^{\mathbf{T}}$ such that for any $D \in \mathcal{B}(\mathbb{R}^{\mathbf{T}})$,

$$Q(D) = P\{\omega \in \Omega : \xi(\cdot,\omega) \in D\}.$$

Usually, a stochastic process $\xi$ is defined by conditions on finitely many values of time (as in the definition of the Brownian motion above), i.e., by properties of $\xi(t_j)$, $j = 1,\ldots,k$ and not by its distribution on the infinite dimensional (in general) space of paths $\mathbb{R}^{\mathbf{T}}$ which is more difficult to describe. Namely, let for all positive integers $k$ and any $k$-tuples $t_1,\ldots,t_k \in \mathbf{T}$ of distinct numbers probability measures $Q_{t_1,\ldots,t_k}$ on $(\mathbb{R}^k, \mathcal{B}^k)$, called finite dimensional distributions, be defined. We will discuss next the conditions which ensure the existence of a probability measure $Q$ on $(\mathbb{R}^{\mathbf{T}}, \mathcal{B}(\mathbb{R}^{\mathbf{T}}))$ such that for any cylinder set $C_{t_1,\ldots,t_k;B}$,

$$Q(C_{t_1,\ldots,t_k;B}) = Q_{t_1,\ldots,t_k}(B). \tag{7.1.8}$$

Once such probability measure $Q$ on $\mathbb{R}^{\mathbf{T}}$ is obtained we can construct a stochastic process $\xi$ by taking $\Omega = \mathbb{R}^{\mathbf{T}}$, $\mathcal{F} = \mathcal{B}(\mathbb{R}^{\mathbf{T}})$, $P = Q$ and setting $\xi(t,\omega) = \omega(t)$ for each $\omega \in \mathbb{R}^{\mathbf{T}}$ which is a real valued function on $\mathbf{T}$. Then

$$P\{\omega \in \Omega : \xi(\cdot,\omega) \in C_{t_1,\ldots,t_k;B}\} = Q(C_{t_1,\ldots,t_k;B}) = Q_{t_1,\ldots,t_k}(B),$$

i.e., $Q_{t_1,\ldots,t_k}$ become finite dimensional distributions of $\xi$ while $Q$ is its infinite dimensional distribution on $\mathbb{R}^{\mathbf{T}}$.

Observe that $C_{t_1,\ldots,t_k;B_1 \times \cdots \times B_k} = C_{t_{i_1},\ldots,t_{i_k};B_{i_1} \times \cdots \times B_{i_k}}$ for any permutation $\left(\begin{smallmatrix} 1 & 2 & \cdots & k \\ i_1 & i_2 & \cdots & i_k \end{smallmatrix}\right)$ where $B_j \in \mathcal{B}(\mathbb{R})$, $j = 1,\ldots,k$. Hence, if a probability distribution $Q$ on $\mathbb{R}^{\mathbf{T}}$ satisfying (7.1.8) exists then we must have that

$$Q_{t_1,\ldots,t_k}(B_1 \times \cdots \times B_k) = Q_{t_{i_1},\ldots,t_{i_k}}(B_{i_1} \times \cdots \times B_{i_k}) \tag{7.1.9}$$

for any $B_1,\ldots,B_k \in \mathcal{B}(\mathbb{R})$ and a permutation $\left(\begin{smallmatrix} 1 & 2 & \cdots & k \\ i_1 & i_2 & \cdots & i_k \end{smallmatrix}\right)$. Since, clearly, $C_{t_1,\ldots,t_k;B} = C_{t_1,\ldots,t_k,t_{k+1};B\times\mathbb{R}}$ for any $B \in \mathcal{B}(\mathbb{R}^k)$ then we also must have

$$Q_{t_1,\ldots,t_k,t_{k+1}}(B \times \mathbb{R}) = Q_{t_1,\ldots,t_k}(B). \tag{7.1.10}$$

The conditions (7.1.9) and (7.1.10) are called the consistency conditions.

**Theorem 7.1.2.** *(Kolmogorov) If finite dimensional distributions $Q_{t_1,\ldots,t_k}$ satisfy the consistency conditions (7.1.9) and (7.1.10) then there exists a probability measure $Q$ on $\mathbb{R}^{\mathbf{T}}$ which satisfies (7.1.8).*

**Proof.** First, we define $Q$ on cylinder sets by (7.1.8). It is easy to see by the consistency conditions that this definition is legitimate in the sense that it does not depend on a specific representation of a cylinder set. This yields a finitely additive probability measure on the algebra of cylinder sets. Next, we will employ the Carathéodory extension theorem (see, for instance, Theorem 3.1 in [1] or Theorem 3.1.4 in [7]) which plays also a crucial role in the construction of the Lebesgue measure. This theorem says that if a finitely additive probability measure on an algebra of sets is, in fact, $\sigma$-additive there then it can be extended as a $\sigma$-additive probability measure to the minimal $\sigma$-algebra containing this algebra. Denote by $\mathcal{C}$ the algebra of cylinder sets and let $C_1, C_2, \ldots$ be disjoint sets in $\mathcal{C}$ with $C = \cup_{j=1}^\infty C_j$. Set $D_k = C \setminus \cup_{j=1}^k C_j$. Then $D_k \downarrow \emptyset$ as $k \uparrow \infty$ and the $\sigma$-additivity of $Q$ will follow if

$$Q(D_k) \downarrow 0 \quad \text{as} \quad k \uparrow \infty. \qquad (7.1.11)$$

Suppose that, on the contrary, (7.1.11) does not hold true, i.e., $Q(D_n) > \varepsilon > 0$ for some $\varepsilon > 0$ and all $n = 1, 2, \ldots$ Since $D_n \in \mathcal{C}$, we can renumber these sets and choose a sequence $t_1, t_2, \ldots \in \mathbf{T}$ of distinct numbers so that $D_n = \{x \in \mathbb{R}^\mathbf{T} : (x(t_1), \ldots, x(t_n)) \in B_n\}$ for some $B_n \in \mathcal{B}(\mathbb{R}^n)$. Since any probability measure on $\mathbb{R}^n$ is regular (see, for instance, Theorem 12.3 in [1] or Theorem 7.1.4 in [7]), it follows that there exist compact sets $K_n \subset B_n$, $n = 1, 2, \ldots$ such that $Q_{t_1,\ldots,t_n}(B_n \setminus K_n) < \frac{\varepsilon}{2^n}$. Set $F_n = \{x \in \mathbb{R}^\mathbf{T} : (x(t_1), \ldots, x(t_n)) \in K_n\}$ and $G_n = \cap_{j=1}^n F_j$. Then $G_n \subset F_n \subset D_n$. Moreover, $G_n = \{x \in \mathbb{R}^\mathbf{T} : (x(t_1), \ldots, x(t_n)) \in L_n\}$ where $L_n = \cap_{j=1}^n (K_j \times \mathbb{R}^{n-j})$ under the convention $K_n \times \mathbb{R}^0 = K_n$.

Next,

$$Q(G_n) = Q(D_n) - Q(D_n \setminus G_n) = Q(D_n) - Q(\cup_{j=1}^n (D_n \setminus F_j))$$
$$> \varepsilon - \sum_{j=1}^n Q(D_n \setminus F_j) \geq \varepsilon - \sum_{j=1}^n Q(D_j \setminus F_j)$$
$$= \varepsilon - \sum_{j=1}^n Q_{t_1,\ldots,t_n}(B_j \setminus K_j) > \frac{\varepsilon}{2^n} > 0.$$

Hence, $Q_{t_1,\ldots,t_n}(L_n) = Q(G_n) > 0$, and so $L_n \neq \emptyset$ for all $n = 1, 2, \ldots$ Now, $G_n \downarrow$ as $n \uparrow$ and we claim that $\cap_{n=1}^\infty G_n \neq \emptyset$. Indeed, since $L_n \neq \emptyset$, we can choose $(x_1^{(n)}, \ldots, x_n^{(n)}) \in L_n$, $n = 1, 2, \ldots$. Since the sequence $\{x_1^{(n)}\}_{n=1}^\infty$ belongs to a compact set $K_1$, it must have a convergent subsequence $\{x_1^{(n_j)}\}_{j=1}^\infty$ with a limit $x_1 \in K_1$. The sequence $\{x_1^{(n_j)}, x_2^{(n_j)}\}_{j=2}^\infty$ is contained in $K_2$, and so it has a convergent subsequence with a limit $(x_1, x_2)$. Continuing this process, we can construct a sequence $(x_1, x_2, \ldots)$ such that $(x_1, \ldots, x_n) \in K_n$ for each $n$. It follows that

$\{x \in \mathbb{R}^{\mathbf{T}} : x(t_j) = x_j, j = 1, 2, ...\} \subset G_n$ for all $n$, and so $\cap_{n=1}^{\infty} G_n \neq \emptyset$. Hence, $\cap_{n=1}^{\infty} D_n \supset \cap_{n=1}^{\infty} G_n \neq \emptyset$ which is a contradiction proving the theorem. □

Next, we will show how to obtain the existence of the Brownian motion $W$ relying on the above theorem. Let $0 = t_0 < t_1 < ... < t_n < \infty$ and set $b(t_1, ..., t_n) = (W(t_1), ..., W(t_n))$ and $\hat{b}(t_1, ..., t_n) = (W(t_1), W(t_2) - W(t_1), ..., W(t_n) - W(t_{n-1}))$. Then $b(t_1, ..., t_n) = A\hat{b}(t_1, ..., t_n)$ where $A = (a_{ij})_{i,j=1}^{n}$ satisfies $a_{ij} = 1$ if $i \geq j$ and $a_{ij} = 0$ if $i < j$. Since $\hat{b}(t_1, ..., t_n)$ is a Gaussian random vector as a collection of independent Gaussian random variables, it follows that $b(t_1, ..., t_n)$ is a Gaussian random vector as well. Applying the change of variables formula in multiple integrals with $(x_1, ..., x_n) = A(y_1, ..., y_n)$, $x_k = y_1 + \cdots + y_k$, $k = 1, ..., n$ and taking into account that $\det A = 1$, we obtain that for any Borel $B \subset \mathbb{R}^n$,

$$P\{b(t_1, ..., t_n) \in B\} = P\{\hat{b}(t_1, ..., t_n) \in A^{-1}B\}$$

$$= \int_{A^{-1}B} p(t_1, y_1) p(t_2 - t_1, y_2) \cdots p(t_n - t_{n-1}, y_n) dy_1 \cdots dy_n$$

$$= \int_B p(t_1, x_1) p(t_2 - t_1, x_2 - x_1) \cdots p(t_n - t_{n-1}, x_n - x_{n-1}) dx_1 \cdots dx_n$$

where $p(t, z) = (2\pi t)^{-1/2} \exp(-\frac{z^2}{2t})$ and we used independence of increments of the Brownian motion appearing in its definition. Thus, the multidimensional distributions $Q_{t_1, ..., t_n}$ of the Brownian motion must have the form

$$Q_{t_1, ..., t_n}(B)$$

$$= \int_B p(t_1, x_1) p(t_2 - t_1, x_2 - x_1) \cdots p(t_n - t_{n-1}, x_n - x_{n-1}) dx_1 \cdots dx_n$$

whenever $0 < t_1 < ... < t_n$ and $B \in \mathcal{B}(\mathbb{R}^n)$. On the other hand, if a stochastic process has such finite dimensional distributions, then taking $B = A(B_1 \times \cdots \times B_n)$ for Borel sets $B_1, ..., B_n \in \mathbb{R}^n$ and reversing the above formula we obtain that this process must have independent Gaussian increments, i.e., the properties (ii) and (iii) from the definition of the Brownian motion will be satisfied and a stochastic process with such finite dimensional distributions will be the Brownian motion. Clearly, the consistency condition (7.1.10) holds true for $Q_{t_1, ..., t_n}$ while we also obtain (7.1.9) just by setting $Q_{t_{i_1}, ..., t_{i_n}}(B_{i_1} \times \cdots \times B_{i_n}) = Q_{t_1, ..., t_n}(B_1 \times \cdots \times B_n)$ for any $B_1, ..., B_n \in \mathcal{B}(\mathbb{R})$ and a permutation $\left( \begin{smallmatrix} 1 & 2 & \cdots & k \\ i_1 & i_2 & \cdots & i_k \end{smallmatrix} \right)$. By Theorem 7.1.2 these finite dimensional distributions can be extended to a probability measure on $(\mathbb{R}^{[0,\infty)}, \mathcal{B}(\mathbb{R}^{[0,\infty)}))$ so that the corresponding stochastic process defined by $W(t, \omega) = \omega(t)$, $t \geq 0$, $\omega \in \mathbb{R}^{[0,\infty)}$ will satisfy the conditions (i)–(iii) of the one-dimensional Brownian motion. As before, we obtain the $d$-dimensional

Brownian motion by taking $d$ independent copies of the one-dimensional Brownian motion.

**Remark 7.1.2.** Even existence of infinite sequences of independent random variables requires arguments similar to Theorem 7.1.2 since this amounts to an extension of finite product probability measures to a probability measure on the infinite product space. Recall, that our first construction of the Brownian motion employed a sequence of independent normal random variables, and so, in fact, it relied on such existence argument.

Theorem 7.1.2 yields the Brownian motion with paths which are not necessarily continuous. In fact, it is easy to see that $\mathcal{B}(\mathbb{R}^{[0,\infty)})$ is the $\sigma$-algebra of sets $\Gamma$ having the form $\Gamma = \{x \in \mathbb{R}^{[0,\infty)} : (x(t_1), x(t_2), ...) \in B\}$ for some countable set $t_1, t_2, ... \in [0, \infty)$ of distinct numbers and a set $B \in \mathcal{B}(\mathbb{R} \times \mathbb{R} \times \cdots)$ (see, for instance, Section II.2.5 in [67]). But, it is not possible to decide whether a function is continuous or not by knowing its values on only a countable set. Hence, the space of continuous functions $\mathcal{C}(\mathbb{R}^{[0,\infty)})$ on the half line does not belong to $\mathcal{B}(\mathbb{R}^{[0,\infty)})$, i.e., it is not measurable with respect to this $\sigma$-algebra. Still, observe that the definition of the Brownian motion imposes conditions only on finite dimensional distributions, and so any modification (as defined in Lecture 11) of the Brownian motion will also be a Brownian motion. The following theorem due to Kolmogorov also ensures existence of Brownian motions not only with continuous but also with Hölder continuous paths.

**Theorem 7.1.3.** *Let $\xi_t = \xi_t(\omega)$, $t \in [0, S]$, $S > 0$ be a stochastic process defined on a probability space $(\Omega, \mathcal{F}, P)$ and satisfying*

$$E|\xi_t - \xi_s|^\alpha \leq C|t - s|^{1+\beta}, \ 0 \leq s, t \leq S \qquad (7.1.12)$$

*for some constants $\alpha, \beta, C > 0$. Then there exists a modification $\tilde{\xi}_t$ of $\xi_t$ which is Hölder continuous with an exponent $\gamma$ when $\gamma \in (0, \beta/\alpha)$ in the sense that*

$$P\left\{ \sup_{0<t-s<h,\, s,t\in[0,S]} \frac{|\tilde{\xi}_t - \tilde{\xi}_s|}{|t-s|^\gamma} \leq \delta \right\} = 1 \qquad (7.1.13)$$

*for some constant $\delta > 0$ and a random variable $h > 0$.*

**Proof.** Without loss of generality it suffices to consider the case $S = 1$. By (7.1.12) for any $\varepsilon > 0$,

$$P\{|\xi_t - \xi_s| \geq \varepsilon\} \leq \frac{E|\xi_t - \xi_s|^\alpha}{\varepsilon^\alpha} \leq C\varepsilon^{-\alpha}|t - s|^{1+\beta},$$

and so $\xi_s \to \xi_t$ in probability as $s \to t$. Take $t = \frac{k}{2^n}$, $s = \frac{k-1}{2^n}$, $\varepsilon = 2^{-\gamma n}$ with $0 < \gamma < \beta/\alpha$. Then

$$P\{|\xi_{k/2^n} - \xi_{(k-1)/2^n}| \geq 2^{-\gamma n}\} \leq C2^{-n(1+\beta-\alpha\gamma)},$$

and so

$$P\{\max_{1\leq k\leq 2^n} |\xi_{k/2^n} - \xi_{(k-1)/2^n}| \geq 2^{-\gamma n}\}$$
$$\leq P\{\cup_{k=1}^{2^n}\{|\xi_{k/2^n} - \xi_{(k-1)/2^n}| \geq 2^{-\gamma n}\}\}$$
$$\leq \sum_{k=1}^{2^n} P\{|\xi_{k/2^n} - \xi_{(k-1)/2^n}| \geq 2^{-\gamma n}\} \leq C2^{-n(\beta-\alpha\gamma)}.$$

By the Borel–Cantelli lemma there exists $\Omega^* \in \mathcal{F}$ with $P(\Omega^*) = 1$ and a random variable $n^*$ such that $n^*(\omega) < \infty$ for each $\omega \in \Omega^*$ and for any $n \geq n^*(\omega)$,

$$\max_{1\leq k\leq 2^n} |\xi_{k/2^n}(\omega) - \xi_{(k-1)/2^n}(\omega)| < 2^{-\gamma n}. \tag{7.1.14}$$

Set $D_n = \{k/2^n, k = 0, 1, ..., 2^n\}$ and let $D = \cup_{n=1}^{\infty} D_n$ be the set of all dyadic numbers on $[0,1]$. We fix $\omega \in \Omega^*$ and $n \geq n^*(\omega)$ and show by induction that for any $m > n$,

$$|\xi_t(\omega) - \xi_s(\omega)| \leq 2 \sum_{j=n+1}^{m} 2^{-\gamma j} \tag{7.1.15}$$

whenever $s,t \in D_m$ and $0 < t - s < 2^{-n}$. If $m = n+1$ then $t = k/2^m$, $s = (k-1)/2^m$, and so in this case (7.1.15) follows from (7.1.14). Now, suppose that (7.1.15) holds true for all $m = n+1, ..., M-1$. Take $s < t$, $s,t \in D_M$ and set $t' = \max\{u \in D_{M-1} : u \leq t\}$ and $s' = \min\{u \in D_{M-1} : u \geq s\}$. Then $s \leq s' \leq t' \leq t$ and $s' - s \leq 2^{-M}$, $t - t' \leq 2^{-M}$. It follows by (7.1.14) that

$$|\xi_{s'}(\omega) - \xi_s(\omega)| < 2^{-\gamma M}, \ |\xi_t(\omega) - \xi_{t'}(\omega)| < 2^{-\gamma M}$$

and by (7.1.15) with $m = M-1$ (induction hypothesis),

$$|\xi_{t'}(\omega) - \xi_{s'}(\omega)| < 2 \sum_{j=n+1}^{M-1} 2^{-\gamma j}.$$

Thus, (7.1.15) holds true also for $m = M$ completing the induction step.

Next, we show that $\{\xi_t(\omega), t \in D\}$ is uniformly continuous in $t$ for each $\omega \in \Omega^*$. Indeed, set $h(\omega) = 2^{-n^*(\omega)}$ and for any $s,t \in D$ with $0 < t - s < h(\omega)$ choose $n \geq n^*(\omega)$ so that $2^{-(n+1)} \leq t - s < 2^{-n}$. Then, it follows from (7.1.15) that

$$|\xi_t(\omega) - \xi_s(\omega)| \leq 2 \sum_{j=n+1}^{\infty} 2^{-\gamma j} \leq \delta|t - s|^{\gamma}$$

where $\delta = 2/(1 - 2^{-\gamma})$. This yields Hölder continuity of $\xi$ on $D$. Define $\tilde{\xi}_t$ by

$$\tilde{\xi}_t(\omega) = \begin{cases} \xi_t(\omega) & \text{if } \omega \in \Omega^* \text{ and } t \in D, \\ 0 & \text{if } \omega \notin \Omega^* \text{ and } 0 \le t \le 1, \\ \lim_{s \to t, s \in D} \xi_s(\omega) & \text{if } \omega \in \Omega^* \text{ and } t \in [0,1] \setminus D \end{cases}$$

where the last limit exists by the Cauchy criterion since $\xi$ is uniformly continuous on $D$. We obtain that $\tilde{\xi}$ is Hölder continuous as required and $\tilde{\xi}_t = \xi_t$ a.s. when $t \in D$. If $t \notin D$ then $\xi_s \to \xi_t$ as $s \to t$ in probability and $\xi_s \to \tilde{\xi}_t$ a.s. when $s \to t$, $s \in D$, and so in this case $\tilde{\xi}_t = \xi_t$ a.s. as well. Thus, $\tilde{\xi}$ is the required modification of $\xi$, completing the proof of the theorem. $\qquad\square$

In order to apply this result to the Brownian motion we compute integrating by parts successively for any $t > s \ge 0$ and an integer $k > 0$,

$$E(W(t) - W(s))^{2k} = (2\pi(t-s))^{-1/2} \int_{-\infty}^{\infty} x^{2k} \exp(-\tfrac{x^2}{2(t-s)})dx$$

$$= (t-s)(2k-1)(2\pi(t-s))^{-1/2} \int_{-\infty}^{\infty} x^{2(k-1)} \exp(-\tfrac{x^2}{2(t-s)})dx$$

$$= (t-s)(2k-1)E(W(t) - W(s))^{2(k-1)} = \cdots = (t-s)^k \prod_{j=1}^{k}(2j-1).$$

Thus there exist Hölder continuous modifications of $W$ with any exponent $\gamma < \frac{k-1}{2k} = \frac{1}{2} - \frac{1}{2k}$, and since $k$ is arbitrary there exist such modifications for any exponent less than $1/2$. These modifications are Brownian motions by themselves, so we obtain Hölder continuous Brownian motions for any exponent less than $1/2$.

We will conclude this section showing that with probability one the Brownian motion cannot have Hölder continuous paths with an exponent bigger than $1/2$ (leaving the case of the exponent equal $1/2$ for exercises) and, in particular, with probability one the paths of the Brownian motion are not differentiable.

**Theorem 7.1.4.** *With probability one paths $W(t,\omega)$, $t \in [0,\infty)$ of the Brownian motion are not Hölder continuous at any $t \ge 0$ with an exponent bigger than $1/2$, i.e., if $\gamma > 1/2$ then for almost all $\omega \in \Omega$ there exist no $t = t(\omega) \ge 0$, $\delta = \delta(\omega) > 0$ and $C = C(\omega) > 0$ such that $|W(t,\omega) - W(s,\omega)| \le C|t-s|^{\gamma}$ for all $s \ge 0$ satisfying $|t-s| \le \delta$.*

**Proof.** For any $k, l \ge 1$ set

$$X_{nk} = X_{nk}^{(l)} = \max_{0 \le j \le l-1} \left\{ |W(\tfrac{k+j}{2^n}) - W(\tfrac{k+j-1}{2^n})| \right\}.$$

From independency of increments and the Brownian scaling it follows that

$$P\{X_{nk}^{(l)} \leq \varepsilon\} = \left(P\{W(2^{-n}) \leq \varepsilon\}\right)^l$$

$$= \left(P\{|W(1)| \leq \varepsilon 2^{\frac{n}{2}}\}\right)^l = \left((2\pi)^{-1/2} \int_{-\varepsilon 2^{n/2}}^{\varepsilon 2^{n/2}} e^{-\frac{z^2}{2}} dz\right)^l \leq (\varepsilon 2^{\frac{n}{2}})^l.$$

Set $Y_n = \min_{1 \leq k \leq n 2^n} X_{nk}$. Then

$$P\{Y_n \leq \varepsilon\} \leq n 2^n (\varepsilon 2^{\frac{n}{2}})^l.$$

Let $\gamma > \frac{1}{2}$ and define

$$A_\gamma = \{\omega \in \Omega : \text{ there exist } t = t(\omega) \geq 0, \, \delta = \delta(\omega) > 0 \text{ and}$$
$$C = C(\omega) > 0 \text{ such that } |W(t, \omega) - W(s, \omega)| \leq C|t - s|^\gamma$$
$$\text{whenever } s \geq 0 \text{ and } |t - s| \leq \delta\}.$$

Suppose that $\omega \in A_\gamma$ and for some $t = t(\omega) \geq 0, \delta = \delta(\omega) > 0$ and $C = C(\omega) > 0$,

$$|W(t, \omega) - W(s, \omega)| \leq C|t - s|^\gamma \text{ whenever } s \geq 0 \text{ and } |t - s| \leq \delta.$$

Choose $n > \max(t, \log_2(l\delta^{-1}), 2Cl^\gamma)$ and for such an $n$ pick up $k \leq n 2^n$ such that $(k - 1)2^{-n} \leq t < k 2^{-n}$. Then $|t - j 2^{-n}| \leq l 2^{-n} < \delta$ for all $j = k - 1, k, ..., k + l - 1$. Hence,

$$X_{nk} \leq 2C(l 2^{-n})^\gamma = 2Cl^\gamma 2^{-n\gamma} \leq n 2^{-n\gamma} \text{ and } k \leq 2^n t \leq n 2^n,$$

and so

$$Y_n \leq n 2^{-n\gamma}.$$

It follows that if $\omega \in A_\gamma$ then $\omega \in A^{(n)} = \{Y_n \leq n 2^{-n\gamma}\}$ for all $n$ large enough, and so

$$A_\gamma \subset \cup_{k=1}^\infty \cap_{n=k}^\infty A^{(n)} = \liminf_{n \to \infty} A^{(n)}.$$

On the other hand,

$$P(A^{(n)}) \leq n 2^n (2^{\frac{n}{2}} n 2^{-n\gamma})^l = n^{l+1} 2^n 2^{nl(\frac{1}{2} - \gamma)}.$$

Taking into account that $\gamma > 1/2$ choose $l > (\gamma - \frac{1}{2})^{-1}$. Then $P(A^{(n)}) \to 0$ as $n \to \infty$, and so $P(A_\gamma) = 0$ (as we always assume that the probability space is complete, i.e., subsets of sets with zero probability are measurable, and so they also have zero probability). □

## 7.2 Lecture 14: Stochastic integrals

In this lecture we will define and study first the stochastic integrals with respect to the Brownian motion and then with respect to martingales.

### 7.2.1 *Stochastic integrals with respect to the Brownian motion*

Since $W(t)$ is not differentiable in $t$ the straightforward definition of the stochastic integral $\int_0^t f(s,\omega)dW(s,\omega)$ by $\int_0^t f(s,\omega)\frac{dW(s,\omega)}{ds}ds$ does not work. As usual, we will try to approximate the integral by integral sums $\sum_{i=0}^k f(u_i)(W(s_{i+1}) - W(s_i))$ where $0 = s_0 < s_1 < ... < s_{k+1} = t$ and $s_i \leq u_i \leq s_{i+1}$. Unlike the classical case the choice of the intermediate point $u_i$ turns out to be crucial. In this lecture we will discuss first the Itô stochastic integrals which together with other parts of the Itô calculus played a very important role in the development of probability in the second half of the 20th century.

Fix a complete probability space $(\Omega, \mathcal{F}, P)$ and let $W(t)$, $t \geq 0$ be the one-dimensional Brownian motion with continuous paths. Denote by $\mathcal{B}^+$ the $\sigma$-field of Borel subsets of $[0, \infty)$ and let $\mathcal{A}_t$, $t \geq 0$ be a filtration of complete $\sigma$-fields such that $\mathcal{A}_t \supset \mathcal{F}_t = \sigma\{W(u), u \leq t\}$ for all $t \geq 0$ and $\mathcal{A}_s$ is independent of the $\sigma$-field $\mathcal{F}_s^+ = \sigma\{W(t+s) - W(s), t \geq 0\}$ for each $s \geq 0$. A random function (process) $f(t) = f(t, \omega)$ is called non-anticipating or adapted to the filtration $\mathcal{A}_t$, $t \geq 0$ if $f$ as a function on $[0, \infty) \times \Omega$ is $\mathcal{B}^+ \times \mathcal{F}$-measurable (which is called: jointly measurable) and $f(t)$ as a function on $\Omega$ is $\mathcal{A}_t$ measurable for each $t \geq 0$. In this chapter we will usually deal with a finite time interval $[0, T]$, $0 < T < \infty$ but most of the results and constructions can be easily extended under appropriate conditions to the infinite time interval $[0, \infty)$ just by letting $T \to \infty$. We will fix $T > 0$ and define $\int_0^t f(s)dW(s)$, simultaneously for all $t \in [0, T]$ and almost all Brownian paths, for all non-anticipating functions $f$ satisfying the condition

$$E \int_0^T f^2(s)ds < \infty. \tag{7.2.1}$$

In a way reminiscent of constructions of deterministic integrals of Riemann and Lebesgue we will define, first, stochastic integrals for "simple" functions and then extend them to general ones. Later we will extend stochastic integrals to integrands $f$ under a more general than (7.2.1) condition.

A non-anticipating function $f$ on $[0, \infty]$ is called simple if there exist numbers $0 = t_0 < t_1 < ... < t_m < T$ such that $f(t, \omega) = f(t_{k-1}, \omega)$ whenever $t_{k-1} \leq t < t_k$, $k = 1, 2, ..., m+1$ where $t_{m+1} = T$. Given such $f$, define the stochastic integral of $f$ for each $t \in [0, T]$ by

$$\int_0^t f(u)dW(u) = \sum_{1 \leq k \leq l} f(t_{k-1})(W(t_k) - W(t_{k-1})) + f(t_l)(W(t) - W(t_l))$$

(7.2.2)

provided $t_l \leq t < t_{l+1}$. Observe that this definition does not depend on the choice of a partition $0 = t_0 < t_1 < ... < t_m < T$ in the representation of a simple function $f$ but only on the function itself. Furthermore, the stochastic integral is linear

$$\int_0^t (a_1 f_1(u) + a_2 f_2(u))dW(u) = a_1 \int_0^t f_1(u)dW(u) + a_2 \int_0^t f_2(u)dW(u)$$

for any simple functions $f_1$ and $f_2$ and constants $a_1, a_2$. Since we chose a continuous in time version of the Brownian motion then it is clear from the formula (7.2.2) that the integral $\int_0^t f(u)dW(u)$ is continuous in $t$ too. It turns out that the stochastic integral is a martingale with respect to the filtration $\mathcal{A}_u$, $u \geq 0$ which is crucial for its numerous applications. Indeed, $\int_0^t f(u)dW(u)$ is clearly measurable with respect to $\mathcal{A}_t$ and if $s < t$ then $E(\int_0^t f(u)dW(u)|\mathcal{A}_s) = E(\int_0^s f(u)dW(u))$ which follows by applying to (7.2.2) successively the formula

$$E\big(f(u)(W(v) - W(u))|\mathcal{A}_u\big) = f(u)E(W(v) - W(u)) = 0$$

which holds true for $T \geq v \geq u \geq 0$ since $f(u)$ is $\mathcal{A}_u$-measurable and $W(v) - W(u)$ is independent of $\mathcal{A}_u$. In particular, $E\int_0^t f(u)dW(u) = 0$. Another important property of the stochastic integral is, so called, Itô's isometry (whose meaning will be explained later on),

$$E(\int_0^t f(u)dW(u))^2 = E\int_0^t f^2(u)du$$

and, more generally, for $0 \leq s \leq t \leq T$,

$$E\big((\int_s^t f(u)dW(u))^2|\mathcal{A}_s\big) = \int_s^t E(f^2(u)|\mathcal{A}_s)du = E\big(\int_s^t f^2(u)du|\mathcal{A}_s\big).$$

The latter equality is obtained from (7.2.2) by using that for $v \geq u \geq w$,

$$E\big(f^2(u)(W(v) - W(u))^2|\mathcal{F}_w\big)$$
$$= E\big(f^2(u)E((W(v) - W(u))^2|\mathcal{A}_u)|\mathcal{A}_w\big) = E(f^2(u)|\mathcal{A}_w)(v - u)$$

and that for $t \geq s \geq v \geq u \geq w$,

$$E\big(f(u)(W(v) - W(u))f(s)(W(t) - W(s))|\mathcal{A}_w\big)$$
$$= E\big(f(u)(W(v) - W(u))f(s)E((W(t) - W(s))|\mathcal{A}_s)|\mathcal{A}_w\big) = 0.$$

In order to extend the definition of stochastic integrals to all non-anticipating $f$ satisfying (7.2.1) we will need the following crucial result.

**Lemma 7.2.1.** *Suppose that a non-anticipating function $f = f(t, \omega)$, $t \in [0, T]$, $\omega \in \Omega$ satisfies (7.2.1). Then there exists a sequence of (non-anticipating) simple functions $f_n = f_n(t, \omega)$, $t \in [0, T]$, $\omega \in \Omega$, $n = 1, 2, ...$ such that*

$$\lim_{n \to \infty} E \int_0^T (f(s) - f_n(s))^2 ds = 0. \tag{7.2.3}$$

**Proof.** We will proceed by slightly modified arguments from Section 4.2 of [53]. First, observe that by the dominated convergence theorem

$$E \int_0^T (f(s) - f(s) \mathbb{I}_{|f(s)| \le N})^2 ds = \int_0^T ds \int_{|f(s)| > N} f^2(s) dP \to 0 \text{ as } N \to \infty$$

which implies that it suffices to prove the lemma for a bounded $f$, i.e., we assume from now on that $|f(t, \omega)| \le C < \infty$ for some constant $C > 0$ and all $t \in [0, T]$, $\omega \in \Omega$.

For convenience we extend $f$ to all real $t$ setting $f(t, \omega) \equiv 0$ if $t \notin [0, T]$. Let

$$\psi_m(t) = \frac{j}{2^m} \text{ when } \frac{j}{2^m} < t \le \frac{j+1}{2^m}, \quad j = 0, \pm 1, \pm 2, ....$$

Then $f_{m,s}(t, \omega) = f(\psi_m(t-s)+s, \omega)$ is a (non-anticipating) simple function of $(t, \omega)$ for each fixed $s$ since $f_{m,s}(t, \omega) = f(\frac{j}{2^m} + s, \omega)$ when $\frac{j}{2^m} + s < t \le \frac{j+1}{2^m} + s$. Hence, the lemma will follow if we show the existence of $\bar{s}$ such that $f_n = f_{m_n, \bar{s}}$ will satisfy (7.2.3) for some subsequence $m_n \to \infty$ as $n \to \infty$.

We will prove existence of an appropriate $\bar{s}$ relying on the fact that

$$\lim_{h \to 0} \int_0^\infty (f(s+h, \omega) - f(s, \omega))^2 ds = 0 \quad \text{a.s.} \tag{7.2.4}$$

which will be established later on. Clearly, (7.2.4) implies that for any $t \ge -T$,

$$\lim_{h \to 0} \int_0^\infty (f(s+t+h, \omega) - f(s+t, \omega))^2 ds$$
$$= \lim_{h \to 0} \int_t^\infty (f(s+h, \omega) - f(s, \omega))^2 ds$$
$$\le \lim_{h \to 0} \int_0^\infty (f(s+h, \omega) - f(s, \omega))^2 ds + \lim_{h \to 0} \int_0^{|h|} f^2(s, \omega) ds = 0$$

where we use that $f$ is bounded and $f(t, \omega) = 0$ for $t \notin [0, T]$. Writing $\psi_m(t) = t + (\psi_m(t) - t)$ we derive from here that for any $t \ge -T$,

$$\lim_{m \to \infty} \int_0^\infty (f(s + \psi_m(t), \omega) - f(s+t, \omega))^2 ds = 0.$$

Again, since $f$ is bounded and it vanishes outside $[0, T]$, this together with the dominated convergence theorem yields that

$$\lim_{m \to \infty} \int_0^\infty \int_0^\infty (f(s + \psi_m(t), \omega) - f(s + t, \omega))^2 ds dt$$
$$= \lim_{m \to \infty} \int_{-T}^T \int_0^T (f(s + \psi_m(t), \omega) - f(s + t, \omega))^2 ds dt = 0.$$

Employing Chebyshev's inequality and the Borel–Cantelli lemma it follows that there exists a subsequence $m_n \to \infty$ as $n \to \infty$ such that for almost all $(s, t, \omega) \in [0, T] \times [0, T] \times \Omega$ with respect to the measure $\ell_{[0,T]} \times \ell_{[-T,T]} \times P$ (where $\ell_{[a,b]}$ is the normalized Lebesgue measure on $[a, b]$),

$$f(s + \psi_{m_n}(t), \omega) - f(s + t, \omega) \to 0 \quad \text{as} \quad n \to \infty.$$

Set $u = s + t$ then for almost all $(s, u, \omega) \in [0, T] \times [0, T] \times \Omega$ with respect to the measure $\ell_{[0,T]} \times \ell_{[0,T]} \times P$,

$$f(s + \psi_{m_n}(u - s), \omega) - f(u, \omega) \to 0 \quad \text{as} \quad n \to \infty. \tag{7.2.5}$$

It follows that (7.2.5) holds true for some fixed $s = \bar{s}$ (in fact, for $\ell_{[0,T]}$-almost all $s$) and $\ell_{[0,T]} \times P$-almost all $(u, \omega)$. Hence, by the dominated convergence theorem

$$E \int_0^\infty (f(\bar{s} + \psi_{m_n}(u - \bar{s}), \omega) - f(u, \omega))^2 du \to 0 \quad \text{as} \quad n \to \infty$$

and (7.2.3) follows with $f_n = f_{m_n, \bar{s}}$.

It remains to prove (7.2.4) which will follow if for each $\varepsilon > 0$ we construct a measurable in $(t, \omega)$ and a.s. continuous in $t$ function $f_\varepsilon = f_\varepsilon(t, \omega)$ such that $f_\varepsilon(t, \omega) = 0$ when $t \notin [-S, S]$ for some $0 < S < \infty$, $|f_\varepsilon(t, \omega)| \leq D$ for some $0 < D < \infty$ and

$$E \int_0^\infty (f_\varepsilon(s, \omega) - f(s, \omega))^2 ds \leq \varepsilon^2. \tag{7.2.6}$$

Indeed, by the Minkowsky inequality (which is here just the triangle inequality for the $L^2$-norm) together with (7.2.6),

$$\limsup_{h \to 0} \left( E \int_0^\infty (f(s + h, \omega) - f(s, \omega))^2 ds \right)^{1/2}$$
$$\leq \limsup_{h \to 0} \left( E \int_0^\infty (f_\varepsilon(s + h, \omega) - f_\varepsilon(s, \omega))^2 ds \right)^{1/2} + 2\varepsilon.$$

Since $f_\varepsilon(t, \omega)$ is continuous in $t$, $f_\varepsilon(s + h, \omega) - f_\varepsilon(s, \omega) \to 0$ as $h \to 0$, and so by the dominated convergence theorem $\limsup_{h \to 0}$ in the right hand side above is 0 which implies (7.2.4) as $\varepsilon > 0$ is arbitrary.

Finally, in order to construct the functions $f_\varepsilon$ needed above we set

$$\varphi_m(t, \omega) = m \int_{(t - \frac{1}{m})^+}^t f(s, \omega) ds.$$

Since the Lebesgue integral $F(t,\omega) = \int_0^t f(s,\omega)ds$ is continuous in $t$, it follows that $\varphi_m(t,\omega) = m(F(t,\omega) - F((t-\frac{1}{m})^+,\omega))$ is continuous in $t$ as well. By the fundamental theorem of calculus for the Lebesgue integral for $\ell_{[0,T]}$-almost all $t \in [0,T]$,

$$f(t,\omega) = F'(t,\omega) = \lim_{m\to\infty} m(F(t,\omega) - F((t-\frac{1}{m})^+,\omega)) = \lim_{m\to\infty} \varphi_m(t,\omega).$$

Hence, $\lim_{m\to\infty} \varphi_m(t,\omega) = f(t,\omega)$ for $\ell_{[0,T]} \times P$-almost all $(t,\omega) \in [0,T] \times \Omega$, and so by the dominated convergence theorem

$$E \int_0^T (\varphi_m(t,\omega) - f(t,\omega))^2 dt \to 0 \quad \text{as} \quad m \to \infty.$$

It follows that for any $\varepsilon > 0$ we can choose $m_\varepsilon$ such that $f_\varepsilon = \varphi_{m_\varepsilon}$ satisfies (7.2.6), completing the proof of the lemma. $\qquad\square$

Now, relying on the above lemma we will complete the construction of the stochastic integral for any non-anticipating function $f = f(t,\omega)$, $t \in [0,T]$, $\omega \in \Omega$ satisfying (7.2.1). Relying on Lemma 7.2.1 we can choose a sequence of simple functions $f_n$, $n = 1, 2, \ldots$ such that

$$E \int_0^T (f_n(t) - f(t))^2 dt \leq 2^{-(3n+2)},$$

and so

$$E \int_0^T (f_{n+1}(t) - f_n(t))^2 dt \leq 2^{-3n}.$$

By the martingale inequality from Lecture 11,

$$P\{\sup_{0 \leq t \leq T} |\int_0^t (f_{n+1}(s) - f_n(s))dW(s)| > 2^{-n}\}$$
$$\leq E \int_0^T (f_{n+1}(t) - f_n(t))^2 dt \leq 2^{-n}.$$

This together with the Borel–Cantelli lemma yields that there exists $N = N(\omega) < \infty$ a.s. such that

$$\sup_{0 \leq t \leq T} |\int_0^t (f_{n+1}(s) - f_n(s))dW(s)| \leq 2^{-n} \quad \text{for all} \quad n \geq N.$$

It follows that the series $f_1 + \sum_{n=1}^{\infty} \int_0^t (f_{n+1}(s) - f_n(s))dW(s)$ converges with probability one uniformly in $t \in [0,T]$, and so we can define the stochastic integral of $f$ by

$$\int_0^t f(s)dW(s) = \lim_{n\to\infty} \int_0^t f_n(s)dW(s), \quad t \in [0,T]. \tag{7.2.7}$$

In view of the uniform convergence, the integral is continuous in $t$ and it is defined simultaneously for all $t \in [0,T]$ and almost all Brownian paths. It follows also from the martingale inequality employed above that the limit in (7.2.7) will remain the same no matter which sequence of simple functions $f^{(n)}$ satisfying $\lim_{n\to\infty} E \int_0^T (f(t) - f^{(n)}(t))^2 dt = 0$ is chosen. Since

$$E\left( \int_0^t f_n(s)dW(s) - \int_0^t f_m(s)dW(s)\right)^2$$
$$= E \int_0^t \left( f_n(s) - f_m(s)\right)^2 ds \leq 2^{-\min(m,n)},$$

the sequence $\int_0^t f_n(s)dW(s)$, $n = 1,2,...$ is fundamental in the complete space $L^2(\Omega, P)$, and so the convergence in (7.2.7) is also in $L^2(\Omega, P)$. This completes the construction of the stochastic integral.

**Remark 7.2.1.** If $T = \infty$ and $E \int_0^\infty f^2(t)dt < \infty$ then we can choose simple functions $f_n = f_n(t)$ which are zero for $t > t_n$ for some $t_n \to \infty$ and $E \int_0^\infty (f(t) - f_n(t))^2 dt \leq 2^{-3n}$. Similarly to the above we conclude that $\sup_{t\geq 0} | \int_0^t (f_{n+1}(s) - f_n(s))dW(s)|$ tends to 0 geometrically fast, and so $\int_0^t f_n(s)dW(s)$ converges as $n \to \infty$ uniformly in $t \geq 0$ and the limit denoted $\int_0^t f(s)dW(s)$ is called the stochastic integral of $f$ and it is defined and continuous now for all $t \geq 0$.

**Theorem 7.2.1.** *(Some properties of stochastic integrals). Let $f$ be a non-anticipating function satisfying (7.2.1).*

*(i)(linearity) If $g$ is another non-anticipating function satisfying (7.2.1) and $a,b$ are constants then $\int_0^t(af(s) + bg(s))dW(s) = a \int_0^t f(s)dW(s) + b \int_0^t g(s)dW(s)$;*

*(ii)(martingale) $\int_0^t f(s)dW(s)$, $t \in [0,T]$ is a continuous martingale with respect to the filtration $\mathcal{A}_t$, $t \in [0,T]$. In particular, $\int_0^t f(s)dW(s)$ is $\mathcal{A}_t$-measurable and $E \int_0^t f(s)dW(s) = \int_0^0 f(s)dW(s) = 0$ and, moreover, for $0 \leq s \leq t \leq T$,*

$$E\left( \int_s^t f(u)dW(u)|\mathcal{A}_s\right) = 0;$$

*(iii)(Itô's isometry) $E(\int_0^T f(s)dW(s))^2 = E \int_0^T f^2(s)ds$ and, more generally, for $0 \leq s \leq t \leq T$,*

$$E\left(\left( \int_s^t f(u)dW(u)\right)^2|\mathcal{A}_s\right) = \int_s^t E(f^2(u)|\mathcal{A}_s)du = E\left( \int_s^t f^2(u)du|\mathcal{A}_s\right);$$

*(iv) If $\tau \leq T$ is a stopping time then for each $t \in [0,T]$,*

$$\int_0^{t\wedge\tau} f(s)dW(s) = \int_0^t f(s)\mathbb{I}_{\tau \geq s}dW(s) \quad a.s.$$

**Proof.** (i)–(iii) follow approximating $f$ by simple functions and using the corresponding properties of stochastic integrals of simple functions obtained above. The assertion (iv) is proved first for simple $f$ and stopping times $\tau_n \downarrow \tau$ taking on finitely many values and then approximating $f$ by simple functions (see Chapter II in [28]). $\qquad\square$

Observe that Itô's isometry exhibits an $L^2$ isometry between the space of stochastic integrals as random variables and the space of adapted square integrable functions on $[0, T] \times \Omega$.

**Example** Next, we compute directly a particular stochastic integral showing that

$$\int_0^t W(s)dW(s) = \frac{1}{2}(W^2(t) - t)$$

which exhibits the difference between Itô's stochastic integrals or, more generally, between Itô's calculus and the standard calculus.

Define a sequence of simple non-anticipating functions $f_n(s) = W(2^{-n}[2^n s])$, $n = 1, 2, ....$ Then by Itô's isometry

$$E\big(\int_0^t (W(s) - f_n(s))dW(s)\big)^2 = \int_0^t E(W(s) - f_n(s))^2 ds$$
$$\leq \sum_{k=0}^{[2^n t]} \int_{k2^n}^{(k+1)2^{-n}} E(W(s) - W(k2^{-n}))^2 ds \leq 2^{-n}(t+1).$$

Applying the Chebyshev inequality and the Borel–Cantelli lemma we conclude from here that $\int_0^t (W(s) - f_n(s))^2 dW(s)$ tends to 0 a.s. as $n \to \infty$ for every $t$. Thus, it suffices to show that with probability one,

$$\lim_{n \to \infty} \int_0^t f_n(s)dW(s) = \frac{1}{2}(W^2(t) - t).$$

Set $\Delta_{kn} = W(k2^{-n}) - W((k-1)2^{-n})$ and $l = l_n(t) = [2nt]$. Then, as $n \to \infty$,

$$2\int_0^t f_n(s)dW(s) = 2(\sum_{k\leq l} W((k-1)2^{-n})\Delta_{kn}) + 2W(l2^{-n})(W(t)$$
$$-W(l2^{-n})) = \sum_{k\leq l}(W^2(k2^{-n}) - W^2((k-1)2^{-n})) - \sum_{k\leq l}\Delta_{kn}^2 + o(1)$$
$$= W^2(t) - \sum_{k\leq l}\Delta_{kn}^2 + o(1).$$

It remains to show that with probability one $\sum_{k\leq l}\Delta_{kn}^2 \to t$ as $n \to \infty$. In order to prove this consider

$$\zeta_n(t) = \sum_{k\leq l_n(t)} \Delta_{kn}^2 + (W(t) - W(l_n(t)2^{-n}))^2 - t.$$

It is immediate to check that $\zeta_n(t)$ is a continuous martingale with respect to the Brownian filtration $\mathcal{F}_t$, $t \geq 0$, and so by the martingale inequality from Lecture 11,

$$P\{ \sup_{0 \leq s \leq t} |\zeta_n(s)| \geq 2^{-n/2}n\} \leq 2^n n^{-2} E\zeta_n^2(t)$$

and

$$E\zeta_n^2(t) = \sum_{k \leq l} E(\Delta_{kn}^2 - 2^{-n})^2 + E\big((W(t) - W(l2^{-n}))^2$$
$$-(t - l2^{-n})\big)^2 \leq (t+1)2^{-n}.$$

This together with the previous inequality and the Borel–Cantelli lemma yields that $\zeta_n \to 0$ a.s. with probability one implying the required convergence and completing the proof of the formula above.

**Remark 7.2.2.** Consider other possibilities to define the stochastic integral
  a) $\int_0^t W(s)dW(s) = \lim_{n \to \infty} \sum_{k \leq 2^n t} W(k2^{-n})\Delta_{kn}$, with the same $\Delta_{kn}$ as above, and
  b) $\int_0^t W(s)dW(s) = \lim_{n \to \infty} \sum_{k \leq 2^n t} \frac{1}{2}(W(k2^{-n}) + W((k-1)2^{-n}))\Delta_{kn}$.
  In the same way as in the example above we compute easily that $\int_0^t W(s)dW(s) = \frac{1}{2}(W^2(t) + t)$ in the case a) and $\int_0^t W(s)dW(s) = \frac{1}{2}W^2(t)$ in the case b). The definition a) leads to an integral which has no good properties, and so it is not studied. In the case b) we obtain the, so called, Stratonovich integral which is not a martingale as the Itô integral but it has an advantage to lead to formulas having the same form as in the standard calculus which is especially convenient in the study of stochastic calculus on Riemannian manifolds.

### 7.2.2 *Progressive measurability and extension of stochastic integrals*

**Definition 7.2.1.** Denote by $\mathcal{B}_{[0,t]}$ the $\sigma$-field of Borel subsets of $[0,t]$. A random function (process) $f : [0,T] \times \Omega \to \mathbb{R}$ is called progressively measurable on $[0,T]$, $T > 0$ with respect to a filtration $\{\mathcal{F}_t, t \in [0,T]\}$ if for each $t \in [0,T]$ the restriction of $f$ to $[0,t] \times \Omega$, i.e., the map $f : [0,t] \times \Omega \to \mathbb{R}$, is measurable with respect to the $\sigma$-field $\mathcal{B}_{[0,t]} \times \mathcal{F}_t$. If this holds true only for $t = T$ then the function (process) $f$ is called (jointly) measurable.

Let $\mathcal{BF}_T$ be the collection of all sets $\Gamma \subset [0,T] \times \Omega$ such that for each $t \in [0,T]$ the intersection $\Gamma \cap ([0,t] \times \Omega)$ is $\mathcal{B}_{[0,t]} \times \mathcal{F}_t$-measurable. It is immediate to check that $\mathcal{BF}_T$ is a $\sigma$-algebra and it is called progressive.

**Proposition 7.2.1.** *Progressively measurable functions on $[0,T]$ are exactly $\mathcal{BF}_T$-measurable functions.*

**Proof.** It suffices to show that progressively measurable indicator functions are exactly indicator functions of $\mathcal{BF}_T$-measurable sets. Now, if $\chi$ takes on only values 0 and 1 and it is progressively measurable on $[0,T]$ then $\{(s,\omega) : \chi(s,\omega) = 1\} \cap ([0,t] \times \Omega)$ is a $\mathcal{B}_{[0,t]} \times \mathcal{F}_t$-measurable by definition, i.e., $\{(s,\omega) : \chi(s,\omega) = 1\} \in \mathcal{BF}_T$. In the other direction, consider $\mathbb{I}_\Gamma$ with $\Gamma \in \mathcal{BF}_T$. Then $\Gamma = \{(s,\omega) : \mathbb{I}_\Gamma(s,\omega) = 1\}$ and $\Gamma \cap ([0,t] \times \Omega) \in \mathcal{B}_{[0,t]} \times \mathcal{F}_t$ implying that $\mathbb{I}_\Gamma$ is progressively measurable since $\{\mathbb{I}_\Gamma = 0\}$ is the complement set, and so its intersection with $([0,t] \times \Omega)$ also belongs to $\mathcal{B}_{[0,t]} \times \mathcal{F}_t$. $\qquad\square$

**Proposition 7.2.2.** *(i) Any progressively measurable function is non-anticipating (adapted) with respect to the same filtration;*

*(ii) If $f = f(t,\omega)$ is right or left-continuous and non-anticipating then it is progressively measurable;*

*(iii) If $f$ is progressively measurable with respect to the filtration $\{\mathcal{F}_t\}$ and $\tau < \infty$ a.s. is a stopping time then the stopped process $g(t,\omega) = f(\tau(\omega) \wedge t, \omega)$ is progressively measurable and $f(\tau(\omega), \omega)$ is $\mathcal{F}_\tau$-measurable;*

*(iv) Any predictable process is progressively measurable.*

**Proof.** (i) If $f$ is progressively measurable on $[0,T]$ then each map $f : [0,t] \times \Omega \to \mathbb{R}$ is $\mathcal{B}_{[0,t]} \times \mathcal{F}_t$-measurable. Then the map $f(t,\cdot) : \Omega \to \mathbb{R}$ is $\mathcal{F}_t$-measurable. Indeed, for each number $a$ the set $\{\omega \in \Omega : f(t,\omega) \le a\}$ is $\mathcal{F}_t$-measurable as the $t$-section of the set $\{(s,\omega) \in [0,t] \times \Omega : f(s,\omega) \le a\} \in \mathcal{B}_{[0,t]} \times \mathcal{F}_t$.

(ii) Let $f$ be right-continuous in $t$ and adapted. Fix $t \in [0,T]$ and define for each $s \in [0,t]$ and $n = 1, 2, ...,$

$$f_n(s,\omega) = \begin{cases} f(\frac{k+1}{2^n}t, \omega) & \text{if} \quad s \in [\frac{k}{2^n}t, \frac{k+1}{2^n}t), 0 \le k < 2^n \\ f(t,\omega) & \text{if} \quad s = t. \end{cases}$$

Since each $f(\frac{k+1}{2^n}t, \omega)$ is $\mathcal{F}_{\frac{k+1}{2^n}t}$-measurable, we see that for any Borel set $\Gamma \subset \mathbb{R}$,

$$\{(s,\omega) \in [0,t] \times \Omega : f_n(s,\omega) \in \Gamma\}$$
$$= \bigcup_{0 \le k < 2^n} \left([\tfrac{k}{2^n}t, \tfrac{k+1}{2^n}t) \times \{f(\tfrac{k+1}{2^n}t, \omega) \in \Gamma\}\right) \cup (\{t\} \times \{f(t,\omega) \in \Gamma\})$$
$$\in \mathcal{B}_{[0,t]} \times \mathcal{F}_t,$$

and so $f_n$ is $\mathcal{B}_{[0,t]} \times \mathcal{F}_t$-measurable. Since $\lim_{n\to\infty} f_n(s,\omega) = f(s,\omega)$ for any $\omega \in \Omega$ and $s \in [0,t]$ and limits of measurable functions are measurable, we conclude that $f$ restricted to $[0,t] \times \Omega$ is $\mathcal{B}_{[0,t]} \times \mathcal{F}_t$-measurable, as required. For left-continuous $f$ the proof is similar defining

$$f_n(s,\omega) = \begin{cases} f(\frac{k}{2^n}t,\omega) & \text{if } s \in (\frac{k}{2^n}t, \frac{k+1}{2^n}t], \ 0 \le k < 2^n \\ f(0,\omega) & \text{if } s = 0. \end{cases}$$

(iii) First, observe that $f(\tau(\omega) \wedge t, \omega) = f(\varphi(t,\omega))$ where $\varphi(u,\omega) = (\tau(\omega) \wedge u, \omega)$. Since $f$ is progressively measurable, for any Borel set $\Gamma \subset \mathbb{R}$,

$$G = ([0,t] \times \Omega) \cap f^{-1}(\Gamma) \in \mathcal{B}_{[0,t]} \times \mathcal{F}_t.$$

In order to prove that $g(t,\omega) = f(\tau(\omega) \wedge t, \omega)$ is progressively measurable it remains to show that the map $\varphi$ of $([0,t] \times \Omega, \mathcal{B}_{[0,t]} \times \mathcal{F}_t)$ to itself is measurable for each $t$. Let $A = [a,b]$, $b \le t$, $B \in \mathcal{F}_t$ and $G = A \times B$. Then

$$\varphi^{-1}(G) = \{(s,\omega) : \tau(\omega) \wedge s \in A, \ \omega \in B\}$$
$$= [a,b] \times (B \cap \{\omega : \tau(\omega) \ge a\}) \cup (b,t] \times (B \cap \{\omega : \tau(\omega) \in [a,b]\})$$
$$\in \mathcal{B}_{[0,t]} \times \mathcal{F}_t.$$

The same holds true if $A = (a,b]$, $A = [a,b)$ or $A = (a,b)$ and this extends to complements and finite disjoint unions of sets $G = A \times B$ of this form. The property $\varphi^{-1}(G) \in \mathcal{B}_{[0,t]} \times \mathcal{F}_t$ is preserved under monotone limits of sets, and so it holds true for any $G \in \mathcal{B}_{[0,t]} \times \mathcal{F}_t$. In particular, $f(\tau(\omega) \wedge t, \omega)$ is $\mathcal{F}_t$-measurable, and so for any Borel set $\Gamma \subset \mathbb{R}$,

$$\{\omega : f(\tau(\omega),\omega) \in \Gamma\} \cap \{\tau \le t\} = \{\omega : f(\tau(\omega) \wedge t, \omega) \in \Gamma\} \cap \{\tau \le t\} \in \mathcal{F}_t.$$

Thus, $\{\omega : f(\tau(\omega),\omega) \in \Gamma\} \in \mathcal{F}_\tau$, and so $f(\tau(\omega),\omega)$ is $\mathcal{F}_\tau$-measurable.

(iv) The predictable $\sigma$-algebra is generated by left-continuous processes and the latter are progressively measurable by (ii). By Proposition 7.2.1 the progressive $\sigma$-algebra is generated by progressively measurable processes, and so the progressive $\sigma$-algebra contains the predictable $\sigma$-algebra. Since by definition the latter is generated also by predictable processes, it follows that predictable processes are progressively measurable. $\quad\square$

**Corollary 7.2.1.** *If $f = f(t,\omega)$ is progressively measurable on $[0,T]$ with respect to a filtration $\mathcal{F}_t$, $t \in [0,T]$ and $\int_0^T |f(t,\omega)|dt < \infty$ a.s. then $F(t,\omega) = \int_0^t f(s,\omega)ds$ is also progressively measurable on $[0,T]$ with respect to the same filtration.*

**Proof.** Since $f$ restricted to $[0,t] \times \Omega$ is $\mathcal{B}_{[0,t]} \times \mathcal{F}_t$ measurable, it follows by the Fubini theorem that $F(t,\omega) = \int_0^t f(s,\omega)ds$ is $\mathcal{F}_t$ measurable. Since the Lebesgue integral $\int_0^t f(s,\omega)ds$ is continuous in $t$ we obtain from Proposition 7.2.2(iii) that $F(t,\omega)$ is progressively measurable. $\quad\square$

The following example taken from an internet course by G. Zitkovic shows not only that adapted processes may not be progressively measurable but also that their integrals in time may not be adapted unlike the progressively measurable case considered in the corollary above.

**Example** Let $\Omega$ be the set of all lower semicontinuous functions $\omega :$ $[0,1] \to \mathbb{R}$ (i.e., all preimages $\omega^{-1}(c,\infty)$, $c \in \mathbb{R}$ are open in $[0,1]$) satisfying $\int_0^1 |\omega(t)|dt < \infty$. For any $0 < a \leq b < 1$ define the map $I_{a,b} : \Omega \to \mathbb{R}$ by $I_{a,b}(\omega) = \inf_{a \leq t \leq b} \omega(t)$. Observe that $I_{a,b}(\omega) > -\infty$ since $\liminf_{s \to t} \omega(s) \geq \omega(t)$ for any $t \in (0,1)$. Let $\mathcal{F} = \mathcal{F}_1$ be generated by all $I_{a,b}$, $0 < a \leq b < 1$. For any $0 \leq t_1 < t_2 < ... < t_n < 1$ define cylinder sets $D_{t_1,...,t_n}(B_n) = \{\omega : (\omega(t_1),...,\omega(t_n)) \in B_n\}$ where $B_n$ is a Borel set in $\mathbb{R}^n$. We define the filtration $\mathcal{F}_t$, $0 \leq t < 1$ so that $\mathcal{F}_t$ is generated by all cylinder sets $D_{t_1,...,t_n}(B_n)$ with $t_n \leq t$. It is easy to see that for any $\Gamma \in \mathcal{F}_t$ there exists a countable set $t_i \in [0,1)$, $i = 1,2,...$ and a Borel set $B$ in the infinite product $\mathbb{R}^\infty = \mathbb{R} \times \mathbb{R} \times \cdots$ such that $\omega \in \Gamma$ if and only if $(\omega(t_1), \omega(t_2),...) \in B$, i.e., it is possible to decide whether $\omega$ belongs to $\Gamma$ or not knowing only values $\omega(t_1), \omega(t_2), ....$ Observe that the situation here is similar to what we had in the construction of the Brownian motion via the Kolmogorov theorem in Lecture 13. Clearly, the coordinate process $H_t(\omega) = \omega(t)$, $t \in [0,1]$ is adapted to the above filtration. It is also measurable in $(t,\omega)$ with respect to $\mathcal{B}_{[0,1]} \times \mathcal{F}$ since

$$\{(t,\omega) \in [0,1] \times \Omega : H_t(\omega) > c\} = \cup\{[p,q] \times \{\omega : I_{p,q}(\omega) > c\}\},$$

where the union is taken over all pairs of rational numbers $0 < p < q < 1$. In view of the corollary above, in order to prove that $H$ is not progressively measurable it suffices to show that the process $X_t = \int_0^t H_s ds$, $t \in [0,1)$ is not adapted. Let $t \in (0,1)$ and suppose that $X_t$ is $\mathcal{F}_t$-measurable. Then $\Gamma = \{\omega : X_t(\omega) \geq t/2\}$ is $\mathcal{F}_t$-measurable and nonempty since $\omega_0 \equiv 1$ belongs to $\Gamma$. As explained above there exists a countable set $S \subset [0,t]$ such that if $\omega|_S = \omega_0|_S$ then $\omega \in \Gamma$. Let $U \subset [0,1)$ be an open set with the Lebesgue measure less than $t/2$ and such that $U \supset S$ which obviously exists. Set $\omega(t) = \mathbb{I}_U$. Then $\omega \in \Omega$, $\omega|_S = \omega_0|_S$ but $\omega \notin \Gamma$ since $\int_0^t H_s(\omega)ds < t/2$, arriving at a contradiction with the assumption that $X_t$ is $\mathcal{F}_t$-measurable.

In view of Proposition 7.2.1 it is easier to construct stochastic integrals for progressively measurable integrands than for general non-anticipating ones. Namely, fix $T \leq \infty$ and consider the space $L^2([0,T] \times \Omega, \mathcal{BF}, \ell \times P)$ of functions $f$ with the norm

$$\|f\|^2 = E \int_0^T f^2(s,\omega)ds,$$

where $\ell$ denotes the Lebesgue measure on $[0, T]$. Now we can define stochastic integrals of simple functions from $L^2([0, T] \times \Omega, \mathcal{BF}, \ell \times P)$ in the same way as above and relying on the usual approximation of functions in $L^2$-spaces by simple ones together with the Itô isometry we can extend the definition of the stochastic integral to the whole $L^2([0, T] \times \Omega, \mathcal{BF}, \ell \times P)$. Here we remain in the same $L^2$-space and do not have to take care about additional measurability conditions like before when we had to ensure that approximating simple functions are non-anticipating.

It is often useful and turns out to be possible to define and study stochastic integrals under the less restrictive condition

$$P\{\int_0^T f^2(s, \omega)ds < \infty\} = 1 \qquad (7.2.8)$$

in place of (7.2.1). This can be done directly as in [54] or using the above construction under (7.2.1) in the following way. First, consider the case of progressively measurable integrands $f$. Define

$$\tau_n = T \wedge \inf\{t \geq 0 : \int_0^t f^2(s)ds \geq n\}.$$

Since $f^2$ is also progressively measurable, it follows by the corollary above that

$$\{\omega : \tau_n(\omega) > t\} = \{\omega : \int_0^t f^2(s, \omega)ds < n\} \in \mathcal{F}_t, \qquad (7.2.9)$$

and so $\tau_n$ is a stopping time and, clearly, $\tau_n \uparrow T$ a.s. as $n \uparrow \infty$. Moreover, $\Omega_n = \{\tau_n = T\} \uparrow \tilde{\Omega}$ as $n \uparrow \infty$ where $P(\tilde{\Omega}) = 1$. Define $f_n(s, \omega) = f(s, \omega)\mathbb{I}_{\tau_n > s}$ for all $s \in [0, T]$. Then

$$E\int_0^T f_n^2(s)ds = E\int_0^T f^2(s)\mathbb{I}_{\tau_n > s}ds = E\int_0^{\tau_n} f^2(s)ds \leq n,$$

and so (7.2.1) holds true for $f_n$ in place of $f$. Hence, the stochastic integrals $\int_0^t f_n(s)dW(s)$, $t \in [0, T]$, $n = 1, 2, ...$ are defined via the previous construction. Now, for each $t \in [0, T]$ set

$$\int_0^t f(s)dW(s) = \lim_{n \uparrow \infty} \int_0^t f_n(s)dW(s),$$

and observe that taking into account Theorem 7.2.1(iv) the limit exists here since $\tau_n$ increases with $n$, and so $f_m(s) = f(s)$ remains the same for all $m \geq n$ if $\tau_n > s$.

Another approach taken in Section 4.2 of [53] proceeds as follows. Using Lemma 7.2.1 we approximate each $f_n$ by simple functions in the sense of

(7.2.3) which yields a sequence of simple functions $g_m$, $m \geq 1$ satisfying (7.2.1) such that $\int_0^T (f(s) - g_m(s))^2 ds \to 0$ as $m \to \infty$ a.s. It is not difficult to derive from here that the stochastic integrals $\int_0^T g_m(s) dW(s)$ converge in probability as $m \to \infty$ and their limit is designated as the stochastic integral $\int_0^T f(s) dW(s)$.

The stochastic integral $\int_0^t f(s) dW(s)$ constructed this way is, in general, not a martingale as any stochastic integral obtained under the condition (7.2.1) but it is, so called, a local martingale since all $\int_0^{t \wedge \tau_n} f(s) dW(s)$, $n = 1, 2, \ldots$ are martingales. Moreover, the stochastic integral $\int_0^t f(s) dW(s)$ is continuous in $t$ with probability one since each stochastic integral $\int_0^{t \wedge \tau_n} f(s) dW(s)$, $n = 1, 2, \ldots$ is continuous in $t$ and $\tau_n \uparrow \infty$ as $n \uparrow \infty$.

An extension of stochastic integrals to all adapted integrands $f$ satisfying only (7.2.8) requires additional arguments. Here we have to rely on the theorem saying that an adapted to a completed filtration $\mathcal{F}_t$, $t \in [0, T]$ and measurable with respect to $\mathcal{B}_{[0,T]} \times \mathcal{F}_T$ random function (process) $g = g(t, \omega)$, $t \in [0, T]$, $\omega \in \Omega$ has a progressively measurable modification $\tilde{g} = \tilde{g}(t, \omega)$, $t \in [0, T]$, $\omega \in \Omega$ (see Section 4.3 in [55] and a shorter more recent proof in [61]). Set $\chi_s(\omega) = \mathbb{I}_{\omega:\, g(s,\omega) \neq \tilde{g}(s,\omega)}$ which is $\mathcal{B}_{[0,T]} \times \mathcal{F}_T$-measurable since both $g$ and $\tilde{g}$ are. By the Fubini theorem $E \int_0^T \chi_s(\omega) ds = \int_0^T P\{g(s,\omega) \neq \tilde{g}(s,\omega)\} ds = 0$. Hence, $\int_0^T \chi_s(\omega) ds = 0$ a.s., and so $G(t, \omega) = \int_0^t g(s, \omega) ds = \tilde{G}(t, \omega) = \int_0^t \tilde{g}(s, \omega) ds$ a.s. for each $t \in [0, T]$ since the $\sigma$-algebras $\mathcal{F}_t$ are complete. Hence, $\int_0^t \tilde{g}(s, \omega) ds$ is $\mathcal{F}_t$ measurable and taking $g = f^2$ we can construct the stochastic integral of $f$ in the same way as above for progressively measurable functions.

**Remark 7.2.3.** In the example above we dealt with a filtration without specifying any probability while the last argument shows that if our filtration consists of $\sigma$-algebras completed with respect to some probability then time integrals of adapted processes form an adapted process, and then the example above will not serve its purpose.

### 7.2.3   *Stochastic integrals with respect to martingales*

Let $M(t)$, $t \in [0, T]$ be a cádlág martingale on a probability space $(\Omega, \mathcal{F}, P)$ with respect to a filtration $\mathcal{F}_t$, $t \in [0, T]$ satisfying usual conditions and assume that $EM^2(T) < \infty$ where we fix a positive $T < \infty$. As a corollary of the Cauchy (or, more generally, Jensen's) inequality it follows that $M^2(t)$, $t \in [0, T]$ is a positive cádlág submartingale, and so $EM^2(t) \leq EM^2(T) < \infty$ for any $t \in [0, T]$. The space of such square integrable cádlág martingales $M$ with $M(0) = 0$ will be denoted by $\mathcal{M}_2 = \mathcal{M}_2[0, T]$ while the subspace of those martingales which are continuous will be denoted by $\mathcal{M}_2^c = \mathcal{M}_2^c[0, T]$. Since $T < \infty$ each $M \in \mathcal{M}_2$ belongs to the class $D$. Indeed, we have to show that $M^2(\tau)$, $\tau \in \mathcal{T}_{0,T}$ is uniformly integrable, where $\mathcal{T}_{0,T}$ is the set of all stopping times taking values between 0 and $T$. But, for any $\tau \in \mathcal{T}_{0,T}$ by the optional sampling theorem,

$$\int_{\{M^2(\tau)>K\}} M_\tau^2 dP \leq \int_{\{M^2(\tau)>K\}} M_T^2 dP \to 0 \text{ as } K \to \infty$$

uniformly in $\tau \in \mathcal{T}_{0,T}$ in view of the absolute continuity of the Lebesgue integral and the fact that

$$P\{M^2(\tau) > K\} \leq K^{-1}EM_\tau^2 \leq K^{-1}EM_T^2 \to 0 \text{ as } K \to \infty.$$

It follows that when $M \in \mathcal{M}_2[0, T]$ the submartingale $M^2(t)$, $t \in [0, T]$ satisfies the conditions of the Doob–Meyer decomposition theorem, and so there exists a unique predictable increasing cádlág process $A(t)$, $t \in [0, T]$ such that $M^2(t) - A(t)$, $t \in [0, T]$ is a martingale. The standard notation for such process $A$ is $\langle M, M \rangle = \langle M \rangle$ and it is called the quadratic variation of the martingale $M$. If $M, N \in \mathcal{M}_2$ then both $(M + N)^2 - \langle M + N \rangle$ and $(M - N)^2 - \langle M - N \rangle$ are martingales, and so their difference $4MN - (\langle M + N \rangle - \langle M - N \rangle)$ is a martingale as well. We set $\langle M, N \rangle(t) = \frac{1}{4}(\langle M + N \rangle(t) - \langle M - N \rangle(t))$, $t \in [0, T]$, which is called the cross-variation process of $M$ and $N$, and observe that $MN - \langle M, N \rangle$ is a martingale.

Our goal is to define integrals

$$I_M(\Phi)(t) = \int_0^t \Phi(s, \omega) dM(s)$$

for predictable with respect to the filtration $\mathcal{F}_s$, $s \in [0, T]$ processes (functions) $\Phi = \Phi(s, \omega)$, $s \in [0, T]$, $\omega \in \Omega$ satisfying

$$(\|\Phi\|_T^M)^2 = E\left(\int_0^T \Phi^2(s, \omega) d\langle M \rangle(s)\right) < \infty,$$

where the integral $\int_0^T$ in the right hand side is the path-wise Lebesgue-Stieltjes integral with respect to the increasing, and so bounded variation, process $\langle M \rangle$. The space of such square integrable processes $\Phi$ will be denoted by $\mathcal{L}_2^M$ which is an $L^2$ space with the norm $\|\cdot\|_T^M$. More precisely, if we introduce the measure

$$\mu_M(\Gamma) = E \int_0^T \mathbb{I}_\Gamma(t, \omega) d\langle M \rangle(t), \ \Gamma \in \mathcal{B}_{[0,T]} \times \mathcal{F}$$

then $\mathcal{L}_2^M$ is the $L^2$-space of functions on the measure space $([0,T] \times \Omega, \mathcal{B}_{[0,T]} \times \mathcal{F}, \mu)$. As in the Brownian motion case we will define first the integral for simple processes (functions) which have the form

$$\Phi(t,\omega) = f_0(\omega)\mathbb{I}_{t=0}(t) + \sum_{i=0}^{m} f_i(\omega)\mathbb{I}_{(t_i,t_{i+1}]}(t) \qquad (7.2.10)$$

where $0 = t_0 < t_1 < ... < t_m < t_{m+1} = T$, $f_i$ is $\mathcal{F}_{t_i}$-measurable and $\max_{0 \le i \le m} \|f_i\|_\infty < \infty$ ($\|\cdot\|_\infty$ is the $L^\infty$-norm). Denote the space of such simple processes by $\mathcal{L}_0$ and observe that $\mathcal{L}_0 \subset \mathcal{L}_2^M$. In order to extend the integral to the whole $\mathcal{L}_2^M$ we will need the following result.

**Lemma 7.2.2.** *The space $\mathcal{L}_0$ is dense in $\mathcal{L}_2^M$ with respect to the norm $\|\cdot\|_T^M$.*

**Proof.** (See also Ch. II in [28] and Ch. 3 in [44]). Let $\Phi \in \mathcal{L}_2^M$ and set $\Phi^K(t,\omega) = \Phi(t,\omega)\mathbb{I}_{[-K,K]}(\Phi(t,\omega))$ where $K > 0$ is a number. Then $\|\Phi - \Phi^K\|_T^M \to 0$ as $K \to \infty$, and so it suffices to approximate by simple processes only bounded processes from $\mathcal{L}_2^M$. Let $\hat{\mathcal{L}} = \{\Phi \in \mathcal{L}_2^M : \Phi$ is bounded, there exists a sequence $\Phi_n \in \mathcal{L}_0$ such that $\|\Phi - \Phi_n\|_T^M \to 0$ as $n \to \infty\}$. Clearly, $\hat{\mathcal{L}}$ is linear and if $\Phi_n \in \hat{\mathcal{L}}$, $|\Phi_n| < K$ for some $K > 0$ and all $n = 1, 2, ...$ and $\Phi_n \uparrow \Phi$ as $n \uparrow \infty$, then $\Phi \in \hat{\mathcal{L}}$. Suppose that the process $\Phi$ is bounded, left-continuous and adapted with respect to the filtration $\{\mathcal{F}_t, t \ge 0\}$. Set $\Phi_n(t,\omega) = \Phi(\frac{k}{2^n},\omega)$ when $t \in (\frac{k}{2^n}, \frac{k+1}{2^n}]$, $k = 0, 1, ...$ and $\Phi_n(0,\omega) = \Phi(0,\omega)$. Then $\Phi_n \in \mathcal{L}_0$ and $\|\Phi_n - \Phi\|_T^M \to 0$ as $n \to \infty$ by the dominated convergence theorem. It follows that $\hat{\mathcal{L}}$ contains all bounded predictable processes. $\qquad \square$

We will also need the following result where, in fact, we are talking about the spaces of classes of indistinguishable martingales, i.e., such that $P\{M(t) = N(t) \,\forall t \in [0,T]\} = 1$, in other words, we identify such martingales. This slight abuse of precision is similar to saying that we are dealing, for instance, with an $L^2$ space of functions while, in fact, we have to talk about space of classes of almost everywhere equal functions.

**Lemma 7.2.3.** *The space $\mathcal{M}_2$ of square integrable cádlág martingales is a complete metric space with respect to the distance $d(M,N) = \||M-N|\|_T = (E(M(T)-N(T))^2)^{1/2}$, $M, N \in \mathcal{M}_2$ and $\mathcal{M}_2^c$ is the closed subspace of $\mathcal{M}_2$.*

**Proof.** If $\||M-N|\|_T = 0$ then $M = N$ a.s. since $(M-N)^2$ is a submartingale, and so $0 = E(M(T)-N(T))^2 \ge E(M(t)-N(t))^2$ for $t \in [0,T]$. Thus, with probability one $M(r) = N(r)$ for all rational $r \in [0,T]$ and for

$r = T$, so by right continuity this holds true, in fact, for all $t \in [0, T]$. Next, let $M^{(n)}$, $n = 1, 2, \ldots$ be a Cauchy (fundamental) sequence, i.e., $\lim_{n,m\to\infty} |||M^{(n)} - M^{(m)}|||_T^2 = 0$. Then, by the Doob martingale inequality

$$P\{ \sup_{0 \le t \le T} |M^{(n)}(t) - M^{(m)}(t)| \ge \varepsilon \} \le \varepsilon^{-2} |||M^{(n)} - M^{(m)}|||_T^2.$$

It follows that there exists $M = M(t), t \in [0, T]$ such that $\sup_{0 \le t \le T} |M^{(n)}(t) - M(t)| \to 0$ in probability as $n \to 0$. In fact, $M^{(n)}(t)$ is a Cauchy sequence in $L^2(\Omega, \mathcal{F}, P)$ for each $t \in [0, T]$ and since the latter space is complete, this sequence converges and its limit must be $M(t)$. Hence, $E(M^{(n)}(t) - M(t))^2 \to 0$ as $n \to \infty$. It follows from here that $M$ is a square integrable martingale. Since the convergence in probability is uniform in $t \in [0, T]$, we can choose a subsequence which converges almost surely uniformly in $t \in [0, T]$, and so the cádlág property is inherited by $M$ from $M^{(n)}$'s while if the latter are continuous martingales then $M$ is a continuous martingale as well, i.e., $\mathcal{M}_2^c$ is a closed subspace of $\mathcal{M}_2$ as required. $\qquad\square$

Now, we are ready to define stochastic integrals with respect to martingales and to study their properties. For each $\Phi \in \mathcal{L}_0$ having the form (7.2.10) for each $t \le T$ and $\omega \in \Omega$ we set

$$I^M(\Phi)(t, \omega)) = \int_0^t \Phi(s, \omega) dM(s, \omega) \qquad (7.2.11)$$
$$= \sum_{i=0}^m f_i(\omega) \big( M(t \wedge t_{i+1}, \omega) - M(t \wedge t_i, \omega) \big)$$
$$= \sum_{i=0}^{n-1} f_i(\omega) \big( M(t_{i+1}, \omega) - M(t_i, \omega) \big) + f_n(\omega)(M(t, \omega) - M(t_n, \omega))$$

where $t_n \le t < t_{n+1}$. It is clear that the integral $I^M(\Phi)$ defined by (7.2.11) is right-continuous since $M$ is so, and if $M$ is continuous then the same holds true for $I^M(\Phi)$. Clearly, the integral $I^M$ is linear $I^M(a\Phi + b\Psi) = aI^M(\Phi) + bI^M(\Psi)$ for any constants $a, b$ and $\Phi, \Psi \in \mathcal{L}_0$. Next, $I^M(\Phi)$ is a martingale. Indeed, if $t_i \le s \le t$ then

$$E\big(f_i(M(t \wedge t_{i+1}) - M(t \wedge t_i))|\mathcal{F}_s\big) = f_i E\big(M(t \wedge t_{i+1}) - M(t \wedge t_i)|\mathcal{F}_s\big)$$
$$= f_i(M(s \wedge t_{i+1}) - M(s \wedge t_i)).$$

If $t \ge t_i \ge s$ then

$$E\big(f_i(M(t \wedge t_{i+1}) - M(t \wedge t_i))|\mathcal{F}_s\big)$$
$$= E\big(f_i E(M(t \wedge t_{i+1}) - M(t \wedge t_i)|\mathcal{F}_{t_i})|\mathcal{F}_s\big)$$
$$= 0 = f_i(M(s \wedge t_{i+1}) - M(s \wedge t_i))$$

implying that $E(I^M(\Phi)(t)|\mathcal{F}_s) = I^M(\Phi)(s)$, i.e., $I^M$ is a martingale.

We claim next, that for any $t \in [0, T]$,

$$||I^M(\Phi)||_T = ||\Phi||_T^M. \qquad (7.2.12)$$

Indeed, first, if $i < j$ then $T \geq t_j \geq t_{i+1} > t_i$, and so,

$$E\big(f_i f_j (M(t_{i+1}) - M(t_i))(M(t_{j+1}) - M(t_j))\big)$$
$$= E\big(f_i (M(t_{i+1}) - M(t_i)) f_j E(M(t_{j+1}) - M(t_j)|\mathcal{F}_{t_j})\big) = 0.$$

Secondly, using the representation $M^2 = N + \langle M \rangle$, where $N$ is a martingale, we obtain

$$E\big(f_i^2 (M(t_{i+1}) - M(t_i))^2\big) = E\big(f_i^2 E((M(t_{i+1}) - M(t_i))^2 | \mathcal{F}_{t_i})\big)$$
$$= E\big(f_i^2 E(M^2(t_{i+1}) - M^2(t_i) | \mathcal{F}_{t_i})\big) = E\big(f_i^2 E((\langle M \rangle(t_{i+1})$$
$$- \langle M \rangle(t_i)) | \mathcal{F}_{t_i})\big) = E\big(f_i^2 (\langle M \rangle(t_{i+1}) - \langle M \rangle(t_i))\big).$$

It follows that

$$(||I^M(\Phi)||_M)^2 = E(I^M(\Phi)(T))^2 = E\big(\sum_{i=0}^m f_i^2 (M(t_{i+1}) - M(t_i))\big)^2$$
$$= \sum_{i=0}^m E\big(f_i^2 (M(t_{i+1}) - M(t_i))^2\big) = \sum_{i=0}^m E\big(f_i^2 (\langle M \rangle(t_{i+1})$$
$$- \langle M \rangle(t_i))\big) = E \int_0^T \Phi^2(t) d\langle M \rangle(t) = (||\Phi||_T^M)^2$$

proving (7.2.12).

Now, let $\Phi \in \mathcal{L}_2^M$. By Lemma 7.2.2 there exists a sequence $\Phi_n \in \mathcal{L}_0$ such that $||\Phi_n - \Phi||_2^M \to 0$ as $n \to \infty$. Hence, $\{\Phi_n, n \geq 1\}$ is a Cauchy sequence. Since $||\Phi_n - \Phi_m||_2^M = |||I^M(\Phi_n) - I^M(\Phi_m)|||_T$ by (7.2.12), the sequence $I(\Phi_n)$, $n \geq 1$ is a Cauchy sequence in the space $\mathcal{M}_2$ which is complete by Lemma 7.2.3. Hence, there exists $X = \lim_{n \to \infty} I(\Phi_n) \in \mathcal{M}_2$ where the convergence is in the norm $||| \cdot |||_T$. Clearly, $X$ does not depend on a particular choice of a sequence $\Phi_n$, $n \geq 1$ converging to $\Phi$ and we denote the limit $X$ by $I^M(\Phi)$ calling it the stochastic integral $\int_0^T \Phi dM$ of $\Phi$ with respect to a martingale $M \in \mathcal{M}_2$. Since $I^M(\Phi_n) - I^M(\Phi_m)$ is a martingale and $(I^M(\Phi_n) - I^M(\Phi_m))^2$ is a submartingale, we have that $|||I^M(\Phi_n) - I^M(\Phi_m)|||_T \geq E(I^M(\Phi_n)(t) - I^M(\Phi_m)(t))^2$ for any $t \in [0, T]$, and so $I^M(\Phi)(t)$ is also a Cauchy sequence whose limit we denote by $I^M(\Phi)(t) = \int_0^t \Phi dM$. In fact,

$$E\big(I^M(\Phi_n(T)) - I^M(\Phi_m(T))\big)^2$$
$$\geq E\big(E(I^M(\Phi_n(T))|\mathcal{F}_t) - E(I^M(\Phi_m(T))|\mathcal{F}_t)\big)^2,$$

and so $I^M(\Phi_n)(t) = E(I^M(\Phi_n)(T)|\mathcal{F}_t)$, $n \geq 1$ is a Cauchy sequence in $L^2(\Omega, \mathcal{F}, P)$ and it converges to $E(I^M(\Phi)(T)|\mathcal{F}_t)$ which means that $I^M(\Phi)(t) = E(I^M(\Phi)(T)|\mathcal{F}_t)$ a.s., i.e., $I^M(\Phi)(t)$, $t \in [0, T]$ is a martingale.

Next, we collect the main properties of stochastic integrals $I^M(\Phi)(t) = \int_0^t \Phi dM$.

**Proposition 7.2.3.** *Let $M \in \mathcal{M}_2[0,T]$ and $\Phi \in \mathcal{L}_2^M$. Then with probability one, (i) $I^M(\Phi)(0) = 0$;*
*(ii) For any $0 \le s \le t \le T$,*

$$E\big(I^M(\Phi)(t) - I^M(\Phi)(s)|\mathcal{F}_s\big) = 0,$$

*in particular, $I^M(\Phi)(t)$, $t \in [0,T]$ is a martingale. Furthermore,*

$$E\big((I^M(\Phi)(t) - I^M(\Phi)(s))^2|\mathcal{F}_s\big) = E\big(\int_s^t \Phi^2(u)d\langle M\rangle(u)|\mathcal{F}_s\big);$$

*(iii) Let $\sigma, \tau \le T$ be stopping times such that $\tau \ge \sigma$ a.s. Then for any $t \in [0,T]$,*

$$E\big(I^M(\Phi)(t \wedge \tau) - I^M(\Phi)(t \wedge \sigma)|\mathcal{F}_\sigma\big) = 0 \quad and$$

$$E\big((I^M(\Phi)(t \wedge \tau) - I^M(\Phi)(t \wedge \sigma))^2|\mathcal{F}_s\big) = E\big(\int_{t \wedge \sigma}^{t \wedge \tau} \Phi^2(u)d\langle M\rangle(u)|\mathcal{F}_\sigma\big);$$

*(iv) More generally, if $\Phi, \Psi \in \mathcal{L}_2^M$, $0 \le s \le t \le T$ and $0 \le \sigma \le \tau \le T$ are stopping times, then*

$$E\big((I^M(\Phi)(t) - I^M(\Phi)(s))(I^M(\Psi)(t) - I^M(\Psi)(s))|\mathcal{F}_s\big)$$
$$= E\big(\int_s^t (\Phi\Psi)(u)d\langle M\rangle(u)|\mathcal{F}_s\big) \quad and$$

$$E\big((I^M(\Phi)(t \wedge \tau) - I^M(\Phi)(t \wedge \sigma))(I^M(\Psi)(t \wedge \tau) - I^M(\Psi)(t \wedge \sigma))|\mathcal{F}_{t \wedge \sigma}\big)$$
$$= E\big(\int_{t \wedge \sigma}^{t \wedge \tau} (\Phi\Psi)(u)d\langle M\rangle(u)|\mathcal{F}_{t \wedge \sigma}\big);$$

*(v) If $t \in [0,T]$ and $0 \le \sigma \le T$ is a stopping time, then*

$$I^M(\Phi)(t \wedge \sigma) = I^M(\Phi')(t) \quad where \quad \Phi'(t) = \mathbb{I}_{\sigma \ge t}\Phi(t);$$

*(vi) Let $M, N \in \mathcal{M}_2$ and $0 \le s \le t \le T$. Then*

$$E\big((I^M(\Phi)(t) - I^M(\Phi)(s))(I^N(\Psi)(t) - I^N(\Psi)(s))|\mathcal{F}_s\big)$$
$$= E\big(\int_s^t (\Phi\Psi)(u)d\langle M,N\rangle(u)|\mathcal{F}_s\big).$$

**Proof.** All these properties are proved by verifying them first for simple $\Phi, \Psi \in \mathcal{L}_0$ and then approximating $\Phi, \Psi \in \mathcal{L}_2^M$ (or $\Phi \in \mathcal{L}_2^M$, $\Psi \in \mathcal{L}_2^N$) by simple $\Phi_n, \Psi_n \in \mathcal{L}_0$. The equalities with stopping times $\sigma, \tau$ are proved by considering first stopping times $\sigma_n, \tau_n$ taking on only finitely many values and such that $\sigma_n \downarrow \sigma$ and $\tau_n \downarrow \tau$ as $n \uparrow \infty$ (see [46] for more details). $\square$

Similarly to the case of the Brownian motion $M = W$ it is easy to see that if $M \in \mathcal{M}_2^c[0,T]$ and $\Phi \in \mathcal{L}_2^M$ then $I^M(\Phi) \in \mathcal{M}_2^c[0,T]$.

**Remark 7.2.4.** It is possible to extend the above setup to integrals with respect to semi-martingales, i.e., stochastic processes which are sums of a martingale $M$ and a process with bounded variation $A$. Such integral is defined as a sum of the stochastic integral with respect to $M$ and the pathwise Riemann–Stieltjes integral with respect to $A$.

For several applications which will be discussed later it will be important to extend the notion of stochastic integrals to the framework of local martingales.

**Definition 7.2.2.** A real stochastic process $X = (X(t),\ t \geq 0)$ on a complete probability space $(\Omega, \mathcal{F}, P)$ is called a local martingale with respect to a filtration $\mathcal{F}_t,\ t \geq 0$ if $X$ is adapted to the filtration and there exists a sequence of stopping times $\tau_n < \infty$, $\tau_n \uparrow \infty$ a.s. as $n \uparrow \infty$ such that each $X_n(t) = X(t \wedge \tau_n),\ t \geq 0$ is a martingale with respect to $\mathcal{F}_t,\ t \geq 0$. If $X_n,\ n = 1, 2, \ldots$ are square integrable martingales then $X$ is called a locally square integrable martingale and the space of them will be denoted by $\mathcal{M}_2^{loc}$. If the objects above are restricted to a time interval $[0,T]$ then we denote the corresponding space by $\mathcal{M}_2^{loc}[0,T]$.

The above definition of stochastic integrals is extended in a straightforward way to the case of local martingales. Namely, let $M, N \in \mathcal{M}_2^{loc}$. Then there exists a sequence of stopping times $\tau_n \uparrow \infty$ a.s. as $n \uparrow \infty$ such that $M^{\tau_n}(t) = M(t \wedge \tau_n)$ and $N^{\tau_n} = N(t \wedge \tau_n),\ t \geq 0$ are in $\mathcal{M}_2$. By uniqueness of the quadratic variation process, if $m < n$ then

$$\langle M^{\tau_m}, N^{\tau_m} \rangle(t) = \langle M^{\tau_n}, N^{\tau_n} \rangle(t \wedge \tau_m).$$

Hence, there exists a unique predictable process $\langle M, N \rangle$ such that $\langle M, N \rangle(t \wedge \tau_n) = \langle M^{\tau_n}, N^{\tau_n} \rangle(t)$ for all $n \geq 1$ and $t \geq 0$ and, as before, we set $\langle M \rangle$ for $\langle M, M \rangle$.

Let $M \in \mathcal{M}_2^{loc}[0,T]$ and denote by $\mathcal{L}_2^{M,loc} = \mathcal{L}_2^{M,loc}[0,T]$ the space of real predictable (with respect to the above filtration) processes (functions) $\Phi = \Phi(t),\ t \in [0,T]$ such that there exists a sequence of stopping times $\tau_n \uparrow \infty$ a.s. as $n \uparrow \infty$ with

$$E \int_0^{T \wedge \tau_n} \Phi^2(t, \omega) d\langle M \rangle(t) < \infty. \tag{7.2.13}$$

Actually, replacing $\tau_n$ by $T \wedge \tau_n$ we can talk about $\tau_n \uparrow T$ a.s. and then can write $\int_0^{\tau_n}$ in place of $\int_0^{T \wedge \tau_n}$.

Let $M \in \mathcal{M}_2^{loc}[0,T]$ and $\Phi \in \mathcal{L}_2^{M,loc}[0,T]$. Then we can choose a sequence of stopping times $\tau_n \uparrow T$ a.s. such that $M^{\tau_n} \in \mathcal{M}_2[0,T]$ where $M^{\tau_n}(t) = M(t \wedge \tau_n)$, $t \in [0,T]$ and (7.2.13) holds true. It follows that we can define the stochastic integral $I^{M_n}(\Phi_n)$ for $M_n = M^{\sigma_n}$ and $\Phi_n$ defined by $\Phi_n(t,\omega) = \mathbb{I}_{\tau_n(\omega) \geq t} \Phi(t,\omega)$. Since $I^{M_m}(\Phi_m)(t) = I^{M_n}(\Phi_n)(t \wedge \tau_m)$ for $m < n$ and $t \in [0,T]$, there exists a unique process $I^M(\Phi)(t)$ such that $I^{M_n}(\Phi_n)(t) = I^M(\Phi)(t \wedge \tau_n)$, $t \in [0,T]$, $n = 1,2,\ldots$. Clearly, $I^M(\Phi) \in \mathcal{M}_2^{loc}$. The process $I^M(\Phi)$ is called, again, the stochastic integral of $\Phi$ with respect to $M$ and it is denoted by $\int_0^t \Phi(s)dM(s)$, $t \in [0,T]$.

## 7.3 Lecture 15: Itô formula

### 7.3.1 *Itô formula with Brownian motion*

A $d$-dimensional stochastic process $\xi(t) = (\xi_1(t),\ldots,\xi_d(t))$, $t \in [0,T]$ on a probability space $(\Omega, \mathcal{F}, P)$ is said to have the stochastic (Itô) differential

$$d\xi_i(t) = \sum_{j=1}^r f_{ij}(t,\omega)dW_j(t) + g_i(t,\omega)dt, \ i = 1,\ldots,d, \ t \in [0,T]$$

if

$$\xi_i(t) = \xi_i(0) + \sum_{j=1}^r \int_0^t f_{ij}(s,\omega)dW_j(s) + \int_0^t g_i(s,\omega)ds, \ i = 1,\ldots,d, \ t \in [0,T]$$

where $W_1,\ldots,W_r$ are independent Brownian motions, $f_{ij}, g_i, i = 1,\ldots,d, j = 1,\ldots,r$ are functions adapted to the filtration $\{\mathcal{F}_t = \sigma\{W_1(s),\ldots,W_r(s), s \leq t\}, t \in [0,T]\}$ and satisfying

$$\int_0^T (f_{ij}^2(s) + |g_i(s)|)ds < \infty \quad \text{a.s. for all} \quad i = 1,\ldots,d, \ j = 1,\ldots,r. \quad (7.3.1)$$

If we denote by $f(t,\omega)$ the matrix $(f_{ij}(t,\omega))_{i=1,\ldots,d;j=1,\ldots,r}$, by $g(t,\omega)$ the vector $(g_1(t,\omega),\ldots,g_d(t,\omega))$ and by $W(t,\omega)$ the vector $(W_1(t,\omega),\ldots,W_r(t,\omega))$ then we can write the above in a more compact vector form

$$d\xi(t) = f(t,\omega)dW(t) + g(t,\omega)dt \quad \text{and} \quad (7.3.2)$$
$$\xi(t) = \xi(0) + \int_0^t f(s,\omega)dW(s) + \int_0^t g(s,\omega)ds.$$

**Theorem 7.3.1.** *(Itô formula).* *Let* $\xi(t), t \in [0,T]$ *be an adapted $d$-dimensional stochastic process having the stochastic differential (7.3.2) with coefficients $f$ and $g$ satisfying (7.3.1) and let $F = F(t, x_1,\ldots,x_d)$ be a function on $[0,T] \times \mathbb{R}^d$ which is once differentiable in the time variable $t$ and* *twice differentiable in the space variables $x_i$, $i = 1, ..., d$ with continuous partial derivatives $F_0 = \frac{\partial F}{\partial t}$, $F_i = \frac{\partial F}{\partial x_i}$ and $F_{ij} = \frac{\partial^2 F}{\partial x_i \partial x_j}$. Then the process $F(t, \xi(t))$, $t \in [0, T]$ has also the stochastic differential having the form*

$$dF(t, \xi_t) = \sum_{j=1}^{r} \left( \sum_{i=1}^{d} F_i(t, \xi_t) f_{ij}(t, \omega) \right) dW_j(t) + \left( F_0(t, \xi_t) \right) \text{ (7.3.3)}$$
$$+ \sum_{i=1}^{d} F_i(t, \xi_t) g_i(t, \omega) + \frac{1}{2} \sum_{i,j=1}^{d} F_{ij}(t, \xi_t) \sum_{k=1}^{r} f_{ik}(t, \omega) f_{jk}(t, \omega) \right) dt.$$

**Proof.** To avoid notational complications we will consider first the case $r = d = 1$ and will describe at the end additional arguments required for the multidimensional case. Thus, we have now

$$d\xi(t) = f(t, \omega) dW(t) + g(t, \omega) dt$$

with adapted to the Brownian filtration $f$ and $g$ satisfying

$$\int_0^T (f^2(s) + |g(s)|) ds < \infty \quad \text{a.s.} \tag{7.3.4}$$

and ought to show that for each $t \in [0, T]$ with probability one,

$$F(t, \xi(t)) - F(0, \xi(0)) = \int_0^t \frac{\partial F}{\partial x}(s, \xi(s)) f(s) dW(s) \tag{7.3.5}$$
$$+ \int_0^t \left( \frac{\partial F}{\partial s}(s, \xi(s)) + \frac{\partial F}{\partial x}(s, \xi(s)) g(s) + \frac{1}{2} \frac{\partial^2 F}{\partial x^2}(s, \xi(s)) f^2(s) \right) ds$$

where $F = F(t, x)$ is a function on $[0, T] \times \mathbb{R}$ continuously differentiable once in $t$ and twice in $x$.

Observe first that $H(t, \omega) = \frac{\partial F}{\partial x}(t, \xi(t)) f(t, \omega)$ satisfies the condition

$$P\left\{ \int_0^T H^2(t, \omega) dt < \infty \right\} = 1. \tag{7.3.6}$$

Indeed, since $\xi(t)$ is given by the sum of the stochastic and the Lebesgue integral from 0 to $t$, it is a time continuous stochastic process, and so $\sup_{0 \le t \le T} |\xi(t)| < \infty$ a.s. This together with our assumption that $\frac{\partial F}{\partial x}(t, x)$ is continuous in $t, x$ gives that $\sup_{0 \le t \le T} |\frac{\partial F}{\partial x}(t, \xi(t))| < \infty$ a.s., and so taking into account (7.3.4) we obtain (7.3.6). On the other hand, the condition $E \int_0^T H^2(t, \omega) dt < \infty$ may not be satisfied (even if we assume that $E \int_0^T (f^2(s) + |g(s)|) ds < \infty$) since $\frac{\partial F}{\partial x}(t, x)$ can grow too fast when $x \to \infty$ and though $|\xi(t)|$ can be very large only with small probability the expectation above may still be infinite. Therefore, in order to have Theorem 7.3.1 in the generality stated above we have to consider the extension of stochastic integrals under the condition of the form (7.3.6) as described in Lecture 14.

Next, introduce stopping times

$$\tau_n = T \wedge \inf\{t \ge 0 : |\xi_t| \ge n\}$$

and set $f_n(s,\omega) = f(s,\omega)\mathbb{I}_{\tau_n>s}$, $g_n(s,\omega) = g(s,\omega)\mathbb{I}_{\tau_n>s}$. Then the stopped process $\xi_n(t) = \xi(t \wedge \tau_n)$ has the stochastic differential

$$d\xi_n(t) = f_n(t,\omega)dW(t) + g_n(t,\omega)dt.$$

If the equality

$$F(t,\xi_n(t)) - F(0,\xi_n(0)) = \int_0^t \frac{\partial F}{\partial x}(s,\xi_n(s))f_n(s)dW(s)$$
$$+ \int_0^t \left(\frac{\partial F}{\partial s}(s,\xi_n(s)) + \frac{\partial F}{\partial x}(s,\xi_n(s))g_n(s) + \frac{1}{2}\frac{\partial^2 F}{\partial x^2}(s,\xi_n(s))f_n^2(s)\right)ds$$

holds true for each $t \in [0,T]$ then (7.3.5) is satisfied for all $t \in [0,T]$ on the set $\Omega_n = \{\omega : \tau_n(\omega) = T\}$. Letting $n \uparrow \infty$ we obtain (7.3.5) on the set $\tilde{\Omega}$ where $\Omega_n \uparrow \tilde{\Omega}$ as $n \uparrow \infty$ and $P(\tilde{\Omega}) = 1$. It follows that, without loss of generality, we can assume that $F$ has a compact support, and so it is bounded together with its derivatives $\frac{\partial F}{\partial t}$, $\frac{\partial F}{\partial x}$ and $\frac{\partial^2 F}{\partial x^2}$.

By the definition of stochastic integrals it suffices to prove (7.3.5) for simple functions $f^{(n)}$ and $g^{(n)}$ and then to pass to the limit approximating $f$ and $g$ by simple functions. Indeed, let $f^{(n)}$ and $g^{(n)}$ be simple functions such that

$$E\int_0^T (f^{(n)}(s) - f(s))^2 ds \to 0$$
$$\text{and} \quad E\int_0^T |g^{(n)}(s) - g(s)|ds \to 0 \text{ as } n \to \infty.$$

Then, setting

$$\xi^{(n)}(t) = \int_0^t f^{(n)}(s)dW(s) + \int_0^t g^{(n)}(s)ds$$

we obtain by the martingale inequality for stochastic integrals that

$$E\sup_{0\le t\le T} |\xi^{(n)}(t) - \xi(t)| \le E\sup_{0\le t\le T} |\int_0^t (f^{(n)}(s) - f(s))dW(s)|$$
$$+E\int_0^T |g^{(n)}(s) - g(s)|ds \le 2\left(E\int_0^T (f^{(n)}(s) - f(s))^2 ds\right)^{1/2}$$
$$+E\int_0^T |g^{(n)}(s) - g(s)|ds.$$

Hence,

$$\sup_{0\le t\le T} |\xi^{(n)}(t) - \xi(t)| \to 0 \text{ as } n \to \infty \text{ in } L^1(\Omega, P),$$

and so this convergence is also in probability. Since $\frac{\partial F}{\partial x}$ is now bounded and continuous

$$\frac{\partial F}{\partial x}(t,\xi^{(n)}(t))f^{(n)}(t) \to \frac{\partial F}{\partial x}(t,\xi(t))f(t) \text{ as } n \to \infty \text{ in } L^2([0,T] \times \Omega, \ell \times P)$$

where $\ell$ is the Lebesgue measure on $\mathbb{R}$. Since $\frac{\partial F}{\partial t}$, $\frac{\partial F}{\partial x}$ and $\frac{\partial^2 F}{\partial x^2}$ are bounded and continuous, it follows that as $n \to \infty$,

$\frac{\partial F}{\partial t}(t, \xi^{(n)}(t)) \to \frac{\partial F}{\partial t}(t, \xi(t))$, $\frac{\partial F}{\partial x}(t, \xi^{(n)}(t))g^{(n)}(t) \to \frac{\partial F}{\partial x}(t, \xi(t))g(t)$ and
$\frac{\partial^2 F}{\partial x^2}(t, \xi^{(n)}(t))(f^{(n)}(t))^2 \to \frac{\partial^2 F}{\partial x^2}(t, \xi(t))f^2(t)$ in $L^1([0,T] \times \Omega, \ell \times P)$.

Thus, it remains to verify (7.3.5) for simple functions

$$f(t) = \sum_{1 \le k \le m+1} f(t_{k-1})\mathbb{I}_{[t_{k-1}, t_k)}(t) \text{ and } g(t) = \sum_{1 \le k \le m+1} g(t_{k-1})\mathbb{I}_{[t_{k-1}, t_k)}(t)$$

where $0 = t_0 < t_1 < ... < t_m < t_{m+1} = T$, $Ef^2(t_i) < \infty$ and $E|g(t_i)| < \infty$ for all $i = 0, 1, ..., m+1$. In this case the right hand side in (7.3.5) is the sum of integrals $\int_{t_{k-1}}^{t_k}$ and the left hand side there is the sum of differences $F(t_k, \xi(t_k)) - F(t_{k-1}, \xi(t_{k-1}))$. Hence, it suffices to verify the equality (7.3.5) on each interval $[t_{k-1}, t_k)$ where $f \equiv f(t_{k-1})$ and $g \equiv g(t_{k-1})$. We conclude from here that it suffices to show that for any $0 \le s \le t \le T$,

$$F(t, \xi_n(t)) - F(s, \xi(s)) = \int_s^t \frac{\partial F}{\partial x}(u, \xi(u))X dW(u) \qquad (7.3.7)$$
$$+ \int_s^t \left(\frac{\partial F}{\partial u}(u, \xi(u)) + \frac{\partial F}{\partial x}(u, \xi(u))Y\right)du + \frac{1}{2}\int_s^t \frac{\partial^2 F}{\partial x^2}(u, \xi(u))X^2 du$$

where $\xi(t) = \xi(s) + X(W(t) - W(s)) + Y(t-s)$ and $X, Y$ are $\mathcal{F}_s$-measurable random variables with $EX^2 + E|Y| < \infty$.

Now set $u_k = u_k^{(n)} = s + k2^{-n}(t-s)$, $k = 0, 1, ..., 2^n$. Then using the Taylor formula with two terms we have

$$F(t, \xi(t)) - F(s, \xi(s)) \qquad (7.3.8)$$
$$= \sum_{k=1}^{2^n} \frac{\partial F}{\partial x}(u_{k-1}, \xi(u_{k-1}))X(W(u_k) - W(u_{k-1}))$$
$$+ \sum_{k=1}^{2^n} \left(\frac{\partial F}{\partial u}(u_{k-1}, \xi(u_{k-1})) + \frac{\partial F}{\partial x}(u_{k-1}, \xi(u_{k-1}))Y\right)(u_k - u_{k-1})$$
$$+ \frac{1}{2}\sum_{k=1}^{2^n} \frac{\partial^2 F}{\partial x^2}(u_{k-1}, \xi(u_{k-1}))X^2(W(u_k) - W(u_{k-1}))^2$$
$$+ o(1) + o\left(\sum_{k=1}^{2^n}(W(u_k) - W(u_{k-1}))^2\right).$$

As $n \to \infty$ the first sum in (7.3.8) converges to the stochastic integral in (7.3.7) while the second sum in (7.3.8) converges to the second integral in (7.3.7). Since $E\sum_{k=1}^{2^n}(W(u_k) - W(u_{k-1}))^2 = t - s$ we see that $o\left(\sum_{k=1}^{2^n}(W(u_k) - W(u_{k-1}))^2\right) = o(1)$. Hence, it remains to show that the third sum in (7.3.8) converges to the third integral in (7.3.7). Approximating the latter integral by integral sums we can write

$$\int_s^t \frac{\partial^2 F(u, \xi(u))}{\partial x^2}X^2 du = X^2 \lim_{n\to\infty} \sum_{k=1}^{2^n} \frac{\partial^2 F}{\partial x^2}(u_{k-1}, \xi(u_{k-1}))(u_k - u_{k-1}).$$

Since $X$ is a fixed random variable, it suffices to prove that

$$\sum_{k=1}^{2^n} \frac{\partial^2 F}{\partial x^2}(u_{k-1}, \xi(u_{k-1}))\big((W(u_k) - W(u_{k-1}))^2 - (u_k - u_{k-1})\big) \quad (7.3.9)$$
$$\to 0 \text{ as } n \to \infty.$$

Set $Z_k = \frac{\partial^2 F}{\partial x^2}(u_k, \xi(u_k))$ and observe that $Z_k$ is $\mathcal{F}_{u_k}$-measurable and

$$\max_{0 \le k \le 2^n} |Z_k| \le C = \sup_{u,x} |\frac{\partial^2 F}{\partial x^2}(u, x)| < \infty.$$

It follows that

$$E\big(\sum_{k=1}^{2^n} Z_{k-1}((W(u_k) - W(u_{k-1}))^2 - (u_k - u_{k-1}))\big)^2 \quad (7.3.10)$$
$$= \sum_{k=1}^{2^n} E\big(Z_{k-1}^2 E\big((W(u_k) - W(u_{k-1}))^2 - (u_k - u_{k-1})\big)^2 \big| \mathcal{F}_{u_{k-1}}\big)\big)$$
$$= \sum_{k=1}^{2^n} (u_k - u_{k-1})^2 E Z_{k-1}^2 E(W^2(1) - 1)^2 \le C 2^{-(n-1)},$$

and so the left hand side of (7.3.9) tends to 0 as $n \to \infty$ in $L^2(\Omega, P)$. In (7.3.10) we used that $W(u_k) - W(u_{k-1})$ is independent of $\mathcal{F}_{u_{k-1}}$, $W(u_k) - W(u_{k-1})$ is distributed as $W(u_k - u_{k-1})$ which, in turn, is distributed as $\sqrt{u_k - u_{k-1}} W(1)$ and that for $k < l$,

$$E\big(Z_{k-1} Z_{l-1}((W(u_k) - W(u_{k-1}))^2 - (u_k - u_{k-1}))((W(u_l) - W(u_{l-1}))^2$$
$$-(u_l - u_{l-1}))\big) = E\big(Z_{k-1} Z_{l-1}((W(u_k) - W(u_{k-1}))^2$$
$$-(u_k - u_{k-1}))E((W(u_l) - W(u_{l-1}))^2 - (u_l - u_{l-1})|\mathcal{F}_{u_{l-1}})\big) = 0.$$

This completes the proof of Itô's formula in the one-dimensional case.

In the multidimensional case we proceed with the same arguments and the only terms which are slightly different have the form

$$\sum_{k=1}^{2^n} \frac{\partial^2 F}{\partial x_i \partial x_j}(u_{k-1}, \xi(u_{k-1}))\big(\sum_{a=1}^r X_{ia}(W_a(u_k) - W_a(u_{k-1}))$$
$$\times \big(\sum_{b=1}^r X_{ib}(W_b(u_k) - W_b(u_{k-1}))\big)\big) = \sum_{a,b=1}^r X_{ia} X_{jb} \sum_{k=1}^{2^n}$$
$$\frac{\partial^2 F}{\partial x_i \partial x_j}(u_{k-1}, \xi(u_{k-1}))(W_a(u_k) - W_a(u_{k-1}))(W_b(u_k) - W_b(u_{k-1})).$$

Setting $Z_k(i,j) = \frac{\partial^2 F}{\partial x_i \partial x_j}(u_k, \xi(u_k))$ we will have to show that as $n \to \infty$ both

$$\sum_{k=1}^{2^n} Z_{k-1}(i,j)((W_a(u_k) - W_a(u_{k-1})^2 - (u_k - u_{k-1})) \to 0 \quad (7.3.11)$$

and for $a \ne b$,

$$I_n(a,b) = \sum_{k=1}^{2^n} Z_{k-1}(i,j)(W_a(u_k) - W_a(u_{k-1}))(W_b(u_k) - W_b(u_{k-1})) \to 0$$
$$(7.3.12)$$

where the convergence is in $L^2(\Omega, P)$. The convergence in (7.3.11) holds true in the same way as for the similar expression in the one-dimensional case. Concerning (7.3.12), we take into account that $W_1, ..., W_r$ are independent Brownian motions and that each $W_a(u_k) - W_a(u_{k-1})$, $a = 1, ..., r$ is independent of $\mathcal{F}_{u_{k-1}}$ which yields that as $n \to \infty$,

$$EI_n^2(a, b) = \sum_{k=1}^{2^n} EZ_{k-1}^2(i, j) E\big((W_a(u_k) - W_a(u_{k-1}))^2$$
$$\times (W_b(u_k) - W_b(u_{k-1}))^2\big) \leq 3\tilde{C}^2 2^{-n} \to 0$$

where $\tilde{C} = \max_{i,j} \sup_{u \in [0,T], x \in \mathbb{R}^d} |\frac{\partial^2 F}{\partial x_i \partial x_j}(u, x)|$ which is finite after the reduction to the case when $F$ has a compact support. This completes the proof of Theorem 7.3.1. $\qquad\qquad\square$

### 7.3.2    *Itô formula with martingales*

Let $M_i(t)$, $t \in [0, T]$, $i = 1, ..., d$ be continuous local martingales from the space $\mathcal{M}_2^{c,loc} = \mathcal{M}_2^{c,loc}[0, T]$ (see Lecture 14) defined on a (complete) probability space $(\Omega, \mathcal{F}, P)$ with a filtration $\mathcal{F}_t$, $t \in [0, T]$ satisfying usual conditions and let $A_i(t)$, $i = 1, ..., d$ be continuous adapted stochastic processes with bounded variation on $[0, T]$. Recall, that the (total, path-wise) variation of a stochastic process (in particular, of a function) $A$ on an interval $[0, t] \subset [0, T]$ is

$$V_{0t}(A) = \sup_{0 = t_0 < t_1 < ... < t_{m-1} < t_m = t, \ m \geq 0} \sum_{i=1}^{m} |A(t_i) - A(t_{i-1})|.$$

Introduce stochastic processes $X_i(t) = X_i(0) + M_i(t) + A_i(t)$, $i = 1, ..., d$ on $[0, T]$ and observe that they are generalizations of expressions $\xi_i(t) = \xi_i(0) + \sum_{j=1}^{r} \int_0^t f_{ij}(s, \omega) dW_j(s) + \int_0^t g_i(s, \omega) ds$ considered above since under (7.3.1) the stochastic integrals are continuous local martingales and the other (Lebesgue) integral is a process with variation bounded by $\int_0^t |g_i(s, \omega)| ds < \infty$. Set $X(t) = (X_1(t), ..., X_d(t))$.

**Theorem 7.3.2.** *Let $F = F(t, x)$ be a function on $[0, T] \times \mathbb{R}^d$, $T < \infty$ once differentiable in $t$ and twice in $x = (x_1, ..., x_d) \in \mathbb{R}^d$. Then for each $t \in [0, T]$ with probability one,*

$$F(t, X(t)) - F(0, X(0)) \qquad\qquad (7.3.13)$$
$$= \int_0^t \frac{\partial F}{\partial s}(s, X(s)) ds + \sum_{i=1}^{d} \int_0^t \frac{\partial F}{\partial x_i}(s, X(s)) dM_i(s)$$
$$+ \sum_{i=1}^{d} \int_0^t \frac{\partial F}{\partial x_i}(s, X(s)) dA_i(s) + \frac{1}{2} \sum_{i,j=1}^{d} \int_0^t \frac{\partial^2 F}{\partial x_i \partial x_j}(s, X(s)) d\langle M_i, M_j \rangle(s).$$

**Proof.** As in the case of the Brownian motion, in order to avoid writing too many indexes, we consider in detail the case $d = 1$ and indicate at the end some additional arguments needed for the multidimensional extension. Thus, we have to prove the above formula for a function $F$ on $[0, T] \times \mathbb{R}$, i.e., that a.s.,

$$F(t, X(t)) - F(0, X(0)) \tag{7.3.14}$$
$$= \int_0^t \frac{\partial F}{\partial s}(s, X(s))ds + \int_0^t \frac{\partial F}{\partial x}(s, X(s))dM(s)$$
$$+ \int_0^t \frac{\partial F}{\partial x}(s, X(s))dA(s) + \frac{1}{2}\int_0^t \frac{\partial^2 F}{\partial x^2}(s, X(s))d\langle M\rangle(s).$$

Set $\tau_n = 0$ if $|X(0)| > n$ and $\tau_n = T \wedge \inf\{t \geq 0 : \max(|M(t)|, V_{0t}(A), |\langle M\rangle(t)|) > n\}$ if $|X(0)| \leq n$. Then $\tau_n \to T$ as $n \to \infty$ a.s., and so arguing in the same way as before we conclude that the theorem can be proved, first, for each $X(t \wedge \tau_n)$ and letting $n \to \infty$ the equality (7.3.14) will follow. Hence, we can restrict ourselves to the case when $M(t)$, $t \in [0, T]$ is a martingale, $X(0)$, $M(t)$, $V_{0T}(A)$ and $A(t)$ are bounded on $[0, T] \times \Omega$ and $F$ has a compact support, in particular, it is bounded together with one derivative in $t$ and two derivatives in $x$.

Let $\Delta = \{0 = t_0 < t_1 < ... < t_m = t\}$ be a partition of the interval $[0, t]$. Using the Taylor formula we write

$$F(t, X(t)) - F(0, X(0))$$
$$= \sum_{i=1}^m \left(F(t_k, X(t_k)) - F(t_{k-1}, X(t_{k-1}))\right) = I_0^\Delta + I_1^\Delta + I_2^\Delta$$

where

$$I_0^\Delta = \sum_{i=1}^m \frac{\partial F}{\partial t}(\eta_k, X(t_k))(t_k - t_{k-1}), \ t_{k-1} \leq \eta_k \leq t_k,$$

$$I_1^\Delta = \sum_{i=1}^m \frac{\partial F}{\partial x}(t_{k-1}, X(t_{k-1}))(X(t_k) - X(t_{k-1})),$$

$$I_2^\Delta = \frac{1}{2}\sum_{i=1}^m \frac{\partial^2 F}{\partial x^2}(t_{k-1}, \xi_k)(X(t_k) - X(t_{k-1}))^2$$

and $X(t_k) \wedge X(t_{k-1}) \leq \xi_k \leq X(t_k) \vee X(t_{k-1})$. It is easy to see that

$$I_0^\Delta \to \int_0^t \frac{\partial F}{\partial s}(s, X(s))ds \text{ as } |\Delta| = \max_{1 \leq k \leq m} |t_k - t_{k-1}| \to 0.$$

Next, write $I_1^\Delta = I_3^\Delta + I_4^\Delta$ where

$$I_3^\Delta = \sum_{i=1}^m \frac{\partial F}{\partial x}(t_{k-1}, X(t_{k-1}))(M(t_k) - M(t_{k-1})) \text{ and}$$

$$I_4^{\Delta} = \sum_{i=1}^{m} \frac{\partial F}{\partial x}(t_{k-1}, X(t_{k-1}))(A(t_k) - A(t_{k-1})).$$

As we are dealing with a Riemann–Stieltjes integral it is easy to see that

$$I_4^{\Delta} \to \int_0^t \frac{\partial F}{\partial x}(s, X(s))dA(s) \text{ as } |\Delta| \to 0.$$

Set

$$\Phi^{\Delta}(s, \omega) = \mathbb{I}_{s=0}(s)\frac{\partial F}{\partial x}(0, X(0)) + \sum_{k=1}^{m} \mathbb{I}_{(t_{k-1}, t_k]}(s)\frac{\partial F}{\partial x}(t_{k-1}, X(t_{k-1}))$$

and $\Phi(s, \omega) = \frac{\partial F}{\partial x}(s, X(s))$. Then $\Phi^{\Delta} \in \mathcal{L}_0$ (the space of simple processes, see Lecture 14) and as $|\Delta| \to 0$,

$$\|\Phi^{\Delta} - \Phi\|_t^M = \left(E\int_0^t (\Phi^{\Delta}(s, \omega) - \Phi(s, \omega))^2 d\langle M\rangle(s)\right)^{1/2} \to 0.$$

It follows that in $L^2(\Omega, P)$,

$$I_3^{\Delta} = \int_0^t \Phi^{\Delta}(s, \omega)dM(s) \to \int_0^t \frac{\partial F}{\partial x}(s, X(s))dM(s) \text{ as } |\Delta| \to 0.$$

It remains to deal with $I_2^{\Delta}$ which we write as $I_2^{\Delta} = I_5^{\Delta} + I_6^{\Delta} + I_7^{\Delta}$ where

$$I_5^{\Delta} = \frac{1}{2}\sum_{i=1}^{m} \frac{\partial^2 F}{\partial x^2}(t_{k-1}, \xi_k)(A(t_k) - A(t_{k-1}))^2,$$

$$I_6^{\Delta} = \sum_{i=1}^{m} \frac{\partial^2 F}{\partial x^2}(t_{k-1}, \xi_k)(M(t_k) - M(t_{k-1}))(A(t_k) - A(t_{k-1}))$$

and $I_7^{\Delta} = \frac{1}{2}\sum_{i=1}^{m} \frac{\partial^2 F}{\partial x^2}(t_{k-1}, \xi_k)(M(t_k) - M(t_{k-1}))^2.$

It is easy to see that $I_5^{\Delta} \to 0$ and $I_6^{\Delta} \to 0$ as $|\Delta| \to 0$. Indeed, as $|\Delta| \to 0$,

$$|I_5^{\Delta}| \leq \sup_{s\in[0,T], x\in\mathbb{R}} |\frac{\partial^2 F}{\partial x^2}(s, x)| \max_{1\leq k\leq m} |A(t_k) - A(t_{k-1})|V_{0T}(A) \to 0$$

and

$$|I_6^{\Delta}| \leq \sup_{s\in[0,T], x\in\mathbb{R}} |\frac{\partial^2 F}{\partial x^2}(s, x)| \max_{1\leq k\leq m} |M(t_k) - M(t_{k-1})|V_{0T}(A) \to 0$$

since $\sum_{k=1}^{m} |A(t_k) - A(t_{k-1})| \leq V_{0T}(A).$

It remains to show that in $L^1(\Omega, P)$,

$$I_7^\Delta \to \frac{1}{2} \int_0^t \frac{\partial^2 F}{\partial x^2}(s, x) d\langle M \rangle(s) \text{ as } |\Delta| \to 0. \tag{7.3.15}$$

Set $V_l^\Delta = \sum_{k=1}^l (M(t_k) - M(t_{k-1}))^2$ where $l = 1, 2, ..., m$. In order to obtain (7.3.15) we will need the estimate

$$E(V_m^\Delta)^2 \leq 12C^4 \tag{7.3.16}$$

provided $\sup_{0 \leq s \leq t} |M(s)| \leq C$. Indeed,

$$(V_m^\Delta)^2 = \sum_{k=1}^m (M(t_k) - M(t_{k-1}))^4 + 2 \sum_{k=1}^m (V_m^\Delta - V_k^\Delta)(M(t_k) - M(t_{k-1}))^2$$

and

$$E(V_m^\Delta - V_k^\Delta | \mathcal{F}_{t_k}) = E\left( \sum_{l=k+1}^m (M(t_l) - M(t_{l-1}))^2 | \mathcal{F}_{t_k} \right)$$
$$= E((M(t) - M(t_k))^2 | \mathcal{F}_{t_k}) \leq (2C)^2.$$

Hence,

$$E\left( \sum_{k=1}^m (V_m^\Delta - V_k^\Delta)(M(t_k) - M(t_{k-1}))^2 \right)$$
$$\leq (2C)^2 E V_m^\Delta = (2C)^2 E(M(t))^2 \leq 4C^4$$

where we use that $E(M(t_k) - M(t_{k-1}))^2 = EM^2(t_k) - EM^2(t_{k-1})$ since $EM(t_k)M(t_{k-1}) = EM^2(t_{k-1})$ for any square integrable martingale $M$. This together with

$$E\left( \sum_{k=1}^m (M(t_k) - M(t_{k-1}))^4 \right) \leq (2C)^2 E V_m^\Delta \leq 4C^4$$

yields (7.3.16).

Next, set

$$I_8^\Delta = \frac{1}{2} \sum_{i=1}^m \frac{\partial^2 F}{\partial x^2}(t_{k-1}, X(t_{k-1}))(M(t_k) - M(t_{k-1}))^2.$$

Then, as $|\Delta| \to 0$,

$$E|I_7^\Delta - I_8^\Delta| \leq \frac{1}{2} \left( E \max_{1 \leq k \leq m} |\frac{\partial^2 F}{\partial x^2}(t_{k-1}, \xi_k) \right.$$
$$\left. - \frac{\partial^2 F}{\partial x^2}(t_{k-1}, X(t_{k-1}))|^2 \right)^{1/2} (E(V_m^\Delta)^2)^{1/2}$$
$$\leq \sqrt{3}C^2 \left( E \max_{1 \leq k \leq m} |\frac{\partial^2 F}{\partial x^2}(t_{k-1}, \xi_k) \right.$$
$$\left. - \frac{\partial^2 F}{\partial x^2}(t_{k-1}, X(t_{k-1}))|^2 \right)^{1/2} \to 0$$

by the bounded convergence theorem for Lebesgue integrals. Let

$$I_9^\Delta = \frac{1}{2} \sum_{k=1}^m \frac{\partial^2 F}{\partial x^2}(t_{k-1}, X(t_{k-1}))(\langle M\rangle(t_k) - \langle M\rangle(t_{k-1})).$$

Then, by the integral bounded convergence theorem, as $|\Delta| \to 0$,

$$E|I_9^\Delta - \frac{1}{2}\int_0^t \frac{\partial^2 F}{\partial x^2}(s,x)d\langle M\rangle(s)| \to 0.$$

It remains to compare $I_8^\Delta$ and $I_9^\Delta$. We have

$$E|I_8^\Delta - I_9^\Delta|^2 = \tfrac{1}{4}E\left( \sum_{k=1}^m \frac{\partial^2 F}{\partial x^2}(t_{k-1}, X(t_{k-1}))\left((M(t_k)-M(t_{k-1}))^2 \right.\right.$$
$$\left.\left. -(\langle M\rangle(t_k)-\langle M\rangle(t_{k-1}))\right)\right)^2.$$

This together with

$$E\left((M(t_k)-M(t_{k-1}))^2 - (\langle M\rangle(t_k)-\langle M\rangle(t_{k-1}))|\mathcal{F}_{t_{k-1}}\right) = 0$$

yields that

$$E|I_8^\Delta - I_9^\Delta|^2 \le \tfrac{1}{2}\max_{s\in[0,T],x\in\mathbb{R}}|\frac{\partial^2 F}{\partial x^2}(s,x)|$$
$$\times E\sum_{k=1}^m \left((M(t_k)-M(t_{k-1}))^4 + (\langle M\rangle(t_k)-\langle M\rangle(t_{k-1}))^2\right)$$
$$\le \tfrac{1}{2}\max_{s\in[0,T],x\in\mathbb{R}}|\frac{\partial^2 F}{\partial x^2}(s,x)|$$
$$\times E\left( \max_{1\le k\le m}(M(t_k)-M(t_{k-1}))^2 V_m^\Delta \right.$$
$$+ \max_{1\le k\le m}(\langle M\rangle(t_k)-\langle M\rangle(t_{k-1}))\langle M\rangle(t))$$
$$\le \tfrac{1}{2}\max_{s\in[0,T],x\in\mathbb{R}}|\frac{\partial^2 F}{\partial x^2}(s,x)|$$
$$\times\left((E(V_m^\Delta)^2)^{1/2}(E\max_{1\le k\le m}(M(t_k)-M(t_{k-1}))^4)^{1/2}\right.$$
$$+ \max_{1\le k\le m}(\langle M\rangle(t_k)-\langle M\rangle(t_{k-1}))\langle M\rangle(t)).$$

Taking into account the bound (7.3.16) we conclude that the last sum tends to 0 as $|\Delta| \to 0$ proving (7.3.15). Thus, the Itô formula holds true for each fixed $t$ a.s. and since both sides of the formula are a.s. continuous we obtain its validity for all $t \in [0,T]$ a.s. simultaneously, completing the proof in the one-dimensional case.

In the multidimensional case the proof is similar to the above except that we will have to show that in $L^1(\Omega, P)$,

$$\sum_{k=1}^m \frac{\partial^2 F}{\partial x_i \partial x_j}(t_{k-1}, X(t_{k-1}))(M_i(t_k)-M_i(t_{k-1}))(M_j(t_k)-M_j(t_{k-1}))$$
$$\longrightarrow \int_0^t \frac{\partial^2 F}{\partial x_i \partial x_j}(s, X(s))d\langle M_i, M_j\rangle(s) \quad \text{as} \quad |\Delta| \to 0.$$

This is verified in the same way as above taking into account that

$$E\big((M_i(t_k) - M_i(t_{k-1}))(M_j(t_k) - M_j(t_{k-1}))$$
$$-(\langle M_i, M_j\rangle(t_k) - \langle M_i, M_j\rangle(t_{k-1}))|\mathcal{F}_{t_{k-1}}\big)$$
$$= E\big(M_i(t_k)M_j(t_k) - \langle M_i, M_j\rangle(t_k)$$
$$-(M_i(t_{k-1})M_j(t_{k-1}) - \langle M_i, M_j\rangle(t_{k-1}))|\mathcal{F}_{t_{k-1}}\big) = 0$$

since $M_iM_j - \langle M_i, M_j\rangle$ is a martingale. $\qquad\qquad\qquad\square$

## 7.4 Lecture 16: Lévy's characterization of the Brownian motion and Girsanov theorem

### 7.4.1 *Lévy's characterization of the Brownian motion*

**Theorem 7.4.1.** *Let* $M(t) = (M_1(t), ..., M_d(t))$, $t \geq 0$, $M(0) = 0$ *be a continuous d-dimensional local martingale on a complete probability space* $(\Omega, \mathcal{F}, P)$ *with respect to a filtration* $\mathcal{F}_t$, $t \geq 0$, *i.e., each* $M_k$, $k = 1, ..., d$ *is a continuous local martingale with respect to this filtration, and suppose that* $\langle M_k, M_l\rangle(t) = \delta_{kl}t$ *for all* $t \geq 0$, *where* $\delta_{kl} = 1$ *if* $k = l$ *and* $= 0$ *if* $k \neq l$. *Then* $\{M(t), \mathcal{F}_t\}_{t\geq 0}$ *is the d-dimensional Brownian motion.*

**Proof.** It suffices to show that for any $t \geq s$ and a real $d$-vector $u = (u_1, ..., u_d)$,

$$E\big(e^{i(u, M(t)-M(s))}|\mathcal{F}_s\big) = e^{\frac{1}{2}|u|^2(t-s)}, \qquad\qquad (7.4.1)$$

where $i = \sqrt{-1}$, $(u, v) = \sum_{k=1}^{d} u_k v_k$ denotes the inner product of $u, v \in \mathbb{R}^d$ and $|u|^2 = (u, u)$, since by the properties of characteristic functions this would imply that $M(t) - M(s)$ is independent of $\mathcal{F}_s$, $M_k(t) - M_k(s)$, $k = 1, ..., d$ are independent and all of them have the normal distribution with zero mean and the variance $t - s$. Set $f(x) = e^{i(u,x)}$, $x = (x_1, ..., x_d)$. Then $\frac{\partial f}{\partial x_k}(x) = iu_k f(x)$ and $\frac{\partial^2 f}{\partial x_k \partial x_l}(x) = -u_k u_l f(x)$. Hence, taking into account that $\langle M_k, M_l\rangle(t) = \delta_{kl}t$, it follows by the multidimensional Itô formula (applied separately to the real and imaginary parts of $f$) that

$$e^{i(u, M(t))} = e^{i(u, M(s))} + i\sum_{k=1}^{d} u_k \int_s^t e^{i(u, M(r))}dM_k(r)$$
$$-\tfrac{1}{2}\sum_{k=1}^{d} u_k^2 \int_s^t e^{i(u, M(r))}dr.$$

We multiply this equality by $e^{-i(u, M(s))}\mathbb{I}_\Gamma$ where $\Gamma \in \mathcal{F}_s$ and take the expectation. Setting $\varphi_\Gamma(r) = E\big(e^{i(u, M(r)-M(s))}\mathbb{I}_\Gamma\big)$ we obtain

$$\varphi_\Gamma(t) = P(\Gamma) + i\sum_{k=1}^{d} u_k E\big(e^{-i(u, M(s))}\mathbb{I}_\Gamma E(\int_s^t e^{i(u, M(r))}dM_k(r)|\mathcal{F}_s)\big)$$
$$-\tfrac{1}{2}|u|^2 \int_s^t \varphi_\Gamma(r)dr = P(\Gamma) - \tfrac{1}{2}|u|^2 \int_s^t \varphi_\Gamma(r)dr$$

where we take into account that the conditional expectation above is zero. This yields the differential equation $\varphi'_\Gamma(t) = -\frac{1}{2}|u|^2\varphi_\Gamma(t)$ with the initial condition $\varphi_\Gamma(s) = P(\Gamma)$ and solving it we obtain $\varphi_\Gamma(t) = P(\Gamma)e^{-\frac{1}{2}|u|^2(t-s)}$ implying (7.4.1) since $\Gamma \in \mathcal{F}_s$ is arbitrary. $\qquad\square$

**Remark 7.4.1.** Differentiating the integral $\int_s^t \varphi_\Gamma(r)dr$ in $t$ is legitimate since $\varphi_\Gamma$ is continuous as the result of our assumption that $M$ is a continuous martingale. To see that this condition cannot be dropped consider the Poisson process $N(t)$, $t \geq 0$ with the parameter 1, i.e., a process with independent increments such that $P\{N(t) - N(s) = k\} = e^{-(t-s)}\frac{(t-s)^k}{k!}$, $t \geq s \geq 0$, $k = 0, 1, 2, ...$ Then it is easy to verify that $M(t) = N(t) - t$ is a martingale and $(N(t)-t)^2 - t$ is a martingale as well, i.e., $\langle M \rangle(t) = t$. Still, since $M$ is not continuous (though, right-continuous) it does not satisfy the conditions of the above theorem.

### 7.4.2 *Exponential martingales*

**Theorem 7.4.2.** *Let $M \in \mathcal{M}_2^c[0,T]$, $T < \infty$, $M(0) = 0$ be a square integrable continuous martingale on a complete probability space $(\Omega, \mathcal{F}, P)$ with respect to a filtration $\{\mathcal{F}_t\}_{t\in[0,T]}$. Set*

$$X(t) = \exp\left(M(t) - \frac{1}{2}\langle M \rangle(t)\right), \ t \in [0,T].$$

*Then $X(t)$, $t \in [0,T]$ is a martingale if and only if*

$$EX(t) = 1 \quad \text{for all} \quad t \in [0,T] \tag{7.4.2}$$

*and the latter holds true provided*

$$E\exp\left(\frac{1}{2}\langle M \rangle(T)\right) < \infty \quad \text{(Novikov's condition)} . \tag{7.4.3}$$

**Proof.** By the Itô formula

$$X(t) = 1 + \int_0^t X(u)dM(u), \ t \in [0,T].$$

Let $\tau_n = T \wedge \inf\{u \geq 0 : X(u) > n\}$ and set

$$X_n(t) = X(\tau_n \wedge t) = 1 + \int_0^{\tau_n \wedge t} X(u)dM(u) = 1 + \int_0^t X(u)\mathbb{I}_{\{\tau_n > u\}}dM(u). \tag{7.4.4}$$

Hence, $X_n(t)$, $t \in [0,T]$ is a martingale (as a stochastic integral) while $X(t)$, $t \in [0,T]$ is a local martingale. Now, for $0 \le s < t$ by the Fatou lemma (for conditional expectations),

$$E(X(t)|\mathcal{F}_s) = E(\liminf_{n\to\infty} X_n(t)|\mathcal{F}_s) \le \liminf_{n\to\infty} E(X_n(t)|\mathcal{F}_s)$$
$$= \liminf_{n\to\infty} X_n(s) = X(s),$$

and so $X(t)$, $t \in [0,T]$ is a supermartingale. If $X(t)$, $t \in [0,T]$ is a martingale then $EX(t) = EX(0) = 1$ for all $t \in [0,T]$. On the other hand, if $EX(T) = 1$ then for $0 \le s \le t \le T$,

$$0 = E(X(0) - X(T)) \ge E(X(s) - X(T))$$
$$\ge E(X(s) - X(t)) = E\big(X(s) - E(X(t)|\mathcal{F}_s)\big).$$

But $X(s) \ge E(X(t)|\mathcal{F}_s)$ a.s., since $X$ is a supermartingale, and so by the above inequality $X(s) = E(X(t)|\mathcal{F}_s)$ a.s., i.e., in fact, $X$ is a martingale.

It remains to prove (7.4.2) under (7.4.3) knowing that always

$$EX(t) \le 1 \quad \text{for all} \quad t \in [0,T] \tag{7.4.5}$$

since $X$ is a supermartingale. In the proof we will follow [41]. For each $\varepsilon \in (0,1)$ set $M_\varepsilon(t) = (1-\varepsilon)M(t)$ and $X_\varepsilon(t) = \exp(M_\varepsilon(t) - \frac{1}{2}\langle M_\varepsilon\rangle(t))$, $t \in [0,T]$. We will show first that

$$EX_\varepsilon(t) = 1 \quad \text{for each} \quad t \in [0,T]. \tag{7.4.6}$$

Observe that by (7.4.3),

$$E \exp\big(\frac{1}{2}(1+\varepsilon)^2\langle M_\varepsilon\rangle(T)\big) = E \exp\big(\frac{1}{2}(1-\varepsilon^2)^2\langle M\rangle(T)\big) < \infty. \tag{7.4.7}$$

Since $(1+\varepsilon)M_\varepsilon(t) = (1-\varepsilon^2)M(t) = M_{\varepsilon^2}(t)$, we have in the same way as in (7.4.5) that

$$EX_{\varepsilon^2}(t) \le 1 \quad \text{for all} \quad t \in [0,T]. \tag{7.4.8}$$

Hence, by (7.4.3) and the Cauchy–Schwarz inequality,

$$Ee^{\frac{1}{2}M_{\varepsilon^2}(t)} = E\big(e^{\frac{1}{2}(M_{\varepsilon^2}(t) - \frac{1}{2}\langle M_{\varepsilon^2}\rangle(t))}e^{\frac{1}{4}\langle M_{\varepsilon^2}\rangle(t)}\big) \tag{7.4.9}$$
$$\le (EX_{\varepsilon^2}(t))^{1/2}(Ee^{\frac{1}{2}M_{\varepsilon^2}(t)})^{1/2} \le Ee^{\frac{1}{2}(1-\varepsilon^2)^2\langle M\rangle(T)} < \infty.$$

Next, we are going to apply martingale inequalities from Lecture 11 to positive local martingales $X_\varepsilon(t)$, $t \in [0,T]$. This is possible since if for $r > 1$,

$$E\big(\sup_{t\in[0,T\wedge\tau_n]} X_\varepsilon^r(t)\big) \le \big(\frac{r}{r-1}\big)^r EX_\varepsilon^r(T)$$

then letting $n \uparrow \infty$ (and so $\tau_n \uparrow T$) we obtain

$$E(\sup_{t\in[0,T]} X_\varepsilon^r(t)) \le (\frac{r}{r-1})^r EX_\varepsilon^r(T). \qquad (7.4.10)$$

Now, set $r = \frac{(1+\varepsilon)^2}{1+2\varepsilon} > 1$, $p = 1 + 2\varepsilon$ and $q = p(p-1)^{-1}$. Then by (7.4.8)–(7.4.10) and the Hölder inequality

$$E(\sup_{t\in[0,T]} X_\varepsilon^r(t)) \le (\tfrac{r}{r-1})^r EX_\varepsilon^r(T)$$

$$= (\tfrac{r}{r-1})^r E\big(e^{(\frac{r}{p})^{1/2}M_\varepsilon(T)-\frac{r}{2}\langle M_\varepsilon\rangle(T)}e^{(r-(\frac{r}{p})^{1/2})M_\varepsilon(T)}\big)$$

$$\le (\tfrac{r}{r-1})^r \big(Ee^{\sqrt{rp}M_\varepsilon(T)-\frac{1}{2}rp\langle M_\varepsilon\rangle(T)}\big)^{1/p}\big(Ee^{(p-1)^{-1}\sqrt{rp}(\sqrt{rp}-1)M_\varepsilon(T)}\big)^{1/q}$$

$$= (\tfrac{r}{r-1})^r(EX_{\varepsilon^2}(T))^{1/p}Ee^{\frac{1}{2}M_{\varepsilon^2}(T)} \le (\tfrac{r}{r-1})^r Ee^{\frac{1}{2}M_{\varepsilon^2}(T)} < \infty.$$

Hence, also $E(\sup_{t\in[0,T]} X_\varepsilon(t)) < \infty$.

Let $\tau_n^\varepsilon = T \wedge \inf\{u \ge 0 : X_\varepsilon(u) > n\}$. Then, using the Itô formula as in (7.4.4) we conclude that $X_{\varepsilon,n}(t) = X_\varepsilon(t \wedge \tau_n)$, $t \in [0,T]$ is a martingale for each $n = 1, 2, ...$, and so

$$EX_{\varepsilon,n}(t) = 1 \quad \text{for all} \quad t \in [0,T]. \qquad (7.4.11)$$

Since $\sup_{s\in[0,T]} X_\varepsilon(s) \ge X_{\varepsilon,n}(t)$ for all $t \in [0,T]$, we obtain from (7.4.11) and the dominated convergence theorem that

$$EX_\varepsilon(t) = E\lim_{n\to\infty} X_{\varepsilon,n}(t) = \lim_{n\to\infty} EX_{\varepsilon,n}(t) = 1.$$

Finally, by the Hölder inequality

$$1 = EX_\varepsilon(t) = E\big(e^{(1-\varepsilon)(M(t)-\frac{1}{2}\langle M\rangle(t))}e^{\frac{1}{2}(1-\varepsilon)\varepsilon\langle M\rangle(t)}\big)$$

$$\le (EX(t))^{1-\varepsilon}(Ee^{\frac{1}{2}(1-\varepsilon)\langle M\rangle(t)})^\varepsilon.$$

Letting $\varepsilon \downarrow 0$ we obtain that $EX(t) \ge 1$ which together with (7.4.5) completes the proof of (7.4.2) yielding that $X(t)$, $t \in [0,T]$ is a martingale. $\square$

**Corollary 7.4.1.** *Let $M(t) = \int_0^t f(s,\omega)dW(s)$, $t \in [0,T]$ be a stochastic integral with respect to the Brownian motion $W$ where $f$ is a non-anticipating function satisfying $E\exp(\frac{1}{2}\int_0^T f^2(s)ds) < \infty$. Then*

$$X(t) = \exp\big(\int_0^t f(s)dW(s) - \frac{1}{2}\int_0^t f^2(s)ds\big), \ t \in [0,T]$$

*is a martingale and $EX(t) = 1$ for each $t \in [0,T]$. The result remains true in the multidimensional case, i.e., when $M(t) = \sum_{l=1}^d \int_0^t f_l(s,\omega)dW_l(s)$, $E\exp(\frac{1}{2}\sum_{l=1}^k \int_0^T f_l^2(s)ds) < \infty$ and $W_1, ..., W_d$ are independent Brownian motions. Then*

$$X(t) = \exp\big(\sum_{l=1}^d \int_0^t f_l(s)dW_l(s) - \frac{1}{2}\sum_{l=1}^d \int_0^t f_l^2(s)ds\big), \ t \in [0,T]$$

*is a martingale and $EX(t) = 1$ for each $t \in [0,T]$.*

**Proof.** Since for $0 \le s < t \le T$,

$$E\left(\left(\int_s^t f(u)dW(u)\right)^2 | \mathcal{F}_s\right) = E\left(\int_s^t f^2(u)du | \mathcal{F}_s\right)$$

we obtain that

$$M(t) = \left(\int_0^t f(u)dW(u)\right)^2 - \int_0^t f^2(u)du$$

satisfies

$$E(M(t)|\mathcal{F}_s) = E\left(\left(\int_s^t f(u)dW(u)\right)^2 | \mathcal{F}_s\right) + 2\int_0^s f(u)dW(u)$$

$$\times E\left(\int_s^t f(u)dW(u)|\mathcal{F}_s\right) + \left(\int_0^s f(u)dW(u)\right)^2 - \int_0^t f^2(u)du = M(s).$$

Hence, $M(t)$, $t \in [0,T]$ is a martingale, and so

$$\langle M \rangle(t) = \int_0^t f^2(u)du.$$

Similarly, it is easy to see that in the multidimensional version $\langle M \rangle(t) = \sum_{l=1}^d \int_0^t f_l^2(u)du$. Thus, the assertion of the corollary follows from the above theorem. $\qquad\square$

### 7.4.3  *Girsanov theorem*

Suppose that a $d$-dimensional stochastic process $\xi(t) = (\xi_1(t), ..., \xi_d(t))$, $t \in [0,T]$ on a complete probability space $(\Omega, \mathcal{F}, P)$ has the stochastic differential

$$d\xi(t) = f(t,\omega)dW(t) + g(t,\omega)dt, \ t \in [0,T] \qquad (7.4.12)$$

where $W = (W_1, ..., W_r)$ is the $r$-dimensional Brownian motions, $f = (f_{ij})_{i=1,...,d, \, j=1,...,r}$ and $g = (g_i)_{i=1,...,d}$ are $d \times r$-matrix and $d$-vector functions, respectively, adapted to the filtration $\{\mathcal{F}_t = \sigma\{W_1(s), ..., W_r(s), s \le t\}, t \in [0,T]\}$ and satisfying $\int_0^T (f_{ij}^2(s) + |g_i(s)|)ds < \infty$ for all $i = 1, ..., d$, $j = 1, ..., r$ a.s.

**Theorem 7.4.3.** *Suppose that* $u(t,\omega) = (u_1(t,\omega), ..., u_r(t,\omega))$ *and* $h(t,\omega) = (h_1(t,\omega), ..., h_d(t,\omega))$, $t \in [0,T]$ *are adapted to the above filtration vector functions such that* $\int_0^T (u_i^2(s) + |h_j(s)|)ds < \infty$ *for all* $i = 1, ..., r$, $j = 1, ..., d$ *a.s. and*

$$f(t,\omega)u(t,\omega) = g(t,\omega) - h(t,\omega). \qquad (7.4.13)$$

*Set*

$$M_t = \exp\left(-\sum_{i=1}^r \int_0^t u_i(s)dW_i(s) - \frac{1}{2}\sum_{i=1}^r \int_0^t u_i^2(s)ds\right), \ t \in [0,T]$$

*and define the new measure $Q$ on $(\Omega, \mathcal{F}_T)$ by*

$$dQ = M_T dP \quad \text{i.e.,} \quad Q(\Gamma) = \int_\Gamma M_T dP \quad \text{for any} \quad \Gamma \in \mathcal{F}_T.$$

*If $M_t$, $t \in [0, T]$ is a martingale on $(\Omega, \mathcal{F}, P)$ with respect to the filtration $\mathcal{F}_t$, $t \in [0, T]$ then $Q$ is a probability measure and the process*

$$\hat{W}(t) = \int_0^t u(s)ds + W(t), \ t \in [0, T]$$

*is a Brownian motion with respect to the probability $Q$ and we can write*

$$d\xi(t) = f(t, \omega)d\hat{W}(t) + h(t, \omega)dt. \tag{7.4.14}$$

*In particular, if $r = d$, the matrix $f(t, \omega)$ is nondegenerate for all $(t, \omega)$, and $u = f^{-1}(g - h)$ satisfies the Novikov condition $E \exp(\frac{1}{2} \int_0^T |u|^2(s)ds) < \infty$ then $Q$ is a probability measure and $\hat{W}$ defined above is a Brownian motion with respect to $Q$.*

**Proof.** Assuming that $M$ is a martingale we have $Q(\Omega) = E_Q(1) = E_P M(T) = E_P M(0) = 1$ where $E_Q$ and $E_P$ are the expectations with respect to $Q$ and $P$, respectively. In order to show that $\hat{W}$ is a Brownian motion with respect to $Q$ we observe, first, that $\hat{W}$ is a continuous process and according to the Lévy theorem above we have only to verify that $\hat{W}_i$, $i = 1, ..., r$ and $\hat{W}_i(t)\hat{W}_j(t) - \delta_{ij}t$, $i, j = 1, ..., r$ are local martingales with respect to the probability $Q$. The latter condition means that $\langle \hat{W}_i(t), \hat{W}_j(t) \rangle = \delta_{ij}t$ with $\langle \cdot, \cdot \rangle$ considered on the probability space $(\Omega, \mathcal{F}, Q)$.

Set $X(t) = -\sum_{i=1}^r \int_0^t u_i(s)dW_i(s) - \frac{1}{2}\sum_{i=1}^r \int_0^t u_i^2(s)ds$, $Y(t) = \hat{W}(t)$ and $F(X(t), Y(t)) = e^{X(t)}Y(t) = M(t)Y(t)$ where $F(x, y) = e^x y$, $x \in \mathbb{R}$, $y = (y_1, ..., y_r) \in \mathbb{R}^r$ is a vector function on $\mathbb{R} \times \mathbb{R}^r$. Write $F_i(x, y) = e^x y_i$ and apply the multidimensional Itô formula to each component of $F$. Namely, we consider the new $(r+1)$-dimensional stochastic process $\eta(t) = (X(t), Y(t)) = (X(t), Y_1(t), ..., Y_r(t))$ which has the stochastic differential $d\eta(t) = a(t)W(t) + b(t)dt$ where $a(t) = (a_{ij})_{i=1,...,r+1, j=1,...,r}$ is a matrix and $b(t) = (b_i(t))_{i=1}^{r+1}$ is a vector such that $a_{i,i-1}(t) = 1$ and $b_i(t) = u_i(t)$ for $i = 2, ..., r+1$, $a_{1,j}(t) = -u_j(t)$ for $j = 1, ..., r$, $a_{ij}(t) = 0$ for all other indexes $i, j$ and $b_1(t) = -\frac{1}{2}\sum_{i=1}^r u_i^2$.

Now we can write

$$F(\eta(t)) = F(X(t), Y(t)) = (F_1(X(t), Y(t)), ..., F_r(X(t), Y(t)))$$
$$= (F_1(\eta(t)), ..., F_r(\eta(t))).$$

Next, we compute partial derivatives

$$\frac{\partial F_i}{\partial x}(\eta(t)) = F_i(\eta(t)), \ \frac{\partial F_i}{\partial y_i}(\eta(t)) = M(t), \ \frac{\partial F_i}{\partial y_j}(\eta(t)) = 0 \text{ if } i \neq j,$$

$$\frac{\partial^2 F_i}{\partial x^2}(\eta(t)) = F_i(\eta(t)), \ \frac{\partial^2 F_i}{\partial x \partial y_i}(\eta(t)) = M(t), \ \frac{\partial^2 F_i}{\partial x \partial y_j}(\eta(t)) = 0 \text{ if } i \neq j$$

$$\text{and } \frac{\partial^2 F_i}{\partial y_j \partial y_k}(\eta(t)) = 0 \text{ for any } j, k = 1, ..., r.$$

Taking into account cancelations we obtain by the Itô formula that

$$dF_i(\eta(t)) = -F_i(\eta(t)) \sum_{k=1}^{r} u_k(t) dW_k(t) + M(t) dW_i.$$

Hence, $F_i(\eta(t)) = M(t)Y_i(t)$, $i = 1, ..., r$ are stochastic integrals, and so they are local martingales with respect to the probability $P$. Hence, by the generalized Bayes formula from Lecture 7 for $s \leq t \leq T$,

$$E_Q(Y_i(t \wedge \tau_n)|\mathcal{F}_s) = \frac{1}{M(s)} E_P(M(t)Y_i(t \wedge \tau_n)|\mathcal{F}_s) = Y_i(s \wedge \tau_n),$$

where $\tau_n$, $n \geq 1$ is the localizing sequence of stopping times, i.e., $Y_i(t) = \hat{W}_i(t)$, $t \in [0, T]$ is a local martingale with respect to $Q$ for each $i = 1, ..., r$, as required.

Next, write $G_{ij}(t, X(t), Y(t)) = e^{X(t)}(Y_i(t)Y_j(t) - \delta_{ij}t)$ where $G(t, x, y) = e^x(y_i y_j - \delta_{ij}t)$ is a function on $[0, T] \times \mathbb{R} \times \mathbb{R}^r$. Relevant for the Itô formula partial derivatives of $G_{ij}$ have the form $\frac{\partial G_{ij}}{\partial t}(t, \eta(t)) = -M(t)\delta_{ij}$, $\frac{\partial G_{ij}}{\partial x}(t, \eta(t)) = G_{ij}(t, \eta(t))$, $\frac{\partial G_{ij}}{\partial y_k}(t, \eta(t)) = M(t)Y_l(t)$ if $k = i \neq j$ and $l = j$ or $k = j \neq i$ and $l = i$, $\frac{\partial G_{ij}}{\partial y_i}(t, \eta(t)) = 2M(t)Y_i(t)$, $\frac{\partial G_{ij}}{\partial y_k}(t, \eta(t)) = 0$ if $k \neq i$ and $k \neq j$, $\frac{\partial^2 G_{ij}}{\partial x^2}(t, \eta(t)) = G_{ij}(t, \eta(t))$, $\frac{\partial^2 G_{ij}}{\partial x \partial y_k}(t, \eta(t)) = M(t)Y_l(t)$ if $k = i \neq j$ and $l = j$ or $k = j \neq i$ and $l = i$, $\frac{\partial^2 G_{ij}}{\partial x \partial y_i}(t, \eta(t)) = 2M(t)Y_i(t)$, $\frac{\partial^2 G_{ij}}{\partial x \partial y_k}(t, \eta(t)) = 0$ if $k \neq i$ and $k \neq j$, $\frac{\partial^2 G_{ij}}{\partial y_i \partial y_j}(t, \eta(t)) = M(t)$ if $i \neq j$, $\frac{\partial^2 G_{ii}}{\partial y_i^2}(t, \eta(t)) = 2M(t)$ and $\frac{\partial^2 G_{ij}}{\partial y_k \partial y_l}(t, \eta(t)) = 0$ unless $k = i$ and $l = j$ or $k = i$ and $l = j$. Again, taking into account cancelations we obtain from the Itô formula that

$$dG_{ij}(t, \eta(t)) = -G_{ij}(t, \eta(t)) \sum_{k=1}^{r} u_k(t) dW_k$$
$$+ M(t)(Y_j(t) dW_i(t) + Y_i(t) dW_j(t)).$$

Hence, $G_{ij}(t, \eta(t)) = M(t)(Y_i(t)Y_j(t) - \delta_{ij}t)$, $i, j = 1, ..., r$ are stochastic integrals, and so they are local martingales with respect to the probability $P$. Hence, by the generalized Bayes formula from Lecture 7 for $s \leq t \leq T$,

$$E_Q(Y_i(t \wedge \tau_n)Y_j(t \wedge \tau_n) - \delta_{ij}t \wedge \tau_n|\mathcal{F}_s) = \frac{1}{M(s)} E_P(M(t)$$
$$\times (Y_i(t \wedge \tau_n)Y_j(t \wedge \tau_n) - \delta_{ij}t \wedge \tau_n)|\mathcal{F}_s) = Y_i(s \wedge \tau_n)Y_j(s \wedge \tau_n) - \delta_{ij}s \wedge \tau_n,$$

where, again, $\tau_n$, $n \geq 1$ is the localizing sequence of stopping times, i.e., $Y_i(t)Y_j(t) - \delta_{ij}t = \hat{W}_i(t)\hat{W}_j(t) - \delta_{ij}t$, $t \in [0,T]$ is a local martingale with respect to $Q$ for any $i,j = 1,...,r$. Therefore, by the Lévy theorem above $\hat{W}$ is the Brownian motion with respect to $Q$. Finally, substituting $dW(t) = d\hat{W}(t) - u(t)dt$ into (7.4.12) and using (7.4.13) we obtain (7.4.14), completing the proof of the theorem. □

## 7.5   Lecture 17: Stochastic differential equations

### 7.5.1   *Existence and uniqueness of solutions*

We will use here the Euclidean norms $|b| = (\sum_{i=1}^{d} b_i^2)^{1/2}$ and $|\sigma| = (\sum_{1 \leq i \leq d, 1 \leq j \leq r} \sigma_{ij}^2)^{1/2}$ for each vector $b = (b_1,...,b_d)$ and a matrix $\sigma = (\sigma_{ij})_{1 \leq i \leq d, 1 \leq j \leq r}$. Let $W(t) = (W_1(t),...,W_r(t))$ be the $r$-dimensional continuous Brownian motion and $\sigma(t)$, $t \in [0,T]$ be a $d \times r$ adapted to the Brownian filtration random matrix-function. Assume that

$$\sum_{1 \leq i \leq d, 1 \leq j \leq r} E \int_0^T \sigma_{ij}^2(s)ds < \infty. \tag{7.5.1}$$

Then, in the same way as for the multidimensional Itô formula, we compute that for each $t \in [0,T]$,

$$E\left| \int_0^t \sigma(s)dW(s)\right|^2 = E\sum_{i=1}^d \left( \int_0^t \sum_{j=1}^r \sigma_{ij}(s)dW_j(s)\right)^2 \tag{7.5.2}$$

$$= E\sum_{i=1}^d \sum_{j=1}^r \int_0^t \sigma_{ij}^2(s)ds = \sum_{i=1}^d \sum_{j=1}^r \int_0^t E\sigma_{ij}^2(s)ds = \int_0^t E|\sigma(s)|^2 ds.$$

Observe also that $|\int_0^t \sigma(s)dW(s)|^2$ is a continuous submartingale. Indeed, for $0 \leq u \leq t$,

$$E\left(|\int_0^t \sigma(s)dW(s)|^2|\mathcal{F}_u\right) = E\left(|\int_u^t \sigma(s)dW(s)|^2|\mathcal{F}_u\right)$$

$$+2\left(E(\int_u^t \sigma(s)dW(s)|\mathcal{F}_u), \int_0^u \sigma(s)dW(s)\right) + |\int_0^u \sigma(s)dW(s)|^2$$

$$= E\left(|\int_u^t \sigma(s)dW(s)|^2|\mathcal{F}_u\right) + |\int_0^u \sigma(s)dW(s)|^2 \geq |\int_0^u \sigma(s)dW(s)|^2.$$

Hence, by (7.5.2) and the submartingale inequality from Lecture 11,

$$E\left( \sup_{0 \leq s \leq t} |\int_0^s \sigma(u)dW(u)|^2\right) \leq 4\int_0^t E|\sigma(s)|^2 ds. \tag{7.5.3}$$

The following result deals with solutions of stochastic differential equations.

**Theorem 7.5.1.** *Let $b = b(t,x) = (b_1(t,x),...,b_d(t,x))$ and $\sigma = \sigma(t,x) = (\sigma_{ij}(t,x)$, $i = 1,...,d$, $j = 1,...,r)$ be measurable vector and matrix functions, respectively, on $[0,T] \times \mathbb{R}^d$, $T > 0$ such that for some $C > 0$, each $t \in [0,T]$ and all $x,y \in \mathbb{R}^d$,*

$$|b(t,x)| + |\sigma(t,x)| \leq C(1 + |x|) \tag{7.5.4}$$

*and the Lipschitz condition*

$$|b(t,x) - b(t,y)| + |\sigma(t,x) - \sigma(t,y)| \le C|x-y| \quad (7.5.5)$$

*hold true. Then the stochastic differential equation*

$$dX(t) = b(t, X(t))dt + \sigma(t, X(t))dW(t), \ t \in [0,T], \quad (7.5.6)$$

*with the $\mathcal{F}_0$-measurable initial condition $X(0)$ satisfying*

$$E|X(0)|^2 < \infty, \quad (7.5.7)$$

*has a unique continuous in t solution $X(t,\omega) = (X_1(t,\omega), ..., X_d(t,\omega)), t \in [0,T]$ such that $X$ is adapted to the Brownian filtration and*

$$\int_0^T E|X(t)|^2 dt < \infty. \quad (7.5.8)$$

**Proof.** Recall, that stochastic differentials should be understood in the stochastic integral form, and so in order to solve (7.5.6) we have to solve the equivalent stochastic integral equation

$$X(t) = X(0) + \int_0^t b(s, X(s))ds + \int_0^t \sigma(s, X(s))dW(s), \ t \in [0,T]. \quad (7.5.9)$$

As it is used for similar purposes in ordinary differential equations we will employ the method of successive approximations. Namely, set $Y^{(0)}(t) \equiv X(0)$, $t \in [0,T]$ and define for $k = 0, 1, 2, ...,$

$$Y^{(k+1)}(t) = X(0) + \int_0^t b(s, Y^{(k)}(s))ds + \int_0^t \sigma(s, Y^{(k)}(s))dW(s). \quad (7.5.10)$$

First, we will prove by induction that for any $k \ge 0$,

$$E \sup_{s\in[0,t]} |Y^{(k)}(s)|^2 \le (1 + E|X(0)|^2) \sum_{j=0}^k \frac{(8(C+1)(t+1))^{2(j+1)}}{(j+1)!} \quad (7.5.11)$$

which together with (7.5.3) and (7.5.4) implies that the integrals in (7.5.10) exist, and so $Y^{(k+1)}$ is well defined by (7.5.10). In view of (7.5.7) the inequality (7.5.11) is trivially true for $k = 0$. Assume that (7.5.11) holds true for all $k = 0, 1, .., n-1$ and prove it for $k = n$. We have

$$E \sup_{s\in[0,t]} |Y^{(n)}(s)|^2 \le 3E|X(0)|^2 + 3A_n(t) + 3B_n(t) \quad (7.5.12)$$

where by (7.5.4), the Cauchy–Schwarz inequality and the induction hypothesis

$$A_n(t) = E\sum_{i=1}^d \left(\int_0^t |b_i(s, Y^{(n-1)}(s))|ds\right)^2$$
$$\le t\int_0^t E|b(s, Y^{(n-1)}(s))|^2 ds \le 2tC^2(t + \int_0^t E|Y^{(n-1)}(s)|^2 ds)$$
$$\le 2C^2 t^2 + 2C^2 t(1 + E|X(0)|^2)\sum_{j=0}^{n-1} \frac{(8(C+1))^{2(j+1)}(t+1)^{2j+3}}{(j+1)!(2j+3)}$$

and by (7.5.3) together with the induction hypothesis

$$B_n(t) = E \sup_{s \in [0,t]} \left| \int_0^s \sigma(u, Y^{(n-1)}(u)) dW(u) \right|^2$$

$$\leq 4 \int_0^t E |\sigma(s, Y^{(n-1)}(s))|^2 ds \leq 8C^2 (t + \int_0^t E |Y^{(n-1)}(s)|^2 ds)$$

$$\leq 8C^2 t + 8C^2 (1 + E|X(0)|^2) \sum_{j=0}^{n-1} \frac{(8(C+1))^{2(j+1)}(t+1)^{2j+3}}{(j+1)!(2j+3)}$$

which together with (7.5.12) yields (7.5.11) for $k = n$.

Next, we will show by induction that for all $k \geq 0$ and $t \in [0, T]$,

$$R_k(t) = E \sup_{s \in [0,t]} |Y^{(k+1)}(s) - Y^{(k)}(s)|^2 \qquad (7.5.13)$$

$$\leq \frac{(4(C+1)(t+1))^{2(k+1)}}{(k+1)!} (1 + E|X(0)|^2).$$

For $k = 0$,

$$R_0(t) = E \sup_{s \in [0,t]} |Y^{(1)}(s) - Y^{(0)}(s)|^2 \leq 2D_0(t) + 2F_0(t),$$

where by (7.5.4) and the Cauchy–Schwarz inequality,

$$D_0(t) = E \sum_{i=1}^d \left( \int_0^t |b_i(s, X(0))| ds \right)^2$$

$$\leq t \int_0^t E|b(s, X(0)|^2 ds \leq 2t^2 C^2 (1 + E|X(0)|^2)$$

and by (7.5.3),

$$F_0(t) = E \sup_{s \in [0,t]} \left| \int_0^s \sigma(u, X(0)) dW(u) \right|^2$$

$$\leq 4 \int_0^t E|\sigma(s, X(0))|^2 ds \leq 8tC^2(1 + E|X(0)|^2)$$

yielding (7.5.13) for $k = 0$.

Suppose now that (7.5.13) holds true for all $k = 0, 1, ..., n-1$ and prove it for $k = n$. We have

$$R_n(t) = E \sup_{s \in [0,t]} |Y^{(n+1)}(s) - Y^{(n)}(s)| \leq 2D_n(t) + 2F_n(t),$$

where by (7.5.5), the Cauchy–Schwarz inequality and the induction hypothesis,

$$D_n(t) = E \sum_{i=1}^d \left( \int_0^t |b_i(s, Y^{(n)}(s)) - b_i(s, Y^{(n-1)}(s))| ds \right)^2$$

$$\leq t \int_0^t E|b(s, Y^{(n)}(s)) - b_i(s, Y^{(n-1)}(s))|^2 ds \leq tC^2 \int_0^t R_{n-1}(s) ds$$

$$\leq \frac{tC^2 (4(C+1))^{2n}(t+1)^{2n+1}}{(n+1)!(2n+1)} (1 + E|X(0)|^2)$$

and by (7.5.3),

$$F_n(t) = E \sup_{s \in [0,t]} \left| \int_0^s (\sigma(u, Y^{(n)}(u)) - \sigma(u, Y^{(n-1)}(u))) dW(u) \right|^2$$

$$\leq 4 \int_0^t E|\sigma(s, Y^{(n)}(s)) - \sigma(s, Y^{(n-1)}(s))|^2 ds \leq 4 \int_0^t R_{n-1}(s) ds$$

$$\leq \frac{4(4(C+1))^{2n}(t+1)^{2n+1}}{(n+1)!(2n+1)} (1 + E|X(0)|^2)$$

and (7.5.13) follows for $k = n$ as well.

Now, by (7.5.13) and the Chebyshev inequality

$$P\{\sup_{t\in[0,T]} |Y^{(k+1)}(t) - Y^{(k)}(t)| > 2^{-k}\} \qquad (7.5.14)$$

$$\leq \frac{(32(C+1)^2(T+1)^2)^{k+1}}{(k+1)!}(1 + E|X(0)|^2).$$

Since the right hand sides of (7.5.14) form a convergent series, we conclude by the Borel–Cantelli lemma that there exists $k_0 = k_0(\omega) < \infty$ a.s. such that

$$\sup_{t\in[0,T]} |Y^{(k+1)}(t) - Y^{(k)}(t)| \leq 2^{-k} \quad \text{for} \quad k \geq k_0.$$

Therefore, the sequence

$$Y^{(n)}(t) = Y^{(0)}(t) + \sum_{k=0}^{n-1}(Y^{(k+1)}(t) - Y^{(k)}(t))$$

is uniformly convergent on $[0,T]$ with probability one. Set $X(t) = \lim_{n\to\infty} Y^{(n)}(t)$ defining $X(t)$ to be 0 on the zero probability set where there is no convergence. Observe also that $X(t)$, $t \in [0,T]$ is continuous in $t$ with probability one as a uniform limit of continuous functions. Clearly, $X(t)$ is $\mathcal{F}_t$-measurable since each $Y^{(n)}(t)$ is.

Observe that

$$\left| \sup_{t\in[0,T]} |Y^{(k)}(t)| - \sup_{t\in[0,T]} |X(t)| \right| \leq \sup_{t\in[0,T]} |Y^{(k)}(t) - X(t)|,$$

and so in view of the uniform convergence of $Y^{(k)}$ to $X$,

$$\lim_{k\to\infty} \sup_{t\in[0,T]} |Y^{(k)}(t)| = \sup_{t\in[0,T]} |X(t)| \quad \text{a.s..}$$

Let $D$ be the limit of the right hand side of (7.5.11) as $k \to \infty$. Then, by the Fatou lemma,

$$E\sup_{t\in[0,T]} |X(t)|^2 = E\liminf_{k\to\infty} \sup_{t\in[0,T]} |Y^{(k)}(t)|^2 \quad (7.5.15)$$

$$\leq \liminf_{k\to\infty} E\sup_{t\in[0,T]} |Y^{(k)}(t)|^2 \leq D < \infty$$

which is even stronger than (7.5.8).

For each random vector $Z$ on $(\Omega, \mathcal{F}, P)$ set $\|Z\| = (E|Z|^2)^{1/2}$ which is the $L^2$-norm of vector functions. Hence, by (7.5.13) and the norms triangle inequality for each $t \in [0,T]$ and $m \geq 1$,

$$\|Y^{(n+m)}(t) - Y^{(n)}(t)\| \leq \sum_{k=n}^{\infty} \|Y^{(k+1)}(t) - Y^{(k)}(t)\| \to 0 \quad \text{as} \quad n \to \infty.$$

It follows that $Y^{(n)}(t)$, $n \geq 0$ is a Cauchy sequence and by completeness of $L^2$ spaces it has a limit. But, $Y^{(n)}(t) \to X(t)$ a.s., and so for each $t \in [0, T]$,
$$\lim_{n \to \infty} E|Y^{(n)}(t) - X(t)|^2 = 0.$$
This together with (7.5.11), (7.5.15) and the bounded convergence theorem yields that

$$\lim_{n \to \infty} \int_0^T E|Y^{(n)}(t) - X(t)|^2 dt = 0. \qquad (7.5.16)$$

Now we can show that $X(t)$, $t \in [0, T]$ is the solution of the stochastic integral equation (7.5.9). Indeed, the left hand side of (7.5.10) converges to $X(t)$ as $k \to \infty$. As for the right hand side of (7.5.10) observe first that by (7.5.5), (7.5.16) and the Cauchy–Schwarz inequality

$$E\left| \int_0^t \left( b(s, Y^{(k)}) - b(s, X(s)) \right) ds \right|^2 \leq T \int_0^T E|b(s, Y^{(k)}) - b(s, X(s))|^2 ds$$

$$\leq TC^2 \int_0^T E|Y^{(k)}(s) - X(s)|^2 ds \to 0 \quad \text{as} \quad k \to \infty.$$

Concerning the stochastic integrals we obtain by (7.5.2), (7.5.5) and (7.5.16) that

$$E\left| \int_0^t \left( \sigma(s, Y^{(k)}(s)) - \sigma(s, X(s)) \right) dW(s) \right|^2$$

$$= \int_0^t E|\sigma(s, Y^{(k)}(s)) - \sigma(s, X(s))|^2 ds$$

$$\leq C^2 \int_0^T E|Y^{(k)}(s) - X(s)|^2 ds \to 0 \quad \text{as} \quad n \to \infty.$$

Thus, the right hand side of (7.5.10) converges as $k \to \infty$ to the right hand side of (7.5.9) for each fixed $t$ which implies (7.5.9) for each fixed $t$ with probability one. Since both parts of (7.5.9) are continuous in $t$ a.s. we conclude that, in fact, (7.5.9) holds true with probability one simultaneously for all $t \in [0, T]$.

It remains to establish uniqueness of the solution of (7.5.9). Suppose that a random adapted vector function $Z$ satisfies

$$Z(t) = X(0) + \int_0^t b(s, Z(s)) ds + \int \sigma(s, Z(s)) dW(s), \quad s \in [0, T]$$

and $\int_0^T E|Z(s)|^2 ds < \infty$. Then by (7.5.2), (7.5.5) and the Cauchy–Schwarz inequality similarly to the above estimates we derive

$$E|X(t) - Z(t)|^2 \leq 2C^2(T+1) \int_0^t E|X(s) - Z(s)|^2 ds.$$

Recall, that the Gronwall inequality (appearing in every textbook on ordinary differential equations) says that if $g$ is a continuous function such that $g(t) \leq A + B \int_0^t g(s) ds$ for some $A, B \geq 0$ and all $t \in [0, T]$ then $g(t) \leq Ae^{Bt}$ for such $t$'s. This means in our situation that $X(t) = Z(t)$ a.s. for each $t \in [0, T]$. Since both $X(t)$ and $Z(t)$ are continuous in $t \in [0, T]$ as the integrals defining them are, we obtain that $X(t) = Z(t)$ for all $t \in [0, T]$ a.s., completing the proof of the theorem. $\qquad \square$

### 7.5.2   Weak and strong solutions

The solution found by successive approximations above is called a strong (path-wise) solution since the version $W(t)$, $t \geq 0$ of the Brownian motion is given in advance and the solution $X(t)$, $t \geq 0$ constructed above is adapted to a given filtration. If we are given only functions $b(t, x)$ and $\sigma(t, x)$ and look for a pair of processes $(X(t), W(t))$, $t \geq 0$ together with a filtration $\mathcal{A}_t$, $t \geq 0$ such that $X$ is adapted to it while $W$ is a Brownian motion with respect to it and the equation (7.5.9) holds true then the pair $(X, W)$ is called a weak solution of (7.5.9) (or of (7.5.6)). We proved the strong (path-wise) uniqueness. The weak uniqueness simply means that any two solutions (weak or strong) are identical in law, i.e., have the same finite dimensional distributions. If $b$ and $\sigma$ satisfy the conditions of the above theorem then any two solutions $X$, $\tilde{X}$ constructed as above with respect to two Brownian motions $W$, $\tilde{W}$, respectively, are identical in law. This follows by observing that for each $k$ the corresponding processes $Y^{(k)}$, $\tilde{Y}^{(k)}$ defined by successive approximations are identical in law at each step of approximation if we start with $Y^{(0)} = \tilde{Y}^{(0)} \equiv X(0)$. Hence, by the strong uniqueness, the limiting solutions must also have the same finite dimensional distributions, i.e., a solution (weak or strong) of a stochastic differential equation with coefficients satisfying the conditions of the above theorem is weakly unique. For more information on weak and strong solutions of stochastic differential equations we refer the reader to Chapter 5 in [46].

We will apply weak uniqueness of solutions of stochastic differential equations in order to discuss time homogeneous diffusions.

**Definition 7.5.1.** A time homogeneous Itô diffusion is a stochastic process $X(t) = X_{x,s}(t, \omega) \in \mathbb{R}^d$ satisfying a stochastic differential equation of the form

$$dX(t) = b(X(t))dt + \sigma(X(t))dW(t), \quad \text{for} \quad t > s \quad and \quad X(s) = x \tag{7.5.17}$$

where $W$ is the $r$-dimensional Brownian motion, $b(x) \in \mathbb{R}^d$ is a vector function on $\mathbb{R}^d$ and $\sigma(x)$ is a $d \times r$ matrix function on $\mathbb{R}^d$ satisfying the conditions of the existence and uniqueness theorem above.

Observe that it does not matter that instead of an initial condition at the time 0 we take it now at the time $s$ since setting $Y(t) = X(t + s)$ we obtain the same equation for $Y$ with the initial condition taken now at the time 0. Denote the unique solution of the equation (7.5.17) by

$X(t) = X_{x,s}(t)$, $t \geq s$. Then

$$X_{x,t}(t+h) = x + \int_t^{t+h} b(X_{x,t}(u))du + \int_t^{t+h} \sigma(X_{x,t}(u))dW(u)$$
$$= x + \int_0^h b(X_{x,t}(t+v))dv + \int_0^h \sigma(X_{x,t}(t+v))d\tilde{W}(v)$$

where $\tilde{W}(v) = W(t+v) - W(t)$, $v \geq 0$ is the new Brownian motion independent of $W(t)$ (and even of $\mathcal{F}_t = \sigma\{W(u), 0 \leq u \leq t\}$). On the other hand, we can write

$$X_{x,0}(h) = x + \int_0^h b(X_{x,0}(v))dv + \int_0^h \sigma(X_{x,0}(v))dW(v).$$

Since $\{W(v)\}_{v \geq 0}$ and $\{\tilde{W}(v)\}_{v \geq 0}$ have the same finite dimensional distributions, it follows by weak uniqueness of the solution of the stochastic differential equation

$$dX(t) = b(X(t))dt + \sigma(X(t))dW(t), \quad X(0) = x$$

that $\{X_{x,t}(t+h)\}_{h \geq 0}$ and $\{X_{x,0}(h)\}_{h \geq 0}$ have the same finite dimensional distributions and this property is called time homogeneity of the process $X$.

**Remark 7.5.1.** Set $P(t, x, \Gamma) = P\{X_{x,0}(t) \in \Gamma\}$ where $X$ is a time homogeneous Itô diffusion, $t \geq 0$, $x \in \mathbb{R}^d$ and $\Gamma \subset \mathbb{R}^d$ is a Borel set. Then for any $t, h \geq 0$, $x \in \mathbb{R}^d$ and a Borel $\Gamma \subset \mathbb{R}^d$ with probability one

$$P\{X_{x,0}(t+h) \in \Gamma | \mathcal{F}_t\} = P(h, X_{x,0}(t), \Gamma)$$

which is a form of the Markov property of Itô diffusions while $P(t, x, \cdot)$ is called the transition probability of the corresponding Markov process. This relation remains true if a fixed $t$ is replaced by a stopping time $\tau$ which is called then the strong Markov property. For the proofs of these relations and other properties of Itô's diffusions we refer the reader to [60].

### 7.5.3  *Diffusions and partial differential equations*

Let $X(t) = X_{x,s}(t, \omega) \in \mathbb{R}^d$ be a solution of the stochastic differential equation

$$dX(t) = b(X(t))dt + \sigma(X(t))dW(t), \, t \geq s \quad \text{with the condition } X(s) = x,$$
$$(7.5.18)$$

where $W(t) = (W_1(t), ..., W_d(t))$ is the standard $d$-dimensional Brownian motion, $b(x) = (b_1(x), ..., b_d(x)) \in \mathbb{R}^d$ is a vector function on $\mathbb{R}^d$

and $\sigma(x) = (\sigma_{i,j}(x))_{i,j=1,\ldots,d}$ is a $d \times d$ matrix function on $\mathbb{R}^d$ satisfying the conditions of the existence and uniqueness theorem above. Introduce the new matrix function $a(x) = \sigma(x)\sigma^*(x) = (a_{i,j}(x))_{i,j=1,\ldots,d}$, $a_{ij}(x) = \sum_{k=1}^d \sigma_{ik}(x)\sigma_{jk}(x)$ and the (elliptic) differential operator $L$ acting on $C^2$ functions $f$ on $\mathbb{R}^d$ by

$$Lf(x) = \frac{1}{2} \sum_{i,j=1}^d a_{ij}(x)\frac{\partial^2 f(x)}{\partial x_i \partial x_j} + \sum_{i=1}^d b_i(x)\frac{\partial f(x)}{\partial x_i}.$$

Let $u = u(t,x)$ be a bounded function on $\mathbb{R}_+ \times \mathbb{R}^d$ with two bounded derivatives in $x$ and one bounded derivative in $t$ and, in addition, let $c$ be a bounded continuous function on $\mathbb{R}^d$. For $0 \le s \le t$ set

$$Y(t) = Y_s(t) = \exp\left(\int_s^t c(X(\alpha))d\alpha\right) \text{ and } F(t,X(t),Y(t)) = u(t,X(t))Y(t).$$

Since $dY(t) = c(X(t))Y(t)dt$ we can employ the $(d+1)$-dimensional Itô formula from Lecture 15 which yields

$$dF(t,X(t),Y(t)) = Y(t)\left(\frac{\partial u(t,X(t))}{\partial t} + L_x u(t,X(t))\right) \quad (7.5.19)$$
$$+c(X(t))u(t,X(t)))dt + Y(t)\sum_{i,j=1}^d \frac{\partial u(t,X(t))}{\partial x_i}\sigma_{i,j}(X(t))dW_j(t)$$

where $L_x$ is the application of the operator $L$ in the variable $x$ of the function $u$. Suppose that $u$ satisfies the (parabolic) partial differential equation

$$\frac{\partial u(t,x)}{\partial t} + L_x u(t,x) + c(X(t))u(t,x) = 0. \quad (7.5.20)$$

Then we obtain from (7.5.19) that for any $0 \le s \le t$,

$$F(t,X(t),Y(t)) = F(s,X(s),Y(s)) \quad (7.5.21)$$
$$+\int_s^t Y(\alpha)\sum_{i,j=1}^d \frac{\partial u(\alpha,X(\alpha))}{\partial x_i}\sigma_{i,j}(X(\alpha))dW_j(\alpha).$$

Hence,

$$EF(t,X(t),Y(t)) = EF(s,X(s),Y(s))$$

since the expectation of the stochastic integral is zero.

Recall that $X(s) = X_{x,s}(s) = x$ and $Y(s) = 1$, and so $F(s,X(s),Y(s)) = u(s,x)$. Therefore, for $0 \le s \le t$,

$$u(s,x) = E\left(u(t,X_{x,s}(t))\exp\left(\int_s^t c(X_{x,s}(\alpha))d\alpha\right)\right). \quad (7.5.22)$$

Set also $v(s,x) = u(t-s,x)$. Then

$$\frac{\partial v(t,x)}{\partial t} = L_x v(t,x) + c(X(t))v(t,x) = 0 \quad (7.5.23)$$

and

$$v(t,x) = E\big(v(0, X_{x,0}(t)) \exp(\int_0^t c(X_{x,0}(\alpha))d\alpha)\big). \qquad (7.5.24)$$

Let $t = T$ then the final time data $u(T,x) = f(x)$ for $u$, as it is usual in mathematical finance, corresponds to the initial condition data $v(0,x) = f(x)$ for $v$ which is standard in parabolic partial differential equations. We have shown that solutions $u$ and $v$ of the equations (7.5.20) and (7.5.23) have probabilistic representations (7.5.22) and (7.5.24), respectively. The existence of solutions of such equations is usually proved by analytic methods which are beyond the scope of this book (see, for instance, [20]). Here it is usually assumed that the operator $L$ is strictly elliptic, i.e., that the matrix $a(x) = (a_{ij})$ is (strictly) positive definite.

Observe that we can integrate in (7.5.21) up to a stopping time $\tau$ instead of a fixed time $t$. Let, for instance, $G$ be a bounded domain in $\mathbb{R}^d$ with a smooth boundary $\partial G$ and $\tau = \inf\{t \geq 0 : X(t) \notin G\}$. Suppose that $u = u(x)$ does not depend on $t$, $c \equiv 0$, and so $u$ satisfies the (elliptic) equation $Lu(x) = 0$ in $G$. Then (7.5.22) becomes $u(x) = Eu(X(\tau)) = Ef(X(\tau))$ where $u|_{\partial G} = f$ is the boundary condition (which is called the Dirichlet boundary value problem).

## 7.6   Lecture 18: Integral representation of martingales

Let $W(t)$, $t \geq 0$, $W(0) = 0$ be the continuous Brownian motion and $\mathcal{F}_t$, $t \geq 0$ be the filtration generated by it complemented by all events of zero probability. As in Lecture 14 we will fix here $T > 0, T < \infty$ and denote by $\mathcal{M}_2$ the space of cádlág martingales $M$ with respect to the Brownian filtration $\mathcal{F}_t$, $t \geq 0$ such that $EM^2(T) < \infty$, $M(0) = 0$ while those martingales which are in addition continuous will form a subspace $\mathcal{M}_2^c$ of $\mathcal{M}_2$. Denote by $\mathcal{M}_2^{loc}$ the space of local martingales with respect to the filtration $\mathcal{F}_t$, $t \geq 0$ such that if $M \in \mathcal{M}_2^{loc}$ and $\tau_n \uparrow \infty$ is a localizing sequence of stopping times for $M$ then $M^{\tau_n} \in \mathcal{M}_2$ for each $n$ where $M^{\tau_n}(t) = M(t \wedge \tau_n)$. By $\mathcal{L}_2^{loc}$ we will denote the space of all real measurable processes $\Phi = \{\Phi(t,\omega)\}$, $t \in [0,T]$ adapted to the Brownian filtration such that $\int_0^T \Phi^2(s)ds < \infty$ a.s. while $\mathcal{L}_2$ will denote the subspace of $\mathcal{L}_2^{loc}$ containing processes $\Phi$ satisfying $\|\Phi\|_2 = E\int_0^T \Phi^2(s)ds < \infty$. The following representation theorem is the main result of this lecture.

**Theorem 7.6.1.** *Let $M \in \mathcal{M}_2$. Then there exists $\Phi \in \mathcal{L}_2$ such that for all*

$t \in [0, T]$,

$$M(t) = \int_0^t \Phi(s) dW(s). \tag{7.6.1}$$

*For any $M \in \mathcal{M}_2^{loc}$ the representation (7.6.1) also holds true with some $\Phi \in \mathcal{L}_2^{loc}$. In particular, such $M$ has a continuous modification. If also $M(t) = \int_0^t \Phi'(s) dW(s)$ with another $\Phi' \in \mathcal{L}_2^{loc}$ then $\int_0^T |\Phi(s) - \Phi'(s)|^2 ds = 0$ a.s.*

**Proof.** The uniqueness part follows from the properties of stochastic integrals. Indeed, if we have two representations of $M \in \mathcal{M}_2^{loc}$ with $\Phi$ and $\Phi'$ then $0 = \int_0^T (\Phi(s) - \Phi'(s)) dW(s)$ which implies that $E \int_0^T |\Phi(s) - \Phi'(s)|^2 \mathbb{I}_{\tau_n \geq s} ds = 0$ where $\tau_n \uparrow \infty$ is a localizing sequence. Hence, $\int_0^T |\Phi(s) - \Phi'(s)|^2 \mathbb{I}_{\tau_n \geq s} ds = 0$ a.s. for each $n \geq 1$ and letting $n \uparrow \infty$ we obtain $\int_0^T |\Phi(s) - \Phi'(s)|^2 ds = 0$ a.s.

Observe that the extension to $M \in \mathcal{M}_2^{loc}$ follows if we consider the representation (7.6.1) for each $M_{t \wedge \tau_m}$, $t \in [0, T]$, $n = 1, 2, \ldots$ where $\tau_n \uparrow \infty$ is a localizing sequence, and so it suffices to consider only the case $M \in \mathcal{M}_2$. In the existence part we will follow [28] observing that the required result is a corollary of

$$\mathcal{M}_2 = \mathcal{M}_2^* = \{ M \in \mathcal{M}_2^c : M(T) = \int_0^T \Phi(s) dW(s) \quad \text{for some} \quad \Phi \in \mathcal{L}_2 \} \tag{7.6.2}$$

since if $M(T) = \int_0^T \Phi(s) dW(s)$ for $M \in \mathcal{M}_2^c$ then $M(t) = E(M(T)|\mathcal{F}_t) = E(\int_0^T \Phi(s) dW(s)|\mathcal{F}_t) = \int_0^t \Phi(s) dW(s)$ a.s. for each $t \in [0, T]$. We will show first that each $M \in \mathcal{M}_2$ has a unique representation in the form

$$M(t) = M_1(t) + M_2(t), \, t \in [0, T] \tag{7.6.3}$$

where $M_1 \in \mathcal{M}_2^*$ and $M_2 \in \mathcal{M}_2$ satisfies $\langle M_2, N \rangle = 0$ for any $N \in \mathcal{M}_2^*$.

Set $\mathcal{H} = \{ M_1(T) : M_1 \in \mathcal{M}_2^* \}$ which is a closed subspace of $L_2(\Omega, P)$. Let $M \in \mathcal{M}_2$. Since $M(T) \in L^2(\Omega, P)$ we have the orthogonal decomposition

$$M(T) = H_1 + H_2,$$

where $H_1 \in \mathcal{H}$ and $H_2$ belongs to the orthogonal complement $\mathcal{H}^\perp$ of $\mathcal{H}$ in $L_2(\Omega, P)$. Then $H_1 = \int_0^T \Phi(s) ds$ for some $\Phi \in \mathcal{L}_2$. Let $M_2(t)$ be the right-continuous modification of the martingale $E(H_2|\mathcal{F}_t)$, $t \in [0, T]$. Then the equality (7.6.3) holds true with $M_1(t) = \int_0^t \Phi(s) dW(s)$ and it remains to show that $\langle M_2, N \rangle(t) = 0$ for any $t \in [0, T]$ and all $N \in \mathcal{M}_2^*$. Namely, it suffices to show that $M_2(t) N(t)$, $t \in [0, T]$ is a martingale.

The latter would follow if we show that

$$E\big(M_2(\sigma)N(\sigma)\big) = 0 \qquad (7.6.4)$$

for any stopping time $\sigma \leq T$. Indeed, $E\big(M_2(T)N(T)|\mathcal{F}_t\big) = M_2(t)N(t)$ if $E(M_2(T)N(T)\mathbb{I}_\Gamma) = E(M_2(t)N(t)\mathbb{I}_\Gamma)$ for any $\Gamma \in \mathcal{F}_t$ which is equivalent to $E(M_2(\sigma)N(\sigma)) = E(M_2(T)N(T)) = 0$ where $\sigma = t\mathbb{I}_\Gamma + T\mathbb{I}_{\Omega\setminus\Gamma}$ is a stopping time. In order to obtain (7.6.4) observe that if $N(t) = \int_0^t \Psi(s)dW(s)$, then $N^\sigma(t) = N(t \wedge \sigma) = \int_0^t \Psi(s)\mathbb{I}_{s\leq\sigma}dW(s) \in \mathcal{M}_2^*$. Hence,

$$E(N(\sigma)M_2(\sigma)) = E\big(N(\sigma)E(M_2(T)|\mathcal{F}_\sigma)\big)$$
$$= E(N(\sigma)M_2(T)) = E(N^\sigma(T)H_2) = 0$$

yielding the decomposition (7.6.3).

To prove the theorem, it suffices to show that there exists a dense subspace $\mathcal{N} \subset \mathcal{M}_2$ such that for any $M \in \mathcal{N}$ the $\mathcal{M}_2$-part in (7.6.3) vanishes. Indeed, we will have then that $\mathcal{N} \subset \mathcal{M}_2^*$ and since $\mathcal{M}_2^* \subset \mathcal{M}_2^c$ is closed, it will follow that $\mathcal{M}_2^c = \mathcal{M}_2^*$. We claim that the set $\tilde{\mathcal{N}} = \{M \in \mathcal{M}_2 : M$ is bounded $\}$ is dense in $\mathcal{M}_2$. Indeed, $\mathcal{H}_0 = \{F \in \mathcal{H} : F$ is bounded $\}$ is dense in $\mathcal{H} = \{M(T) : M \in \mathcal{M}_2\} = L_2(\Omega, \mathcal{F}_T, P)$ and the norm in $\mathcal{M}_2$ is defined by the $L^2$-norm in $\mathcal{H}$. Let $M \in \tilde{\mathcal{N}}$ and $M = M_1 + M_2$ be the decomposition (7.6.3). Since $M_1$ is a continuous martingale, there exists a sequence of stopping times $\sigma_n \in [0, T]$, $\sigma_n \uparrow T$ as $n \uparrow \infty$ such that $M_1^{\sigma_n}(t) = M_1(t \wedge \sigma_n)$, $t \in [0, T]$ is a bounded martingale for each $n = 1, 2, ...$ We have that $M_1^{\sigma_n} \in \mathcal{M}_2^*$ since $M_1 \in \mathcal{M}_2^*$. Furthermore, $M^{\sigma_n} = M_1^{\sigma_n} + M_2^{\sigma_n}$ is the decomposition (7.6.3) for $M^{\sigma_n}$ since for any $N \in \mathcal{M}_2^*$,

$$\langle N, M_2^{\sigma_n}\rangle = \langle N^{\sigma_n}, M_2^{\sigma_n}\rangle = \langle N, M_2\rangle^{\sigma_n} = 0$$

where we use that $NM_2^{\sigma_n} - N^{\sigma_n}M_2^{\sigma_n}$ is a martingale because

$$E(N(t)M_2^{\sigma_n}(t) - N^{\sigma_n}(t)M_2^{\sigma_n}(t)|\mathcal{F}_{s\wedge\sigma_n})$$
$$= E\big(E(N(t)|\mathcal{F}_{s\wedge\sigma_n})M_2^{\sigma_n}(s) - N^{\sigma_n}(s)M_2^{\sigma_n}(s)\big) = 0.$$

Set

$$\mathcal{N} = \{M^{\sigma_n} : M \in \tilde{\mathcal{N}}, n = 1, 2, ...\}.$$

It is easy to see that $\mathcal{N}$ is dense in $\mathcal{M}_2^c$ and if $M = M_1 + M_2$ is the decomposition (7.6.3) for $M \in \mathcal{N}$ then both $M_1$ and $M_2$ are bounded. It suffices to show that $M_2 = 0$ which is a corollary of the following lemma.

**Lemma 7.6.1.** *Let $M \in \mathcal{M}_2$ be bounded and suppose that $\langle M, N\rangle = 0$ for every $N \in \mathcal{M}_2^*$. Then $M = 0$.*

**Proof.** Observe that the condition $\langle M, N \rangle = 0$ for every $N \in \mathcal{M}_2^*$ is equivalent to the condition $\langle M, W \rangle = 0$ since $\langle M, N \rangle(t) = \int_0^t \Phi(s) d\langle M, W \rangle(s)$ if $N(t) = \int_0^t \Phi(s) dW(s)$. Let $|M(t)| \leq \alpha < \infty$ for all $t \in [0, T]$ and set $D = 1 + \frac{1}{2\alpha} M(T)$. Then $D \geq 1/2$ and $ED = 1$. Introduce a new probability measure $\tilde{P}$ on $\mathcal{F}_T$ by

$$\tilde{P}(\Gamma) = E(D\mathbb{I}_\Gamma), \ \Gamma \in \mathcal{F}_T.$$

Then for any stopping time $\sigma \leq T$,

$$E_{\tilde{P}}(W(\sigma)) = E(DW(\sigma)) = E\big(E(D|\mathcal{F}_\sigma)W(\sigma)\big)$$
$$= EW(\sigma) + \frac{1}{2\sigma} E(M(\sigma)W(\sigma)) = 0$$

since $\langle M, W \rangle = 0$, and so $MW$ is a martingale. Similarly, $E_{\tilde{P}}(W^2(\sigma) - \sigma) = 0$ since $W^2(t) - t = 2\int_0^t W(s) dW(s) \in \mathcal{M}_2^*$, and so $\langle W^2(t) - t, M(t) \rangle = 0$. Hence, both $W(t)$ and $W^2(t) - t$ are continuous martingales with respect to the probability $\tilde{P}$ and the filtration $\mathcal{F}_t$, $t \geq 0$. By the Lévy theorem from Lecture 16 we conclude that $W$ is the Brownian motion also with respect to the probability $\tilde{P}$ which, clearly, implies that $P = \tilde{P}$ on $\mathcal{F}_T$, and so $D = 1$, i.e., $M = 0$ a.s. This completes both the proof of the latter lemma and the whole theorem for $M \in \mathcal{M}_2$. $\square$

For $M \in \mathcal{M}_2^{loc}$ we obtain the representation (7.6.1) by the first part of the theorem for each $M^{\tau_n}$, where $\tau_n \uparrow \infty$ is a localizing sequence of $M$, and then let $n \uparrow \infty$. $\square$

By essentially the same proof we obtain the following multidimensional generalization of the representation theorem.

**Theorem 7.6.2.** *Let $W(t) = (W_1(t), ..., W_r(t))$ be the $r$-dimensional Brownian motion with $W(0) = 0$ and the filtration $\mathcal{F}_t = \sigma\{W(s), s \leq t\}$, $t \in [0, T]$, $T < \infty$. Suppose that $M$ is a martingale with respect to the filtration $\mathcal{F}_t$, $t \in [0, T]$ such that $EM^2(T) < \infty$. Then there exist $\Phi_i \in \mathcal{L}_2$, $i = 1, ..., r$ such that for each $t \in [0, T]$ with probability one,*

$$M(t) = \sum_{i=1}^r \int_0^t \Phi_i(s) dW_i(s).$$

*This representation remains true for any $M \in \mathcal{M}_2^{loc}$ with some $\Phi_i \in \mathcal{L}_2^{loc}$, $i = 1, ..., r$. If $M = (M_1, ..., M_d)$ is a vector (or local) martingale on $[0, T]$ with respect to the above filtration satisfying $EM_i^2(T) < \infty$, $i = 1, ..., d$ then this result applied to each $M_i$ yields the existence of $\Phi_{ij} \in \mathcal{L}_2$, $i =$*

$1, ..., d,\ j = 1, ..., r$ *(or $\Phi_{ij} \in \mathcal{L}_2^{loc}$) such that for each $t \in [0, T]$ with proba-bility one,*

$$M_i(t) = \sum_{j=1}^{r} \int_0^t \Phi_{ij}(s)dW_j(s),\ \ i = 1, ..., d.$$

## 7.7 Exercises

1) Let $W(t)$, $t \geq 0$ be a $d$-dimensional Brownian motion with continuous paths. Prove that $\int_0^1 W(t)dt$ is a Gaussian random vector and find the expectations of its components and the covariance matrix.

2) Prove that for each fixed $t \geq 0$ with probability one the Brownian motion $W(t)$ cannot be Hölder continuous at $t$ with the exponent $1/2$. Observe that this is a weaker statement than we proved with respect to the Hölder continuity with an exponent $\gamma > 1/2$.

3) Let $f(t, \omega)$, $t \in [0, T]$ be a bounded simple adapted (to the Brownian filtration) function, i.e., $f(t, \omega) = f(t_{k-1}, \omega)$ when $t_{k-1} \leq t < t_k$, $k = 1, 2, ..., m+1$ where $0 = t_0 < t_1 < ... < t_m < t_{m+1} = T$. Prove directly (i.e., not using the general theorem on exponential martingales from Lecture 16) that $M(t) = \exp(\int_0^t f(s)dW(s) - \frac{1}{2}\int_0^t f^2(s)ds)$, $t \in [0, T]$ is a martingale and $EM(t) = 1$.

4) Find $E\cos W(t)$ where $W(t), t \geq 0$ is the standard one-dimensional Brownian motion (with $W(0) = 0$).

5) Prove that

$$\int_0^t \int_0^{t_1} ... \int_0^{t_{n-2}} W(t_n)dW(t_1)...dW(t_{n-1}) = H_n(t, W(t))$$

where the left hand side is the multiple stochastic integral and in the right hand side stands the $n$-th Hermit polynomial

$$H_n(t, x) = \frac{(-t)^n}{n!}\exp(\frac{x^2}{2t})\frac{\partial^n}{\partial x^n}\exp(-\frac{x^2}{2t}).$$

# Chapter 8

# Derivatives in the Black–Scholes Market

## 8.1 Lecture 19: Black–Scholes market model

### 8.1.1 Geometric Brownian motion stock model and martingale measures

The standard Black–Scholes market model consists of two types of assets (or securities) which evolve and can be traded continuously in time. The first asset is riskless and it is commonly called a bond but due to its evolution a more appropriate name for it may be a bank account. The price $B_t$ at the time $t \geq 0$ of this asset is given by the (deterministic) formula $B_t = B_0 e^{rt}$ where $r \geq 0$ is viewed as an interest rate. The second asset, commonly called a stock, is a risky one and its price fluctuates randomly. Namely, we assume that the randomness of the market is described by a complete probability space $(\Omega, \mathcal{F}, P)$ which admits a standard continuous one-dimensional Brownian motion $W = (W(t), t \geq 0)$ defined on it. We denote also by $\{\mathcal{F}_t, t \geq 0\}$ the Brownian filtration, i.e., the right-continuous filtration generated by the Brownian motion complemented by events of zero probability. The $\sigma$-algebra $\mathcal{F}_t$ plays the role of information available in the market up to the time $t$. The price $S_t$ of the stock at the time $t \geq 0$ is prescribed to be $S_t = S_0 \exp\left((\mu - \frac{1}{2}\kappa^2)t + \kappa W(t)\right)$ which is called a geometric Brownian motion. Here $\mu$ and $\kappa > 0$ are (constant) parameters, the latter is called the volatility of the stock. In order to understand better the rational behind this formula we apply the Itô formula to $S_t$ to obtain the stochastic differential equation

$$dS_t = S_t(\mu dt + \kappa dW(t)). \tag{8.1.1}$$

Thus, we may view the stock evolution model as a stochastic perturbation of the deterministic growth (if $\mu > 0$) or decay (if $\mu < 0$) model $\frac{dS_t}{dt} = \mu S_t$.

As in the discrete time case an important role in the theory play martingale measures . Again, we call a probability measure $P^*$ on $(\Omega, \mathcal{F})$ a martingale measure if it is equivalent to $P$ and the discounted stock price $\frac{S_t}{B_t}$, $t \geq 0$ is a martingale with respect to $P^*$. Set

$$W^*(t) = \frac{\mu - r}{\kappa} t + W(t).$$

By the Girsanov theorem from Lecture 16 for any $t > 0$ the process $W^*(s)$, $s \in [0, t]$ is the Brownian motion with respect to the probability $P^*$ on $\mathcal{F}_t$ defined by its Radon–Nikodym derivative

$$N_t = \frac{dP_t^*}{dP_t} = \exp\left(-\frac{\mu - r}{\kappa} W(t) - \frac{1}{2}\left(\frac{\mu - r}{\kappa}\right)^2 t\right) \qquad (8.1.2)$$

where we consider restrictions $P_t$ and $P_t^*$ of $P$ and $P^*$ to $\mathcal{F}_t$.

To see directly that $W^*$ is the Brownian motion under $P^*$ rather than to rely on the general Girsanov theorem we show first that $W^*(t)$ has a normal distribution under $P^*$. Indeed,

$$P^*\{W^*(t) \leq a\} = E\mathbb{I}_{W^*(t) \leq a} N_t = E\left(\mathbb{I}_{W(t) \leq a - \frac{\mu-r}{\kappa} t} \exp\left(-\frac{\mu-r}{\kappa} W(t)\right.\right.$$

$$\left.\left. -\frac{1}{2}\left(\frac{\mu-r}{\kappa}\right)^2 t\right)\right) = \frac{1}{\sqrt{2\pi t}} \int_{-\infty}^{a - \frac{\mu-r}{\kappa} t} \exp\left(-\frac{x^2}{2t} - \frac{\mu-r}{\kappa} x - \frac{1}{2}\left(\frac{\mu-r}{\kappa}\right)^2 t\right) dx$$

$$= \frac{1}{\sqrt{2\pi t}} \int_{-\infty}^{a} \exp\left(-\frac{1}{2t} x^2\right) dx = P\{W(t) \leq a\}.$$

We can see also directly that $W^*(t) - W^*(s)$ is independent of $\mathcal{F}_s$ under $P^*$ when $0 \leq s < t$. Indeed, by the generalized Bayes formula from Lecture 7, for any bounded Borel function $f$ and $t > s \geq 0$,

$$E_{P^*}\left(f(W^*(t) - W^*(s))|\mathcal{F}_s\right)$$

$$= \frac{1}{N_s} E_P\left(f(W(t) - W(s) + \frac{\mu-r}{\kappa}(t - s)) N_t|\mathcal{F}_s\right)$$

$$= E_P\left(f(W(t) - W(s) + \frac{\mu-r}{\kappa}(t - s)) \exp\left(-\frac{\mu-r}{\kappa}(W(t) - W(s))\right.\right.$$

$$\left.\left. -\frac{1}{2}\left(\frac{\mu-r}{\kappa}\right)^2(t - s)\right)|\mathcal{F}_s\right)$$

$$= E_P\left(f(W(t - s) + \frac{\mu-r}{\kappa}(t - s)) \exp\left(-\frac{\mu-r}{\kappa} W(t - s) - \frac{1}{2}\left(\frac{\mu-r}{\kappa}\right)^2(t - s)\right)\right)$$

$$= E_P\left(f(W(t - s) + \frac{\mu-r}{\kappa}(t - s)) N_{t-s}\right) = E_{P^*}(f(W^*(t - s))).$$

It follows that $W^*(t) - W^*(s)$ is independent of $\mathcal{F}_s$ under $P^*$ since for all bounded Borel $f$ the above conditional expectation is not random and, moreover, we see from here that $W^*(t) - W^*(s)$ has the same distribution as $W^*(t - s)$. Hence, $W^*$ is the Brownian motion under $P^*$.

Now, observe that

$$R_t = \frac{S_t}{B_t} = \frac{S_0}{B_0} \exp\left((\mu - r - \frac{\kappa^2}{2})t + \kappa W(t)\right) = \frac{S_0}{B_0} \exp\left(-\frac{\kappa^2}{2} t + \kappa W^*(t)\right).$$

Since $W^*$ is the Brownian motion with respect to $P^*$ we conclude by Lecture 16 that $R_t$, $t \geq 0$ is an exponential martingale with respect to the measure $P^*$, and so $P^*$ is a martingale measure for the Black–Scholes market model above.

**8.1.2   Uniqueness of a martingale measure**

Next, we will show that the martingale measure $P^*$ is unique. Let $Q$ be a martingale measure. Then the Radon–Nikodym derivative $L_t = \frac{dQ_t}{dP_t}$, $t \geq 0$ is a martingale with respect to $P$ and the Brownian filtration where $Q_t$ and $P_t$ are restrictions of $Q$ and $P$ to $\mathcal{F}_t$. Indeed, by the generalized Bayes formula from Lecture 7,

$$1 = E_Q(1|\mathcal{F}_s) = E_P(\frac{L_t}{L_s}|\mathcal{F}_s), \ t > s, \ \text{i.e.,} \ E_P(L_t|\mathcal{F}_s) = L_s \ \text{a.s.}$$

The martingale $L_t$, $t \geq 0$ may not be square integrable but $L$ is a local square integrable martingale with the localizing sequence $\tau_n = \inf_{t \geq 0}\{L_t > n\}$. Thus we can use the representation theorem from Lecture 18 which yields that

$$L_t = 1 + \int_0^t \psi_s dW(s) = 1 + \int_0^t L_s \phi_s dW(s) \qquad (8.1.3)$$

where $\phi_s = \frac{\psi_s}{L_s}$ and $\psi$ is an adapted to the filtration $\mathcal{F}_t$, $t \geq 0$ process satisfying $P\{\int_0^t \psi_s^2 ds < \infty\} = 1$ for any $t \geq 0$.

Since $Q$ is a martingale measure, $e^{-rt}S_t$, $t \geq 0$ is a martingale with respect to $Q$. Hence, for $0 \leq s < t$ by the generalized Bayes formula from Lecture 7,

$$e^{-rs}S_s = E_Q(e^{-rt}S_t|\mathcal{F}_s) = E_P(\frac{L_t}{L_s}e^{-rt}S_t|\mathcal{F}_s),$$

and so

$$E_P(L_t e^{-rt}S_t|\mathcal{F}_s) = L_s e^{-rs}S_s,$$

i.e., $L_t e^{-rt}S_t$, $t \geq 0$ is a martingale with respect to $P$. It follows that $L_t e^{-rt}S_t$ can be represented as a stochastic integral with respect to the Brownian motion. On the other hand, by (8.1.1), (8.1.3) and the Itô formula

$$d(L_t e^{-rt}S_t) = L_t e^{-rt}S_t(\mu - r - \kappa\phi_t)dt + L_t e^{-rt}S_t(\phi_t + \kappa)dW(t),$$

and so the coefficient in $dt$ must vanish and since $L_t e^{-rt}S_t > 0$ we conclude that $\mu - r + \kappa\phi_t \equiv 0$ a.s., i.e., that $\phi_t \equiv -\frac{\mu-r}{\kappa}$ a.s. Here we use that if $\int_0^t f_s dW(s) = \int_0^t g_s ds$, $t \in [0,T]$ where $f,g$ are adapted and $\int_0^T (f_s^2 + |g_s|)ds < \infty$ a.s., then both integrals are zero with probability one for each $t \in [0,T]$. This holds true because contrary to Lebesgue integrals nontrivial stochastic integrals have infinite total variation which follows in the same way as the proof of the uniqueness part of the Doob–Meyer decomposition theorem in Lecture 11.

Now, solving the stochastic integral equation (8.1.3) with $\phi_t \equiv -\frac{\mu-r}{\kappa}$ we obtain

$$L_t = \exp\left(-\frac{1}{2}(\frac{\mu-r}{\kappa})^2 t - (\frac{\mu-r}{\kappa})W(t)\right),$$

and so $Q$ coincides with $P^*$ defined by (8.1.2).

### 8.1.3 Trading strategies

**Definition 8.1.1.** A trading strategy defined on a time interval $[0, T]$, $T <$ $\infty$ (with $T$ called a horizon) is a pair $\pi = (\beta, \gamma)$ of stochastic processes $\beta_t, \gamma_t$, $t \in [0, T]$ adapted to the Brownian filtration $\mathcal{F}_t$, $t \in [0, T]$ with the map $(\beta, \gamma) : [0, T] \times \Omega \to \mathbb{R}^2$ measurable with respect to $\mathcal{B}_{[0,T]} \times \mathcal{F}_T$ (where $\mathcal{B}_{[0,T]}$ is the Borel $\sigma$-algebra on $[0, T]$) and such that

$$\int_0^T |\beta_t| dt < \infty \quad \text{and} \quad \int_0^T \gamma_t^2 dt < \infty \quad \text{a.s.} \tag{8.1.4}$$

**Definition 8.1.2.** For a trading strategy $\pi = (\beta, \gamma)$ the corresponding portfolio value at time $t$ is given by

$$X_t^\pi = \beta_t B_t + \gamma_t S_t. \tag{8.1.5}$$

A trading strategy $\pi = (\beta, \gamma)$ and the corresponding portfolio $X$ are called self-financing if

$$X_t^\pi = X_0 + \int_0^t \beta_u dB_u + \int_0^t \gamma_u dS_u \tag{8.1.6}$$

which written in the differential form

$$dX_t^\pi = \beta_t dB_t + \gamma_t dS_t$$

represents the requirement that changes in the portfolio value can result only from changes in prices of assets and no external infusion or spending of capital is permitted.

By (8.1.4) and the formulas for $B_t$ and $S_t$ we see that

$$\int_0^T |\beta_t| B_t dt < \infty \quad \text{and} \quad \int_0^T \gamma_t^2 S_t^2 dt < \infty \quad \text{a.s.,} \tag{8.1.7}$$

and so the integrals in (8.1.6) exist.

Next,

$$dS_t = S_t(\mu dt + \kappa dW(t)) = S_t(r dt + \kappa dW^*(t))$$

where, recall, $W^*(t) = \frac{\mu - r}{\kappa} t + W(t)$ is the Brownian motion with respect to $P^*$. Hence, for any self-financing trading strategy $\pi = (\beta, \gamma)$,

$$dX_t^\pi = \beta_t dB_t + \gamma_t dS_t = \beta_t r B_t dt + \gamma_t S_t(r dt + \kappa dW^*(t))$$
$$= r X_t^\pi dt + \kappa \gamma_t S_t dW^*(t).$$

Set $M_t^\pi = e^{-rt} X_t^\pi$ which is (up to $B_0$) the discounted portfolio value. Then

$$dM_t^\pi = -r M_t^\pi dt + e^{-rt} dX_t^\pi = -r M_t^\pi dt + r e^{-rt} X_t^\pi dt + \kappa e^{-rt} \gamma_t S_t dW^*(t)$$
$$= \kappa e^{-rt} \gamma_t S_t dW^*(t),$$

and so

$$M_t^\pi = M_0^\pi + \int_0^t \kappa e^{-ru} \gamma_u S_u dW_u^*. \tag{8.1.8}$$

Since $\gamma$ satisfies (8.1.4) and $e^{-ru} S_u$ is continuous in $u$, it follows that

$$\int_0^T e^{-2rt} \gamma_t^2 S_t^2 dt < \infty \quad \text{a.s.}$$

which together with (8.1.8) yields that the discounted portfolio values $M_t^\pi$, $t \in [0, T]$ form a continuous local martingale (see Lecture 14) for any self-financing trading strategy $\pi$ with the initial portfolio value $M_0^\pi = X_0^\pi$ being an integrable random variable, in particular, $X_0^\pi$ can be a constant. If we impose a stronger requirement

$$E \int_0^T \gamma_t^2 S_t^2 dt < \infty$$

then $M_t^\pi$, $t \in [0, T]$ will be a continuous martingale.

The following result will be useful in pricing of derivative securities.

**Lemma 8.1.1.** *Let $M_t$, $t \in [0, T]$ be a local martingale with respect to a filtration $\mathcal{F}_t$, $t \in [0, T]$ and let $\{\Xi_t, t \in [0, T]\}$ be a stochastic process such that $\Xi_\tau$ is measurable for each $\tau$ from the set $\mathcal{T}_{0T}$ of all stopping times with values in $[0, T]$ and that the family $\{\Xi_\tau, \tau \in \mathcal{T}_{0T}\}$ is uniformly integrable. If with probability one $M_t \geq -|\Xi_t|$ (or $M_t \leq |\Xi_t|$) simultaneously for all $t \in [0, T]$ then $M_t$, $t \in [0, T]$ is a supermartingale (submartingale). Consequently, if with probability one $|M_t| \leq |\Xi_t|$ for all $t \in [0, T]$, then $M_t$, $t \in [0, T]$ is a martingale. If $M$ and $\Xi$ are either both right or both left-continuous then it suffices to require the above bounds only for each rational $t \in [0, T]$ a.s.*

**Proof.** Let $\tau_n \uparrow \infty$ be a localizing sequence for $M$. Since $M_{t \wedge \tau_n}$, $t \in [0, T]$ is a martingale, it follows that $M_0 = M_{0 \wedge \tau_n}$ is integrable. Set $M_t^+ = \max(M_t, 0)$ and $M_t^- = -\min(M_t, 0)$. Then $M_t = M_t^+ - M_t^-$, and so with probability one $|\Xi_t| \geq M_t^- \geq 0$ for all $t$. In particular, $|\Xi_{t \wedge \tau_n}| \geq M_{t \wedge \tau_n}^- \geq 0$ a.s., and so $M_{t \wedge \tau_n}^-$ is integrable. Since $M_{t \wedge \tau_n}$, $t \in [0, T]$ is a martingale for each $n \geq 1$, we conclude from here and the Fatou lemma that

$$EM_0 = \liminf_{n \uparrow \infty} EM_{t \wedge \tau_n} \geq \liminf_{n \uparrow \infty} EM_{t \wedge \tau_n}^+$$
$$- \limsup_{n \uparrow \infty} EM_{t \wedge \tau_n}^- \geq E \liminf_{n \uparrow \infty} M_{t \wedge \tau_n}^+ - \sup_{\tau \in \mathcal{T}_{0T}} E|\Xi_\tau|.$$

Since $\liminf_{n \uparrow \infty} M_{t \wedge \tau_n}^+ = M_t^+$, it follows that

$$EM_t^+ \leq EM_0 + \sup_{\tau \in \mathcal{T}_{0T}} E|\Xi_\tau| < \infty,$$

and so $M_t^+$, as well as $M_t = M_t^+ - M_t^-$, are integrable. Next, we can apply the Fatou lemma again to obtain that whenever $0 \leq s < t \leq T$,

$$M_s = \liminf_{n\uparrow\infty} M_{s\wedge\tau_n} = \liminf_{n\uparrow\infty} E(M_{t\wedge\tau_n}|\mathcal{F}_s)$$
$$\geq \liminf_{n\uparrow\infty} E\big((M_{t\wedge\tau_n} + |\Xi_{t\wedge\tau_n}|)\big|\mathcal{F}_s\big) - \limsup_{n\uparrow\infty} E\big(|\Xi_{t\wedge\tau_n}|\big|\mathcal{F}_s\big)$$
$$\geq E\big(\liminf_{n\uparrow\infty}(M_{t\wedge\tau_n} + |\Xi_{t\wedge\tau_n}|)\big|\mathcal{F}_s\big) - E\big(|\Xi_t|\big|\mathcal{F}_s\big) = E(M_t|\mathcal{F}_s).$$

Hence, $M_t$, $t \in [0,T]$ is a supermartingale. If with probability one $M_t \leq |\Xi_t|$ for all $t \in [0,T]$, then applying the above argument to $-M_t \geq -|\Xi_t|$ we obtain that $-M_t$, $t \in [0,T]$ is a supermartingale, and so $M_t$, $t \in [0,T]$ is a submartingale. If with probability one $|M_t| \leq |\Xi_t|$ for all $t \in [0,T]$, then $M_t$, $t \in [0,T]$ is both super and submartingale, and so it is a martingale. If $M$ and $\Xi$ are either both left or both right-continuous and the required bound holds true for each rational $t \in [0,T]$ a.s. then these bounds will be satisfied with probability one simultaneously for all rational $t \in [0,T]$ which will remain true for all $t \in [0,T]$ in view of left or right continuity. $\square$

## 8.2 Lecture 20: Pricing of derivatives in the Black–Scholes market

In this lecture we will discuss fair pricing of three types of derivative securities in the Black–Scholes market model.

### 8.2.1 *European contingent claims*

**Definition 8.2.1.** A European contingent claim in the Black–Scholes market model described in the previous lecture is a contract between its buyer and seller so that the latter accepts an obligation to pay an amount $R = R_T$ at time $T < \infty$ to the buyer where $R \geq 0$ is an $\mathcal{F}_T$-measurable random variable.

**Example 8.1.** The payoffs $R = (K - S_T)^+$ and $R = (S_T - K)^+$ correspond to put and call options, respectively. Here $K$ is called the strike price . A put (call) European option enables (but not obliges) its owner to sell (buy) a stock for the price $K$ at a fixed expiration time $T$. The amount $R$ is considered as a payoff by the seller of the option to its buyer since, if $K > S_T$ in the put option case, the seller of the contract has to pay to its buyer the amount $K$ for the stock which can be sold then by the market price $S_T$. Thus, the actual payment amounts to $K - S_T$. If $K \leq S_T$ then it does not make sense for the buyer to use the contract, and so the payment by the seller is zero. In the call option case the argument is similar.

**Definition 8.2.2.** (hedging) A self-financing trading strategy $\pi = (\beta, \gamma)$ is called hedging or a hedge (and the corresponding portfolio is called a hedging portfolio) in an European contingent claim case if $X_T^\pi \geq R_T$ a.s., i.e., at the expiration time the portfolio value is sufficient to cover the payment obligation. We will call a hedging strategy $\pi$ admissible if $X_t^\pi \geq 0$ a.s for all $t \in [0, T]$.

**Definition 8.2.3.** The fair price $V^*$ of a contingent claim is the infimum of initial capitals of all admissible hedging (self-financing) portfolios, i.e., $V^* = \inf\{x : \text{there exists an admissible hedging trading strategy } \pi \text{ such that } X_0^\pi = x\}$.

Sometimes this notion is called the fair price from the seller's point of view. The admissibility condition means that the portfolio value is not allowed to become negative which would require the investor to borrow funds (to trade on margin). But taking a loan requires a collateral made of other assets which is not compatible with the essence of self-financing. In fact, the admissibility is imposed here by technical reasons in order to be able to employ the lemma from the previous lecture and it suffices to require only, for instance, that at all time the portfolio value remains bounded from below by an integrable with respect to the martingale measure $P^*$ random variable. Moreover, it suffices to require directly that the discounted portfolio values form a supermartingale with respect to the martingale measure. On the other hand, some admissibility condition is necessary to avoid arbitrage, i.e., a riskless gain, as the following example (which appeared in [47], [60] and [72]) shows.

**Example 8.2.** Let $B_t \equiv 1$ and $S_t = \exp(-\frac{t}{2} + W(t))$, i.e., $dS_t = S_t dW(t)$, and fix the horizon $T = 1$. Observe that $S_t,\, t \geq 0$ is already a martingale with respect to the basic (market) probability $P$, and so $P$ itself is a martingale measure. Set

$$I(t) = \int_0^t (1-s)^{-\frac{1}{2}} dW(s) \quad \text{and} \quad \tau_M = \inf\{t \in [0,1) : I(t) = M\} \wedge 1$$

where $M$ is an arbitrary positive number. For any $t \in [0,1)$ the stochastic integral is well defined and it has the quadratic variation

$$\langle I(t) \rangle = \int_0^t (1-s)^{-1} ds = -\ln(1-t), \quad t \in [0,1).$$

Set $\theta_t = 1 - e^{-t}$. Then $\langle I(\theta_t) \rangle = t$. Since $\theta_t$ is increasing in $t$, $J(t) = I(\theta_t)$ is a continuous martingale for $t \geq 0$ and $\langle J(t) \rangle = t$, it follows from the Lévy characterization theorem from Lecture 16 that $J(t)$, $t \geq 0$ is a Brownian motion. Observe that $\tau_M = \inf\{\theta_t : J(t) = M\} < 1$ a.s. since the continuous Brownian motion $J(t)$ will reach $M$ at some time $t \geq 0$ with probability one. Set $\pi = (\beta_t, \gamma_t)_{t \in [0,1]}$ where

$$\gamma_t = \begin{cases} (1-t)^{-\frac{1}{2}} S_t^{-1} \mathbb{I}_{\{t \leq \tau_M\}} & \text{if } \tau_M < 1 \\ 0 \text{ if } \tau_M \geq 1 \end{cases}$$

and

$$\beta_t = \begin{cases} I(t \wedge \tau_M) - \gamma_t S_t & \text{if } \tau_M < 1 \\ 0 \text{ if } \tau_M \geq 1. \end{cases}$$

Then $\pi$ is a trading strategy with the portfolio value at time $t \in [0,1]$ equal with probability one to

$$X_t^\pi = \beta_t B_t + \gamma_t S_t = I(t \wedge \tau_M) = \int_0^{t \wedge \tau_M} (1-u)^{-\frac{1}{2}} dW(u)$$
$$= \int_0^t \gamma_u S_u dW(u) = \int_0^t \gamma_u dS_u$$

which shows that $\pi$ is self-financing. Since $X_0^\pi = 0$ and $X_1^\pi = I(\tau_M) = M$ a.s., it follows that $\pi$ provides an opportunity for arbitrage and, actually, such strategy enables us to start with zero initial capital and to gain any amount at the exercise time. Clearly, $\pi$ is not admissible since $I(t \wedge \tau_M) = J(-\ln(1 - t \wedge \tau_M))$ is not bounded from below taking into account that $J$ is the Brownian motion.

Next, we proceed to pricing of contingent claims.

**Theorem 8.2.1.** *A European contingent claim with a horizon $T > 0$, $T < \infty$ and an $\mathcal{F}_T$-measurable payoff function $R = R_T \geq 0$ satisfying $E^* R < \infty$ has the fair price*

$$V^* = E^*(e^{-rT} R)$$

*where $E^* = E_{P^*}$ is the expectation with respect to the martingale measure $P^*$. Furthermore, there exists a hedging trading strategy with the initial capital $V^*$.*

**Proof.** Let $\pi$ be an admissible hedging strategy. Then, as we saw in the previous lecture, the discounted portfolio values $e^{-rt} X_t^\pi$, $t \in [0,T]$ form a local martingale which together with the admissibility condition yields by the lemma in the previous lecture that $e^{-rt} X_t^\pi$, $t \in [0,T]$ is a supermartingale with respect to $P^*$. Assuming that $X_0^\pi = x$ is constant we obtain from this together with hedging that

$$x = X_0^\pi \geq E^*(e^{-rT} X_T^\pi) \geq E^*(e^{-rT} R) = V^*.$$

Hence, the initial capital of any admissible hedging strategy must be at least $V^*$, i.e., the fair price of this contingent claim cannot be less than $V^*$.

We will complete the proof constructing an admissible hedging (replicating) strategy with the initial capital exactly equal to $V^*$. Since $R$ is integrable with respect to $P^*$ we can introduce the (Doob) martingale

$$M_t = E^*(e^{-rT}R|\mathcal{F}_t), \quad t \in [0,T].$$

This is a martingale with respect to the Brownian filtration and from the beginning we will take its cádlág modification, denoted by the same letter, which is possible as explained in Lecture 11. Clearly, $M \in \mathcal{M}_2^{loc}$ as we can take the localizing sequence $\tau_n = \inf\{s \geq 0 : |M_s| \geq n\}$. Hence, according to Lecture 18 there exists a representation

$$M_t = M_0 + \int_0^t \kappa e^{-ru}\gamma_u S_u dW^*(u), \; t \in [0,T] \quad \text{with} \quad M_0 = e^{-rT}E^*R,$$

where the adapted process $\gamma_u$, $t \in [0,T]$ satisfying $\int_0^T \gamma_u^2 du < \infty$ a.s. is defined by this formula. Next, we set

$$\beta_t = (M_t - e^{-rt}\gamma_t S_t)B_0^{-1}, \quad \pi = (\beta, \gamma) \quad \text{and} \quad X_t^\pi = e^{rt}M_t, \; t \in [0,T].$$

We obtain from here that $X_t^\pi = \beta_t B_t + \gamma_t S_t$ and

$$\begin{aligned}
dX_t^\pi &= rX_t^\pi dt + e^{rt}dM_t = rX_t^\pi dt + \kappa\gamma_t S_t dW^* \\
&= rX_t^\pi dt + \gamma_t(dS_t - rS_t dt) = \beta_t dB_t + \gamma_t dS_t,
\end{aligned}$$

where we use that $dS_t = S_t(rdt + \kappa dW^*)$, which means that the trading strategy $\pi = (\beta_t, \gamma_t)_{t \in [0,T]}$ is self-financing. It is also admissible since conditional expectations of a nonnegative random variable are nonnegative, and so $X_t^\pi \geq 0$ a.s. Moreover, $\pi$ is hedging and even replicating since $X_T^\pi = e^{rT}M_T = R$. Since the initial portfolio value $X_0^\pi$ equals exactly $M_0 = V^*$, this completes the proof of the theorem. $\qquad\square$

Observe that in the European call option case when $R = R_T = (S_T - K)^+ = \left(\exp\left((r - \frac{\kappa^2}{2})T + \kappa W^*(T)\right) - K\right)^+$ we can write the fair price in the form

$$V^* = e^{-rT}E^*R = \frac{1}{e^{rT}\sqrt{2\pi T}}\int_{-\infty}^{\infty}\left(e^{(r-\frac{\kappa^2}{2})T+\kappa x} - K\right)^+ e^{-\frac{x^2}{2T}}\,dx$$

which is a version of the Black–Scholes formula. A similar expression can be written in the European put option case where $R = R_T = (K - S_T)^+$.

## 8.2.2 American contingent claims

**Definition 8.2.4.** An American contingent claim in the Black–Scholes market model described in the previous lecture is a contract between its buyer and seller so that the latter accepts an obligation to pay an amount $R_t$ to the buyer provided the latter exercises the contract at time $t \in [0, T]$. Here $R_t$, $t \in [0, T]$ is an adapted to the Brownian filtration $\{\mathcal{F}_t, t \in [0, T]\}$ nonnegative cádlág stochastic process and $T$ is the contract's expiration time (horizon). If the contract is not exercised before $T$, then it is stopped automatically at $T$ with the payment $R_T$ by the seller to the buyer. The buyer can exercise the contract at any stopping time between 0 and $T$ which represents the requirement that her/his decision should be based only on the information available up to the present moment (no clairvoyance).

**Definition 8.2.5.** (hedging) A self-financing trading strategy $\pi = (\beta, \gamma)$ is called hedging or a hedge (and the corresponding portfolio is called a hedging portfolio) in an American contingent claim case if $X_t^{\pi} \geq R_t$ for all $t \in [0, T]$ a.s. where $T$ is the horizon, i.e., at any time the portfolio value is kept sufficient to cover the payment obligation.

Observe, that we do not have to require nonnegativity of portfolio values (admissibility) here since this follows automatically from hedging of a nonnegative payoff.

**Definition 8.2.6.** The fair price $V^*$ of an American contingent claim is the infimum of initial capitals of all hedging (self-financing) portfolios, i.e., $V^* = \inf\{x : \text{there exists a hedging trading strategy } \pi \text{ such that } X_0^{\pi} = x\}$.

**Theorem 8.2.2.** *An American contingent claim with a horizon $T \in (0, \infty)$ and an adapted nonnegative cádlág payoff process $R_t$, $t \in [0, T]$ satisfying $E^* \sup_{t \in [0,T]} R_t < \infty$ has the fair price*

$$V^* = \sup_{\tau \in \mathcal{T}_{0T}} E^* \left( e^{-r\tau} R_\tau \right)$$

*where, as before, $\mathcal{T}_{tT}$ denotes the set of stopping times with values between $t$ and $T$. Furthermore, there exists a hedging strategy with the initial capital $V^*$.*

**Proof.** First, observe that since the payoff process $R$ is a cádlág, $R_\tau$ is measurable for any stopping time. Let $\pi = (\beta_t, \gamma_t)_{t \in [0,T]}$ be a hedging trading strategy with an initial capital $X_0^{\pi} = x$. Since $\pi$ is self-financing, according to the previous lecture its discounted portfolio value $e^{-rt} X_t^{\pi}$, $t \in$

$[0, T]$ forms a continuous local martingale with respect to $P^*$. The hedging condition $X_t^\pi \geq R_t$ for all $t \in [0, T]$ a.s. implies that $X_t^\pi \geq 0$ for all $t \in [0, T]$ a.s., and so by the lemma in the previous lecture $e^{-rt} X_t^\pi$, $t \in [0, T]$ forms a continuous supermartingale with respect to $P^*$. Hence, by the optional stopping theorem and by hedging for any $\tau \in \mathcal{T}_{0T}$,

$$x = X_0^\pi \geq E^*(e^{-r\tau} X_\tau^\pi) \geq E^*(e^{-r\tau} R_\tau).$$

Taking the supremum over $\tau \in \mathcal{T}_{0,T}$ we obtain $x \geq V^*$, i.e., the fair price cannot be lower than $V^*$.

In order to complete the proof we will construct a hedging (replicating) strategy with the initial capital exactly equal to $V^*$. Set

$$V_t = ess \sup_{\tau \in \mathcal{T}_{tT}} E^*(e^{-r\tau} R_\tau | \mathcal{F}_t)$$

which according to Lecture 12 is a supermartingale and

$$E^* V_t = \sup_{\tau \in \mathcal{T}_{tT}} E^*(e^{-r\tau} R_\tau). \tag{8.2.1}$$

According to Lecture 11 there exists a cádlág modification of $V_t$, $t \in [0, T]$ in $t$ provided $E^* V_t$ is right-continuous. To show this observe that if $t, t_n \in [0, T]$ and $t_n \downarrow t$ as $n \uparrow \infty$, then

$$E^* V_t \geq \lim_{n \uparrow \infty} E^* V_{t_n} \tag{8.2.2}$$

since $V_t$, $t \in [0, T]$ is a supermartingale.

For the inequality in the other direction take an arbitrary $\varepsilon > 0$, set $\tilde{R}_t = e^{-rt} R_t$ and using (8.2.1) choose a stopping time $\tau = \tau(\varepsilon) \in \mathcal{T}_{tT}$ such that

$$E^* V_t \leq E^* \tilde{R}_\tau + \varepsilon \quad \text{and} \quad P\{\tau > t\} = 1$$

which is possible since $\tilde{R}$ is right-continuous and nonnegative. For each $n \geq 1$ define stopping times $\tau_n \in \mathcal{T}_{t_n T}$ by

$$\tau_n = \begin{cases} \tau \text{ if } \tau \geq t_n \\ T \text{ if } \tau < t_n. \end{cases}$$

Then

$$|E^* \tilde{R}_\tau - E^* \tilde{R}_{\tau_n}| \leq E^*(\tilde{R}_\tau + \tilde{R}_T) \mathbb{I}_{\tau < t_n} \to 0 \text{ as } n \to \infty.$$

Since $\tau_n \geq t_n$, and so by (8.2.1), $E^* V_{t_n} \geq E^* \tilde{R}_{\tau_n}$, it follows that

$$E^* V_t \leq \lim_{n \to \infty} E^* \tilde{R}_{\tau_n} + \varepsilon \leq \lim_{n \to \infty} E^* V_{t_n} + \varepsilon.$$

Since $\varepsilon > 0$ is arbitrary, we obtain that $E^*V_t \leq \lim_{n\uparrow\infty} E^*V_{t_n}$ which together with (8.2.2) yields the right-continuity of $E^*V_t$ in $t \in [0,T]$. Thus, there exists a cádlág modification of $V_t$, $t \in [0,T]$ which we denote by the same letter and will work with from now on.

Recall that $Q = \sup_{t \in T} R_t$ is integrable by the assumption. Since $V_\tau \leq E^*(Q|\mathcal{F}_\tau)$ for any stopping time $\tau \in \mathcal{T}_{0T}$, we obtain

$$\int_{\{V_\tau > N\}} V_\tau dP^* \leq \int_{\{E^*(Q|\mathcal{F}_\tau) > N\}} E^*(Q|\mathcal{F}_\tau) dP^* = \int_{\{E^*(Q|\mathcal{F}_\tau) > N\}} Q dP^*.$$
(8.2.3)

Since $P^*\{E^*(Q|\mathcal{F}_\tau) > N\} \leq N^{-1}E^*Q$ we obtain from the absolute continuity of the Lebesgue integral that the right hand side of (8.2.3) tends to zero as $N \to \infty$ uniformly in $\tau \in \mathcal{T}_{0T}$ implying that $V_t$, $t \in [0,T]$ belongs to the class D. Thus, the conditions for the Doob–Meyer decomposition theorem from Lecture 11 are satisfied and we obtain that there exist a cádlág martingale $M_t$, $t \in [0,T]$ with $M_0 = V_0$ and a predictable increasing cádlág process $A_t$, $t \in [0,T]$ with $A_0 = 0$ such that

$$V_t = M_t - A_t \quad \text{a.s.} \quad \text{for each} \quad t \in [0,T].$$

In the same way as in the European option case we rely on the martingale representation from Lecture 18 to write

$$M_t = M_0 + \int_0^t \kappa e^{-ru} \gamma_u S_u dW^*(u), \ t \in [0,T] \quad \text{with}$$
$$M_0 = V_0 = \sup_{\tau \in \mathcal{T}_{[0,T]}} E^*(e^{-r\tau} R_\tau),$$

where the adapted process $\gamma_u$, $t \in [0,T]$ satisfying $\int_0^T \gamma_u^2 du < \infty$ a.s. is defined by this formula. Again, we set

$$\beta_t = (M_t - e^{-rt}\gamma_t S_t)B_0^{-1}, \ \pi = (\beta, \gamma) \quad \text{and} \quad X_t^\pi = e^{rt}M_t, \ t \in [0,T]$$

which yields that $X_t^\pi = \beta_t B_t + \gamma_t S_t$ and in the same way as in the European option case we see that the trading strategy $\pi = (\beta_t, \gamma_t)_{t \in [0,T]}$ is self-financing. To see that $\pi$ is hedging we write

$$X_t^\pi = e^{rt}M_t = e^{rt}(V_t + A_t) \geq e^{rt}V_t = e^{rt}\operatorname*{ess\,sup}_{\tau \in \mathcal{T}_{tT}} E^*(e^{-r\tau} R_\tau|\mathcal{F}_t) \geq R_t$$

where for the last inequality we take the specific stopping time $\tau \equiv t$ which, of course, also belongs to $\mathcal{T}_{tT}$. Thus, $\pi$ is hedging and since $X_0^\pi = M_0 = V_0 = V^* = \sup_{\tau \in \mathcal{T}_{0T}} E^*(e^{-r\tau} R_\tau)$, this completes the proof of the theorem.

$\square$

### 8.2.3   Israeli (game) contingent claims

**Definition 8.2.7.** An Israeli (game) contingent claim (option or other derivative) in the Black–Scholes market model is a contract between its buyer and seller so that the latter accepts an obligation to pay an amount $R(s,t) = Y_s \mathbb{I}_{s<t} + Z_t \mathbb{I}_{s \geq t}$ provided the seller cancels the contract at time $s$ and the buyer exercises it at time $t$. Here, $\{Y_t\}_{t \geq 0}$ and $\{Z_t\}_{t \geq 0}$ are nonnegative adapted to the filtration $\{\mathcal{F}_t\}_{t \geq 0}$ cádlág processes such that $Y_t \geq Z_t$ a.s. for all $t$. Again, there is a horizon $T$ so that the contract is stopped automatically at time $T$ if it was not cancelled or exercised before in which case the seller pays to the buyer the amount $R(T,T) = Y_T = Z_T$. The amount $\delta_t = Y_t - Z_t \geq 0$ is interpreted as a penalty the seller has to pay to the buyer for cancellation of the contract at time $t$. The buyer and the seller can exercise and cancel the contract at any stopping time between $0$ and $T$ which represents the requirement that their decisions should be based only on the information available up to the present moment.

**Definition 8.2.8.** An investment strategy of the seller in the game contingent claim case is a pair $(\pi, \sigma)$ where $\pi = (\beta, \gamma)$ is a trading strategy and $\sigma$ is the cancellation of the contract stopping time. Such an investment strategy $(\pi, \sigma)$ is called hedging if $\pi$ is self-financing and the corresponding portfolio value satisfies $X^{\pi}_{\sigma \wedge t} \geq R(\sigma, t)$ a.s. for all $t \in [0, T]$.

Again, we will not have to require nonnegativity of portfolio values (admissibility) here since this will follow automatically from hedging of a nonnegative payoff. Observe that if $(\pi, \sigma)$ is a hedging investment strategy then

$$X^{\pi}_{\sigma \wedge \tau} \geq R(\sigma, \tau) \quad \text{a.s. for any} \quad \tau \in \mathcal{T}_{0T}. \tag{8.2.4}$$

Indeed, let $\tau_n \in \mathcal{T}_{0T}$, $n = 1, 2, \ldots$ be stopping times taking on only countably many values and such that $\tau_n \downarrow \tau$ as $n \uparrow \infty$. Then $X^{\pi}_{\sigma \wedge \tau_n} \geq R(\sigma, \tau_n)$ a.s. for each $n$. Since $Y \geq Z$ and both are right-continuous, it follows that $\liminf_{n \uparrow \infty} R(\sigma, \tau_n) \geq R(\sigma, \tau)$. Taking into account that the discounted self-financing portfolio values $e^{-rt} X^{\pi}_t$, $t \geq 0$ form a continuous local martingale we obtain that $\lim_{n \uparrow \infty} X^{\pi}_{\sigma \wedge \tau_n} = X^{\pi}_{\sigma \wedge \tau}$, and so (8.2.4) holds true.

**Definition 8.2.9.** The fair price $V^*$ of a game contingent claim is the infimum of initial capitals of all hedging (self-financing) portfolios, i.e.,
$V^* = \inf\{x : \text{there exists a hedging investment strategy } (\pi, \sigma) \text{ such that } X^{\pi}_0 = x\}$.

**Theorem 8.2.3.** *An Israeli contingent claim with a horizon $T \in (0,\infty)$ and the payoff process $R(s,t) = Y_s \mathbb{I}_{s<t} + Z_t \mathbb{I}_{s \geq t}$, where $Y_t \geq Z_t$, $t \in [0,T]$ are adapted nonnegative cádlág processes satisfying $E^* \sup_{t \in [0,T]} Y_t < \infty$, has the fair price*

$$V^* = \inf_{\sigma \in \mathcal{T}_{0T}} \sup_{\tau \in \mathcal{T}_{0T}} E^* \big( e^{-r(\sigma \wedge \tau)} R(\sigma, \tau) \big).$$

*If, in addition, $Y_t$, $Z_t$, $t \in [0,T]$ are left-continuous then there exists a hedging investment strategy $(\pi^*, \sigma^*)$ with the initial capital $V^*$.*

**Proof.** First, observe that for any stopping times $\sigma, \tau \in \mathcal{T}_{0T}$ the random variables $Y_\sigma$ and $Z_\tau$ are measurable since the processes $Y$ and $Z$ are cádlágs. Since the random variables $\mathbb{I}_{\sigma < \tau}$ and $\mathbb{I}_{\sigma \geq \tau}$ are measurable too, we conclude that $R(\sigma, \tau)$ is measurable. Next, let $(\pi, \sigma)$ be a hedging investment strategy with an initial capital $X_0^\pi = x$. Since $\pi$ is self-financing, according to the previous lecture its discounted portfolio value $e^{-rt} X_t^\pi$, $t \in [0,T]$ forms a continuous local martingale with respect to $P^*$. Hence, for any stopping time $\sigma \in \mathcal{T}_{0,T}$ the stopped process $e^{-r(\sigma \wedge t)} X_{\sigma \wedge t}^\pi$, $t \in [0,T]$ also forms a continuous local martingale with respect to $P^*$ and by the hedging condition this local martingale is nonnegative. Thus, by the lemma in the previous lecture it forms a continuous supermartingale with respect to $P^*$. Hence, if $(\pi, \sigma)$ is hedging, then by the optional stopping theorem for any $\tau \in \mathcal{T}_{0T}$,

$$x = X_0^\pi \geq E^* (e^{-r(\sigma \wedge \tau)} X_{\sigma \wedge \tau}^\pi) \geq E^* (e^{-r(\sigma \wedge \tau)} R(\sigma, \tau)).$$

Taking the supremum over $\tau \in \mathcal{T}_{0,T}$ and infimum over $\sigma \in \mathcal{T}_{0,T}$ we obtain $x \geq V^*$, i.e., the fair price cannot be lower than $V^*$.

   In order to complete the proof we will show that for any $\varepsilon > 0$ there exists a hedging strategy with the initial capital not exceeding $V^* + \varepsilon$. Set $Q_t^\sigma = e^{-r(\sigma \wedge t)} R(\sigma, t)$ and observe that $Q_t^\sigma$ is $\mathcal{F}_{\sigma \wedge t}$-measurable while $Q_\tau^\sigma$ is a measurable random variable as explained above. Hence, we can define

$$V_t^\sigma = \text{ess} \sup_{\tau \in \mathcal{T}_{tT}} E^* \big( Q_\tau^\sigma | \mathcal{F}_t \big)$$

which according to Lecture 12 is a supermartingale and

$$E^* V_t^\sigma = \sup_{\tau \in \mathcal{T}_{tT}} E^* Q_\tau^\sigma. \tag{8.2.5}$$

According to Lecture 11 there exists a cádlág modification of $V_t^\sigma$, $t \in [0,T]$ in $t$ provided $E^* V_t^\sigma$ is right-continuous. To show this observe that if $t, t_n \in [0,T]$ and $t_n \downarrow t$ as $n \uparrow \infty$, then

$$E^* V_t^\sigma \geq \lim_{n \uparrow \infty} E V_{t_n}^\sigma \tag{8.2.6}$$

since $V_t^\sigma$, $t \in [0, T]$ is a supermartingale.

For the inequality in the other direction observe first that

$$\lim_{s \downarrow t} Q_s^\sigma = e^{-r\sigma \wedge t}(Y_\sigma \mathbb{I}_{\sigma \leq t} + Z_t \mathbb{I}_{t < \sigma}) \geq Q_t^\sigma \qquad (8.2.7)$$

since $Y_v \geq Z_v$, $v \in [0, T]$ and both $Y$ and $Z$ are right-continuous. Now we proceed in the same way as in the American options case. Namely, take an arbitrary $\varepsilon > 0$ and using the definition of $V^\sigma$ choose a stopping time $\tau = \tau(\varepsilon) \in \mathcal{T}_{tT}$ such that

$$E^* V_t^\sigma \leq E^* Q_\tau^\sigma + \varepsilon \quad \text{and} \quad P\{\tau > t\} = 1$$

which is possible in view of (8.2.5) and (8.2.7). For each $n \geq 1$ define stopping times $\tau_n \in \mathcal{T}_{t_n T}$ by

$$\tau_n = \begin{cases} \tau & \text{if } \tau \geq t_n \\ T & \text{if } \tau < t_n. \end{cases}$$

Then

$$|E^* Q_\tau^\sigma - E^* Q_{\tau_n}^\sigma| \leq E^*(Q_\tau^\sigma + Q_T^\sigma) \mathbb{I}_{\tau < t_n} \to 0 \text{ as } n \to \infty.$$

Since $\tau_n \geq t_n$, and so by (8.2.5), $E^* V_{t_n}^\sigma \geq E^* Q_{\tau_n}^\sigma$, it follows that

$$E^* V_t^\sigma \leq \lim_{n \to \infty} E^* Q_{\tau_n}^\sigma + \varepsilon \leq \lim_{n \to \infty} E V_{t_n}^\sigma + \varepsilon.$$

Since $\varepsilon > 0$ is arbitrary, we obtain that $E^* V_t^\sigma \leq \lim_{n \uparrow \infty} E V_{t_n}^\sigma$ which together with (8.2.6) yields the right-continuity of $E V_t^\sigma$ in $t \in [0, T]$. Thus, there exists a cádlág modification of $V_t^\sigma$, $t \in [0, T]$ which we denote by the same letter and will work with from now on.

Recall, that $\sup_{t \in [0, T]} Y_t$ is integrable by the assumption, and so $\sup_{s, t \in [0, T]} R(s, t) \leq 2 \sup_{t \in [0, T]} Y_t$ is integrable as well. Using this we conclude in the same way as in the American contingent claim case above that the cádlág supermartingale $V_t^\sigma$, $t \in [0, T]$ belongs to the class D. Thus, the conditions for the Doob–Meyer decomposition theorem from Lecture 11 are satisfied and we obtain that there exists a cádlág martingale $M_t^\sigma$, $t \in [0, T]$ with $M_0^\sigma = V_0^\sigma$ and a predictable increasing cádlág process $A_t^\sigma$, $t \in [0, T]$ with $A_0^\sigma = 0$ such that

$$V_t^\sigma = M_t^\sigma - A_t^\sigma \quad \text{a.s.} \quad \text{for each} \quad t \in [0, T].$$

In the same way as in the American contingent claim case we rely on the martingale representation from Lecture 18 to write

$$M_t^\sigma = M_0^\sigma + \int_0^t \kappa e^{-ru} \gamma_u^\sigma S_u dW^*(u), \ t \in [0, T] \quad \text{with}$$

$$M_0 = V_0^\sigma = \sup_{\tau \in \mathcal{T}_{0T}} E^*(e^{-r(\sigma \wedge \tau)} R(\sigma, \tau)),$$

where the adapted process $\gamma_u^\sigma$, $t \in [0,T]$ satisfying $\int_0^T (\gamma_u^\sigma)^2 du < \infty$ a.s. is defined by this formula. Again, we set

$$\beta_t^\sigma = (M_t^\sigma - e^{-rt}\gamma_t^\sigma S_t)B_0^{-1}, \quad \pi^\sigma = (\beta^\sigma, \gamma^\sigma) \text{ and } X_t^{\pi^\sigma} = e^{rt}M_t^\sigma, \quad t \in [0,T]$$

which yields that $X_t^{\pi^\sigma} = \beta_t^\sigma B_t + \gamma_t^\sigma S_t$ and in the same way as in the American contingent claim case we see that the trading strategy $\pi^\sigma = (\beta_t^\sigma, \gamma_t^\sigma)_{t \in [0,T]}$ is self-financing. To see that $\pi^\sigma$ is hedging we write

$$X_t^{\pi^\sigma} = e^{rt}M_t^\sigma = e^{rt}(V_t^\sigma + A_t^\sigma) \geq e^{rt}V_t^\sigma$$
$$= e^{rt}ess\sup\nolimits_{\tau \in \mathcal{T}_{tT}} E^*(e^{-r\sigma\wedge\tau}R(\sigma,\tau)|\mathcal{F}_t) \geq R(\sigma,t)$$

where for the last inequality we take the specific stopping time $\tau \equiv t$ which, of course, also belongs to $\mathcal{T}_{tT}$. Thus, $(\pi^\sigma, \sigma)$ is a hedging investment strategy with the initial capital

$$X_0^{\pi^\sigma} = M_0^\sigma = V_0^\sigma = \sup_{\tau \in \mathcal{T}_{0T}} E^*(e^{-r\sigma\wedge\tau}R(\sigma,\tau)).$$

Next, we apply the theorem about Dynkin's games from Lecture 12 to the payoff $\tilde{R}(s,t) = e^{-rs}Y_s\mathbb{I}_{s<t} + e^{-rt}Z_t\mathbb{I}_{s\geq t}$. This theorem implies that for any $\varepsilon > 0$ there exists $\sigma^\varepsilon \in \mathcal{T}_{0,T}$ such that for any $\tau \in \mathcal{T}_{0,T}$,

$$E^*(e^{r(\sigma^\varepsilon\wedge\tau)}R(\sigma^\varepsilon,\tau)) \leq V^* + \varepsilon$$

and if $Y$ and $Z$ are also left-continuous then there exists $\sigma^* \in \mathcal{T}_{0,T}$ such that for any $\tau \in \mathcal{T}_{0,T}$,

$$E^*(e^{r(\sigma^*\wedge\tau)}R(\sigma^*,\tau)) \leq V^*.$$

For such $\sigma^\varepsilon$ (or $\sigma^*$) we can construct by the above hedging investment strategy $(\pi^{\sigma^\varepsilon}, \sigma^\varepsilon)$ (or $(\pi^*, \sigma^*)$ where $\pi^* = \pi^{\sigma^*}$) with the initial capital not exceeding $V^* + \varepsilon$ (or not exceeding $V^*$). Since $\varepsilon > 0$ is arbitrary, it follows that the fair price should not exceed $V^*$. If $\sigma^*$ exists then the initial capital of the hedging investment strategy $(\pi^*, \sigma^*)$ does not exceed $V^*$ and since by the first part of the proof the initial capital of any hedging investment strategy cannot be less than $V^*$, we obtain the equality completing the proof of the theorem.  $\square$

### 8.2.4  *Prices of contingent claims from a buyer's point of view*

As in the discrete time case the fair price of a contingent claim from a buyer's point of view is based on taking a bank loan in an amount $x$ in order to pay for the contract and then handling a self-financing portfolio with the initial capital $-x$ so that at an exercise time it will be possible to settle the debt after receiving the contract's payoff. We will see that in the Black–Scholes market model both prices from seller's and buyer's point of view coincide which is, essentially, the consequence of uniqueness of the martingale measure.

**Definition 8.2.10.** A self-financing trading strategy $\pi = (\beta_t, \gamma_t)_{t \in [0,T]}$ is called hedging from a buyer's point of view for an European contingent claim with a payoff $R = R_T \geq 0$ at an expiration time $T < \infty$ if the corresponding portfolio values $X^\pi$ satisfy $X_T^\pi + R \geq 0$ a.s.

**Definition 8.2.11.** A pair $(\pi, \tau)$ of a self-financing trading strategy $\pi = (\beta_t, \gamma_t)_{t \in [0,T]}$ and a stopping time $\tau \in \mathcal{T}_{0T}$ is called hedging from a buyer's point of view for an American contingent claim with a payoff $R_t \geq 0$, $t \in [0, T]$ and a horizon $T < \infty$ if the corresponding portfolio values $X^\pi$ satisfy $X_\tau^\pi + R_\tau \geq 0$ a.s.

We observe that unlike the seller's point of view case, the buyer here not only chooses a trading strategy buy also an exercise time $\tau$, and so the full strategy of the buyer is the pair $(\pi, \tau)$.

**Definition 8.2.12.** A pair $(\pi, \tau)$ of a self-financing trading strategy $\pi = (\beta_t, \gamma_t)_{t \in [0,T]}$ and a stopping time $\tau \in \mathcal{T}_{0T}$ is called hedging from a buyer's point of view for an Israeli contingent claim with a payoff $R(s,t) \geq 0$, $s, t \in [0, T]$ and a horizon $T < \infty$ if the corresponding portfolio values $X^\pi$ satisfy $X_{s \wedge \tau}^\pi + R(s, \tau) \geq 0$ a.s. for any $s \in [0, T]$.

The self-financing trading strategy $\pi$ (and the portfolio $X^\pi$) will be called here admissible if the corresponding discounted portfolio values $e^{-rt} X_t^\pi$, $t \in [0, T]$ form a supermartingale. In the buyer's approach setup the portfolio values can stay negative but some lower bounds control in order to secure the repayment of the loan can be justified from the financial point of view. As shown in the lemma of the previous lecture appropriate lower bounds on portfolio values make them a supermartingale.

**Definition 8.2.13.** The fair price $U^*$ of a contingent claim from the buyer's point of view is given by $U^* = \sup\{x :$ there exists an admissible hedging trading strategy $\pi$ (in the European contingent claim case) or a hedging pair $(\pi, \tau)$ with an admissible $\pi$ and a stopping time $\tau$ (in the American and Israeli contingent claims cases) such that the initial capital of the corresponding portfolio $X^\pi$ equals $-x \}$.

**Theorem 8.2.4.** *The fair price $U^*$ from the buyer's point of view of a European, American and Israeli contingent claims with a horizon $T \in (0, \infty)$ coincides with the corresponding fair price $V^*$ from the seller's point of view provided that the payoffs satisfy the same conditions as in the corresponding theorems above. In the European case there exists a hedging strategy with the initial capital $-U^*$. In the American and Israeli cases hedging strategies with the initial capital $-U^*$ exist if the corresponding payoff functions (R in the American case and Y, Z in the Israeli case) are also left-continuous (i.e., they are continuous since the right continuity is the basic assumption).*

**Proof.** (i) In the European contingent claim case with the payoff $R = R_T$ let $\pi$ be an admissible hedging strategy with an initial capital $-x$. Then $e^{-rt}X_t^\pi$, $t \in [0, T]$ is a supermartingale, and so

$$-x = X_0^\pi \geq E^*(e^{-rT}X_T^\pi) \geq -E^*(e^{-rT}R) = -V^*.$$

Hence, $x \leq V^*$ holds true for any $\pi$ and $x$, and so $U^* \leq V^*$.

To obtain the other bound we will construct an admissible hedging portfolio with the initial capital $-V^*$. Define the martingale

$$\tilde{M}_t = -M_t = -E^*(e^{-rT}R|\mathcal{F}_t), \ t \in [0, T].$$

Here $M_t$, $t \in [0, T]$ is the martingale appearing in the above theorem on the fair price of an European contingent claim from the seller's point of view. Using the constructions and notations there we define

$$X_t^\pi = e^{rt}\tilde{M}_t = -e^{rt}M_t = \tilde{\beta}_t dB_t + \tilde{\gamma}_t dS_t$$

where $\tilde{\beta}_t = -\beta_t$ and $\tilde{\gamma}_t = -\gamma_t$. Then the self-financing condition follows. The hedging and even replicating of $\pi$ follows as well. Set $\Xi_t = E^*(R|\mathcal{F}_t)$ and observe that the family $\{\Xi_\tau, \tau \in \mathcal{T}_{0T}\}$ is uniformly integrable. Indeed, since $R \geq 0$,

$$\int_{\{|\Xi_\tau|>K\}} |\Xi_\tau| dP^* = \int_{\{|\Xi_\tau|>K\}} R dP^* \tag{8.2.8}$$

and

$$P^*\{|\Xi_\tau| > K\} \leq K^{-1}E^*|\Xi_\tau| = K^{-1}ER.$$

By the absolute continuity of the Lebesgue integral the right hand side of (8.2.8) tends to zero as $K \to \infty$ uniformly in $\tau \in \mathcal{T}_{0T}$. Since with probability one $X_t^\pi \geq -\Xi_t$ for all $t \in [0, T]$ we can apply the lemma from the previous lecture to conclude that $X_t^\pi$, $t \in [0, T]$ is a supermartingale establishing admissibility of $\pi$.

(ii) In the American contingent claim case we assume that the payoff $R_t$, $t \in [0, T]$ is a nonnegative càdlàg such that $R = \sup_{t \in [0,T]} R_t$ is integrable with respect to $P^*$. Let $(\pi, \tau)$ be a hedging pair with an admissible $\pi$. Since $e^{-rt} X_t^\pi$, $t \in [0, T]$ is a supermartingale, we obtain by the optional stopping theorem that

$$-x = X_0^\pi \geq E^*(e^{-r\tau} X_\tau^\pi) \geq -E^*(e^{-r\tau} R_\tau),$$

i.e., $x \leq E^*(e^{-r\tau} R_\tau)$. Thus, the buyer cannot take a loan above $V^* = \sup_{\tau \in \mathcal{T}_{0T}} E^*(e^{-r\tau} R_\tau)$ which can be settled at an exercise time, and so $U^* \leq V^*$.

For the bound in the other direction we define for each $\tau \in \mathcal{T}_{0T}$ a martingale

$$M_t^\tau = -E^*(e^{-r\tau} R_\tau | \mathcal{F}_t), \ t \in [0, T].$$

Again, we use the martingale representation theorem from Lecture 18 to obtain

$$M_t^\tau = M_0^\tau + \int_0^t \kappa e^{-ru} \gamma_u^\tau S_u dW^*(u), \ t \in [0, T]$$

where $\gamma_u^\tau$, $u \in [0, T]$ is defined by this formula. As before, we set for each $t \in [0, T]$,

$$\beta_t^\tau = (M_t^\tau - e^{-rt} \gamma_t^\tau S_t) B_0^{-1}, \ \pi^\tau = (\beta^\tau, \gamma^\tau), \ X_t^{\pi^\tau} = e^{rt} M_t^\tau$$

but now we define the strategy $\pi^\tau = (\beta^\tau, \gamma^\tau)$ and the corresponding portfolio $X_t^{\pi^\tau} = e^{rt} M_t^\tau$ only for $t \leq \tau$ since the contract is exercised at the time $\tau$ and the portfolio is not managed after this time. As before, we see that $\pi^\tau$ is self-financing. Now, $X_\tau^{\pi^\tau} = e^{r\tau} M_\tau^\tau = -R_\tau$, i.e., $X_\tau^{\pi^\tau} + R_\tau = 0$ which means that the pair $(\pi^\tau, \tau)$ is hedging. In the same way as in the European contingent claim case above we see here that the strategy $\pi^\tau$ is admissible since for any $t \leq \tau$,

$$X_t^{\pi^\tau} \geq -E^*(R_\tau | \mathcal{F}_t)$$

where the right hand side forms a uniformly integrable family, and so the lemma from the previous lecture is applicable. The initial value of the portfolio $X^{\pi^\tau}$ is $X_0^{\pi^\tau} = M_0^\tau = -E^*(e^{-r\tau} R_\tau)$. For each $\varepsilon > 0$ there exists $\tau_\varepsilon \in \mathcal{T}_{0T}$ such that

$$E^*(e^{-r\tau_\varepsilon} R_{\tau_\varepsilon}) \geq \sup_{\tau \in \mathcal{T}_{0T}} E^*(e^{-r\tau} R_\tau) - \varepsilon = V^* - \varepsilon,$$

and so by the definition $U^* \geq V^*$. This together with the other inequality above yields that $U^* = V^*$. Observe that by Lecture 12 if $R_t$, $t \in [0, T]$ is also left-continuous over stopping times then there exists $\tau_0 \in \mathcal{T}_{0T}$ such that $E^*(e^{-r\tau_0}R_{\tau_0}) = V^*$, and so the hedging strategy $(\pi^{\tau_0}, \tau_0)$ has the initial capital $-U^*$.

(iii) Let $(\pi, \tau)$ be a hedging pair in the Israeli contingent claim case with the payoff $R(s,t) = Y_s\mathbb{I}_{s<t} + Z_t\mathbb{I}_{s\geq t}$. Then for each $s \in [0,T]$,

$$X^\pi_{s\wedge\tau} + R(s,\tau) \geq 0 \quad \text{a.s.}$$

Since $|R(s,t)| \leq 2\sup_{s\in[0,T]}Y_s$ and the right hand side here is integrable we conclude that $R(s,\tau)$, $s \in [0,T]$ is a uniformly integrable family. Now, $e^{-r(s\wedge\tau)}X^\pi_{s\wedge\tau}$, $s \in [0,T]$ is a local martingale and $e^{-r(s\wedge\tau)}X^\pi_{s\wedge\tau} \geq -R(s,\tau)$ which yields that $e^{-r(s\wedge\tau)}X^\pi_{s\wedge\tau}$, $s \in [0,T]$ is a supermartingale by the lemma in the previous lecture. Hence, we obtain the admissibility here from hedging and not as an extra assumption. Thus, relying on the optional stopping theorem we can write for any $\sigma \in \mathcal{T}_{0T}$ that

$$-x = X^\pi_0 \geq E^*(e^{-r(\sigma\wedge\tau)}X^\pi_{\sigma\wedge\tau}) \geq -E^*(e^{-r(\sigma\wedge\tau)}R(\sigma,\tau)).$$

It follows that

$$x \leq \inf_{\sigma\in\mathcal{T}_{0T}} \sup_{\tau\in\mathcal{T}_{0T}} E^*(e^{-r(\sigma\wedge\tau)}R(\sigma,\tau)) = V^*,$$

and so $U^* \leq V^*$.

To obtain the inequality in the other direction, for each $\tau \in \mathcal{T}_{0T}$ define

$$N^\tau_t = ess\sup_{\sigma\in\mathcal{T}_{tT}} E^*\left(-e^{-r(\sigma\wedge\tau)}R(\sigma,\tau)|\mathcal{F}_t\right)$$

which is a supermartingale with respect to $P^*$. Similarly to the case of the Israel contingent claim from a seller's point of view we conclude that $EN^\tau_t$ is right-continuous, and so $N^\tau_t$, $t \in [0,T]$ has a right-continuous modification which we denote by the same letter and work with. Next, we rely on the Doob–Meyer decomposition theorem from Lecture 11 obtaining

$$N^\tau_t = M^\tau_t - A^\tau_t, \ M^\tau_0 = N^\tau_0, \ A^\tau_t \geq 0$$

where $M^\tau_t$, $t \in [0,T]$ is a martingale. Using as before the martingale representation theorem from Lecture 18 we write

$$M^\tau_t = M^\tau_0 + \int_0^t \kappa e^{-ru}\gamma^\tau_u S_u dW^*(u), \ t \in [0,T]$$

where $\gamma^\tau_u$, $u \in [0,T]$ is defined by this equality. Again, for each $t \in [0,T]$ we put

$$\beta^\tau_t = (M^\tau_t - e^{-rt}\gamma^\tau_t S_t)B_0^{-1}, \ \pi^\tau = (\beta^\tau, \gamma^\tau), \ X^{\pi^\tau}_t = e^{rt}M^\tau_t$$

and observe that $X_t^{\pi^\tau}$, $t \in [0,T]$ is a self-financing portfolio. Moreover,

$$X_{s\wedge\tau}^{\pi^\tau} = e^{r(s\wedge\tau)} M_{s\wedge\tau}^\tau \geq e^{r(s\wedge\tau)} N_{s\wedge\tau}^\tau \geq -R(s\wedge\tau,\tau) \geq -R(s,\tau)$$

where in the second inequality we take a specific stopping time $\sigma = s \wedge \tau$ in place of the essential supremum. Thus, the pair $(\pi,\tau)$ is hedging. Using the lemma from the previous lecture we see that this hedging implies admissibility of $\pi^\tau$ since $R(s,\tau)$, $s \in [0,T]$ is a uniformly integrable family.

The initial portfolio value is

$$X_0^{\pi^\tau} = M_0^\tau = N_0^\tau = \sup_{\sigma\in\mathcal{T}_{0T}} E^*\big(-e^{-r(\sigma\wedge\tau)}R(\sigma,\tau)\big)$$
$$= -\inf_{\sigma\in\mathcal{T}_{0T}} E^*\big(e^{-r(\sigma\wedge\tau)}R(\sigma,\tau)\big).$$

For each $\varepsilon > 0$ there exists $\tau_\varepsilon \in \mathcal{T}_{0T}$ such that

$$\inf_{\sigma\in\mathcal{T}_{0T}} E^*\big(e^{-r(\sigma\wedge\tau_\varepsilon)}R(\sigma,\tau_\varepsilon)\big) \geq \sup_{\tau\in\mathcal{T}_{0T}}\inf_{\sigma\in\mathcal{T}_{0T}} E^*\big(e^{-r(\sigma\wedge\tau)}R(\sigma,\tau)\big)-\varepsilon = V^*-\varepsilon$$

where in the last equality we use that the Dynkin game with the payoff

$$e^{-r(s\wedge t)}R(s,t) = Y_s e^{-rs}\mathbb{I}_{s<t} + Z_t e^{-rt}\mathbb{I}_{s\geq t}$$

has a value, and so $\sup_{\tau\in\mathcal{T}_{0T}}\inf_{\sigma\in\mathcal{T}_{0T}} = \inf_{\sigma\in\mathcal{T}_{0T}}\sup_{\sigma\in\mathcal{T}_{0T}}$ (see Lecture 12). Since $\varepsilon > 0$ is arbitrary, it follows that $U^* \geq V^*$ which together with the other inequality proved above yields $U^* = V^*$. If $Y$ and $Z$ appearing in the definition of the payoff are also left-continuous then we conclude relying on Lecture 12 that the hedging strategy $(\pi^{\tau_0},\tau_0)$ has the initial capital $-U^*$, completing the proof of the theorem. $\qquad\square$

### 8.2.5 *Arbitrage arguments*

In the same way as in the discrete time case we say here that a self-financing trading strategy $\pi = (\beta_t,\gamma_t)_{t\in[0,T]}$ with a portfolio $X^\pi$ provides an arbitrage opportunity if $X_0^\pi = 0$, $X_T^\pi \geq 0$ and $P\{X_T^\pi > 0\} > 0$. More generally, an arbitrage opportunity is a possibility for a market participant to make a riskless profit. A financial market is called viable if there are no arbitrage opportunities in it. A mature market usually eliminates quickly any arbitrage opportunities, and so the mathematical finance mostly studies market models without arbitrage opportunities.

If the discounted portfolio values $e^{-rt}X_t^\pi$, $t \in [0,T]$ for a trading strategy $\pi$ form a supermartingale with respect to the martingale measure $P^*$ then $E^*X_0^\pi \geq E^*e^{-rT}X_T^\pi$. If $X_0^\pi = 0$ and $X_T^\pi \geq 0$ a.s. then $X_T^\pi = 0$ a.s., i.e., the strategy $\pi$ does not provide an arbitrage opportunity. Hence, admissible trading strategies do not provide arbitrage opportunities. We recall that, as the example at the beginning of this lecture shows, in the continuous time case some conditions on self-financing trading strategies are needed to prevent arbitrage opportunities.

Observe that $V^* = E^* e^{-rT} R_T$ is the unique no arbitrage price of the European contingent claim with a horizon $T$ and a payoff $R_T$. Indeed, if the seller sold the contract for a price $V > V^*$ then a hedging (replicating) portfolio with the initial capital $V^*$ can be built which will lead to the riskless profit $V - V^*$. If the contract is sold for $V < V^*$ then the buyer can take the loan $V^*$ paying only $V$ for the contract and to build a hedging portfolio with the initial capital $-V$ as described above which will allow to settle the loan at time $T$ after obtaining the contract's payoff. This will lead to the riskless profit $V^* - V$ of the buyer.

### 8.2.6 *A partial differential equation approach*

After deriving formulas for pricing of contingent claims in the Black–Scholes market it is natural to inquire about efficient methods to compute these prices and corresponding hedging strategies with minimal initial capital. In discrete time markets we saw in Lecture 4 that for American and Israeli contingent claims this can be done by a dynamical programming algorithm. We will see in Lecture 22 that prices and hedging strategies in the Black–Scholes market can be approximated by the corresponding objects in a sequence of binomial CRR markets while the latter can be computed as described in Lecture 4. On the other hand, if the payoff functions depend only on the last stock price and they are sufficiently regular then we can apply the following partial differential equations approach which we describe briefly and refer the reader for more details to [49], [47], [60], [27] and references there.

Namely, let the payoff functions $R_T$, $R_t$, $t \in [0,T]$ and $R(s,t) = Y_s \mathbb{I}_{s<t} + Z_t \mathbb{I}_{s \geq t}$ of the European, American and Israeli contingent claims, respectively, satisfy $R_T = R(S_T)$, $R_t = R(S_t)$, $Y_t = Y(S_t)$ and $Z_t = Z(S_t)$ where $S_t$, $t \in [0,T]$ is the stock price at the time $t$ and the functions $R$, $Y$ and $Z$ are twice continuously differentiable. Let $S_t^{x,u}$ be the solution of the stochastic differential equation

$$dS_t = S_t(rdt + \kappa dW^*(t)) \quad \text{with} \quad S_u = S_u^{x,u} = x,$$

$E^*$ be the expectation with respect to the martingale measure $P^*$ and $W^*$ be the corresponding Brownian motion. Introduce the differential operator

$$L = \frac{\kappa^2}{2} x^2 \frac{\partial^2}{\partial x^2} + rx \frac{\partial}{\partial x}.$$

Suppose that $g(t, x)$ has one bounded derivative in $t$ and two in $x$ and it satisfies the equation

$$\frac{\partial g}{\partial t} + Lg - rg = 0 \text{ with the condition } g(T, x) = R(x). \tag{8.2.9}$$

Using the probabilistic representation of solutions of such equations described in Lecture 17 we obtain

$$g(0, x) = E^* \left( e^{-rT} R(S_T^{0,x}) \right) = V(0, x)$$

which is the price of the European contingent claim above provided $S_0 = x$. Thus, $g$ can be obtained if we solve (8.2.9) either analytically or numerically.

Observe that the operator $L$ in (8.2.9) is not strictly elliptic since the coefficient in the second derivative is zero at $x = 0$. Nevertheless, we can consider $U_t = \ln S_t$ which (by the Itô formula) satisfies the stochastic differential equation

$$dU_t = (r - \frac{\kappa^2}{2})dt + \kappa dW^*(t).$$

Let $\hat{L} = \frac{\kappa^2}{2} \frac{\partial^2}{\partial x^2} + (r - \frac{\kappa^2}{2}) \frac{\partial}{\partial x}$ and suppose that $f = f(t, x)$ satisfies the equation

$$\frac{\partial f}{\partial t} + \hat{L}f - rf = 0$$

where now $\hat{L}$ is a strictly elliptic partial differential operator with constant coefficients. Then $g(t, x) = f(t, \ln x)$ will satisfy the equation (8.2.9).

In the American and Israeli contingent claims cases the situation is more complicated. Namely, let

$$g(t, x) = \sup_{\tau \in \mathcal{T}_{tT}} E^* (e^{-r\tau} R(S_\tau^{x,t})) \text{ and}$$

$$g(t, x) = \inf_{\sigma \in \mathcal{T}_{tT}} \sup_{\tau \in \mathcal{T}_{tT}} E^* \left( e^{-r\sigma \wedge \tau} (Y(S_\sigma^{x,t}) \mathbb{I}_{\sigma < \tau} + Z(S_\tau^{x,t}) \mathbb{I}_{\sigma \geq \tau}) \right)$$

be the corresponding value functions, i.e., prices of the contingent claims at the time $t$ if the stock price at this time is $x$. Then it is possible to show that $g$ satisfies (8.2.9) in the domain $C = \{(t, x) : g(t, x) > e^{-rt} R(x)\}$ in the American case and in the domain $C = \{(t, x) : e^{-rt} Y(x) > g(t, x) > e^{-rt} Z(x)\}$ in the Israeli case. The first domain is bounded by one boundary $\partial C = \{(t, x) : g(t, x) = e^{-rt} R(x)\}$ and the second one by two $\partial C_{seller} = \{(t, x) : g(t, x) = e^{-rt} Y(x)\}$ and $\partial C_{buyer} = \{(t, x) : g(t, x) = e^{-rt} Z(x)\}$. Both the domain $C$ and the boundaries are not fixed but depend on the unknown function $g$, and so the problem of finding of a solution $g$ is called a free boundary problem. For details see [40], [62] and references there.

## 8.3   Lecture 21: Multi-asset Black–Scholes market

### 8.3.1   *Market model*

Let $W = \{W(t), t \in [0,T]\}$, $W(t) = (W_1(t), ..., W_n(t))$ be the $n$-dimensional continuous Brownian motion on a probability space $(\Omega, \mathcal{F}, P)$ and $\mathcal{F}_t = \sigma\{W(s), s \in [0,t]\}$, $t \in [0,T]$ be the corresponding Brownian filtration (complemented by the sets of zero probability) where we assume for convenience that $\mathcal{F} = \mathcal{F}_T$. The multi-asset Black–Scholes market consists of a riskless security, called a bond or a money market account, whose value can only grow, though may be randomly, and of $d$ risky securities, called stocks, whose values fluctuate randomly both up and down.

More formally, the price of the bond at a time $t$ is supposed to evolve according to the formula

$$B_t = 1 + \int_0^t r_s B_s ds, \quad \text{i.e.,} \quad B_t = \exp\left(\int_0^t r_s ds\right) \quad \text{a.s.}$$

where the interest rate $r = \{r_t, t \in [0,T]\}$ is a nonnegative adapted process such that the map $r : [0,T] \times \Omega \to [0,\infty)$ acting by $r(t,\omega) = r_t(\omega)$ is measurable with respect to the product $\sigma$-algebra $\mathcal{B}_{[0,T]} \times \mathcal{F}_T$, where $\mathcal{B}_{[0,T]}$ is the Borel $\sigma$-algebra on $[0,T]$. We assume also that

$$\int_0^T r_t dt < \infty \quad \text{a.s.} \tag{8.3.1}$$

By the properties of the Lebesgue integral $\int_0^t r_s ds$ is a.s. continuous in $t$, and so $B_t$ is also continuous in $t$. We assume for simplicity that $B_0 = 1$ but this does not harm generality as we can measure values in the units of $B_0$ (i.e., take $B_0$ as a numeraire). Observe that the process $r_t$, $t \in [0,T]$ can be viewed as a variable random interest rate.

The price evolution at time $t$ of the $i$-th stock, $i = 1, ..., d$ will be given by the formula

$$S_t^i = S_0^i \exp\left(\int_0^t (\mu_s^i - \frac{1}{2}(\kappa_s^i, \kappa_s^i))ds + \int_0^t (\kappa_s^i, dW(s))\right)$$

where $S_0^i$ is a positive constant, $\mu = (\mu_t, t \in [0,T])$, $\mu_t = (\mu_t^1, ..., \mu_t^d)$ and $\kappa = \{\kappa_t, t \in [0,T]\}$, $\kappa_t = (\kappa_t^{ij}, i = 1, ..., d; j = 1, ..., n)$, $\kappa_t^i = (\kappa_t^{i1}, ..., \kappa_t^{in})$ are, respectively, vector and matrix valued adapted processes such that $\mu, \kappa : (t,\omega) \to \mu_t(\omega), \kappa_t(\omega)$ are $\mathcal{B}_{[0,T]} \times \mathcal{F}_T$-measurable maps and

$$\int_0^T |\mu_t| dt < \infty \quad \text{and} \quad \int_0^T |\kappa_t|^2 dt < \infty \quad \text{a.s.} \tag{8.3.2}$$

where, as before, $(\cdot, \cdot)$ denotes the inner product of vectors, $|a|^2 = \sum_{i=1}^{d} a_i^2$ and $|b|^2 = \sum_{i=1}^{d} \sum_{j=1}^{n} b_{ij}^2$ for each vector $a = (a_1, ..., a_d)$ and a matrix $b = (b_{ij}, i = 1, ..., d; j = 1, ..., n)$. Accordingly, for any vector process $c_t = (c_t^1, ..., c_t^n)$ we write $\int_0^t (c_s, dW(s)) = \sum_{j=1}^{n} \int_0^t c_s^j dW_j(s)$. The conditions on $\mu$ and $\kappa$ above enable us to conclude that both Lebesgue and stochastic integrals in the exponent in the definition of $S_t^i$ above are well defined and they are continuous in $t$, and so the stock price processes $S_t^i$, $i = 1, ..., d$ themselves are a.s. continuous in $t$ as well. Observe that this setup generalizes the one from the previous two lectures not only in the number of stocks but also because the interest rate $r_t$, the parameter $\mu_t$ and the volatility matrix $\kappa_t$ are now time dependent and random.

By the multidimensional Itó formula from Lecture 15,

$$S_t^i = S_0^i + \int_0^t \mu_u^i S_u^i du + \int_0^t S_u^i (\kappa_u^i, dW(u)) \tag{8.3.3}$$

or in the differential form

$$dS_t^i = S_t^i \big( \mu_t^i dt + (\kappa_t^i, dW(t)) \big).$$

Since, as explained above, the processes $S_t^i$, $i = 1, ..., d$ are continuous in $t$, $\sup_{t \in [0,T]} |S_t^i(\omega)| < \infty$ with probability one. Hence,

$$\int_0^T |\mu_t| S_t^i dt < \infty \quad \text{and} \quad \int_0^T |\kappa_t|^2 (S_t^i)^2 dt < \infty \quad \text{a.s.,} \tag{8.3.4}$$

and so the integrals in (8.3.3) are well defined and they are continuous in $t$. Rewriting the above formulas for the discounted stock price $\tilde{S}_t^i = S_t^i \exp(- \int_0^t r_s ds)$ we obtain

$$d\tilde{S}_t^i = \tilde{S}_t^i \big( (\mu_t^i - r_t) dt + (\kappa_t^i, dW(t)) \big). \tag{8.3.5}$$

### 8.3.2 Martingale measures

Introduce the vector process $\eta_t = (\eta_t^1, ..., \eta_t^d)$ by $\eta_t^i = \mu_t^i - r_t$, $i = 1, ..., d$. Suppose that there exists another adapted vector process $\zeta_t = (\zeta_t^1, ..., \zeta_t^d)$ measurable as a map $\zeta : (t, \omega) \to \zeta_t(\omega)$ and such that

$$\int_0^T |\zeta_t|^2 dt < \infty \quad \text{a.s. and} \quad \kappa_t \zeta_t = \eta_t \quad \text{a.s. for each } t \in [0, T], \tag{8.3.6}$$

where $\kappa_t \zeta_t$ is the multiplication between the matrix $\kappa_t$ and the vector $\zeta_t$. Set

$$W^\zeta(t) = W(t) + \int_0^t \zeta_s ds.$$

Then

$$\kappa_t dW^\zeta(t) = \kappa_t dW(t) + \kappa_t \zeta_t dt = \kappa_t dW(t) + \eta_t dt,$$

and so for $i = 1, 2, ..., d$,

$$dS_t^i = S_t^i\big(r_t dt + (\kappa_t^i, dW^\zeta(t))\big) \text{ and } d\tilde{S}_t^i = \tilde{S}_t^i(\kappa_t^i, dW^\zeta(t)). \tag{8.3.7}$$

Suppose that

$$\Lambda_t = \exp\big(-\int_0^t (\zeta_s, dW(s)) - \frac{1}{2}\int_0^t |\zeta_s|^2 ds\big), \, t \in [0, T]$$

is a martingale which follows under Novikov's condition

$$E \exp(\frac{1}{2}\int_0^T |\zeta_s|^2 ds) < \infty \tag{8.3.8}$$

as explained in Lecture 16. Then by the Girsanov theorem from Lecture 16, $W^\zeta$ is the Brownian motion with respect to the probability measure $P^\zeta$ defined by

$$P^\zeta(\Gamma) = E(\mathbb{I}_\Gamma \Lambda_T) \text{ for any } \Gamma \in \mathcal{F}(= \mathcal{F}_T), \text{ i.e., } \frac{dP^\zeta}{dP} = \Lambda_T,$$

and so $P^\zeta$ is equivalent to $P$. Taking into account (8.3.4) and (8.3.7) we conclude that

$$\tilde{S}_t^i = \tilde{S}_0^i + \int_0^t \tilde{S}_u^i(\kappa_u^i, dW^\zeta(u)) \tag{8.3.9}$$

is a continuous local martingale with respect to $P^\zeta$ and the latter is called a local martingale measure . Observe that if $\kappa_t$ is invertible with probability one for all $t \in [0, T]$ then $\zeta_t = \kappa_t^{-1}\eta_t$ exists. If, in addition, $\kappa_t$ is continuous in $t$, in particular $\kappa$ is constant a.s., then $\sup_{t\in[0,T]}|\kappa_t^{-1}| < \infty$ a.s. and taking into account (8.3.1) and (8.3.2) we see that the integral boundedness condition on $\zeta$ in (8.3.6) holds true as well. The Novikov condition (8.3.8) will be satisfied if, for instance, $r_t$, $|\mu_t|$ and $|\kappa_t^{-1}|$ are uniformly bounded, in particular constant, and so $|\zeta_t|$ will be uniformly bounded too. In this case $\tilde{S}_t$, $t \in [0, T]$ will be a true martingale with respect to the measure $P^\zeta$ which will be a martingale measure then.

Next, we proceed in the other direction where the argument is similar to the proof of uniqueness of a martingale measure in the setup of Lecture 19. Let $Q$ be a probability measure equivalent to $P$. Then, in the same way as in Lecture 19 using the generalized Bayes formula, we see that the Radon–Nikodym derivatives $L_t = \frac{dQ_t}{dP_t}$, $t \in [0, T]$, where $Q_t$ and $P_t$ are restrictions of $Q$ and $P$ to $\mathcal{F}_t$, form a right-continuous martingale with respect to $P$. This martingale may not be square integrable but $L$ is locally square integrable with the localizing sequence $\tau_n = \inf\{t \geq 0 : L_t > n\}$. Using the multidimensional martingale representation theorem from Lecture 18 we can write

$$L_t = 1 + \int_0^t (\psi_s, dW(s)) = 1 + \int_0^t L_s(\zeta_s, dW(s)) \qquad (8.3.10)$$

where $\zeta_s = \frac{\psi_s}{L_s}$ and $\psi$ is an adapted to the filtration $\mathcal{F}_t$, $t \in [0, T]$ vector process satisfying $\int_0^T |\psi_s|^2 ds < \infty$ a.s. Hence, $L_t$, $t \in [0, T]$ is a.s. continuous in $t$.

Now suppose that $Q$ is a local martingale measure so that the discounted stock values $\tilde{S}_t = S_t \exp(-\int_0^t r_s ds)$ form a vector local martingale with respect to $Q$ with some localizing sequence $\tau_n \uparrow \infty$ as $n \uparrow \infty$. Then, each $\tilde{S}_{t \wedge \tau_n}$, $t \in [0, T]$ is a martingale and by the generalized Bayes formula from Lecture 7,

$$\tilde{S}_{s \wedge \tau_n} = E_Q(\tilde{S}_{t \wedge \tau_n} | \mathcal{F}_s) = E_P\left(\frac{L_t}{L_s} \tilde{S}_{t \wedge \tau_n} | \mathcal{F}_s\right)$$

for each $s \in [0, t)$. Here, the indexes $Q$ and $P$ emphasise the measure with respect to which an expectation is taken though in other places we write just $E$ if the expectation is taken with respect to the basic probability $P$. Since $L_s$ is $\mathcal{F}_s$-measurable, we obtain

$$L_s \tilde{S}_{s \wedge \tau_n} = E_P(L_t \tilde{S}_{t \wedge \tau_n} | \mathcal{F}_s).$$

Since $L$ is a continuous martingale with respect to $P$, by taking the conditional expectation with respect to $\mathcal{F}_{s \wedge \tau_n}$ we obtain that

$$L_{s \wedge \tau_n} \tilde{S}_{s \wedge \tau_n} = E_P\left(E_P(L_t | \mathcal{F}_{t \wedge \tau_n}) \tilde{S}_{t \wedge \tau_n} | \mathcal{F}_{s \wedge \tau_n}\right) = E_P\left(L_{t \wedge \tau_n} \tilde{S}_{t \wedge \tau_n} | \mathcal{F}_{s \wedge \tau_n}\right).$$

But the right hand side here equals

$$E_P\left(L_{t \wedge \tau_n} \tilde{S}_{t \wedge \tau_n} | \mathcal{F}_s\right) \mathbb{I}_{s \leq \tau_n} + E_P\left(L_{\tau_n} \tilde{S}_{\tau_n} | \mathcal{F}_{\tau_n}\right) \mathbb{I}_{s > \tau_n}$$

and

$$E_P\left(L_{\tau_n} \tilde{S}_{\tau_n} | \mathcal{F}_{\tau_n}\right) \mathbb{I}_{s > \tau_n} = L_{\tau_n} \tilde{S}_{\tau_n} \mathbb{I}_{s > \tau_n} = E_P(L_{\tau_n} \tilde{S}_{\tau_n} | \mathcal{F}_s) \mathbb{I}_{s > \tau_n}$$

since $L_{\tau_n} \tilde{S}_{\tau_n} \mathbb{I}_{s > \tau_n}$ is $\mathcal{F}_s$-measurable. Hence,

$$L_{s \wedge \tau_n} \tilde{S}_{s \wedge \tau_n} = E_P\left(L_{t \wedge \tau_n} \tilde{S}_{t \wedge \tau_n} | \mathcal{F}_s\right).$$

It follows that $L_t \tilde{S}_t^i$, $t \in [0, T]$ is a continuous local martingale with respect to $P$ for each $i = 1, ..., d$.

Now, by the multidimensional martingale representation theorem from Lecture 18 we obtain that each $L_t \tilde{S}_t^i$, $i = 1, ..., d$ can be written as a sum of stochastic integrals with respect to the Brownian motions $W_j$, $j = 1, ..., n$. On the other hand, by (8.3.5), (8.3.10) and the Itô formula

$$d\big(L_t S_t^i \exp(-\int_0^t r_s ds)\big)$$

$$= L_t e^{-\int_0^t r_s ds} S_t^i \big((\mu_t^i - r_t - (\kappa_t^i, \zeta_t))dt + (\zeta_t + \kappa_t^i, dW(t))\big).$$

In the same way as in Lecture 19, relying on the arguments in the proof of uniqueness of the Doob–Meyer decomposition, we conclude that the coefficient in $dt$ must vanish. Hence, $\mu_t^i - r_t - (\kappa_t^i, \zeta_t) = 0$ a.s. for each $t \in [0, T]$ and $i = 1, ..., d$ or in vector-matrix notations

$$\kappa_t \zeta_t = \eta_t \quad \text{a.s. where} \quad \eta_t^i = \mu_t^i - r_t, \ i = 1, ..., d. \tag{8.3.11}$$

Solving the stochastic integral equation (8.3.10) we see that

$$L_t = \exp\big(-\int_0^t (\zeta_s, dW(s)) - \frac{1}{2} \int_0^t |\zeta_s|^2 ds\big), \tag{8.3.12}$$

and so $Q = P^\zeta$. Thus, we obtained the following result

**Theorem 8.3.1.** *A probability measure $Q$ is a local martingale measure (i.e., the discounted stock price $\tilde{S}$ is a local martingale under $Q$ and $Q \sim P$) if and only if there exists an adapted vector process $\zeta = (\zeta_t^1, ..., \zeta_t^n)$ such that the map $\zeta : (t, \omega) \to \zeta_t(\omega)$ is measurable, $\int_0^T |\zeta_t|^2 < \infty$ a.s., $L_t, t \in [0, T]$ given by (8.3.12) is a continuous martingale with respect to $P$, $Q = P^\zeta$, i.e., $\frac{dQ_t}{dP_t} = L_t, t \in [0, T]$ and (8.3.11) holds true.*

*If $d = n$ and $\kappa_t$ is invertible a.s. for each $t$, then there exists at most one solution to (8.3.11) which satisfies conditions, and so in this case there exists at most one local martingale measure. If $d = n$, $\kappa_t^{-1}$ exists and $|\kappa_t|$, $|\kappa_t^{-1}|$, $|\mu_t|$, $r_t$ are all uniformly bounded, then $L_t, t \in [0, T]$ is a martingale and there exists a unique martingale measure. If $\kappa_t \equiv \kappa$ is a constant matrix, $\mu_t \equiv \mu$ is a constant vector, $r_t \equiv r$ is constant too, $\text{rank}(\kappa) < n$ and $\eta \in \kappa\mathbb{R}^n$ then there exist infinitely many martingale measures while if $\eta \notin \kappa\mathbb{R}^n$ then there exists none.*

### 8.3.3　Trading strategies

**Definition 8.3.1.** (i) A trading strategy in a multi-asset Black–Scholes market on a time interval $[0, T]$, $T < \infty$ is a pair $\pi = (\beta, \gamma)$ of a scalar and a vector stochastic processes $\beta_t$, $\gamma_t = (\gamma_t^1, ..., \gamma_t^d)$, $t \in [0, T]$ adapted to the filtration $\{\mathcal{F}_t, t \in [0, T]\}$ with the map $(\beta, \gamma) : (t, \omega) \to (\beta_t(\omega), \gamma_t(\omega))$ measurable with respect to $\mathcal{B}_{[0,T]} \times \mathcal{F}_T$ and such that with probability one

$$\int_0^T |\beta_t| dt < \infty \text{ and } \int_0^T \big(|\gamma_t| r_t + |\sum_{i=1}^d \gamma_t^i (\mu_t^i - r_t)| + |\sum_{i=1}^d \gamma_t^i \kappa_t^i|^2\big) dt < \infty. \tag{8.3.13}$$

(ii) For a trading strategy $\pi = (\beta, \gamma)$ the corresponding portfolio value at time $t$ is given by

$$X_t^\pi = \beta_t B_t + (\gamma_t, S_t).$$

(iii) A trading strategy $\pi = (\beta, \gamma)$ and a corresponding portfolio $X^\pi$ are called self-financing if

$$X_t^\pi = X_0^\pi + \int_0^t \beta_u dB_u + \int_0^t (\gamma_u, dS_u), \text{ i.e., } dX_t^\pi = \beta_t dB_t + (\gamma_t, dS_t).$$
(8.3.14)

(iv) A self-financing trading strategy $\pi$ is called admissible if $X_t^\pi \geq 0$ a.s. for all $t \in [0, T]$.

Since $B_t$ and $S_t$ are continuous in $t \in [0, T]$ we obtain from (8.3.1), (8.3.2) and (8.3.13) that with probability one

$$\int_0^T \left( |\beta_t| B_t + |(\gamma_t, S_t)| r_t + \left| \sum_{i=1}^d \gamma_t^i S_t^i (\mu_t^i - r_t) \right| + \left| \sum_{i=1}^d \gamma_t^i S_t^i \kappa_t^i \right|^2 \right) dt < \infty$$
(8.3.15)

which implies that the integrals in (iii) above exist. Observe that if the parameters $r, \mu$ and $\kappa$ are constant then it suffices to require only that

$$\int_0^T \left( |\beta_t| + |\gamma_t|^2 \right) dt < \infty \quad \text{a.s.}$$

As before, we will use admissibility of trading strategies only to apply the lemma from Lecture 19 to make the discounted portfolio value $\tilde{X}_t^\pi$, $t \in [0, T]$ a supermartingale, and so a lower bound on $\tilde{X}^\pi$ by an integrable random variable will already suffice.

Let $P^\varsigma$ be a local martingale measure and $W^\varsigma(t) = W(t) + \int_0^t \varsigma_s ds$ be the corresponding Brownian motion with respect to $P^\varsigma$. Set $\tilde{X}_t^\pi = e^{-\int_0^t r_s ds} X_t^\pi$ and suppose that $\pi = (\beta, \gamma)$ is a self-financing trading strategy. Then by (8.3.5), (8.3.7) and (iii) of the above definition,

$$d\tilde{X}_t^\pi = -r_t \tilde{X}_t^\pi dt + e^{-\int_0^t r_s ds} dX_t^\pi = -r_t \tilde{X}_t^\pi dt + r_t e^{-\int_0^t r_s ds} X_t^\pi dt$$
$$+ e^{-\int_0^t r_s ds} \sum_{i=1}^d \gamma_t^i S_t^i (\kappa_t^i, dW^\varsigma(t)) = e^{-\int_0^t r_s ds} \sum_{i=1}^d \gamma_t^i S_t^i (\kappa_t^i, dW^\varsigma(t)).$$

Hence,

$$\tilde{X}_t^\pi = \tilde{X}_0^\pi + \int_0^t e^{-\int_0^u r_s ds} \sum_{i=1}^d \gamma_u^i S_u^i (\kappa_u^i, dW^\varsigma(u))$$

is a continuous local martingale.

**Lemma 8.3.1.** *(i) Let $\pi = (\beta_t, \gamma_t)_{t \in [0,T]}$ be a trading strategy and $X_t^\pi = \beta_t B_t + (\gamma_t, S_t)$. Then (8.3.14) holds true if and only if the discounted values $\tilde{X}_t^\pi$ and $\tilde{S}_t$ satisfy*

$$d\tilde{X}_t^\pi = \big((\gamma_t \tilde{S}_t), \eta_t\big)dt + \big((\gamma_t \tilde{S}_t), \kappa_t dW(t)\big) \qquad (8.3.16)$$

*where $\gamma_t \tilde{S}_t = ((\gamma_t \tilde{S}_t)^i, i = 1, ..., d)$ with $(\gamma_t \tilde{S}_t)^i = \gamma_t^i \tilde{S}_t^i$ and $\eta_t$ is the same as in (8.3.11).*

*(ii) Let $\gamma_t = (\gamma_t^1, ..., \gamma_t^d)$ satisfy (8.3.13), (8.3.16) and set*

$$\beta_t = X_0 + \int_0^t \big((\gamma_u \tilde{S}_u), \eta_u\big)du + \int_0^t \big((\gamma_u \tilde{S}_u), \kappa_u dW(u)\big) - (\gamma_t, \tilde{S}_t).$$

*Then $\pi = (\beta_t, \gamma_t)_{t \in [0,T]}$ is a self-financing trading strategy with $X_0^\pi = X_0$ and $\tilde{X}_t^\pi$ satisfying (8.3.16).*

**Proof.** (i) If (8.3.14) holds true then using (8.3.3) we obtain by Itô's formula that

$$d\tilde{X}_t^\pi = -r_t \tilde{X}_t^\pi dt + \beta_t r_t dt + \big((\gamma_t \tilde{S}_t), \mu_t\big)dt$$
$$+\big((\gamma_t \tilde{S}_t), \kappa_t dW(t)\big) = \big((\gamma_t \tilde{S}_t), \eta_t\big)dt + \big((\gamma_t \tilde{S}_t), \kappa_t dW(t)\big)$$

taking into account that $\tilde{X}_t^\pi = \beta_t + (\gamma_t, \tilde{S}_t)$. In the other direction, starting from (8.3.16) and using that

$$e^{\int_0^t r_s ds} d(X_t^\pi e^{-\int_0^t r_s ds}) = dX_t^\pi - r_t X_t^\pi dt,$$

we obtain (8.3.14) taking into account (8.3.3) and that $X_t^\pi = \beta_t B_t + (\gamma_t, S_t)$.

(ii) If $\beta_t$ is given by the above formula and (8.3.16) holds true then $\beta_t = \tilde{X}_t^\pi - (\gamma_t, \tilde{S}_t)$, i.e., $X_t^\pi = \beta_t B_t + (\gamma_t, S_t)$. This together with (8.3.16) yields (8.3.14), and so $\pi = (\beta_t, \gamma_t)_{t \in [0,T]}$ is self-financing verifying easily that $\beta$ satisfies the condition (8.3.13). $\qquad \square$

### 8.3.4   Arbitrage

Recall that a self-financing trading strategy $\pi = (\beta_t, \gamma_t)_{t \in [0,T]}$ with a portfolio $X^\pi = X_t^\pi$, $t \in [0,T]$ is said to provide an arbitrage opportunity if $X_0^\pi = 0$, $X_T^\pi \geq 0$ a.s. and $P\{X_T^\pi > 0\} > 0$. Clearly, this is equivalent to $\tilde{X}_0^\pi = 0$, $\tilde{X}_T^\pi \geq 0$ a.s. and $P\{\tilde{X}_T^\pi > 0\} > 0$ for the discounted portfolio values $\tilde{X}_t^\pi = X_t^\pi e^{-\int_0^t r_s ds}$. The following result provides a partial characterization for the existence of arbitrage opportunities in a multi-asset Black–Scholes market.

**Theorem 8.3.2.** *(i) If there exists an adapted vector process $\zeta_t = (\zeta_t^1, ..., \zeta_t^n)$ solving (8.3.11) and satisfying other conditions of Theorem 8.3.1 then admissible trading strategies do not provide arbitrage opportunities;*

*(ii) If no admissible trading strategy provides an arbitrage opportunity in the market above then there exists an adapted vector process $\zeta_t = (\zeta_t^1, ..., \zeta_t^n)$ solving (8.3.11) a.s. for Lebesgue almost all $t \in [0, T]$ (we do not claim that $\zeta$ satisfies other conditions of Theorem 8.3.1).*

**Proof.** (i) If $\pi$ is a self-financing trading strategy then the discounted portfolio values $\tilde{X}_t^\pi$, $t \in [0, T]$ form a continuous local martingale with respect to a local martingale measure $P^\zeta$ which exists under our assumptions. If $\pi$ is admissible then $\tilde{X}_t^\pi$, $t \in [0, T]$ is also a supermartingale by the lemma in Lecture 19. Hence, if $X_0^\pi = 0$ then $0 = \tilde{X}_0^\pi \geq E^\zeta \tilde{X}_T^\pi$ where $E^\zeta$ is the expectation with respect to $P^\zeta$. If $\tilde{X}_T^\pi \geq 0$ a.s. then $\tilde{X}_T^\pi = 0$ a.s., and so $\pi$ does not provide an arbitrage opportunity.

(ii) The matrix $\kappa_t$ can be viewed as a linear map of $\mathbb{R}^n$ to its image $\kappa_t \mathbb{R}^n \subset \mathbb{R}^d$. Each vector $u \in \mathbb{R}^d$ can be represented uniquely in the form $u = \hat{u}_t + u_t^\perp$ where $\hat{u}_t \in \kappa_t \mathbb{R}^n$ and $u_t^\perp$ belongs to the orthogonal complement $(\kappa_t \mathbb{R}^n)^\perp$ of $\kappa_t \mathbb{R}^n$. Set $\Gamma_t = \{\omega : \text{the equation (8.3.11) has no solution}\} = \{\omega : \eta_t(\omega) \notin \kappa_t(\omega)\mathbb{R}^n\}$. Let $v_t(\omega)$ be the orthogonal projection of $\eta_t(\omega)$ to $(\kappa_t \mathbb{R}^n)^\perp$. By the above, $\Gamma_t = \{\omega : v_t(\omega) \neq 0\}$. Observe that

$$(\eta_t, v_t) = (v_t, v_t) \geq 0 \quad \text{and} \quad (\kappa_t^* v_t, z) = (v_t, \kappa_t z) = 0$$

for any $z \in \mathbb{R}^n$, where $\kappa_t^*$ is the conjugate of $\kappa_t$. Since $\kappa_t = \kappa_t(\omega)$ is measurable in $(t, \omega)$, and so both $\kappa_t(\omega)\mathbb{R}^n$ and $(\kappa_t(\omega)\mathbb{R}^n)^\perp$ depend measurably on $(t, \omega)$, it follows that $v_t = v_t(\omega)$ is measurable in $(t, \omega)$, as well as the indicator $\mathbb{I}_{\Gamma_t}(\omega)$ of the set $\Gamma_t$, $t \in [0, T]$. For $i = 1, ..., d$ set

$$\gamma_t^i = \begin{cases} \dfrac{v_t^i(\omega)}{\tilde{S}_t^i(\omega)|v_t(\omega)|} & \text{if} \quad \omega \in \Gamma_t \\ 0 & \text{if} \quad \omega \notin \Gamma_t. \end{cases}$$

In view of (8.3.1) and (8.3.2) it is easy to see that $\gamma_t = (\gamma_t^1, ..., \gamma_t^d)$ satisfies (8.3.13). By the lemma above there exists $\beta_t$, $t \in [0, T]$ satisfying (8.3.13) such that $\pi = (\beta_t, \gamma_t)_{t \in [0,T]}$ is a self-financing trading strategy and $X_0^\pi = 0$. Since $((\gamma_t, \tilde{S}_t), \kappa_t dW(t)) = 0$ (understood in the stochastic integral sense) by the choice of $\gamma$, it follows from (8.3.16) that

$$\tilde{X}_t^\pi = \int_0^t |v_s| \mathbb{I}_{\Gamma_s} ds \geq 0,$$

and so $\pi$ is an admissible trading strategy. Observe that $|v_t| \leq |\eta_t| \leq |\mu_t| + |r_t|$, and so the above integral exists in view of (8.3.1) and (8.3.2). If $\pi$ does not provide an arbitrage opportunity then we must have

$$0 = P\{\tilde{X}_T^\pi > 0\} = P\{\int_0^T |v_s| \mathbb{I}_{\Gamma_s} ds > 0\}.$$

Hence, $\int_0^T |v_s| \mathbb{I}_{\Gamma_s} ds = 0$ a.s. Since $|v_s(\omega)| > 0$ when $\omega \in \Gamma_s$, we see that $\int_0^T \mathbb{I}_{\Gamma_s} ds = 0$ a.s., and so $0 = E \int_0^T \mathbb{I}_{\Gamma_s} ds = \int_0^T P(\Gamma_s) ds$ which implies that $P(\Gamma_s) = 0$ for Lebesgue almost all $s \in [0, T]$, completing the proof of (ii). $\qquad\qquad\qquad\qquad\qquad\qquad\qquad\qquad\qquad\qquad\qquad\qquad\qquad\square$

### 8.3.5   *Pricing of derivatives*

In the same way as in the previous lecture we define European, American and Israeli contingent claims, their hedging, admissible strategies and fair prices from the seller's point of view.

**Theorem 8.3.3.** *Assume that a local martingale measure $P^*$ exists.*

*(i) The fair price $V^*$ from the seller's point of view of a contingent claim with a horizon $T < \infty$ satisfies*

*(a) In the European case with a payoff function $R \geq 0$,*

$$V^* \geq E^*(e^{-\int_0^T r_s ds} R) \tag{8.3.17}$$

*where $E^*$ is the expectation with respect to a local martingale measure $P^*$ and we assume that $E^* R < \infty$;*

*(b) In the American case with an adapted cádlág payoff function $R_t \geq 0$, $t \in [0, T]$,*

$$V^* \geq \sup_{\tau \in \mathcal{T}_{0T}} E^*(e^{-\int_0^\tau r_s ds} R_\tau) \tag{8.3.18}$$

*where we assume that $E^* \sup_{t \in [0,T]} R_t < \infty$;*

*(c) In the Israeli case with a payoff $R(s,t) = Y_s \mathbb{I}_{s<t} + Z_t \mathbb{I}_{s \geq t}$ where $Y_t \geq Z_t \geq 0$, $t \in [0, T]$ are cádlág processes,*

$$V^* \geq \inf_{\sigma \in \mathcal{T}_{0T}} \sup_{\tau \in \mathcal{T}_{0T}} E^*(e^{-\int_0^{\sigma \wedge \tau} r_s ds} R(\sigma, \tau)) \tag{8.3.19}$$

*where we assume that $E^* \sup_{t \in [0,T]} Y_t < \infty$.*

*(ii) Suppose that $n = d$, the matrices $\kappa_t$, $t \in [0, T]$ are invertible with probability one and*

$$\sup_{t \in [0,T]} (|\kappa_t| + |\kappa_t^{-1}| + r_t + |\mu_t|) < \infty \quad a.s. \tag{8.3.20}$$

*Then in (8.3.17)–(8.3.19) we have equalities for fair prices $V^*$ in the corresponding cases.*

**Proof.** (i) As we showed it above the discounted portfolio values $\tilde{X}_t^\pi$, $t \in [0, T]$ for any self-financing trading strategy $\pi$ form a continuous local martingale under each local martingale measure $P^*$. By the lemma from Lecture 19 a proper lower bound on a continuous local martingale makes it a continuous supermartingale. In the European contingent claim case this lower bound comes from the admissibility assumption while in the American and Israeli contingent claims cases the lower bounds come automatically from hedging. Thus, let $\pi$ be an admissible hedging trading strategy, then in the European case,

$$x = \tilde{X}_0^\pi \geq E^*(\tilde{X}_T^\pi) \geq E^*(e^{-\int_0^T r_t dt} R).$$

In the American case by the optional stopping theorem and by hedging, for any $\tau \in \mathcal{T}_{0T}$,

$$x = \tilde{X}_0^\pi \geq E^*(\tilde{X}_\tau^\pi) \geq E^*(e^{-\int_0^\tau r_t dt} R_\tau).$$

In the Israeli case if $(\pi, \sigma)$ is a hedging pair then, again, by the optional stopping theorem and by hedging, for any $\tau \in \mathcal{T}_{0T}$,

$$x = \tilde{X}_0^\pi \geq E^*(\tilde{X}_{\sigma \wedge \tau}^\pi) \geq E^*(e^{-\int_0^{\sigma \wedge \tau} r_t dt} R(\sigma, \tau)),$$

which completes the proof of (i).

Next, we deal with (ii) considering first the case of an European contingent claim. As before, introduce the martingale with respect to $P^*$,

$$M_t = E^*(e^{-\int_0^T r_s ds} R | \mathcal{F}_t), \ t \in [0, T].$$

As in the previous lecture, such martingale may not be square integrable but it is locally square integrable. Hence, we can apply the multi-dimensional martingale representation theorem from Lecture 18 to obtain

$$M_t = M_0 + \int_0^t (\varphi_s, dW^*(s))$$

where $W^*$ is the Brownian motion with respect to $P^*$ and $\varphi_s = (\varphi_s^1, \varphi_s^2, ..., \varphi_s^n)$ is an adapted vector process satisfying $\int_0^T |\varphi_s|^2 ds < \infty$ a.s. Since $\kappa_t$ is invertible for each $t \in [0, T]$ we can define

$$\gamma_t^i = e^{\int_0^t r_s ds} (S_t^i)^{-1} (\varphi_t \kappa_t^{-1})^i$$

where $\varphi_t \kappa_t^{-1}$ is the multiplication of the row vector $\varphi_t$ by the matrix $\kappa_t^{-1}$ from the right. Then

$$\varphi_t^j = e^{-\int_0^t r_s ds} \sum_{i=1}^d \gamma_t^i S_t^i \kappa_t^{ij},$$

and so

$$M_t = M_0 + \int_0^t e^{-\int_0^u r_s ds} \sum_{i=1}^d \gamma_u^i S_u^i \sum_{j=1}^n \kappa_u^{ij} dW_j^*(u).$$

Set

$$\beta_t = (M_t - e^{-\int_0^t r_s ds}(\gamma_t, S_t)), \quad \pi = (\beta, \gamma) \text{ and } X_t^\pi = e^{\int_0^t r_s ds}M_t.$$

It follows from (8.3.20) that the conditions (8.3.15) on $\beta$ and $\gamma$ are satisfied. Now we have $X_t^\pi = \beta_t B_t + \gamma_t S_t$ and

$$dX_t^\pi = r_t X_t^\pi dt + e^{\int_0^t r_s ds} dM_t = r_t X_t^\pi dt + \sum_{i=1}^d \gamma_t^i S_t^i(\kappa_t^i, W^*(t))$$
$$= r_t X_t^\pi dt + \sum_{i=1}^d \gamma_t^i (dS_t^i - r_t S_t^i dt) = \beta_t dB_t + (\gamma_t, dS_t).$$

Hence, $\pi$ is self-financing. Clearly, $\pi$ is hedging and even replicating since $X_T^\pi = R$. Since conditional expectations of nonnegative random variables are nonnegative, $X_t^\pi = e^{\int_0^t r_s ds} M_t \geq 0$, and so $\pi$ is admissible. It follows that in the European contingent claims case under the additional assumptions of (ii) we have the equality in (8.3.17). The proof for the American and Israeli contingent claims is similar to the previous lecture taking into account the modifications above concerning the multidimensional martingale representation and the verification of self-financing. $\qquad\square$

We observe that in the same way as in the previous lecture together with modifications above it follows that under the conditions (ii) of the above theorem, which ensure uniqueness of a martingale measure, the fair prices from the buyer's point of view are the same as from the seller's one described above.

## 8.4 Lecture 22: Binomial approximations of the Black–Scholes market

### 8.4.1 *The setup*

In this lecture we will obtain error estimates for approximations of fair prices of contingent claims in the Black–Scholes (BS) market by fair prices of corresponding contingent claims in the binomial CRR (Cox–Ross–Rubinstein) markets considered in Lecture 5 and we will follow mostly [36] though the proof from there was slightly modified here in order to rely only on results appearing in this book. We observe that binomial approximations for some particular cases of European and American contingent claims were considered in [71] and [50], respectively. As we know from Lecture 5 prices for American and Israeli contingent claims in a CRR market can be obtained via dynamical programming (backward induction) algorithm while in the BS market the discrete time approximation is the only available procedure when payoffs are sufficiently general and depend on the whole history of the stock evolution. As in Lecture 19 we will consider here the standard BS market with one bond and one stock whose prices evolve according to the formulas $B_t = B_0 e^{rt}$ and $S_t = S_0 \exp((\mu - \frac{1}{2}\kappa^2)t + \kappa W(t))$, respectively.

According to Lecture 20, the fair prices $V_E$, $V_A$ and $V_I$ of European, American and Israeli contingent claims, respectively, with nonnegative payoffs $R = R(T)$, $R(t)$ and $R(s,t) = Y_s\mathbb{I}_{s<t} + Z_t\mathbb{I}_{s\geq t}$, $Y \geq Z$ satisfying $ER < \infty$, $E \sup_{0\leq t\leq T} R(t) < \infty$ and $E \sup_{0\leq t\leq T} Y_t < \infty$, are given by the formulas

$$V_E = E^*(e^{-rT}R(T)), \quad V_A = \sup_{\tau\in\mathcal{T}_{0T}} E^*(e^{-r\tau}R(\tau)) \text{ and}$$
$$V_I = \inf_{\tau\in\mathcal{T}_{0T}} \sup_{\tau\in\mathcal{T}_{0T}} E^*(e^{-r\sigma\wedge\tau}R(\sigma,\tau))$$

where $T > 0$ is a horizon, $r$ is the interest rate, $E^*$ is the expectation with respect to the martingale measure and $\mathcal{T}_{0T}$ denotes the set of all stopping times with respect to the Brownian filtration $\mathcal{F}_t$, $t \geq 0$ taking on values between 0 and $T$.

If we replace the original Brownian motion $W(t)$ by the Brownian motion $W^*(t) = \frac{\mu-r}{\kappa}t + W(t)$ with respect to the martingale measure $P^*$ then the stock price at the time $t$ has the form

$$S_t(z) = z \exp\left((r - \frac{1}{2}\kappa^2)t + \kappa W^*(t)\right) = z \exp(rt + \kappa \hat{W}(t)) \qquad (8.4.1)$$

where $\hat{W}(t) = -\frac{\kappa}{2}t + W^*(t)$ and $S_0(z) = z > 0$. We will approximate the latter expression by a sequence

$$S_t^{(n)}(z) = z \exp\left(\sum_{k=1}^{[nt/T]} \left(\frac{rT}{n} + \kappa(\frac{T}{n})^{1/2}\xi_k\right)\right), \quad t \geq T/n \qquad (8.4.2)$$

$$\text{and } S_t^{(n)}(z) = S_0^{(n)}(z) = z > 0, \quad t \in [0, T/n)$$

where $\xi_1$, $\xi_2$, ... are i.i.d. random variables taking on the values 1 and -1 with probabilities $p^{(n)} = (\exp(\kappa\sqrt{\frac{T}{n}})+1)^{-1}$ and $1 - p^{(n)} = (\exp(-\kappa\sqrt{\frac{T}{n}})+1)^{-1}$, respectively. The value $S_t^{(n)}(z)$ is the price at the time $[nt/T]$ of a stock in the CRR market model with the parameters $r = r^{(n)} = \exp(rT/n) - 1$ and $\rho_k = \rho_k^{(n)} = \exp(\frac{rT}{n} + \kappa(\frac{T}{n})^{1/2}\xi_k) - 1$ which have the bond and the stock price at the step $m$ given by the formulas

$$B_m^{(n)} = B_0(1 + r^{(n)})^m \quad \text{and} \quad \tilde{S}_m^{(n)}(z) = z \prod_{k=1}^{m}(1 + \rho_k^{(n)}) = S_{\frac{mT}{n}}^{(n)}. \qquad (8.4.3)$$

Let $P_n^\xi = \{p^{(n)}, 1 - p^{(n)}\}^\mathbb{N}$ be the product measure and $E_n^\xi$ be the corresponding expectation. Since $E_n^\xi \rho_k^{(n)} = r^{(n)}$, it follows that $P_n^\xi$ is the martingale measure for the corresponding CRR market model.

Next, we will write our assumptions concerning the payoffs of contingent claims under consideration. For each $t > 0$ denote by $M[0,t]$ the space of Borel measurable functions on $[0,t]$ with the uniform metric $d_{0t}(v,\tilde{v}) = \sup_{0 \leq s \leq t} |v_s - \tilde{v}_s|$. For each $t > 0$ let $F_t$ and $\Delta_t$ be nonnegative functions on $M[0,t]$ such that for some constant $L \geq 1$ and for any $t \geq s \geq 0$ and $v, \tilde{v} \in M[0,t]$,

$$|F_s(v) - F_s(\tilde{v})| + |\Delta_s(v) - \Delta_s(\tilde{v})| \leq L(s+1)d_{0s}(v,\tilde{v}), \qquad (8.4.4)$$

and

$$|F_t(v) - F_s(v)| + |\Delta_t(v) - \Delta_s(v)| \leq L\big(|t-s|(1+ \sup_{u \in [0,t]} |v_u|) + \sup_{u \in [s,t]} |v_u - v_s|\big).$$
$$\qquad (8.4.5)$$

By (8.4.4), $F_0(v) = F_0(v_0)$ and $\Delta_0(v) = \Delta_0(v_0)$ are functions of $v_0$ only. By (8.4.5),

$$F_t(v) + \Delta_t(v) \leq F_0(v_0) + \Delta_0(v_0) + L(t+2)(1+ \sup_{0 \leq s \leq t} |v_s|) \qquad (8.4.6)$$

The contingent claims will be considered up to a horizon $T > 0$ and we will assume that

$$F_t = G_t = F_T \quad \text{for all} \quad t \geq T,$$

in particular, in the Israeli contingent claim case $\Delta_t \equiv 0$ for all $t \geq T$, and so we require (8.4.5) for $\Delta$ only when $0 \leq s \leq t < T$ (for a modification of the proof without the latter condition see [12]). Now we define the payoff and discounted payoff functions $R_z = R_z(T)$, $Q_z = Q_z(T) = e^{-rT}R_z(T)$, $R_z(t)$, $Q_z(t) = e^{-rt}R_z(t)$, $R_z(s,t)$, $Q_z(s,t) = e^{-rs \wedge t}R_z(s,t)$ of European, American and Israeli contingent claims by

$$R_z = F_T(S(z)), \ R_z(t) = F_t(S(z)) \text{ and} \qquad (8.4.7)$$
$$R_z(s,t) = G_s(S(z))\mathbb{I}_{s<t} + F_t(S(z))\mathbb{I}_{s \geq t},$$

where $G_t = F_t + \Delta_t$ and $S(z) = S(z,\omega) \in M[0,T]$ is the random path of the BS stock price $S_t(z) = S_t(z,\omega)$ on the time interval $[0,T]$ and we extend it for convenience setting $S_t(z) = S_T(z)$ for all $t \geq T$.

Among main examples of options with path-dependent payoff we can consider integral options where $K > 0$ is a strike price,

$$F_t(v) = \Big(\int_0^t f_u(v_u)du - K\Big)^+ \quad \text{(call option case)}$$

or

$$F_t(v) = \Big(K - \int_0^t f_u(v_u)du\Big)^+ \quad \text{(put option case)}.$$

The penalty functional in the Israeli contingent claim case may also have here the integral form

$$\Delta_t(v) = \int_0^t \delta_u(v_u) du.$$

In order to satisfy the conditions (8.4.4) and (8.4.5) we can assume that for some $L > 0$ and all $x, y, u$,

$$|f_u(x) - f_u(y)| + |\delta_u(x) - \delta_u(y)| \le L|x - y|$$

and

$$|f_u(x)| + |\delta_u(x)| \le L|x|.$$

The Asian type (averaged integral) payoffs of the form

$$F_t(v) = \left(\frac{1}{t} \int_0^t f_u(v_u) du - K\right)^+ \text{ or } = \left(K - \frac{1}{t} \int_0^t f_u(v_u) du\right)^+$$

will satisfy our conditions if very small exercise times are prohibited. Anyway, the errors of binomial approximations can be estimated in this case in a similar way considering separately estimates for small stopping times and for stopping times bounded away from zero. As another example of path-dependent payoffs satisfying our conditions we can consider, so called, Russian options where for some $m > 0$,

$$F_t(v) = \max\left(m, \sup_{u \in [0,t]} v_u\right)$$

and in the Israeli contingent claim case $\Delta_t(v) = \delta v_t$.

Of course, the standard options with payoffs depending only on the current stock price satisfy our conditions as well.

Our goal is to estimate errors of approximations of the BS fair contingent claims prices

$$V_E(z) = E^*(Q_z(T)), \quad V_A(z) = \sup_{\tau \in \mathcal{T}_{0T}} E^*(Q_z(\tau)) \qquad (8.4.8)$$
$$\text{and } V_I(z) = \inf_{\sigma \in \mathcal{T}_{0T}} \sup_{\tau \in \mathcal{T}_{0T}} E^*(Q_z(\sigma, \tau))$$

by the corresponding prices in CRR markets

$$V_E^{(n)}(z) = E^{\xi}(Q_z^{(n)}(T)), \quad V_A^{(n)}(z) = \sup_{\eta \in \mathcal{T}_{0n}^{\xi}} E^{\xi}(Q_z^{(n)}(\eta T/n)) \ (8.4.9)$$
$$\text{and } V_I^{(n)}(z) = \min_{\zeta \in \mathcal{T}_{0n}^{\xi}} \max_{\eta \in \mathcal{T}_{0n}^{\xi}} E^{\xi}(Q_z^{(n)}(\zeta T/n, \eta T/n))$$

where

$$Q_z^{(n)} = Q_z^{(n)}(T) = e^{-rT} R_z^{(n)}(T), \quad R_z^{(n)}(T) = F_T(S^{(n)}(z)), \quad (8.4.10)$$
$$Q_z^{(n)}(t) = e^{-rt} R_z^{(n)}(t), \quad R_z^{(n)}(t) = F_t(S^{(n)}(z)),$$
$$Q_z^{(n)}(s, t) = e^{-rt} R_z^{(n)}(s, t), \quad R_z^{(n)}(s, t)$$
$$= G_s(S^{(n)}(z)) \mathbb{I}_{s<t} + F_t(S^{(n)}(z)) \mathbb{I}_{s \ge t},$$

$S^{(n)}(z) = S^{(n)}(z,\omega) \in M[0,T]$ is the random path of the CRR stock price $S_t(z) = S_t(z,\omega)$ on the time interval $[0,T]$ and $\mathcal{T}^\xi_{0n}$ is the set of stopping times with respect to the filtration $\mathcal{F}^\xi_k = \sigma\{\xi_1,...,\xi_k\}$, $k = 0,1,2,...$ taking on values $0,1,...,n$. The following result provides required error estimates.

**Theorem 8.4.1.** *There exists $C > 0$ (which can be estimated from the proof) such that*

$$|V_E(z) - V_E^{(n)}(z)| \le C\delta_n(z),\ |V_A(z) - V_A^{(n)}(z)| \le C\delta_n(z),\quad (8.4.11)$$
$$and\ |V_I(z) - V_I^{(n)}(z)| \le C\delta_n(z)$$

*for all $z, n > 0$, where $\delta_n(z) = (G_0(z) + z + 1)n^{-1/4}(\ln n)^{3/4}$ with $G_0(z) = F_0(z)$ in the European and the American cases.*

**Remark 8.4.1.** As we know from Lecture 5, in the CRR discrete time market a self-financing hedging portfolio with the initial capital equal to the fair price can be constructed by essentially explicit algorithm. This is not the case in the continuous time BS market with path-dependent payoffs. Thus constructions of hedging portfolios with initial capital close to the fair price become important. This also can be done by binomial approximations and we refer the reader to [36] for details.

### 8.4.2 *Skorokhod embedding and strategy of the proof*

In order to compare $V(z)$ and $V^{(n)}(z)$ for three types of contingent claims in question in the case of path-dependent payoffs we have to consider both BS and CRR markets on the same probability space which will enable us to compare stock prices paths $S(z)$ and $S^{(n)}(z)$. The main tool in achieving this goal will be here a simple explicit version of the Skorokhod type embedding (see, for instance, section 37 in [1]). Namely, define recursively stopping times

$$\theta_0^{(n)} = 0,\ \theta_{k+1}^{(n)} = \inf\{t > \theta_k^{(n)} : |\hat{W}(t) - \hat{W}(\theta_k^{(n)})| = \sqrt{\tfrac{T}{n}}\}$$

where $\hat{W}$ is the same as in (8.4.1). This can be written in a slightly different way. Namely, set $W^{(1)}(t) = \hat{W}(t) = \hat{W}^{(1)}(t) = -\tfrac{\kappa}{2}t + W^*(t)$ and $\eta_1^{(n)} = \theta_1^{(n)} = \inf\{t > 0 : |\hat{W}^{(1)}(t)| = \sqrt{\tfrac{T}{n}}\}$. Define the Brownian motion (with respect to $P^*$), $W^{(2)}(t) = W^*(\theta_1^{(n)} + t) - W^*(\theta_1^{(n)})$ which is independent of $\mathcal{F}_{\theta_1^{(n)}}$ (see Lecture 13) and set $\hat{W}^{(2)}(t) = -\tfrac{\kappa}{2}t + W^{(2)}(t)$, $\eta_2^{(n)} = \inf\{t > 0 : |\hat{W}^{(2)}(t)| = \sqrt{\tfrac{T}{n}}\}$ and $\theta_2^{(n)} = \theta_1^{(n)} + \eta_2^{(n)}$. Next, we continue recursively for

$i \geq 2$ introducing Brownian motions $W^{(i+1)}(t) = W^*(\theta_i^{(n)} + t) - W^*(\theta_i^{(n)})$ independent of $\mathcal{F}_{\theta_i^{(n)}}$ and setting $\hat{W}^{(i+1)}(t) = -\frac{\kappa}{2}t + W^{(i+1)}(t)$, $\eta_{i+1}^{(n)} =$ $\inf\{t > 0 : |\hat{W}^{(i+1)}(t)| = \sqrt{\frac{T}{n}}\}$ and $\theta_{i+1}^{(n)} = \theta_i^{(n)} + \eta_{i+1}^{(n)}$. Then we obtain that the pairs $(\hat{W}^{(i)}(t), \eta_i^{(n)}) = (\hat{W}(\theta_{i-1}^{(n)} + t) - \hat{W}(\theta_{i-1}^{(n)}), \theta_i^{(n)} - \theta_{i-1}^{(n)})$, $i = 1, 2, ...$ are independent and identically distributed. In particular, $\theta_m^{(n)} = \sum_{i=1}^m \eta_i$ is a sum of i.i.d. random variables.

Observe that $\exp(\kappa\hat{W}(t))$ is an exponential martingale with respect to the martingale probability $P^*$ (see Lecture 16), and so by the optional stopping theorem $1 = E^* \exp(\kappa\hat{W}(0)) = E \exp(\kappa\hat{W}(\theta_1^{(n)}))$. Hence,

$$1 = e^{\kappa\sqrt{\frac{T}{n}}} P\{\hat{W}(\theta_1^{(n)}) = \sqrt{\frac{T}{n}}\} + e^{-\kappa\sqrt{\frac{T}{n}}}(1 - P\{\hat{W}(\theta_1^{(n)}) = \sqrt{\frac{T}{n}}\}),$$

which yields that

$$P\{\hat{W}(\theta_1^{(n)}) = \sqrt{\frac{T}{n}}\} = (e^{\kappa\sqrt{\frac{T}{n}}} + 1)^{-1} = p^{(n)}.$$

It follows that i.i.d. sequence $\hat{W}^{(i)}(\eta_i^{(n)}) = \hat{W}(\theta_i^{(n)}) - \hat{W}(\theta_{i-1}^{(n)})$, $i = 1, 2, ...$ has the same distribution as the sequence $\sqrt{\frac{T}{n}}\xi_i$, $i = 1, 2, ...$ and, correspondingly, the sum $\sqrt{\frac{T}{n}}\sum_{i=1}^m \xi_i$ has the same distribution as $\hat{W}(\theta_m^{(n)}) = \sum_{i=1}^m \hat{W}^{(i)}(\eta_i^{(n)})$, which exhibits a version of the Skorokhod embedding of sums of i.i.d. random variables (more generally: of martingales) into a Brownian motion (see, for instance, [1]). This hints that it should be possible to replace $S_t^{(n)}(s)$ defined in (8.4.2) by

$$S_t^{W,n}(z) = z \exp\left(\sum_{k=1}^{[nt/T]} \left(\frac{rT}{n} + \kappa(\hat{W}(\theta_k^{(n)}) - \hat{W}(\theta_{k-1}^{(n)}))\right)\right). \tag{8.4.12}$$

This would enable us to consider all relevant processes on the same probability space where the Brownian motion $W^*$ is defined and to deal with stopping times with respect to the same Brownian filtration both for the BS continuous time model and for its discrete time binomial approximation.

Next, we replace $R_z^{(n)}(T)$, $Q_z^{(n)}(T)$, $R_z^{(n)}(t)$, $Q_z^{(n)}(t)$, $R_z^{(n)}(s,t)$, $Q_z^{(n)}(s,t)$ defined in (8.4.10) by

$$R_z^{W,n}(T) = F_T(S^{W,n}(z)), \, Q_z^{W,n} = Q_z^{W,n}(T) = e^{-rT}R_z^{W,n}(T),$$
$$R_z^{W,n}(t) = F_t(S^{W,n}(z)), \, Q_z^{W,n}(t) = e^{-rt}R_z^{W,n}(t), \, R_z^{W,n}(s,t)$$
$$= G_s(S^{W,n}(z))\mathbb{I}_{s<t} + F_t(S^{W,n}(z))\mathbb{I}_{s\geq t}, \, Q_z^{W,n}(s,t) = e^{-rt}R_z^{W,n}(s,t)$$

and introduce the sets $\mathcal{T}^{W,n}$ and $\mathcal{T}^{W,n}_{0n}$ of stopping times with respect to the filtration $\{\mathcal{F}_{\theta_k^{(n)}}\}_{k=0,1,2,\ldots}$ taking on values in $\mathbb{N} = \{0,1,\ldots\}$ and in $\{0,1,\ldots,n\}$, respectively. Define

$$V_E^{W,n}(z) = E^*(Q_z^{W,n}(T)), \ V_A^{W,n}(z) = \sup_{\eta\in\mathcal{T}^{W,n}_{0n}} E^*(Q_z^{W,n}(\tfrac{\eta T}{n}))$$

$$\text{and } V_I^{W,n}(z) = \inf_{\zeta\in\mathcal{T}^{W,n}_{0n}} \sup_{\eta\in\mathcal{T}^{W,n}_{0n}} E^*(Q_z^{W,n}(\tfrac{\zeta T}{n}, \tfrac{\eta T}{n})).$$

Observe that

$$\{\theta_\zeta^{(n)} \le t\} = \cup_{k=0}^\infty \{\theta_k^{(n)} \le t\} \cap \{\zeta = k\} \in \mathcal{F}_t$$

if $\zeta \in \mathcal{T}^{W,n}$ and so $\theta_\zeta^{(n)}$ and $\theta_n^{(n)}$ appearing above are stopping times with respect to the Brownian filtration. Since $S^{W,n}(z)$ has the same distribution as $S^{(n)}(z)$ it is clear that

$$V_E^{W,n}(z) = V_E^{(n)}(z). \tag{8.4.13}$$

The equalities $V_A^{W,n} = V_A^{(n)}$ and $V_I^{W,n} = V_I^{(n)}$ are less clear because of stopping times with respect to different filtrations appearing there but the following result yields that this is still true.

**Lemma 8.4.1.** *For any $z, n > 0$,*

$$V_A^{W,n}(z) = V_A^{(n)}(z) \text{ and } V_I^{W,n}(z) = V_I^{(n)}(z). \tag{8.4.14}$$

**Proof.** We employ the dynamical programming relations for optimal stopping from Lecture 3 which in this case have the form

$$V_A^{(n)}(z) = V_{A,0}^{(n)}(z), \ V_{A,n}^{(n)}(z) = e^{-rT}F_T(S^{(n)}) \tag{8.4.15}$$

$$\text{and for } k = n-1,\ldots,1,$$

$$V_{A,k}^{(n)}(z) = \max\big(e^{-r\frac{kT}{n}}F_{kT/n}(S^{(n)}(z)), E^\xi(V_{A,k+1}^{(n)}(z)|\mathcal{F}_k^\xi)\big);$$

$$V_A^{W,n}(z) = V_{A,0}^{W,n}(z), \ V_{A,n}^{W,n}(z) = e^{-rT}F_T(S^{W,n}) \text{ and for } k = n-1,\ldots,1,$$

$$V_{A,k}^{W,n}(z) = \max\big(e^{-r\frac{kT}{n}}F_{kT/n}(S^{W,n}(z)), E^*(V_{A,k+1}^{W,n}(z)|\mathcal{F}_{\theta_k^{(n)}})\big)$$

and

$$V_I^{(n)}(z) = V_{I,0}^{(n)}(z), \ V_{I,n}^{(n)}(z) = e^{-rT}F_T(S^{(n)}) \text{ and for} \tag{8.4.16}$$

$$k = n-1,\ldots,1, \ V_{I,k}^{(n)}(z) = \min\big(e^{-r\frac{kT}{n}}G_{kT/n}(S^{(n)}(z)),$$

$$\max\big(e^{-r\frac{kT}{n}}F_{kT/n}(S^{(n)}(z)), E^\xi(V_{I,k+1}^{(n)}(z)|\mathcal{F}_k^\xi)\big)\big);$$

$$V_I^{W,n}(z) = V_{I,0}^{W,n}(z), \ V_{I,n}^{W,n}(z) = e^{-rT}F_T(S^{W,n}) \text{ and for } k = n-1,\ldots,1,$$

$$V_{I,k}^{W,n}(z) = \min\big(e^{-r\frac{kT}{n}}G_{kT/n}(S^{W,n}(z)),$$

$$\max\big(e^{-r\frac{kT}{n}}F_{kT/n}(S^{W,n}(z)), E^*(V_{I,k+1}^{W,n}(z)|\mathcal{F}_{\theta_k^{(n)}})\big)\big).$$

For any numbers $x_1, x_2, ..., x_n$ set

$$x_t^{(n)} = x_t^{(n)}(z) = z \exp\left(\sum_{k=1}^{[nt/T]} \left(\frac{rT}{n} + \kappa x_k\right)\right) \text{ if } t \geq T/n$$

and $x_t^{(n)} = x_t^{(n)}(z) = z$ if $t \in [0, T/n)$. In view of (8.4.4) we can write

$$F_{\frac{kT}{n}}(x^{(n)}(z)) = q_k(z, x_1, ..., x_k) \text{ and } \Delta_{\frac{kT}{n}}(x^{(n)}(z)) = r_k(z, x_1, ..., x_k)$$

for some continuous functions $q_k$ and $r_k$ depending only on $z, x_1, ..., x_k$. Next, we show by the backward induction that there exist continuous functions $\Phi_0(z), \Psi_0(z), \Phi_k(z, x_1, ..., x_k)$ and $\Psi_k(z, x_1, ..., x_k)$, $k = 1, 2, ..., n$ such that

$$V_{A,k}^{(n)}(z) = \Phi_k\left(z, \left(\tfrac{T}{n}\right)^{1/2}\xi_1, ..., \left(\tfrac{T}{n}\right)^{1/2}\xi_k\right), \quad V_{A,0}^{(n)}(z) = \Phi_0(z), \quad (8.4.17)$$

$$V_{A,k}^{W,n}(z) = \Phi_k\left(z, \hat{W}(\theta_1^{(n)}), \hat{W}(\theta_2^{(n)}) - \hat{W}(\theta_1^{(n)}), ..., \hat{W}(\theta_k^{(n)}) - \hat{W}(\theta_{k-1}^{(n)})\right),$$

$$V_{A,0}^{W,n}(z) = \Phi_0(z)$$

and

$$V_{I,k}^{(n)}(z) = \Psi_k\left(z, \left(\tfrac{T}{n}\right)^{1/2}\xi_1, ..., \left(\tfrac{T}{n}\right)^{1/2}\xi_k\right), \quad V_{I,0}^{(n)}(z) = \Psi_0(z), \quad (8.4.18)$$

$$V_{I,k}^{W,n}(z) = \Psi_k\left(z, \hat{W}(\theta_1^{(n)}), \hat{W}(\theta_2^{(n)}) - \hat{W}(\theta_1^{(n)}), ..., \hat{W}(\theta_k^{(n)}) - \hat{W}(\theta_{k-1}^{(n)})\right),$$

$$V_{I,0}^{W,n}(z) = \Psi_0(z).$$

Indeed, for $k = n$ set $\Phi_n(z, x_1, ..., x_n) = \Psi_n(z, x_1, ..., x_n) = e^{-rT}q_n(z, x_1, ..., x_n)$. Suppose that we already constructed such $\Phi_k$ and $\Psi_k$ for all $k \geq l+1$. Set

$$\Phi_l(z, x_1, ..., x_l) = \max\left(e^{-r\frac{lT}{n}} q_l(z, x_1, ..., x_l), g_l(z, x_1, ..., x_l)\right),$$

where

$$g_l(z, x_1, ..., x_l) = E^\xi \Phi_{l+1}\left(z, x_1, ..., x_l, (T/n)^{1/2}\xi_{l+1}\right)$$

$$= E^* \Phi_{l+1}\left(z, x_1, ..., x_l, \hat{W}(\theta_{l+1}^{(n)}) - \hat{W}(\theta_l^{(n)})\right),$$

and

$$\Psi_l(z, x_1, ..., x_l) = \min\left(e^{-r\frac{lT}{n}}\left(q_l(z, x_1, ..., x_l) + r_l(z, x_1, ..., x_l)\right),\right.$$

$$\left.\max\left(e^{-r\frac{lT}{n}} q_l(z, x_1, ..., x_l), h_l(z, x_1, ..., x_l)\right)\right),$$

where

$$h_l(z, x_1, ..., x_l) = E^\xi \Psi_{l+1}\left(z, x_1, ..., x_l, (T/n)^{1/2}\xi_{l+1}\right)$$

$$= E^* \Psi_{l+1}\left(z, x_1, ..., x_l, \hat{W}(\theta_{l+1}^{(n)}) - \hat{W}(\theta_l^{(n)})\right).$$

Then (8.4.15) and (8.4.16) will be satisfied for $k = l$ with $V_{A,k}^{(n)}(z)$, $V_{A,k}^{W,n}(z)$ and $V_{I,k}^{(n)}(z)$, $V_{I,k}^{W,n}(z)$ given by (8.4.17) and (8.4.18), respectively, since $\xi_{l+1}$ and $\hat{W}(\theta_{l+1}^{(n)}) - \hat{W}(\theta_l^{(n)})$ are independent of $\mathcal{F}_l^\xi$ and $\mathcal{F}_l^{W,n} = \mathcal{F}_{\theta_l^{(n)}}$, respectively, and so

$$E^\xi\big(\Phi_{l+1}\big(z, (\tfrac{T}{n})^{1/2}\xi_1, ..., (\tfrac{T}{n})^{1/2}\xi_l, (\tfrac{T}{n})^{1/2}\xi_{l+1}\big)\big|\mathcal{F}_l^\xi\big)$$
$$= g_l\big(z, (\tfrac{T}{n})^{1/2}\xi_1, ..., (\tfrac{T}{n})^{1/2}\xi_l\big),$$

$$E^*\big(\Phi_{l+1}(z, \hat{W}(\theta_1^{(n)}), \hat{W}(\theta_2^{(n)})) - \hat{W}(\theta_1^{(n)})),$$
$$..., \hat{W}(\theta_l^{(n)}) - \hat{W}(\theta_{l-1}^{(n)}), \hat{W}(\theta_{l+1}^{(n)}) - \hat{W}(\theta_l^{(n)}))\big|\mathcal{F}_{\theta_l^{(n)}}\big) = h_l\big(z, \hat{W}(\theta_1^{(n)}),$$
$$\hat{W}(\theta_2^{(n)}) - \hat{W}(\theta_1^{(n)}), ..., \hat{W}(\theta_l^{(n)}) - \hat{W}(\theta_{l-1}^{(n)}))$$

and

$$E^\xi\big(\Psi_{l+1}\big(z, (\tfrac{T}{n})^{1/2}\xi_1, ..., (\tfrac{T}{n})^{1/2}\xi_l, (\tfrac{T}{n})^{1/2}\xi_{l+1}\big)\big|\mathcal{F}_l^\xi\big)$$
$$= g_l\big(z, (\tfrac{T}{n})^{1/2}\xi_1, ..., (\tfrac{T}{n})^{1/2}\xi_l\big),$$

$$E^*\big(\Psi_{l+1}(z, \hat{W}(\theta_1^{(n)}), \hat{W}(\theta_2^{(n)})) - \hat{W}(\theta_1^{(n)})),$$
$$..., \hat{W}(\theta_l^{(n)}) - \hat{W}(\theta_{l-1}^{(n)}), \hat{W}(\theta_{l+1}^{(n)}) - \hat{W}(\theta_l^{(n)}))\big|\mathcal{F}_{\theta_l^{(n)}}\big) = h_l\big(z, \hat{W}(\theta_1^{(n)}),$$
$$\hat{W}(\theta_2^{(n)}) - \hat{W}(\theta_1^{(n)}), ..., \hat{W}(\theta_l^{(n)}) - \hat{W}(\theta_{l-1}^{(n)}))$$

completing the induction step. Here we employed the standard fact (see properties of conditional expectations in Lecture 1) that if $\zeta$ is a random vector measurable with respect to a $\sigma$-algebra $\mathcal{G}$, $\eta$ is another random vector independent of $\mathcal{G}$ then $E(f(\zeta, \eta)|\mathcal{G}) = \varphi(\zeta)$ with $\varphi(x) = Ef(x, \eta)$ for any Borel function $f$ such that $E|f(\zeta, \eta)| < \infty$. Finally, applying (8.4.17) and (8.4.18) with $k = 0$ and taking into account (8.4.15) and (8.4.16) we arrive at (8.4.14). □

Next, we formulate a series of lemmas which demonstrate the plan of the proof of Theorem 8.4.1 while the actual proofs of these results will appear later on. First, observe that though $S_t^{W,n}$ defined by (8.4.13) is a piecewise constant approximation of the BS stock price $S_t(z)$ given by (8.4.1) there is certain inconsistency there between times $kT/n$ of jumps of $S_t^{W,n}(z)$ and the Brownian stopping times $\theta_k^{(n)}$ corresponding to the jump $\hat{W}(\theta_k^{(n)}) - \hat{W}(\theta_{k-1}^{(n)})$ in the exponent. In order to pass to the correct time we introduce

$$S_t^{W,\theta,n}(z) = z \exp\big(r\theta_k^{(n)} + \kappa\hat{W}(\theta_k^{(n)})\big) \text{ if } \theta_k^{(n)} \le t < \theta_{k+1}^{(n)}, \quad (8.4.19)$$
$$k = 0, 1, ..., n, \text{ and } S_t^{W,\theta,n}(z) = S_{\theta_n^{(n)}}^{W,\theta,n}(z) \text{ if } t \ge \theta_n^{(n)}.$$

Set

$$R_z^{W,\theta,n} = R_z^{W,\theta,n}(T) = F_T(S^{W,\theta,n}(z)), \; Q_z^{W,\theta,n} = Q_z^{W,\theta,n}(T)$$
$$= e^{-rT}R_z^{W,\theta,n}(T), \; R_z^{W,\theta,n}(t) = F_t(S^{W,\theta,n}(z)), \; Q_z^{W,\theta,n}(t) = e^{-rt}R_z^{W,\theta,n}(t),$$
$$R_z^{W,\theta,n}(s,t) = F_t(S^{W,\theta,n}(z))\mathbb{I}_{s\geq t} + G_s(S^{W,\theta,n}(z))\mathbb{I}_{s<t},$$
$$Q_z^{W,\theta,n}(s,t) = e^{-rs\wedge t}R_z^{W,\theta,n}(s,t)$$

and

$$V_E^{W,\theta,n}(z) = E^*Q_z^{W,\theta,n}, \; V_A^{W,\theta,n}(z) = \sup\nolimits_{\eta\in\mathcal{T}_{0n}^{W,n}} E^*Q_z^{W,\theta,n}(\theta_\eta^{(n)}),$$
$$V_I^{W,\theta,n} = \inf\nolimits_{\zeta\in\mathcal{T}_{0,n}^{W,n}} \sup\nolimits_{\eta\in\mathcal{T}_{0,n}^{W,n}} E^*Q_z^{W,\theta,n}(\theta_\zeta^{(n)},\theta_\eta^{(n)}).$$

In order to compare $V^{W,n}(z)$ and $V^{W,\theta,n}(z)$ we have to be able to compare $S_t^{W,n}(z)$ and $S_t^{W,\theta,n}(z)$ at the same time $t \in [0,T]$. The definitions (8.4.12) and (8.4.19) require us to compare then, in particular, $W(\theta_l^{(n)})$ and $W(\theta_k^{(n)})$ provided $lTn^{-1} \leq t < (l+1)Tn^{-1}$ and $\theta_k^{(n)} \leq t < \theta_{k+1}^{(n)}$. Via certain renewal theory arguments we will conclude that in the average $|k-l|$, for such $k,l \leq n$, is of order $n^{1/2}$, then $|\theta_k^{(n)} - \theta_l^{(n)}|$ is of order $n^{-1/2}$, and so $|W(\theta_k^{(n)}) - W(\theta_l^{(n)})|$ is roughly of order $n^{-1/4}$. The proof of the following result will make these heuristic arguments precise.

**Lemma 8.4.2.** *There exists a constant $C > 0$ such that for all $n, z > 0$,*
$$|V_E^{W,n}(z) - V_E^{W,\theta,n}(z)| \leq C\delta_n, \; |V_A^{W,n}(z) - V_A^{W,\theta,n}(z)| \leq C\delta_n \quad (8.4.20)$$
$$\text{and } |V_I^{W,n}(z) - V_I^{W,\theta,n}(z)| \leq C\delta_n$$
*where $\delta_n$ is the same as in (8.4.11).*

The values $V^{W,\theta,n}(z)$ are still defined for piecewise constant approximations $S^{W,\theta,n}(z)$ of the BS stock prices $S(z)$ given by (8.4.1). Thus on the next step we replace $S^{W,\theta,n}(z)$ by $S(z)$ estimating the corresponding error which turns out to be of smaller order than in other lemmas as we have to compare the Brownian motion here at times $s,t$ such that $|t-s| \leq \theta_k^{(n)} - \theta_{k-1}^{(n)}$ and since the latter is of order $1/n$ the increment $|W^*(t) - W^*(s)|$ is roughly of order $n^{-1/2}$ which is made precise in the following result.

**Lemma 8.4.3.** *For each $\varepsilon > 0$ there exists $C_\varepsilon > 0$ such that for all $z, n > 0$ and $\zeta, \eta \in \mathcal{T}_{0,n}^{W,n}$,*
$$E^*|Q_z^{W,\theta,n}(T) - Q_z(T)| \leq C_\varepsilon z n^{\varepsilon-1/2}, \quad (8.4.21)$$
$$E^*|Q_z^{W,\theta,n}(\theta_\eta^{(n)}) - Q_z(\theta_\eta^{(n)})| \leq E^* \max\nolimits_{0\leq k\leq n} |Q_z^{W,\theta,n}(\theta_k^{(n)}) - Q_z(\theta_k^{(n)})|$$
$$\leq C_\varepsilon z n^{\varepsilon-1/2} \text{ and } E^*|Q_z^{W,\theta,n}(\theta_\zeta^{(n)},\theta_\eta^{(n)}) - Q_z^W(\theta_\zeta^{(n)},\theta_\eta^{(n)})|$$
$$\leq E^* \max\nolimits_{0\leq k,l\leq n} |Q_z^{W,\theta,n}(\theta_k^{(n)},\theta_l^{(n)}) - Q_z^W(\theta_k^{(n)},\theta_l^{(n)})| \leq C_\varepsilon z n^{\varepsilon-1/2}.$$

Lemmas 8.4.2 and 8.4.3 yield already Theorem 8.4.1 for European contingent claims while for American and Israeli contingent claims we will need additional arguments. In the formulas (8.4.8) and (8.4.9) for prices of BS contingent claims the allowed stopping times take values in the interval $[0, T]$, so we have to restrict the stopping times $\theta_k^{(n)}$ (which are not bounded) to this interval. It is not difficult to understand that in the average $|\theta_n^{(n)} - T|$ is of order $n^{-1/2}$ and $\theta_n^{(n)} - \theta_n^{(n)} \wedge T$ is of the same order. Then the absolute value of the increment of the Brownian motion taken at times $\theta_n^{(n)} \wedge T$ and $\theta_n^{(n)}$ is roughly of order $n^{-1/4}$, and so the restriction of embedding times to the interval $[0, T]$ leads to a difference of about that order.

**Lemma 8.4.4.** *There exists a constant $C > 0$ such that for all $z, n > 0$ and $\zeta, \eta \in \mathcal{T}_{0,n}^{W,n}$,*

$$E^* |Q_z^W(\theta_\eta^{(n)}) - Q_z^w(\theta_\eta^{(n)} \wedge T)| \tag{8.4.22}$$
$$\leq E^* \max_{0 \leq k \leq n} |Q_z^W(\theta_k^{(n)}) - Q_z^W(\theta_k^{(n)} \wedge T)| \leq C\tilde{\delta}_n(z) \text{ and}$$
$$E^* |Q_z^W(\theta_\zeta^{(n)}, \theta_\eta^{(n)}) - Q_z^w(\theta_\zeta^{(n)} \wedge T, \theta_\eta^{(n)} \wedge T)|$$
$$\leq E^* \max_{0 \leq k,l \leq n} |Q_z^W(\theta_k^{(n)}, \theta_l^{(n)}) - Q_z^W(\theta_k^{(n)} \wedge T, \theta_l^{(n)} \wedge T)| \leq C\tilde{\delta}_n(z)$$

*where $\tilde{\delta}_n(z) = \delta_n(z)(\ln n)^{-3/4}$.*

Until now we considered only stopping times $\theta_k^{(n)}$ for $k = 0, 1, ..., n$ which may not be enough, in principle, in order to approximate all Brownian stopping times bounded by $T$, so the next result asserts that we can employ the whole sequence $\theta_0^{(n)} = 0, \theta_1^{(n)}, \theta_2^{(n)}, ...$ The estimates of the corresponding difference here are similar to Lemma 8.4.4 and they produce, essentially, the same result.

**Lemma 8.4.5.** *There exists a constant $C > 0$ such that for all $z, n > 0$ and $\zeta, \eta \in \mathcal{T}^{W,n}$,*

$$E^* |Q_z(\theta_\eta^{(n)} \wedge T) - Q_z(\theta_{\eta \wedge n}^{(n)} \wedge T)| \tag{8.4.23}$$
$$\leq E^* \sup_{0 \leq k < \infty} |Q_z(\theta_k^{(n)} \wedge T) - Q_z(\theta_{k \wedge n}^{(n)} \wedge T)| \leq C\tilde{\delta}_n(z) \text{ and}$$
$$E^* |Q_z(\theta_\zeta^{(n)} \wedge T, \theta_\eta^{(n)} \wedge T) - Q_z(\theta_{\zeta \wedge n}^{(n)} \wedge T, \theta_{\eta \wedge n}^{(n)} \wedge T)|$$
$$\leq E^* \sup_{0 \leq k,l < \infty} |Q_z(\theta_k^{(n)} \wedge T, \theta_l^{(n)} \wedge T) - Q_z(\theta_{k \wedge n}^{(n)} \wedge T, \theta_{l \wedge n}^{(n)} \wedge T)|$$
$$\leq C\tilde{\delta}_n(z).$$

Set $\mathcal{T}_T^{W,n} = \{\theta_\zeta^{(n)} \wedge T : \zeta \in \mathcal{T}^{w,n}\}$ and let

$$V_{A,0,T}^{W,n}(z) = \sup_{\tau \in \mathcal{T}_T^{W,n}} E^* Q_z(\tau) \text{ and}$$

$$V_{I,0,T}^{W,n}(z) = \inf_{\sigma \in \mathcal{T}_T^{W,n}} \sup_{\tau \in \mathcal{T}_T^{W,n}} E^* Q_z(\sigma, \tau).$$

Then Lemmas 8.4.3–8.4.5 yield that for some constant $C > 0$,

$$|V_A^{W,\theta,n}(z) - V_{A,0,T}^{W,n}(z)| \leq \sup_{\eta \in \mathcal{T}^{W,n}} E^* (|Q_z^{W,\theta,n}(\theta_{\eta \wedge n}^{(n)}) \quad (8.4.24)$$

$$-Q_z(\theta_{\eta \wedge n}^{(n)})| + |Q_z(\theta_{\eta \wedge n}^{(n)}) - Q_z(\theta_{\zeta \wedge n}^{(n)} \wedge T, \theta_{\eta \wedge n}^{(n)} \wedge T)|$$

$$+|Q_z(\theta_{\eta \wedge n}^{(n)} \wedge T) - Q_z(\theta_\eta^{(n)} \wedge T)|) \leq 3C\tilde{\delta}_n(z) \text{ and}$$

$$|V_I^{W,\theta,n}(z) - V_{I,0,T}^{W,n}(z)| \leq \sup_{\zeta \in \mathcal{T}^{W,n}} \sup_{\eta \in \mathcal{T}^{W,n}} E^W (|Q_z^{W,\theta,n}(\theta_{\zeta \wedge n}^{(n)}, \theta_{\eta \wedge n}^{(n)})$$

$$-Q_z(\theta_{\zeta \wedge n}^{(n)}, \theta_{\eta \wedge n}^{(n)})| + |Q_z(\theta_{\zeta \wedge n}^{(n)}, \theta_{\eta \wedge n}^{(n)}) - Q_z(\theta_{\zeta \wedge n}^{(n)} \wedge T, \theta_{\eta \wedge n}^{(n)} \wedge T)|$$

$$+|Q_z(\theta_{\zeta \wedge n}^{(n)} \wedge T, \theta_{\eta \wedge n}^{(n)} \wedge T) - Q_z(\theta_\zeta^{(n)} \wedge T, \theta_\eta^{(n)} \wedge T)|) \leq 3C\tilde{\delta}_n(z).$$

In the definitions of $V_{A,0,T}^{W,n}(z)$ and $V_{I,0,T}^{W,n}(z)$ above we consider only stopping times of the special form while in (8.4.8) which gives $V_A(z)$ and $V_I(z)$ all Brownian stopping times with values in $[0, T]$ are allowed and the last step in the proof of Theorem 8.4.1 is to estimate the corresponding error which turns out to be of the same order as in Lemma 8.4.3 since, again, we have to estimate increments $|B_t^* - B_s^*|$ when $|t - s| \leq \theta_k^{(n)} - \theta_{k-1}^{(n)}$ though here $k$ runs over all positive integers and not only up to $n$ as in Lemma 8.4.3.

**Lemma 8.4.6.** *For any $\varepsilon > 0$ there exists a constant $C_\varepsilon > 0$ such that for all $z, n > 0$,*

$$|V_A(z) - V_{A,0,T}^{W,n}(z)| \leq C_\varepsilon (F_0(z) + \Delta_0(z) + z + 1)n^{\varepsilon - 1/2} \quad (8.4.25)$$

*and* $|V_I(z) - V_{I,0,T}^{W,n}(z)| \leq C_\varepsilon (F_0(z) + \Delta_0(z) + z + 1)n^{\varepsilon - 1/2}.$

Lemmas 8.4.1–8.4.6 yield the required estimates of $|V_A(z) - V_A^{(n)}(z)|$ and $|V_I(z) - V_I^{(n)}(z)|$ from Theorem 8.4.1.

### 8.4.3 *Proving the estimates*

Set

$$W^{(n)}(t) = -\frac{\kappa t}{2}\sqrt{\frac{T}{n}} + W^*(t) \text{ and } \Theta^{(n)} = \inf\{t > 0 : |W^{(n)}(t)| = 1\}.$$

By the scaling property of the Brownian motion

$$\sqrt{\frac{T}{n}} W^{(n)}(t) \overset{\mathrm{d}}{\sim} \hat{W}(\frac{T}{n} t) \text{ and } \theta_1^{(n)} \overset{\mathrm{d}}{\sim} \frac{T}{n} \Theta^{(n)} \quad (8.4.26)$$

where $\xi \overset{\mathrm{d}}{\sim} \tilde{\xi}$ means that $\xi$ and $\tilde{\xi}$ have the same distribution. Observe that in view of independency of increments $W^*(l) - W^*(l-1)$, $l = 1, 2, \ldots$ for any $n \geq 1$,

$$P^*\{\Theta^{(n)} \geq k\} \leq P^*\{|W^*(l) - W^*(l-1)| \leq 2 + \kappa\sqrt{T} \ \forall l = 1, \ldots, k\} = e^{-b_T k} \tag{8.4.27}$$

where $b_T = -\ln P^*\{|W^*(1)| \leq 2 + \kappa\sqrt{T}\} > 0$. It follows that for any nonnegative $a < b_T$,

$$E^* e^{a\Theta^{(n)}} \leq \sum_{k=0}^{\infty} e^{a(k+1)} P^*\{\Theta^{(n)} \geq k\} \leq e^a(1 - e^{a-b_T})^{-1} < \infty. \tag{8.4.28}$$

The estimates needed for the proof of our results are not trivial only for large $n$, so we will assume that $n$ is sufficiently big, in particular, that various exponential moments of the form $E^* \exp(\frac{aT}{n}\Theta^{(n)})$ are finite, i.e., that $n > aTb_T^{-1}$.

Since $\exp(\kappa\hat{W}(t))$, $t \geq 0$ is a martingale with respect to the probability $P^*$, and so $E^* \exp(\kappa\hat{W}(\theta_1^{(n)})) = 1$, we derive by an easy computation that $\hat{W}(\theta_1^{(n)}) = \sqrt{\frac{T}{n}}$ or $= -\sqrt{\frac{T}{n}}$ and $W^{(n)}(\Theta^{(n)}) = 1$ or $= -1$ with probability $(1 + \exp(\kappa\sqrt{\frac{T}{n}}))^{-1}$ or $(1 + \exp(-\kappa\sqrt{\frac{T}{n}}))^{-1}$, respectively. Set $\alpha_n = E^*\Theta^{(n)}$ so that $E^*\theta_1^{(n)} = \alpha_n\frac{T}{n}$. Since the Brownian motion is a martingale, and so $E^*W^*(\Theta^{(n)}) = 0$, we have that

$$-\tfrac{\kappa}{2}\alpha_n\sqrt{\tfrac{T}{n}} = E^*\left(-\tfrac{\kappa}{2}\Theta^{(n)}\sqrt{\tfrac{T}{n}} + W^*(\Theta^{(n)})\right)$$
$$= (1 + e^{\kappa\sqrt{\frac{T}{n}}})^{-1} - (1 + e^{-\kappa\sqrt{\frac{T}{n}}})^{-1}.$$

This together with an easy estimate show that

$$|\alpha_n - 1| \leq \min\left(2\kappa^{-1}\sqrt{\tfrac{n}{T}}, \tfrac{\kappa^2 T}{2n}|1 - \tfrac{\kappa^2 T}{n}|^{-1}\right) \leq K_1\frac{T}{n} \tag{8.4.29}$$

where

$$K_1 = \min\left(2\kappa^{-1}(2\kappa^2 + T^{-1})^{3/2}, \kappa^2\right).$$

By (8.4.27),

$$E^*|\Theta^{(n)}|^m \leq M_m = \sum_{k=1}^{\infty} k^m e^{-b_T(k-1)} \tag{8.4.30}$$
$$\leq e^{2b_T} \int_0^\infty x^m e^{-b_T x}dx = e^{2b_T} m! b_T^{-(m+1)}.$$

Assuming without loss of generality that $n \geq K_1 T$ we obtain from (8.4.29) that

$$P^*\{|\Theta^{(n)} - \alpha_n|^m \geq k^m\} \leq P^*\{\Theta^{(n)} \geq k - 2\},$$

and so

$$E^* |\Theta^{(n)} - \alpha_n|^m \le \sum_{k=1}^{\infty} k^m e^{-b_T (k-3)} = e^{2b_T} M_m. \qquad (8.4.31)$$

Recall that

$$\theta_k^{(n)} - \alpha_n k \frac{T}{n} = \sum_{i=1}^{k} \left( \theta_i^{(n)} - \theta_{i-1}^{(n)} - \alpha_n \frac{T}{n} \right) = \sum_{i=1}^{k} \left( \eta_i^{(n)} - \alpha_n \frac{T}{n} \right)$$

is a sum of i.i.d. random variables with zero mean with respect to the probability $P^*$. We will need the following estimate for such sums.

**Lemma 8.4.7.** *Let* $\zeta_1, \zeta_2, \ldots$ *be a sequence of i.i.d. random variables such that* $E\zeta_1 = 0$ *and* $D_{2m} = E|\zeta_1|^{2m} < \infty$ *for some integer* $m \ge 1$. *Set* $S_n = \sum_{i=1}^{n} \zeta_i$. *Then for any integer* $n \ge 1$,

$$ES_n^{2m} \le A_m n^m$$

*where we can take* $A_m = \max(1, 2^{-(m-1)} D_{2m}, 2^m m^{-m} (D_{2m}^{\frac{1}{2m}} + 1)^{2m^2})$.

**Proof.** For each $j = 1, 2, \ldots$ we can write

$$S_{j+1}^{2m} = (S_j + \zeta_{j+1})^{2m} = S_j^{2m} + 2m S_j^{2m-1} \zeta_{j+1} + \sum_{l=2}^{2m} \binom{2m}{l} S_j^{2m-l} \zeta_{j+1}^l.$$

Next, we proceed by induction. For $n = 1$ the assertion is clear if we choose $A_m \ge D_{2m}$. Suppose that the assertion holds true for all $n \le k$ and prove it for $n = k + 1$. Summing the above equality in $j$ from 1 to $k$, taking into account that $\zeta_{j+1}$ is independent of $S_j$ and using the induction hypothesis we obtain that

$$ES_{k+1}^{2m} = E\zeta_1^{2m} + \sum_{j=1}^{k} \sum_{l=2}^{2m} \binom{2m}{l} ES_j^{2m-l} E\zeta_{j+1}^l$$

$$\le D_{2m} + \sum_{l=2}^{2m} \binom{2m}{l} \sum_{j=1}^{k} (ES_j^{2m})^{\frac{2m-l}{2m}} D_{2m}^{\frac{l}{2m}}$$

$$\le D_{2m} + \sum_{l=2}^{2m} \binom{2m}{l} A_m^{\frac{2m-l}{2m}} D_{2m}^{\frac{l}{2m}} \sum_{j=1}^{k} j^{m-1}$$

$$\le D_{2m} + \frac{(k+1)^m}{m} \sum_{l=2}^{2m} \binom{2m}{l} A_m^{1-\frac{l}{2m}} D_{2m}^{\frac{l}{2m}}$$

$$= A_m (k+1)^m \left( \frac{D_{2m}}{(k+1)^m A_m} + m^{-1} \sum_{l=2}^{2m} \binom{2m}{l} A_m^{-\frac{l}{2m}} D_{2m}^{\frac{l}{2m}} \right)$$

$$\le A_m (k+1)^m \left( \frac{D_{2m}}{2^m A_m} + A_m^{-\frac{1}{m}} m^{-1} (D_{2m}^{\frac{1}{2m}} + 1)^{2m} \right) \le A_m (k+1)^m$$

where $A_m \ge 1$ is so large that the expression in brackets does not exceed 1, taking, for instance, $A_m = \max(1, 2^{-(m-1)} D_{2m}, 2^m m^{-m} (D_{2m}^{\frac{1}{2m}} + 1)^{2m^2})$ and completing the induction step. $\square$

By (8.4.26) and (8.4.31),

$$D_m = E^* |\theta_1^{(n)} - \alpha_n \frac{T}{n}|^{2m} = (\frac{T}{n})^{2m} E^* (\Theta^{(n)} - \alpha_n)^{2m} \le e^{2b_T} (\frac{T}{n})^{2m} M_{2m},$$

and so by (8.4.29), the martingale inequality (from Lecture 2) and Lemma 8.4.7 for any integer $m \ge 1$,

$$E^* \sup_{\zeta \in T_{0,n}^{W,n}} |\theta_\zeta^{(n)} - \zeta \frac{T}{n}|^{2m} = E^* \max_{0 \le k \le n} |\theta_k^{(n)} - \frac{kT}{n}|^{2m} \quad (8.4.32)$$

$$\le 2^{2m-1} T^{2m} |\alpha_n - 1|^{2m} + 2^{2m-1} E^* \max_{0 \le k \le n} |\theta_k^{(n)} - \alpha_n \frac{kT}{n}|^{2m}$$

$$\le 2^{2m-1} K_1^{2m} T^{4m} n^{-2m} + 2^{2m-1} (\frac{2m}{2m-1})^{2m} E^* |\theta_n^{(n)} - \alpha_n T|^{2m}$$

$$\le K_2^{(m)} T^{4m} n^{-m}$$

where $K_2^{(m)} > 0$ depends only on $m$ but not on $n$ and it can be estimated explicitly from the above inequalities. We will need (8.4.32) mostly with $m = 1$ and in order to simplify notations we set $K_2 = K_2^{(1)}$.

Using the exponential martingales $\exp(aW^*(t) - \frac{1}{2}a^2 t)$ and $\exp(4aW^*(t) - 8a^2 t)$ and applying the martingale inequality (see Lecture 11) together with the Cauchy–Schwarz inequality we obtain that for any finite Brownian stopping time $\tau$ and a number $a$,

$$E^* \sup_{0 \le t \le \tau} \exp(aW^*(t)) \le E^* e^{\frac{1}{2}a^2 \tau} \sup_{0 \le t \le \tau} \exp(aW^*(t) \quad (8.4.33)$$

$$-\tfrac{1}{2}a^2 t) \le (E^* e^{a^2 \tau})^{1/2} (E^* \sup_{0 \le t \le \tau} \exp(2aW^*(t) - a^2 t))^{1/2}$$

$$\le 2 (E^* e^{a^2 \tau})^{1/2} (E^* \exp(2aW^*(\tau) - a^2 \tau))^{1/2} \le 2 (E^* e^{a^2 \tau})^{1/2}$$

$$\times (E^* \exp(4aW^*(\tau) - 8a^2 \tau))^{1/4} (E^* e^{6a^2 \tau})^{1/4}$$

$$= 2 (E^* e^{a^2 \tau})^{1/2} (E^* e^{6a^2 \tau})^{1/4}.$$

We will need also the following estimate which also uses an exponential martingale and a corresponding martingale inequality from Lecture 11,

$$P^* \{\sup_{0 \le t \le u} |W^*(t)| \ge r\} = 2P^* \{\sup_{0 \le t \le u} W^*(t) \ge r\} \quad (8.4.34)$$

$$\le 2P^* \{\sup_{0 \le t \le u} (W^*(t) - \tfrac{\alpha}{2}t) \ge r - \tfrac{\alpha}{2}u\}$$

$$= 2P^* \{\sup_{0 \le t \le u} \exp(\alpha W^*(t) - \tfrac{\alpha^2}{2}t) \ge \exp(\alpha r - \tfrac{\alpha^2}{2}u)\}$$

$$\le 2 \exp(-\alpha r + \tfrac{\alpha^2}{2}u)$$

for any $\alpha > 0$.

Next, we will need the following result.

**Lemma 8.4.8.** *Let* $k_t^{(n)} = \max\{j \le n : \theta_j^{(n)} \le t\}$ *for all* $t \ge 0$ *and* $\ell_t^{(n)} = [nt/T]$ *if* $t \in [0, T]$ *and* $\ell_t^{(n)} = n$ *if* $t > T$. *Then*

$$E^* \sup_{0 \le t \le T} |k_t^{(n)} - \ell_t^{(n)}|^2 \le 2E^* \sup_{0 \le t \le T} |k_t^{(n)} - nt/T|^2 + 2 \le 2(K_2 + 2)n \quad (8.4.35)$$

*and*

$$E^* \sup_{0 \le t \le \theta_n^{(n)}} \left| W^*(\theta_{k_t^{(n)}}^{(n)}) - W^*(\theta_{\ell_t^{(n)}}^{(n)}) \right|^2 \le K_3 n^{-\frac{1}{2}} (\ln n)^{3/2} \quad (8.4.36)$$

*where* $K_3 > 0$ *can be estimated from the proof below.*

**Proof.** Let $\Theta_1^{(n)}, \Theta_2^{(n)}, \dots$ be i.i.d. random variables with the same distribution as $\Theta^{(n)}$. Set $m_u^{(n)} = \max\{j \le n : \sum_{i=1}^{j} \Theta_i^{(n)} \le u\}$ then by (8.4.26) the process $m_{\frac{nt}{T}}^{(n)}$, $t \in [0, T]$ has the same distribution as the process $k_t^{(n)}$, $t \in [0, T]$. Hence

$$E^* \sup_{0 \le t \le T} |k_t^{(n)} - nt/T|^2 = E^* \sup_{0 \le u \le n} |m_u^{(n)} - u|^2.$$

Set $\Psi_t = \sum_{j=1}^{[t]} \Theta_j^{(n)}$ for $t \ge 1$ and $\Psi_t = \Psi_0 = 0$ for $t \in [0, 1)$. It is clear that if $l < n$ then $m_u^{(n)} - u = l - u$ if and only if $l - \Psi_l \ge l - u > l - \Psi_{l+1}$, and so in this case

$$|m_u^{(n)} - u| \le \max \left( |\Psi_l - l|, |\Psi_{l+1} - (l+1)| + 1 \right).$$

If $m_u^{(n)} = n$ and $u \le n$ then $\Psi_n \le u \le n$, and so $|m_u^{(n)} - u| \le |\Psi_n - n|$. Hence,

$$\max_{0 \le u \le n} |m_u^{(n)} - u| \le \max_{0 \le l \le n} |\Psi_l - l| + 1. \quad (8.4.37)$$

Observe that by (8.4.29) for any $l \le n$,

$$|\Psi_l - l| \le |\Psi_l - l\alpha_n| + K_1 T, \quad (8.4.38)$$

and so by the martingale inequality from Lecture 2,

$$E^* \max_{0 \le l \le n} |\Psi_l - l|^2 \le 2E^* \max_{0 \le l \le n} |\Psi_l - l\alpha_n|^2 + 2K_1^2 T^2$$
$$\le 8E^* |\Psi_n - n\alpha_n|^2 + 2K_1^2 T^2 = 8n E^* (\Theta^{(n)} - \alpha_n)^2 + 2K_1^2 T^2$$

and (8.4.35) follows from (8.4.29) and (8.4.31).

Next, we prove (8.4.36) estimating, first,

$$\sup_{0 \le t \le \theta_n^{(n)}} \left| W^*(\theta_{k_t^{(n)}}^{(n)}) - W^*(\theta_{\ell_t^{(n)}}^{(n)}) \right|^2 \quad (8.4.39)$$

$$\le 4 \mathbb{I}_{\sup_{0 \le t \le \theta_n^{(n)}} |k_t^{(n)} - \ell_t^{(n)}| > D\sqrt{n \ln n}} \sup_{0 \le t \le \theta_n^{(n)}} |W^*(t)|^2$$

$$+ 4 \mathbb{I}_{\max_{k, l \le n, |k-l| \le D\sqrt{n \ln n}} |\theta_k^{(n)} - \theta_l^{(n)}| > D^2 \sqrt{n^{-1} \ln n}} \sup_{0 \le t \le \theta_n^{(n)}} |W^*(t)|^2$$

$$+ 4 \mathbb{I}_{\max_{k \le n} \sup_{0 \le t \le D^2 \sqrt{n^{-1} \ln n}} |W^*(\theta_k^{(n)} + t) - W^*(\theta_k^{(n)})| > D^2 n^{-1/4} (\ln n)^{3/4}}$$

$$\times \sup_{0 \le t \le \theta_n^{(n)}} |W^*(t)|^2 + D^4 n^{-\frac{1}{2}} (\ln n)^{3/2}$$

where $D > 0$ will be chosen below. Observe that

$$\sup_{0 \le t \le \theta_n^{(n)}} |k_t^{(n)} - \ell_t^{(n)}| \le 2 \sup_{0 \le t \le T} |k_t^{(n)} - \ell_t^{(n)}| \qquad (8.4.40)$$

since by the definition of $k_t^{(n)}$ and $\ell_t^{(n)}$,

$$\sup_{T \le t \le T \vee \theta_n^{(n)}} |k_t^{(n)} - \ell_t^{(n)}| \le n - k_T^{(n)} = \ell_T^{(n)} - k_T^{(n)}.$$

Since $|k_t^{(n)} - \ell_t^{(n)}| \le |k_t^{(n)} - nt/T| + 1$ and the processes $m_{\frac{nt}{T}}^{(n)}$, $t \in [0, T]$ and $k_t^{(n)}$, $t \in [0, T]$ have the same distribution we derive from (8.4.37), (8.4.38) and (8.4.40) that

$$P^* \left\{ \sup_{0 \le t \le \theta_n^{(n)}} |k_t^{(n)} - \ell_t^{(n)}| > D\sqrt{n \ln n} \right\} \qquad (8.4.41)$$

$$\le P^* \left\{ \max_{0 \le l \le n} |\Psi_l - l\alpha_n| > \tfrac{1}{2} D\sqrt{n \ln n} - 2 \right\}$$

$$\le \sum_{l=0}^{n} \left( P^* \left\{ \Psi_l - l\alpha_n > \tfrac{1}{2} D\sqrt{n \ln n} - 2 \right\} \right.$$

$$\left. + P^* \left\{ l\alpha_n - \Psi_l > \tfrac{1}{2} D\sqrt{n \ln n} - 2 \right\} \right).$$

By (8.4.30), (8.4.31), Chebyshev's inequality and the definition of $\Psi_l$ for any $l \le n$,

$$P^* \left\{ \Psi_l - l\alpha_n > \tfrac{1}{2} D\sqrt{n \ln n} - 2 \right\} \le P^* \left\{ \exp \left( 2\sqrt{n^{-1} \ln n} \right. \right. \qquad (8.4.42)$$

$$\times \sum_{i=1}^{l} (\Theta_i^{(n)} - \alpha_n)) \ge n^D e^{-4} \right\} \le e^4 n^{-D} \left( E^* \exp \left( 2\sqrt{n^{-1} \ln n} (\Theta^{(n)} \right. \right.$$

$$\left. \left. -\alpha_n)) \right)^l \le e^4 n^{-D} \left( 1 + e^{2b_T} \sum_{m=2}^{\infty} \left( \tfrac{4 \ln n}{n} \right)^{m/2} \tfrac{M_m}{m!} \right)^n$$

$$\le \left( 1 + 4n^{-1} e^{4b_T} b_T^{-3} \ln(n(1 - b_T^{-1} \sqrt{4n^{-1} \ln n})^{-1}) \right)^n$$

$$\le e^4 n^{(8e^{4b_T} b_T^{-3} - D)} \le e^4 n^{-2},$$

where we use the inequality $(1 + a/q)^q < e^a$ for $a, q > 0$, choose $D \ge 2 + 8e^{4b_T} b_T^{-3}$ and assume that $n \ge (16/b_T^2)^2$, so that $n^{-1} \ln n \le b_T^2/16$. Similarly, under the same conditions,

$$P^* \left\{ l\alpha_n - \Psi_l > \frac{1}{2} D\sqrt{n \ln n} - 2 \right\} \le e^4 n^{-2}. \qquad (8.4.43)$$

Next, by (8.4.26), (8.4.28) and the Chebyshev inequality

$$P^* \left\{ \max_{k, l \le n, |k-l| \le D\sqrt{n \ln n}} |\theta_k^{(n)} - \theta_l^{(n)}| > D^2 \sqrt{n^{-1} \ln n} \right\} \qquad (8.4.44)$$

$$\le nP^* \left\{ \theta_{[D\sqrt{n \ln n}]}^{(n)} > D^2 \sqrt{n^{-1} \ln n} \right\}$$

$$= nP^* \left\{ anT^{-1} \theta_{[D\sqrt{n \ln n}]}^{(n)} > aD^2 T^{-1} \sqrt{n \ln n} \right\}$$

$$\le n \exp \left( - aD^2 T^{-1} \sqrt{n \ln n} \right) E^* \exp \left( anT^{-1} \theta_{[D\sqrt{n \ln n}]}^{(n)} \right)$$

$$\le n \left( \exp(-aDT^{-1}) E^* e^{a\Theta^{(n)}} \right)^{D\sqrt{n \ln n}} \le n^{-1},$$

(where $[b]$ is the integral part of $b$) if we choose a positive $a < b_T$,

$$D \geq Ta^{-1}\big(1 - \ln(1 - e^{a-b_T})\big) + T + 1, \text{ so that } \big(e^{a(DT^{-1}-1)}(1 - e^{a-b_T})\big)^D \geq e,$$

and assume that $n \geq e^3$, so that $n^{-1}\ln n \leq 1/4$.

Taking into account that $W^*(\theta_k^{(n)} + t) - W^*(\theta_k^{(n)})$, $k = 1, ..., n$ are Brownian motions (see Lecture 13) we obtain from (8.4.34) considered with $r = D^2 n^{-1/4}(\ln n)^{3/4}$, $u = D^2\sqrt{n^{-1}\ln n}$ and $\alpha = (n\ln n)^{1/4}$ that

$$P^*\big\{ \max_{k \leq n} \sup_{\theta_k^{(n)} \leq t \leq \theta_k^{(n)} + D^2\sqrt{n^{-1}\ln n}} \tag{8.4.45}$$

$$|W^*(t) - W^*(\theta_k^{(n)})| > D^2 n^{-1/4}(\ln n)^{3/4}\big\}$$

$$\leq nP^*\big\{ \sup_{0 \leq t \leq D^2\sqrt{n^{-1}\ln n}} |W^*(t)| > D^2 n^{-1/4}(\ln n)^{3/4}\big\}$$

$$\leq 2n^{1-\frac{D^2}{2}} \leq 2n^{-1}$$

provided we choose $D \geq 2$. It remains to show that

$$E^* \sup_{0 \leq t \leq \theta_n^{(n)}} |W^*(t)|^4 \leq C_T < \infty \tag{8.4.46}$$

for some $C_T > 0$ independent of $n$ and we will obtain (8.4.36) from (8.4.39) and (8.4.41)–(8.4.46) together with the Cauchy–Schwarz inequality.

In order to derive (8.4.46) we rely on the martingale inequality from Lecture 11 and take into account that $W^*(\theta_k^{(n)}) - W^*(\theta_{k-1}^{(n)})$, $k = 1, ..., n$ are i.i.d. random variables having the same distribution as $W^*(\theta_1^{(n)})$, which yields

$$E^* \sup_{0 \leq t \leq \theta_n^{(n)}} |W^*(t)|^4 \leq (4/3)^4 E^* |W^*(\theta_n^{(n)})|^4$$

$$= (4/3)^4 E^*\big( \textstyle\sum_{k=1}^n (W^*(\theta_k^{(n)}) - W^*(\theta_{k-1}^{(n)}))\big)^4$$

$$= (4/3)^4 \big( nE^*(W^*(\theta_1^{(n)}))^4 + 4n(n-1)E^*(W^*(\theta_1^{(n)}))^3 E^* W^*(\theta_1^{(n)})$$

$$+ 3n(n-1)(E^*(W^*(\theta_1^{(n)}))^2)^2 + 12n(n-1)(n-2)E^*(W^*(\theta_1^{(n)}))^2$$

$$\times (E^* W^*(\theta_1^{(n)}))^2 + 24n(n-1)(n-2)(n-3)(E^* W^*(\theta_1^{(n)}))^4.$$

recall, that

$$W^*(\theta_1^{(n)}) = \frac{\kappa}{2}\theta_1^{(n)} + \hat{W}(\theta_1^{(n)}), \quad |\hat{W}(\theta_1^{(n)})| = \sqrt{\frac{T}{n}}$$

and by (8.4.26) and (8.4.27) for each integer $m \geq 1$,

$$E^*(\theta_1^{(n)})^m = (\frac{T}{n})^m E(\Theta^{(n)})^m \leq (\frac{T}{n})^m \sum_{k=0}^{\infty} (k+1)^m e^{-b_T k}.$$

It follows that for each integer $m \geq 1$,

$$\sup_{n \geq 1} \left( n^{m/2} E^* |W^*(\theta_1^{(n)})|^m \right) < \infty$$

which suffices to bound terms above with $m = 4, 3, 2$. On the other hand, $\theta_1^{(n)}$ is a stopping time and $W^*$ is a martingale with $W^*(0) = 0$, and so by the optional stopping theorem $E^* W^*(\theta_1^{(n)}) = 0$ which together with the above estimates yields (8.4.46) completing the proof of (8.4.36) and the whole Lemma 8.4.8. $\qquad\square$

Now we are ready to pass directly to the proofs of Lemmas 8.4.2–8.4.6. We will consider in details only the slightly more difficult case of Israeli contingent claims while indicating the simplifications needed to obtain the estimates for American and European contingent claims at the end of the proof of each lemma. By the definition of $V_I^{W,n}$ and $V_I^{W,\theta,n}$ and the equality $\theta_\zeta^{(n)} \wedge \theta_\eta^{(n)} = \theta_{\zeta \wedge \eta}^{(n)}$,

$$|V_I^{W,n}(z) - V_I^{W,\theta,n}(z)| \leq \sup_{\zeta \in \mathcal{T}_{0,n}^{W,n}} \sup_{\eta \in \mathcal{T}_{0,n}^{W,n}} \left( J_1(\zeta,\eta) + J_2(\zeta,\eta) \right) \qquad (8.4.47)$$

where

$$J_1(\zeta,\eta) = E^* \left( \left| e^{-r\frac{T}{n}\zeta \wedge \eta} - e^{-r\theta_{\zeta \wedge \eta}^{(n)}} \right| R_z^{W,n}(\frac{\zeta T}{n}, \frac{\eta T}{n}) \right)$$

and

$$J_2(\zeta,\eta) = E^* \left| R_z^{W,n}(\frac{\zeta T}{n}, \frac{\eta T}{n}) - R_z^{W,\theta,n}(\theta_\zeta^{(n)}, \theta_\eta^{(n)}) \right|.$$

Since $|e^{-ra} - e^{-rb}| \leq r|a-b|$ we obtain by (8.4.32) and the Cauchy–Schwarz inequality

$$J_1(\zeta,\eta) \leq r E^* \left( \left| \theta_{\zeta \wedge \eta}^{(n)} - \frac{T}{n}\zeta \wedge \eta \right| R_z^{W,n}(\frac{\zeta T}{n}, \frac{\eta T}{n}) \right) \leq \frac{\sqrt{K_2} rT}{\sqrt{n}} \left( J_{11}(\eta,\zeta) \right)^{1/2} \qquad (8.4.48)$$

where by (8.4.33),

$$J_{11}(\zeta,\eta) = E^* \left( R_z^{W,n}(\frac{\zeta T}{n}, \frac{\eta T}{n}) \right)^2 \qquad (8.4.49)$$

$$\leq 2E^* \left( \left( F_{\frac{T}{n}\zeta \wedge \eta}(S^{W,n}(z)) \right)^2 + \left( \Delta_{\frac{T}{n}\zeta \wedge \eta}(S^{W,n}(z)) \right)^2 \right) \leq 6 \left( F_0^2(z) + \Delta_0^2(z) \right.$$

$$+ 2L^2(T+2)^2 E^* \left( 1 + \sup_{0 \leq t \leq T} \left( S_t^{W,n}(z) \right)^2 \right) \right) \leq 6 \left( F_0^2(z) + \Delta_0^2(z) \right.$$

$$+ 2L^2(T+2)^2 \left( 1 + z^2 e^{2rT} E^* \sup_{0 \leq t \leq \theta_n^{(n)}} e^{2\kappa W^*(t)} \right) \right) \leq 6 \left( F_0^2(z) + \Delta_0^2(z) \right.$$

$$\left. + 4L^2(T+2)^2 \left( 1 + z^2 e^{2rT} \left( E^* e^{4\kappa^2 \theta_n^{(n)}} \right)^{1/2} \left( E^* e^{24\kappa^2 \theta_n^{(n)}} \right)^{1/4} \right) \right).$$

Since

$$\theta_n^{(n)} = \sum_{k=1}^{n}(\theta_k^{(n)} - \theta_{k-1}^{(n)}) \text{ and } W^*(\theta_n^{(n)}) = \sum_{k=1}^{n}(W^*(\theta_k^{(n)}) - W^*(\theta_{k-1}^{(n)}))$$

are sums of i.i.d. random variables we obtain by (8.4.26)–(8.4.30) and the Taylor formula that for any $a > 0$,

$$E^* e^{a\theta_n^{(n)}} = \left(E^* e^{a\frac{T}{n}\Theta^{(n)}}\right)^n \tag{8.4.50}$$

$$\leq \left(1 + a(K_1\frac{T}{n} + 1)\frac{T}{n} + \frac{a^2 T^2}{n^2}\sum_{m=0}^{\infty}\frac{a^m T^m M_{m+2}}{n^m (m+2)!}\right)^n$$

$$\leq \left(1 + a(K_1\frac{T}{n} + 1)\frac{T}{n} + a^2 T^2 b_T^{-3} n^{-2} e^{2b_T}\frac{nb_T}{nb_T - aT}\right)^n \leq C_a = e^{2a(K_1+1)T}$$

provided $n \geq \frac{2aT}{b_T}\max(1, Tb_T^{-2})$ and we use that $(1 + a/q)^q \leq e^a$ if $a, q > 0$. This together with (8.4.33) give also

$$E^* e^{aW^*(\theta_n^{(n)})} \leq 2\left(E^* e^{a^2\theta_n^{(n)}}\right)^{1/2}\left(E^* e^{6a^2\theta_n^{(n)}}\right)^{1/4} \leq 2(C_{a^2})^{1/2}(C_{6a^2})^{1/4} \tag{8.4.51}$$

assuming that $n \geq \frac{2a^2 T}{b_T}\max(1, Tb_T^{-2})$.

Next, we estimate $J_2(\zeta, \eta)$,

$$J_2(\zeta, \eta) \leq E^*\Bigg(\left|F_{\frac{\zeta T}{n}}(S^{W,n}(z)) - F_{\theta_\zeta^{(n)}}(S^{W,\theta,n}(z))\right| \tag{8.4.52}$$

$$+\left|F_{\frac{\eta T}{n}}(S^{W,n}(z)) - F_{\theta_\eta^{(n)}}(S^{W,\theta,n}(z))\right|$$

$$+\left|\Delta_{\frac{\zeta T}{n}}(S^{W,n}(z)) - \Delta_{\theta_\zeta^{(n)}}(S^{W,\theta,n}(z))\right|\Bigg).$$

For any $\zeta \in \mathcal{T}_{0,n}^{W,n}$ we obtain from (8.4.4) and (8.4.5) that

$$\left|F_{\frac{\zeta T}{n}}(S^{W,n}(z)) - F_{\theta_\zeta^{(n)}}(S^{W,\theta,n}(z))\right| \tag{8.4.53}$$

$$\leq \max_{0 \leq k \leq n}\Bigg(\left|F_{\theta_k^{(n)}}(S^{W,n}(z)) - F_{\theta_k^{(n)}}(S^{W,\theta,n}(z))\right|$$

$$+\left|F_{\frac{kT}{n}}(S^{W,n}(z)) - F_{\theta_k^{(n)}}(S^{W,n}(z))\right|\Bigg)$$

$$\leq L(\theta_n^{(n)} + 1)J_{21} + L\max_{0 \leq k \leq n}|\theta_k^{(n)} - \frac{kT}{n}|(1 + J_{22}) + LJ_{23}$$

where

$$J_{21} = \sup_{0 \leq t \leq \theta_n^{(n)}}\left|S_t^{W,n}(z) - S_t^{W,\theta,n}(z)\right|,$$

$$J_{22} = \sup_{0 \leq t \leq T} S_t^{W,n}(z) \leq ze^{rT}\sup_{0 \leq t \leq \theta_n^{(n)}} e^{\kappa W^*(t)}, \tag{8.4.54}$$

and

$$J_{23} = \max_{0 \le k \le n} \sup_{\frac{kT}{n} \wedge \theta_k^{(n)} \le u \le t \le \frac{kT}{n} \vee \theta_k^{(n)}} \left| S_t^{W,n}(z) - S_u^{W,n}(z) \right|.$$

Set

$$H_1^{(n)}(t) = \left| r \left( \theta_{k_t^{(n)}}^{(n)} - \ell_t^{(n)} \frac{T}{n} \right) + \kappa \left( \hat{W}(\theta_{k_t^{(n)}}^{(n)}) - \hat{W}(\theta_{\ell_t^{(n)}}^{(n)}) \right) \right|$$

and $H_1^{(n)} = \sup_{0 \le t \le \theta_n^{(n)}} H_1^{(n)}(t)$

with $k_t^{(n)}$ and $\ell_t^{(n)}$ defined in Lemma 8.4.8. Then by (8.4.54),

$$J_{21} \le J_{22} \mathbb{I}_{H_1^{(n)} \le 1} \sup_{0 \le t \le \theta_n^{(n)}} \left| e^{H_1^{(n)}(t)} - 1 \right| \tag{8.4.55}$$

$$+ \mathbb{I}_{H_1^{(n)} > 1} \sup_{0 \le t \le \theta_n^{(n)}} \left( S_t^{W,n}(z) + S_t^{W,\theta,n}(z) \right)$$

$$\le 2z e^{rT} \sup_{0 \le t \le \theta_n^{(n)}} e^{\kappa W^*(t)} H_1^{(n)}$$

$$+ 2z \mathbb{I}_{H_1^{(n)} > 1} \left( e^{rT} + e^{r\theta_n^{(n)}} \right) \left( \sup_{0 \le t \le \theta_n^{(n)}} e^{\kappa W^*(t)} \right).$$

Next, by (8.4.40),

$$H_1^{(n)} \le \left| r - \frac{\kappa^2}{2} \right| \max_{0 \le k \le n} \left| \theta_k^{(n)} - k \frac{T}{n} \right| \tag{8.4.56}$$

$$+ 2 \frac{T}{n} \left| r - \frac{\kappa^2}{2} \right| \sup_{0 \le t \le T} \left| k_t^{(n)} - \ell_t^{(n)} \right|$$

$$+ \kappa \sup_{0 \le t \le \theta_n^{(n)}} \left| W^*(\theta_{k_t^{(n)}}^{(n)}) - W^*(\theta_{\ell_t^{(n)}}^{(n)}) \right|.$$

Hence, by (8.4.32), (8.4.35), (8.4.36), (8.4.56), the Cauchy–Schwarz and the Chebyshev inequalities it follows that there exists a constant $\tilde{C} > 0$ such that

$$P^* \{ H_1^{(n)} > 1 \} \le \tilde{C} n^{-\frac{1}{2}} (\ln n)^{3/2}. \tag{8.4.57}$$

This together with (8.4.32), (8.4.33), (8.4.35), (8.4.36), (8.4.50), (8.4.55), (8.4.56) and the Cauchy–Schwarz inequalities yield that there exists a constant $C^{(1)} > 0$ such that

$$E^* J_{21} \le C^{(1)} z n^{-\frac{1}{4}} (\ln n)^{3/4} \tag{8.4.58}$$

and both $\tilde{C}$ and $C^{(1)}$ can be estimated from the above formulas.

In order to estimate $J_{23}$ set

$$H_2^{(n)}(s,t) = \frac{rT}{n} \left( \ell_t^{(n)} - \ell_s^{(n)} \right) + \kappa \left( \hat{W}(\theta_{\ell_t^{(n)}}^{(n)}) - \hat{W}(\theta_{\ell_s^{(n)}}^{(n)}) \right)$$

and

$$H_2^{(n)} = \max_{0 \le k \le n} \sup_{\frac{kT}{n} \wedge \theta_k^{(n)} \le s \le t \le \frac{kT}{n} \vee \theta_k^{(n)}} \left| H_2^{(n)}(s,t) \right|.$$

Then similarly to (8.4.55),

$$J_{23} \leq 2J_{22}(\mathbb{I}_{H_2^{(n)}>1} + H_2^{(n)}). \tag{8.4.59}$$

If $\frac{kT}{n} \wedge \theta_k^{(n)} \leq s \leq t \leq \frac{kT}{n} \vee \theta_k^{(n)}$ then by (8.4.29),

$$\ell_t^{(n)} - \ell_s^{(n)} \leq \frac{n}{T}(t-s) + 1 \leq \frac{n}{T}\max_{0\leq k\leq n}|\theta_k^{(n)} - k\frac{T}{n}| + 1 \tag{8.4.60}$$

$$\leq \frac{n}{T}\max_{0\leq k\leq n}|\theta_k^{(n)} - k\frac{T}{n}\alpha_n| + K_1 T + 1$$

and

$$\theta_{\ell_t^{(n)}}^{(n)} - \theta_{\ell_s^{(n)}}^{(n)} \leq |\theta_{\ell_t^{(n)}}^{(n)} - \frac{\ell_t^{(n)}T}{n}| + |\theta_{\ell_s^{(n)}}^{(n)} - \frac{\ell_s^{(n)}T}{n}| + \frac{T}{n}|\ell_t^{(n)} - \ell_s^{(n)}| \tag{8.4.61}$$

$$\leq 3\max_{0\leq k\leq n}|\theta_k^{(n)} - \frac{kT}{n}| + \frac{T}{n} \leq 3\max_{0\leq k\leq n}|\theta_k^{(n)} - \frac{kT}{n}\alpha_n| + 3K_4 n^{-1}.$$

where $K_4 = T(K_1 T + 1)$. Hence, similarly to (8.4.39) and (8.4.56),

$$H_2^{(n)} \leq |r - \frac{3\kappa^2}{2}|\max_{0\leq k\leq n}|\theta_k^{(n)} - k\frac{T}{n}| + |1 - \frac{\kappa^2}{2}|\frac{T}{n} \tag{8.4.62}$$

$$+ \max_{0\leq k\leq n}\sup_{t\leq\theta_n^{(n)}-\theta_k^{(n)},0\leq t\leq 3\max_{0\leq l\leq n}|\theta_l^{(n)}-\frac{lT}{n}\alpha_n|+3K_4 n^{-1}}$$

$$|W^*(\theta_k^{(n)} + t) - W^*(\theta_k^{(n)})|$$

and

$$\sup_{k\leq n, t\leq\theta_n^{(n)}-\theta_k^{(n)},0\leq t\leq 3\max_{0\leq l\leq n}|\theta_l^{(n)}-\frac{lT}{n}\alpha_n|+3K_4 n^{-1}} \tag{8.4.63}$$

$$|W^*(\theta_k^{(n)} + t) - W^*(\theta_k^{(n)})|$$

$$\leq 2\mathbb{I}_{\max_{0\leq l\leq n}|\theta_l^{(n)}-\frac{lT}{n}\alpha_n|>\frac{1}{3}D\sqrt{n^{-1}\ln n}-K_4 n^{-1}}\sup_{0\leq t\leq\theta_n^{(n)}}|W^*(t)|$$

$$+ Dn^{-1/4}(\ln n)^{3/4}$$

$$+ 2\mathbb{I}_{\max_{0\leq k\leq n}\sup_{0\leq t\leq D\sqrt{n^{-1}\ln n}}|W^*(\theta_k^{(n)}+t)-W^*(\theta_k^{(n)})|>Dn^{-1/4}(\ln n)^{3/4}}$$

$$\sup_{0\leq t\leq\theta_n^{(n)}}|W^*(t)|.$$

Since the sequence $\{\theta_l^{(n)}, l \geq 1\}$ has the same distribution as the sequence $\{\frac{T}{n}\Psi_l, l \geq 1\}$ defined in the proof of Lemma 8.4.8 then in the same way as in (8.4.41) and (8.4.42) we obtain

$$P^*\{\max_{0\leq l\leq n}|\theta_l^{(n)} - \frac{lT}{n}\alpha_n| > \frac{1}{3}D\sqrt{n^{-1}\ln n} - K_4 n^{-1}\} \tag{8.4.64}$$

$$= P^*\{\max_{0\leq l\leq n}|\Psi_l - l\alpha_n| > \frac{D}{3T}\sqrt{n\ln n} - K_4 T^{-1}\} \leq 2e^{3K_4 n^{-1}}$$

provided we choose $D \geq 2 + 18T^2 e^{4b_T} b_T^{-3}$ and assume that $n \geq (36T^2/b_T^2)^2$, so that $n^{-1}\ln n \leq \frac{1}{36}T^{-2}b_T^2$. Hence, by (8.4.32), (8.4.34),(8.4.45),(8.4.62)–(8.4.64), the Cauchy–Schwarz and Chebyshev inequalities it follows that there exists a constant $\tilde{C} > 0$ such that

$$P^*\{H_2^{(n)} > 1\} \leq \tilde{C}n^{-1/2}(\ln n)^{3/2}.$$

This together with (8.4.32)–(8.4.34), (8.4.45), (8.4.50), (8.4.59), (8.4.62)–(8.4.64) and the Cauchy–Schwarz inequality yield that there exists a constant $C^{(2)} > 0$ such that

$$E^* J_{23} \leq C^{(2)} z n^{-1/4} (\ln n)^{3/4} \tag{8.4.65}$$

and both $\tilde{C}$ and $C^{(2)}$ can be estimated explicitly from the above formulas. Finally, estimating the left hand side of (8.4.53) by means of (8.4.32), (8.4.33), (8.4.50), (8.4.53), (8.4.54), (8.4.58), (8.4.65) and estimating the other two terms in the right hand side of (8.4.52) exactly in the same way we obtain that

$$J_2(\zeta, \eta) \leq C^{(3)} z n^{-\frac{1}{4}} (\ln n)^{3/4} \tag{8.4.66}$$

for some $C^{(3)} > 0$ independent of $n$, which together with (8.4.47)–(8.4.51) yield (8.4.20) completing the proof of Lemma 8.4.2 for the Israeli contingent claims case. To obtain the same estimate for the American and European contingent claims cases we can take $\zeta \equiv n$ in place of $\sup_\zeta$ in (8.4.47) for the former and take $\zeta = \eta \equiv n$ in place of $\sup_\zeta \sup_\eta$ for the latter and proceed as above. Observe that we obtain the same order of the error term in all three cases since it comes from the estimate of $J_{21}$ above which appears for all of them. □

Next, in order to prove Lemma 8.4.3 we obtain using (8.4.4) that for any $k, l = 1, 2, ..., n$,

$$\left| Q_z^{W,n,\theta}(\theta_k^{(n)}, \theta_l^{(n)}) - Q_z(\theta_k^{(n)}, \theta_l^{(n)}) \right| \leq \left| R_z^{W,n,\theta}(\theta_k^{(n)}, \theta_l^{(n)}) \right| \tag{8.4.67}$$
$$- R_z(\theta_k^{(n)}, \theta_l^{(n)}) \Big| \leq L \big( (\theta_n^{(n)} + 1) J_3 \big)$$

where

$$J_3 = \sup_{0 \leq t \leq \theta_n^{(n)}} |S_t^{W,\theta,n}(z) - S_t(z)|.$$

Set

$$H_3^{(n,l)} = \max_{1 \leq k \leq l} \sup_{\theta_{k-1}^{(n)} \leq s \leq t \leq \theta_k^{(n)}} \big( r(t-s) + \kappa |\hat{W}(t) - \hat{W}(s)| \big) \text{ and } H_3^{(n)} = H_3^{(n,n)}.$$

Then similarly to (8.4.55),

$$J_3 \leq 2 z e^{r \theta_n^{(n)}} H_3^{(n)} \max_{0 \leq k \leq n} e^{\kappa W^*(\theta_k^{(n)})} + 2 z e^{r \theta_n^{(n)}} \mathbb{I}_{H_3^{(n)} > 1} \max_{0 \leq k \leq n} e^{\kappa W^*(\theta_k^{(n)})}.$$
$$\tag{8.4.68}$$

Since $|W^*(t) - W^*(s)| \leq |W^*(t) - W^*(\theta_{k-1}^{(n)})| + |W^*(s) - W^*(\theta_{k-1}^{(n)})|$ and $(a+b)^{2m} \leq 2^{2m-1}(a^{2m} + b^{2m})$ for $a, b \geq 0$, $m \geq 1/2$ we obtain by (8.4.26), (8.4.30), and (8.4.34) that

$$E^*\big|H_3^{(n)}\big|^{2m} \leq 2^{2m-1}\big|r - \tfrac{\kappa^2}{2}\big|^{2m} E^* \max_{1 \leq k \leq n} |\theta_k^{(n)} - \theta_{k-1}^{(n)}|^{2m} \quad (8.4.69)$$

$$+ 2^{4m-1} \kappa^{2m} E^* \max_{1 \leq k \leq n} \sup_{\theta_{k-1}^{(n)} \leq t \leq \theta_k^{(n)}} |W^*(t) - W^*(\theta_{k-1}^{(n)})|^{2m}$$

$$\leq 2^{2m-1} \sum_{k=1}^n \Big(\big|r - \tfrac{\kappa^2}{2}\big|^{2m} E^* |\theta_k^{(n)} - \theta_{k-1}^{(n)}|^{2m}$$

$$+ 2^{2m} \kappa^{2m} E^* \sup_{\theta_{k-1}^{(n)} \leq t \leq \theta_k^{(n)}} |W^*(t) - W^*(\theta_{k-1}^{(n)})|^{2m}\Big)$$

$$\leq 2^{2m-1} n \Big(\big|r - \tfrac{\kappa^2}{2}\big|^{2m} E^*(\theta_1^{(n)})^{2m} + 2^{2m} \kappa^{2m} \Lambda_m E^*(\theta_1^{(n)})^m\Big) \leq K_5^{(m)} n^{-m+1}$$

where

$$K_5^{(m)} = 2^{2m-1} e^{2bT} b_T^{-(m+1)} T^m \Big(\big|r - \tfrac{\kappa^2}{2}\big|^{2m} (2m)! b_T^{-m} T^m + 2^{2m} \kappa^{2m} \Lambda_m m!\Big).$$

This together with the Chebyshev inequality give that for any integers $m, n \geq 1$,

$$P^*\{H_3^{(n)} > 1\} \leq K_5^{(m)} n^{-m+1}. \quad (8.4.70)$$

Finally, by (8.4.33), (8.4.50), (8.4.67)–(8.4.70), and the Hölder inequality we obtain the assertion of Lemma 8.4.3 for the Israeli contingent claims case. Again, in order to obtain the same estimate for the other two cases we just fix $k = n$ in (8.4.67) in the American case and fix $k = l = n$ in the European one. Proceeding as above we arrive at the required estimates since (8.4.67) holds true for any $k$ and $l$ between 1 and $n$. $\qquad \square$

The proof of Lemma 8.4.4 starts similarly with the estimate

$$\big|Q_z(\theta_k^{(n)}, \theta_l^{(n)}) - Q_z(\theta_k^{(n)} \wedge T, \theta_l^{(n)} \wedge T)\big| \leq E^*(J_4(k, l) + J_5(k, l)) \quad (8.4.71)$$

where by (8.4.4)–(8.4.6),

$$J_4(k, l) = \big|e^{-r\theta_{k \wedge l}^{(n)}} - e^{-r\theta_{k \wedge l}^{(n)} \wedge T}\big| R_z(\theta_k^{(n)} \wedge T, \theta_l^{(n)} \wedge T) \quad (8.4.72)$$

$$\leq rL(T+2)(F_0(z) + \Delta_0(z) + z + 1) e^{rT} \big(|\theta_n^{(n)} - T| \sup_{0 \leq t \leq T} e^{\kappa W^*(t)}\big)$$

and

$$J_5(k, l) \leq \big|R_z(\theta_k^{(n)}, \theta_l^{(n)}) - R_z(\theta_k^{(n)} \wedge T, \theta_l^{(n)} \wedge T)\big| \quad (8.4.73)$$

$$\leq L\big(|\theta_n^{(n)} - T|(1 + e^{r\theta_n^{(n)}} \sup_{0 \leq t \leq \theta_n^{(n)}} e^{\kappa W^*(t)}) + J_{51}\big)$$

with

$$J_{51} = \max_{0 \leq k \leq n} \sup_{\theta_k^{(n)} \wedge T \leq t \leq \theta_k^{(n)}} |S_t(z) - S_{\theta_k^{(n)} \wedge T}(z)|.$$

Here we use that $\Delta_z(t) = 0$ when $t \geq T$. Set

$$H_4^{(n)} = \max_{0 \leq k \leq n} \sup_{\theta_k^{(n)} \wedge T \leq t \leq \theta_k^{(n)}} \left( r(t - \theta_k^{(n)} \wedge T) + \kappa |\hat{W}(t) - \hat{W}(\theta_k^{(n)} \wedge T)| \right).$$

Similarly to (8.4.55) and (8.4.68) we obtain that

$$J_{51} \leq 2ze^{rT} H_4^{(n)} \sup_{0 \leq t \leq T} e^{\kappa W^*(t)} \qquad (8.4.74)$$

$$+ z \mathbb{I}_{H_4^{(n)} > 1} \left( e^{r\theta_n^{(n)}} \sup_{0 \leq t \leq \theta_n^{(n)}} e^{\kappa W^*(t)} + e^{rT} \sup_{0 \leq t \leq T} e^{\kappa W^*(t)} \right).$$

Observe that $\theta_k^{(n)} \wedge T < \theta_k^{(n)}$ if and only if $T < \theta_k^{(n)}$ and since $k \leq n$ we have in this case

$$[\theta_k^{(n)} \wedge T, \theta_k^{(n)}] \subset [T, \theta_n^{(n)}] \subset [T, \theta_n^{(n)} \vee T].$$

Hence,

$$H_4^{(n)} \leq (r + \frac{\kappa^2}{2})|\theta_n^{(n)} - T| + \kappa \sup_{T \leq t \leq \theta_n^{(n)} \vee T} |W^*(t) - W^*(T)|. \qquad (8.4.75)$$

By (8.4.32), the martingale inequality from Lecture 11 and the Cauchy–Schwarz inequality

$$E^* \sup_{T \leq t \leq \theta_n^{(n)} \vee T} |W^*(t) - W^*(T)|^2 \leq 4E^* |W^*(\theta_n^{(n)} \vee T) - W^*(T)|^2$$

$$= 4E^* |\theta_n^{(n)} - T| \leq 4TK_2^{1/2} n^{-1/2}$$

which together with (8.4.32) and (8.4.75) give

$$E^* |H_4^{(n)}|^2 \leq 2(r + \frac{\kappa^2}{2})^2 K^2 T^2 n^{-1} + 8\kappa^2 T K_2^{1/2} n^{-1/2}.$$

This enables us to estimate $P^*\{H_4^{(n)} > 1\}$ by the Chebyshev inequality which together with (8.4.33), (8.4.50) and the Cauchy–Schwarz inequality yields

$$E^* J_{51} \leq C^{(4)} z n^{-\frac{1}{4}}$$

for some $C^{(4)} > 0$ independent of $n$ which can be easily estimated explicitly via the above formulas. Using (8.4.32), (8.4.33) and (8.4.50) together with the Cauchy–Schwarz inequality in order to estimate $E^* J_4(\zeta, \eta)$ and the expectation of the remaining term in $J_5(\zeta, \eta)$ we arrive at (8.4.22), completing the proof of Lemma 8.4.4 in the Israeli contingent claim case. This and the following two lemmas are already not needed for European contingent claims while in order to obtain the required estimate for American ones we just have to take again $k = n$ in (8.4.71) and proceed as above. $\square$

Next, we derive Lemma 8.4.5 using estimates similar to the above. Namely, for any $k, l = 1, 2, \ldots$ we have

$$\left| R_z(\theta_k^{(n)} \wedge T, \theta_l^{(n)} \wedge T) - R_z(\theta_{k \wedge n}^{(n)} \wedge T, \theta_{l \wedge n}^{(n)} \wedge T) \right| \leq E^*(J_6(k, l) + J_7(k, l)) \tag{8.4.76}$$

where by (8.4.4)–(8.4.6) and the equality $\theta_\zeta^{(n)} \wedge \theta_\eta^{(n)} = \theta_{\zeta \wedge \eta}^{(n)}$ similarly to (8.4.68) and (8.4.74),

$$J_6(k, l) = \left( \left| e^{-r\theta_{k \wedge l}^{(n)} \wedge T} - e^{-r\theta_{k \wedge l \wedge n}^{(n)} \wedge T} \right| R_z(\theta_k^{(n)} \wedge T, \theta_l^{(n)} \wedge T) \right) \tag{8.4.77}$$
$$\leq rL(T+2)(F_0(z) + \Delta_0(z) + z + 1)e^{rT}$$
$$\times \max_{n < k < \infty} |\theta_k^{(n)} \wedge T - \theta_n^{(n)} \wedge T| \sup_{0 \leq t \leq T} e^{\kappa W^*(t)}$$

and

$$J_7(k, l) \leq \left| R_z(\theta_k^{(n)} \wedge T, \theta_l^{(n)} \wedge T) - R_z(\theta_{k \wedge n}^{(n)} \wedge T, \theta_{l \wedge n}^{(n)} \wedge T) \right| \tag{8.4.78}$$
$$\leq Lze^{rT}\left( (1 + \sup_{0 \leq t \leq T} e^{\kappa W^*(t)})\left( \max_{n < k < \infty} |\theta_k^{(n)} \wedge T - \theta_n^{(n)} \wedge T| \right. \right.$$
$$\left. \left. + H_5^{(n)} + 2\mathbb{I}_{H_5^{(n)} > 1} \right) \right)$$

with

$$H_5^{(n)} = \max_{n < k < \infty} \sup_{\theta_n^{(n)} \wedge T \leq t \leq \theta_k^{(n)} \wedge T} \left( r(t - \theta_n^{(n)} \wedge T) + \kappa |\hat{W}(t) - \hat{W}(\theta_n^{(n)} \wedge T)| \right).$$

Here we use again that $\Delta_z(t) = 0$ when $t \geq T$. Observe that $\theta_n^{(n)} \wedge T < \theta_k^{(n)} \wedge T$ for $k > n$ if and only if $T > \theta_n^{(n)}$ and then

$$[\theta_n^{(n)} \wedge T, \theta_k^{(n)} \wedge T] \subset [\theta_n^{(n)}, T] \subset [\theta_n^{(n)}, \theta_n^{(n)} \vee T].$$

Hence,

$$|\theta_k^{(n)} \wedge T - \theta_n^{(n)} \wedge T| \leq T \vee \theta_n^{(n)} - \theta_n^{(n)} \leq |T - \theta_n^{(n)}| \tag{8.4.79}$$

and

$$H_5^{(n)} \leq (r + \frac{\kappa^2}{2})|T - \theta_n^{(n)}| + \kappa \sup_{T \leq t \leq \theta_n^{(n)} \vee T} |W^*(t) - W^*(T)|. \tag{8.4.80}$$

The right hand side of (8.4.80) is the same as in (8.4.75), and so we can use the same estimates for $H_5^{(n)}$ as for $H_4^{(n)}$ which together with (8.4.32), (8.4.33), (8.4.50), the Chebyshev and the Cauchy–Schwarz inequalities enable us to estimate $E^* J_7(k, l)$ by $C^{(5)} z n^{-1/4}$ for some $C^{(5)} > 0$ independent of $n$. Finally, using (8.4.32), (8.4.33), (8.4.50), and (8.4.79) together with the Cauchy–Schwarz inequality in order to estimate $E^* J_6(k, l)$ we obtain

(8.4.23) completing the proof of Lemma 8.4.5 for Israeli contingent claims while for American ones take $k = n$ in (8.4.76) and proceed as above. $\quad\square$

In order to complete the proof of Theorem 8.4.1 it remains to establish Lemma 8.4.6. For each $\sigma \in \mathcal{T}_{0,T}$ set $\nu_\sigma = \min\{k \in \mathbb{N} : \theta_k^{(n)} \geq \sigma\}$ which, indeed, defines $\nu_\sigma$ since $\theta_k^{(n)} \to \infty$ with probability one as $k \to \infty$. Observe, that $\nu_\sigma \in \mathcal{T}^{W,n}$ since $\{\nu_\sigma \leq k\} = \{\theta_k^{(n)} \geq \sigma\} \in \mathcal{F}_{\theta_k^{(n)}}$. For any $\sigma \in \mathcal{T}_{0,T}$ we set $\sigma^{(n)} = \theta_{\nu_\sigma}^{(n)} \wedge T$. Since $\mathcal{T}_T^{W,n} \subset \mathcal{T}_{0,T}$ we conclude that

$$V_I(z) \geq \inf_{\sigma \in \mathcal{T}_{0,T}} \sup_{\tau \in \mathcal{T}_T^{W,n}} E^* Q_z(\sigma, \tau). \tag{8.4.81}$$

Then for any $\delta > 0$ there exists $\sigma_\delta \in \mathcal{T}_{0,T}$ such that

$$V_I(z) \geq \sup_{\tau \in \mathcal{T}_T^{W,n}} E^* Q_z(\sigma_\delta, \tau) - \delta. \tag{8.4.82}$$

Hence,

$$V_I(z) \geq \sup_{\tau \in \mathcal{T}_T^{W,n}} E^* Q_z(\sigma_\delta^{(n)}, \tau) - \delta \tag{8.4.83}$$

$$- \sup_{\tau \in \mathcal{T}_T^{W,n}} E^* \big( Q_z(\sigma_\delta^{(n)}, \tau) - Q_z(\sigma_\delta, \tau) \big)$$

$$\geq V_{I,0,T}^{W,n} - \delta - \sup_{\tau \in \mathcal{T}_T^{W,n}} J_8(\sigma_\delta, \tau) - \sup_{\tau \in \mathcal{T}_T^{W,n}} J_9(\sigma_\delta, \tau)$$

where for any $\sigma \in \mathcal{T}_{0,T}$ and $\tau \in \mathcal{T}_T^{W,n}$,

$$J_8(\sigma, \tau) = E^* \big( e^{-r\sigma^{(n)} \wedge \tau} (R_z(\sigma^{(n)}, \tau) - R_z(\sigma, \tau)) \big)$$

and by (8.4.6),

$$J_9(\sigma, \tau) = E^* \big( |e^{-r\sigma^{(n)} \wedge \tau} - e^{-r\sigma \wedge \tau}| R_z(\sigma, \tau) \big) \tag{8.4.84}$$

$$\leq rL(T+2)(F_0(z) + \Delta_0(z) + z)e^{rT} E^* \big( \sup_{0 \leq k < \infty} |\theta_{k+1}^{(n)} \wedge T - \theta_k^{(n)} \wedge T|$$

$$\times (1 + \sup_{0 \leq t \leq T} e^{\kappa W^*(t)}) \big).$$

Since $\sigma^{(n)} \geq \sigma$ it follows that

$$R_z(\sigma, \tau) = F_\sigma(S(z)) + \Delta_\sigma(S(z))$$

whenever

$$R_z(\sigma^{(n)}, \tau) = F_{\sigma^{(n)}}(S(z)) + \Delta_{\sigma^{(n)}}(S(z)).$$

Thus, by (8.4.4) similarly to (8.4.78) for any $\sigma \in \mathcal{T}_{0T}$ and $\tau \in \mathcal{T}_T^{W,n}$,

$$R_z(\sigma^{(n)}, \tau) - R_z(\sigma, \tau) \leq |F_{\sigma^{(n)}}(S(z)) - F_\sigma(S(z))| \tag{8.4.85}$$

$$+ |\Delta_{\sigma^{(n)}}(S(z)) - \Delta_\sigma(S(z))| \leq Lze^{rT}(1 + \sup_{0 \leq t \leq T} e^{\kappa W^*(t)})$$

$$\times \big( \sup_{0 \leq k < \infty} |\theta_{k+1}^{(n)} \wedge T - \theta_k^{(n)} \wedge T| + H_6^{(n)} + 2\mathbb{1}_{H_6^{(n)} > 1} \big)$$

where
$$H_6^{(n)} = \sup_{0 \le k < \infty} \sup_{\theta_k^{(n)} \wedge T \le t \le \theta_{k+1}^{(n)} \wedge T} \left( r(t - \theta_k^{(n)} \wedge T) + \kappa(\hat{W}(t) - \hat{W}(\theta_k^{(n)} \wedge T)) \right).$$

It is clear that
$$|H_6^{(n)}| \le |H_3^{(n)}| + |H_5^{(n)}| \tag{8.4.86}$$

and by (8.4.79) for all $k \ge 0$,
$$|\theta_{k+1}^{(n)} \wedge T - \theta_k^{(n)} \wedge T| \le \max_{0 \le k \le n-1} |\theta_{k+1}^{(n)} - \theta_k^{(n)}| + |T - \theta_n^{(n)}|. \tag{8.4.87}$$

Hence, we can apply to the right hand side of (8.4.85) the estimates of Lemmas 8.4.3–8.4.5 arriving at a bound of order $n^{-1/4}$. In order to obtain a better estimate promised in Lemma 8.4.6 (though it will not help us to improve the estimate of Theorem 8.4.1) we write
$$|H_6^{(n)}| \le |H_3^{(n,2n)}| + \mathbb{I}_{\theta_{2n}^{(n)} < T}(|H_3^{(n)}| + |H_5^{(n)}|) \tag{8.4.88}$$

with $H_3^{(n,l)}$ defined above (8.4.68). In the same way as in (8.4.69) we obtain
$$E^*|H_3^{(n,2n)}|^{2m} \le 2K_5^{(m)} n^{-m+1}. \tag{8.4.89}$$

Next, by (8.4.26) and (8.4.29)–(8.4.31) similarly to (8.4.42) we obtain the following (large deviations) bound valid for $a > 0$ small enough,
$$P^*\{\theta_{2n}^{(n)} < T\} = P^*\{2n\alpha_n - \Psi_{2n} > n(2\alpha_n - 1)\} \tag{8.4.90}$$
$$= P^*\{\textstyle\sum_{j=1}^{2n}(\alpha_n - \Theta_j^{(n)}) > n(2\alpha_n - 1)\} \le e^{-an(2\alpha_n-1)}\left(E^* e^{a(\alpha_n - \Theta^{(n)})}\right)^{2n}$$
$$\le e^{-an(2\alpha_n-1)}\left(1 + e^{4b_T} b_T^{-1} \textstyle\sum_{m=2}^{\infty} a^m b_T^{-m}\right)^{2n}$$
$$= e^{-an(2\alpha_n-1)}\left(1 + a^2 e^{4b_T} b_T^{-3}(1 - ab_T^{-1})^{-1}\right)^{2n}$$
$$\le \exp\left(-an(2\alpha_n - 1 - 2ae^{4b_T} b_T^{-3}(1 - ab_T^{-1})^{-1})\right) \le e^{-\frac{1}{4}an},$$

provided $n \ge 4K_1 T$ and $0 < a \le \frac{1}{2}\min(b_T, \frac{1}{8}e^{-4b_T} b_T^3)$.

Estimating $E^*|T - \theta_n^{(n)}|^m$ by (8.4.32) we obtain by (8.4.34), (8.4.69), (8.4.80), (8.4.88)–(8.4.90) and the Cauchy–Schwarz inequality that for any $m \ge 1$ there exists $K_6^{(m)} > 0$ (which can be explicitly estimated from above formulas) such that
$$E^*|H_6^{(n)}|^{2m} \le K_6^{(m)} n^{-m+1}. \tag{8.4.91}$$

Since $\delta$ in (8.4.83) is arbitrary we conclude by (8.4.32), (8.4.33), (8.4.50), (8.4.80), (8.4.83)–(8.4.85), (8.4.87), and (8.4.91) together with the Chebyshev and Hölder inequalities similarly to Lemma 8.4.3 that for any $\varepsilon > 0$ there exists $C_\varepsilon^{(6)} > 0$ such that for all $n \in \mathbb{N}$,
$$V_I(z) - V_{I,0,T}^{W,n}(z) \ge -C_\varepsilon^{(6)}(F_0(z) + \Delta_0(z) + z + 1)n^{\varepsilon - \frac{1}{2}}. \tag{8.4.92}$$

Since the corresponding Dynkin's game has a value (see Lecture 12) we can represent $V_I(z)$ also as

$$V_I(z) = \sup_{\tau \in \mathcal{T}_{0T}} \inf_{\sigma \in \mathcal{T}_{0T}} E^* Q_z(\sigma, \tau) \le \inf_{\sigma \in \mathcal{T}_T^{W,n}} E^* Q_z(\sigma, \tau_\delta) + \delta$$

for each $\delta > 0$ and some $\tau_\delta \in \mathcal{T}_{0T}$. Introducing $\tau_\delta^{(n)}$ and employing the same arguments as above we obtain that for any $\varepsilon > 0$ there exists $C_\varepsilon^{(7)} > 0$ such that for all $n \in \mathbb{N}$,

$$V_I(z) - V_{I,0,T}^{W,n}(z) \le C_\varepsilon^{(7)}(F_0(z) + \Delta_0(z) + z + 1)n^{\varepsilon - \frac{1}{2}}$$

which together with (8.4.92) yields (8.4.25) and completes the proof of both Lemma 8.4.6 and Theorem 8.4.1 for Israeli contingent claims. For American ones we just take $\sigma_\delta = T$ in (8.4.82), proceed as above and in place of $\inf_\sigma$ fix everywhere $\sigma = T$ which will lead to the required estimate. ☐

## 8.5 Exercises

1) Consider the Black–Scholes market model on a probability space $(\Omega, \mathcal{F}, P)$ with a horizon $T$ and prices of a stock and of a bond at time $t$ given by

$$S_t = S_0 \exp((\mu - \frac{\sigma^2}{2})t + \sigma W(t)) \text{ and } B_t = B_0 e^{rt},$$

respectively, where $W(t)$ is the standard Brownian motion. Make the following computations for European options showing all steps of solutions:

(a) Find the explicit expression for the fair price $V$ when the payoff function is given by $f_T = S_T^2$ (square of the stock price at time $T$), i.e., write and compute explicitly the corresponding integrals.

(b) For the payoff functions $f_T = (S_T - K)^+$, where $K > 0$ is a constant, (call option) and $f_T = (K - S_T)^+$ (put option), assuming $S_0 = 1$, represent the corresponding fair price $V$ by means of the normal cumulative distribution function

$$\Phi(x) = \frac{1}{\sqrt{2\pi}} \int_{-\infty}^x e^{-\frac{u^2}{2}} du.$$

2) For the Black–Scholes market defined in 1) with $r = 1$, $\mu = 0$ and $\sigma = 1$ find the fair price (from seller's point of view) and a replicating self-financing trading strategy for the European option with the payoff $f_T = W^2(T)$.

3) Consider the multi-asset Black–Scholes market with a bond whose price evolves according to $B_t = B_0 e^{2t}$ and two stocks $S_t^1$ and $S_t^2$ whose evolution is described by the stochastic differential equations

$$dS_t^1 = S_t^1(3dt + dW_1(t) - 2dW_2(t))$$
$$dS_t^2 = S_t^2(5dt - 2dW_1(t) + 4dW_2(t))$$

with some positive initial conditions $S_0^1$ and $S_0^2$. Write the expressions for $S_t^1$ and $S_t^2$. Prove the existence of a self-financing admissible strategy which provides an opportunity for arbitrage and construct it.

4) This question has no direct connection to mathematical finance but it is closely related to estimates in Lecture 22. Let $W(t)$, $t \geq 0$ be the standard continuous Brownian motion starting at 0. Define $\Theta_0 = 0$, $\Theta_1 = \inf\{t > 0 : |W(t)| = 1\}$ and recursively $\Theta_{k+1} = \inf\{t > \Theta_k : |W(t) - W(\Theta_k)| = 1\}$.

(a) Prove that $\{\Theta_{k+1} - \Theta_k, \ k = 0, 1, ...\}$ is the sequence of i.i.d. random variables with $E\Theta_1 = 1$;

(b) Prove that

$$\lim_{n\to\infty} n^{-1/2} E|W(n) - W(\Theta_n)|^2 = \sqrt{\frac{10}{3\pi}};$$

(c) Define $M_n = \max_{0 \leq k \leq n} |W(k) - W(\Theta_k)|$ and prove that

$$\limsup_{n\to\infty} n^{-1/2}(\ln n)^{-3/2} EM_n^2 < \infty.$$

It would be interesting to know whether a tighter estimate in c) is possible. Observe that by [33] with probability one,

$$\limsup_{n\to\infty} n^{-1/4}(\ln n)^{-1/2}(\ln \ln n)^{-1/4}|W(n) - W(\Theta_n)| = const < \infty.$$

# PART 3
# Further Topics

# Chapter 9

# Discrete Time Case

## 9.1 Markets with transaction costs

We discussed pricing of derivatives without transaction costs which allows an agent managing the portfolio to buy and to sell the securities without any cost which is not the case in a real market. A big volume investor which buys and sells each time big quantities of each security may as a first approximation disregard transaction costs which are usually a small percentage of the transaction especially for a big investor who gets a discount. But, in general, transaction costs cannot be disregarded. We will describe certain models of markets with transaction costs. More information and references can be found in the book [43] and the papers [65], [37] and [63].

### 9.1.1 *Market model*

We will deal with the market model from [64], [65] and [37] which consists of a finite probability space $\Omega$ with the $\sigma$-field $\mathcal{F} = 2^\Omega$ of all subsets of $\Omega$ and a probability measure $P$ on $\mathcal{F}$ giving a positive weight $P(\omega)$ to each $\omega \in \Omega$. The setup includes also a filtration $\{\emptyset, \Omega\} = \mathcal{F}_0 \subset \mathcal{F}_1 \subset ... \subset \mathcal{F}_T = \mathcal{F}$ where $T$ is a positive integer called the time horizon. It is convenient to denote by $\Omega_t$ the set of atoms in $\mathcal{F}_t$ (which are subsets of $\Omega$) so that any $\mathcal{F}_t$-measurable random variable (vector) $Z$ can be identified with a function (vector function) defined on $\Omega_t$ and its value at $\omega \in \Omega_t$ will be denoted either by $Z(\omega)$ or by $Z^\omega$. The points of $\Omega_t$ can be viewed as vertices of a tree so that an arrow is drawn from $\mu \in \Omega_t$ to $\nu \in \Omega_{t+1}$ if $\nu \subset \mu$.

The market model consists of a risk-free bond and a risky stock. Without loss of generality, we can assume that all prices are discounted so that the bond price equals 1 all the time and a position in bonds is identified with cash holding. On the other hand, the shares of the stock can be traded which involves proportional transaction costs. This will be represented by bid-ask spreads, i.e., shares can be bought at an ask price $S_t^a$ or sold at the bid price $S_t^b$, where $S_t^a \geq S_t^b > 0$, $t = 0, 1, ..., T$ are processes adapted to the filtration $\{\mathcal{F}_t\}_{t=0}^{T}$. We can take, for instance, $S_t^a = (1 + \lambda)S_t$ and $S_t^b = (1 - \mu)S_t$ where $\lambda, \mu > 0$, $\mu < 1$ are constants and $S_t$ is the market ("true") stock price at time $t$, so that buying (or selling) a unit of stock involves also paying $\lambda S_t$ (or $\mu S_t$) as a transaction cost.

The liquidation value at time $t$ of a portfolio $(\beta, \gamma)$ consisting of an amount $\beta$ of cash (or bond) and $\gamma$ shares of the stock equals

$$\theta_t(\beta, \gamma) = \beta + S_t^b \gamma^+ - S_t^a \gamma^- \qquad (9.1.1)$$

which in case $\gamma < 0$ means that a portfolio owner should spend the amount $S_t^a \gamma^-$ in order to close his short position. Similarly, the cost of setting up a portfolio $(\beta, \gamma)$ is

$$-\theta_t(-\beta, -\gamma) = \beta - S_t^b \gamma^+ - S_t^a \gamma^-.$$

Observe that fractional numbers of shares are allowed here so that both $\beta$ and $\gamma$ in a portfolio $(\beta, \gamma)$ could be, in principle, any real numbers. By definition, a self-financing portfolio strategy is a predictable process $(\beta_t, \gamma_t)$ representing positions in cash (or bonds) and stock at time $t$, $t = 0, 1, ..., T$ such that

$$\theta_t(\beta_t - \beta_{t+1}, \gamma_t - \gamma_{t+1}) \geq 0 \quad \forall t = 0, 1, ..., T - 1 \qquad (9.1.2)$$

and the set of all such portfolio strategies will be denoted by $\Phi$.

We will consider here European, American and Israeli type contingent claims. In the first one its seller is committed to deliver at the expiration time $T$ a package of cash amount $Q_T^{(1)}$ and $Q_T^{(2)}$ number of stock shares, both are $\mathcal{F}_T$-measurable random variables, which will be represented by the 2-vector $Q_T = (Q_T^{(1)}, Q_T^{(2)})$. In the American case the buyer can exercise the claim (contract) at any stopping time $\tau$ not exceeding the expiration time $T$ and if $\tau = t$ then the seller is committed to deliver at time $t$ a package of cash amount $Q_t^{(1)}$ and $Q_t^{(2)}$ number of stock shares, both are $\mathcal{F}_t$-measurable random variables, which will be represented by the 2-vector $Q_t = (Q_t^{(1)}, Q_t^{(2)})$. In an Israeli contingent claim case if the seller cancels the contract at time $s$ and the buyer exercises at time $t$ then the former delivers to the latter a package of cash and stock represented by a random 2-vector according to the formula

$$Q_{s,t} = (Q_{s,t}^{(1)}, Q_{s,t}^{(2)}) = Y_s \mathbb{I}_{s<t} + Z_t \mathbb{I}_{t \leq s} \qquad (9.1.3)$$

where $Y_t = (Y_t^{(1)}, Y_t^{(2)})$, $Z_t = (Z_t^{(1)}, Z_t^{(2)})$ are $\mathcal{F}_t$-measurable random 2-vectors with the first and the second coordinates representing the cash amount and the number of stock shares, respectively. The penalty imposed on the seller for cancellation of the contract before buyer's exercise time is represented by the inequality

$$\Delta_t = \theta_t(Y_t^{(1)} - Z_t^{(1)}, Y_t^{(2)} - Z_t^{(2)}) \geq 0. \qquad (9.1.4)$$

A self-financing strategy $\pi = (\beta_t, \gamma_t)_{t=0}^T$ will be called a superhedging strategy for the seller

(i) of a European contingent claim with a payoff $Q_T = (Q_T^{(1)}, Q_T^{(2)})$ if $\theta_T(\beta_T - Q_T^{(1)}, \gamma_T - Q_T^{(2)}) \geq 0$ and

(ii) of an American contingent claim with the payoff $Q_t = (Q_t^{(1)}, Q_t^{(2)})$, $t \in [0, T]$ if $\theta_t(\beta_t - Q_t^{(1)}, \gamma_t - Q_t^{(2)}) \geq 0$ for all $t \in [0, T]$;

(iii) A pair of a stopping time $\sigma \leq T$ and a self-financing strategy $\pi = (\beta_t, \gamma_t)_{t=0}^T$ will be called a superhedging strategy for the seller of an Israeli contingent claim with the payoff $Q_{s,t} = (Q_{s,t}^{(1)}, Q_{s,t}^{(2)})$ if $\theta_{\sigma \wedge t}(\beta_{\sigma \wedge t} - Q_{\sigma,t}^{(1)}, \gamma_{\sigma \wedge t} - Q_{\sigma,t}^{(2)}) \geq 0$ for all $t \in [0, T]$.

The seller's (ask or upper hedging) price $V^a$ of each contingent claim is defined as the infimum of initial amounts required to setting up a super-hedging strategy for the seller. Since in order to get $\beta_0$ amount of cash and $\gamma_0$ shares of stock at time 0 the seller should spend

$$-\theta_0(-\beta_0, -\gamma_0) = \beta_0 + \gamma_0^+ S_0^a - \gamma_0^- S_0^b \qquad (9.1.5)$$

in cash, we can write for the seller's ( ask or upper) European and American contingent claims prices $V^a = V_E^a$ and $V^a = V_A^a$,

$$V^a = \inf_\pi \{-\theta_0(-\beta_0, -\gamma_0) : \pi = (\beta_t, \gamma_t)_{t=0}^T \qquad (9.1.6)$$
$$\text{is a seller's superhedging strategy}\}$$

and for the seller's (or ask) Israeli contingent claim price $V^a = V_I^a$,

$$V^a = \inf_{\sigma, \pi} \{-\theta_0(-\beta_0, -\gamma_0) : (\sigma, \pi), \ \pi = (\beta_t, \gamma_t)_{t=0}^T \qquad (9.1.7)$$
$$\text{is a seller's superhedging strategy}\}.$$

On the other hand, the buyer may borrow from a bank an amount $\theta_0(-\beta_0, -\gamma_0)$ to purchase one of the above contingent claims with the payoff $Q$ and starting with the negative valued portfolio $(\beta_0, \gamma_0)$ to manage a self-financing strategy $\pi = (\beta_t, \gamma_t)_{t=0}^T$ and to choose an exercise time $\tau$ in the American and Israeli contingent claim case so that she (he) could settle the bank loan after receiving the payoff from the seller, i.e.,

$$\theta_T(\beta_T + Q_T^{(1)}, \gamma_T + Q_T^{(2)}) \geq 0 \quad \text{in the European case,} \qquad (9.1.8)$$

$$\theta_\tau(\beta_\tau + Q_\tau^{(1)}, \gamma_\tau + Q_\tau^{(2)}) \geq 0 \quad \text{in the American case,}$$

$$\theta_{s \wedge \tau}(\beta_{s \wedge \tau} + Q_{s,\tau}^{(1)}, \gamma_{s \wedge \tau} + Q_{s,\tau}^{(2)}) \geq 0 \quad \text{for all } s \in [0, T] \text{ in the Israeli case.}$$
$$(9.1.9)$$

In this case $\pi = (\beta_t, \gamma_t)_{t=0}^T$ or $(\tau, \pi)$ will be called a superhedging strategy for the buyer. The buyer's (bid or lower hedging) price $V^b$ of the above contingent claims is defined as the supremum of initial bank loan required to purchase this game option and to manage a superhedging strategy for the buyer. Thus,

$$V^b = \sup_\pi \{\theta_0(-\beta_0, -\gamma_0) : \pi = (\beta_t, \gamma_t)_{t=0}^T \qquad (9.1.10)$$
$$\text{or } (\pi, \tau) \text{ is a buyer's superhedging strategy}\}.$$

### 9.1.2 Price representations and randomized stopping times

Denote by $\mathcal{P}$ the space of pairs $(\tilde{P}, S)$ consisting of a probability measure $\tilde{P}$ on $\Omega$ equivalent to the original (market) probability $P$ and of a martingale $S$ under $\tilde{P}$ such that $S_t^b \leq S_t \leq S_t^a$ for each $t = 0, 1, ..., T$ while we denote by $\bar{\mathcal{P}}$ a bigger space of pairs $(\tilde{P}, S)$ where we drop the condition that $\tilde{P}$ is equivalent to $P$. For instance, if $S_t^a = (1 + \lambda)S_t$ and $S_t^b = (1 - \mu)S_t$, where $S_t$ is the market (discounted) stock price, then assuming that the market with zero transaction costs has no arbitrage we conclude by the first fundamental theorem of asset pricing that $\{S_t\}$ is a martingale with respect to some probability measure $\tilde{P}$ equivalent to $P$, and so $\mathcal{P} \neq \emptyset$ in this case. According to [30] and [45] $\mathcal{P} \neq \emptyset$ is equivalent to the lack of arbitrage in the above model, i.e., to the nonexistence of a self-financing strategy $\pi = (\beta, \gamma)$ such that $-\theta_0(-\beta_0, -\gamma_0) \leq 0$, $\theta_T(\beta_T, \gamma_T) \geq 0$ and $P\{\theta_T(\beta_T, \gamma_T) > 0\} > 0$. The ask and bid prices of a European contingent claim in the market model above are given by the following theorem (see [64]).

**Theorem 9.1.1.** *Suppose that $\mathcal{P} \neq \emptyset$. Then*

$$V^a = V_E^a = \sup_{(\tilde{P}, S) \in \mathcal{P}} E_{\tilde{P}}(Q_T^{(1)} + S_T Q_T^{(2)}) = \max_{(\tilde{P}, S) \in \bar{\mathcal{P}}} E_{\tilde{P}}(Q_T^{(1)} + S_T Q_T^{(2)}).$$

*and*

$$V^b = V_E^b = \inf_{(\tilde{P}, S) \in \mathcal{P}} E_{\tilde{P}}(Q_T^{(1)} + S_T Q_T^{(2)}) = \min_{(\tilde{P}, S) \in \bar{\mathcal{P}}} E_{\tilde{P}}(Q_T^{(1)} + S_T Q_T^{(2)}).$$

In order to represent prices of American and Israeli contingent claims in the model above we will need the notion of a randomized (mixed) stopping time which is defined as a nonnegative adapted process $\chi$ such that $\sum_{t=0}^{T} \chi_t = 1$. The set of all randomized stopping times will be denoted by $\mathcal{X}$ while the set of all usual or pure stopping times will be denoted by $\mathcal{T}$. It is convenient to identify each pure stopping time $\tau$ with a randomized stopping time $\chi^\tau$ such that $\chi_t^\tau = \mathbb{I}_{\{\tau=t\}}$ for any $t = 0, 1, ..., T$, so that we could write $\mathcal{T} \subset \mathcal{X}$. For any adapted process $Z$ and each randomized stopping time $\chi$ the time-$\chi$ value of $Z$ is defined by

$$(Z)_\chi = Z_\chi = \sum_{t=0}^{T} \chi_t Z_t. \qquad (9.1.11)$$

In particular, if $Q$ is the (discounted) payoff of an American contingent claim and the exercise time is a randomized stopping time $\chi$ then the payoff at $\chi$ is equal to

$$Q_\chi = \sum_{t=0}^{T} \chi_t Q_t.$$

For a game option with a payoff given by (9.1.3) we write also

$$Q_{\chi,\tilde\chi} = \sum_{s,t=0}^{T} \chi_s \tilde\chi_t Q_{s,t}$$

which is the seller's payment to the buyer when the former cancels and the latter exercises at randomized stopping times $\chi$ and $\tilde\chi$, respectively. In particular, if $\sigma$ and $\tau$ are pure stopping times then

$$Q_{\chi,\chi^\tau} = \sum_{s=0}^{T} \chi_s Q_{s,\tau} \text{ and } Q_{\chi^\sigma,\chi} = \sum_{t=0}^{T} \chi_t Q_{\sigma,t}.$$

We can also define the "minimum" and the "maximum" of two randomized stopping times $\chi$ and $\tilde\chi$ which are randomized stopping times $\chi \wedge \tilde\chi$ and $\chi \vee \tilde\chi$ given by

$$(\chi \wedge \tilde\chi)_t = \chi_t \sum_{s=t}^{T} \tilde\chi_s + \tilde\chi_t \sum_{s=t+1}^{T} \chi_s \text{ and }$$
$$(\chi \vee \tilde\chi)_t = \chi_t \sum_{s=0}^{t} \tilde\chi_s + \tilde\chi_t \sum_{s=0}^{t-1} \chi_s.$$

In particular, if $\sigma$ and $\tau$ are pure stopping times then

$$\chi^\sigma \wedge \chi^\tau = \chi^{\sigma \wedge \tau}, \ \chi^\sigma \vee \chi^\tau = \chi^{\sigma \vee \tau}$$

and for any adapted process $Z$,

$$Z_{\chi \wedge \chi^\tau} = \sum_{s=0}^{T} \chi_s Z_{s \wedge \tau}, \ Z_{\chi^\sigma \wedge \chi} = \sum_{t=0}^{T} \chi_t Z_{\sigma \wedge t}.$$

272 *Lectures on Mathematical Finance*

and similarly for $\chi \vee \chi^\tau$ and $\chi^\sigma \vee \chi$.

Next, we introduce the notion of an approximate martingale which is defined for any randomized stopping time $\chi$ as a pair $(\tilde{P}, S)$ of a probability measure $\tilde{P}$ on $\Omega$ and of an adapted process $S$ such that for each $t = 0, 1, ..., T$,

$$S_t^b \leq S_t \leq S_t^a \text{ and } \chi_{t+1}^* S_t^b \leq E_{\tilde{P}}(S_{t+1}^{\chi^*}|\mathcal{F}_t) \leq \chi_{t+1}^* S_t^a \qquad (9.1.12)$$

where $E_{\tilde{P}}$ is the expectation with respect to $\tilde{P}$,

$$\chi_t^* = \sum_{s=t}^T \chi_s, \ Z_t^{\chi^*} = \sum_{s=t}^T \chi_s Z_s, \ \chi_{T+1}^* = 0 \text{ and } Z_{T+1}^{\chi^*} = 0. \qquad (9.1.13)$$

Given a randomized stopping time $\chi$ the space of corresponding approximate martingales $(\tilde{P}, S)$ will be denoted by $\bar{\mathcal{P}}(\chi)$ and we denote by $\mathcal{P}(\chi)$ the subspace of $\bar{\mathcal{P}}(\chi)$ consisting of pairs $(\tilde{P}, S)$ with $\tilde{P}$ being equivalent to the original (market) probability $P$.

Now we can formulate the results from [65], [37] and [63] which provide ask and bid price representations for American and Israeli contingent claims.

**Theorem 9.1.2.** *In the above notations,*

$$V^a = V_A^a = \max_{\chi \in \mathcal{X}} \max_{(\tilde{P},S) \in \bar{\mathcal{P}}(\chi)} E_{\tilde{P}}\left(Q_{\cdot}^{(1)} + Q_{\cdot}^{(2)} S_{\cdot}\right)_\chi$$

$$= \max_{\chi \in \mathcal{X}} \sup_{(\tilde{P},S) \in \mathcal{P}(\chi)} E_{\tilde{P}}\left(Q_{\cdot}^{(1)} + Q_{\cdot}^{(2)} S_{\cdot}\right)_\chi,$$

$$V^a = V_I^a = \min_{\sigma \in \mathcal{T}} \max_{\chi \in \mathcal{X}} \max_{(\tilde{P},S) \in \bar{\mathcal{P}}(\sigma \wedge \chi)} E_{\tilde{P}}\left(Q_{\sigma,\cdot}^{(1)} + Q_{\sigma,\cdot}^{(2)} S_{\sigma \wedge \cdot}\right)_\chi$$

$$= \min_{\sigma \in \mathcal{T}} \max_{\chi \in \mathcal{X}} \sup_{(\tilde{P},S) \in \mathcal{P}(\sigma \wedge \chi)} E_{\tilde{P}}\left(Q_{\sigma,\cdot}^{(1)} + Q_{\sigma,\cdot}^{(2)} S_{\sigma \wedge \cdot}\right)_\chi$$

*and*

$$V^a = V_A^a = \max_{\tau \in \mathcal{T}} \min_{(\tilde{P},S) \in \bar{\mathcal{P}}(\tau)} E_{\tilde{P}}\left(Q_{\cdot}^{(1)} + Q_{\cdot}^{(2)} S_{\cdot}\right)_\tau$$

$$= \max_{\tau \in \mathcal{T}} \inf_{(\tilde{P},S) \in \mathcal{P}(\tau)} E_{\tilde{P}}\left(Q_{\cdot}^{(1)} + Q_{\cdot}^{(2)} S_{\cdot}\right)_\tau,$$

$$V^a = V_I^a = \max_{\tau \in \mathcal{T}} \min_{\chi \in \mathcal{X}} \min_{(\tilde{P},S) \in \bar{\mathcal{P}}(\chi \wedge \tau)} E_{\tilde{P}}\left(Q_{\cdot,\tau}^{(1)} + Q_{\cdot,\tau}^{(2)} S_{\cdot \wedge \tau}\right)_\chi$$

$$= \max_{\tau \in \mathcal{T}} \min_{\chi \in \mathcal{X}} \inf_{(\tilde{P},S) \in \mathcal{P}(\chi \wedge \tau)} E_{\tilde{P}}\left(Q_{\cdot,\tau}^{(1)} + Q_{\cdot,\tau}^{(2)} S_{\cdot \wedge \tau}\right)_\chi$$

*where $\cdot$ means that the value of the corresponding function at $t$ is obtained by replacing $\cdot$ by $t$.*

It is important to stress that the above price representations require randomized stopping times and the standard stopping times which we dealt with before are not sufficient here. Relying on some convex analysis notions it is possible to produce dynamical programming algorithms for the computation of the above ask and bid prices as well, as of the optimal hedging self-financing strategies and of corresponding exercise and cancellation times (see [65], [37] and [63]).

## 9.2 Swing options

### 9.2.1 *Preliminaries*

Swing contracts appearing in energy and commodity markets are often modeled by multiple exercising of American style options which leads to multiple stopping problems (see, for instance, [4]). Such models describe options consisting of a package of claims or rights which can be exercised in a prescribed (or in any) order with some restrictions such as a delay time between successive exercises. The peculiarities of multiple exercise options are due only to restrictions such as an order of exercises and a delay time between them since without restrictions the above claims or rights could be considered as separate options which should be dealt with independently.

Attempts to valuate swing options (contingent claims) in multiple exercise models are usually reduced to maximizing the total expected gain of the buyer which is the expected payoff in the corresponding multiple stopping problem deviating from what now became classical and generally accepted methodology of pricing derivatives via hedging and replicating arguments. This digression is sometimes explained by difficulties in using an underlying commodity in a hedging portfolio in view of the high cost of storage, for instance, in the case of electricity. On the other hand, multiple exercise options may appear in their own rights when an investor wants to buy or sell an underlying security in several instalments at times of his choosing and, actually, any usual American or game option can be naturally extended to the multi exercise setup so that they may emerge both in commodities, energy and in different financial markets. For instance, a European car producer (having most expenses in euros) plans to supply autos to USA during a year in several shipments and it buys a multi exercise option which guaranties a favorable dollar — euro exchange rate at time of shipments (of its choosing). The seller of such option can use currencies as underlying risky assets for a hedging portfolio. The study of hedging for multiple exercise options is sufficiently motivated from the financial point of view and it leads to interesting mathematical problems. Following [15] we will assume here that the underlying security can be used for construction of a hedging portfolio without restrictions as in the usual theory of derivatives and we will consider here the game (Israeli) option (contingent claim) setup when both the buyer (holder) and the seller (writer) of the option can exercise or cancel, respectively, the claims (or rights) in a given order but, as before, each cancellation entails a penalty payment by the seller. This requires us, in particular, to extend Dynkin's games machinery to the multiple stopping setup.

A discrete time swing (multiple stopping) game option here is a contract between its seller and the buyer which allows to the seller to cancel (or terminate) and to the buyer to exercise $L$ specific claims or rights in a particular order. Such contract is determined given $2L$ payoff processes $Y_i(n) \geq Z_i(n) \geq 0$, $n = 0, 1, ...$, $i = 1, 2, ..., L$ adapted to a filtration $\mathcal{F}_n$, $n \geq 0$ generated by the stock (underlying risky security) $S_n$, $n \geq 0$ evolution. If the buyer exercises the $k$-th claim $k \leq L$ at the time $n$ then the seller pays to him the amount $Z_k(n)$ but if the latter cancels the claim $k$ at the time $n$ before the buyer he has to pay to the buyer the amount $Y_k(n)$ and the difference $\delta_k(n) = Y_k(n) - Z_k(n)$ is viewed as the cancelation penalty. In addition, we require a delay of one unit of time between successive exercises and cancellations. Observe that unlike some other papers (cf. [4]) we allow payoffs depending on the exercise number so, for instance, our options may change from call to put and vice versa after different exercises.

We will define the notion of a perfect hedge for swing game options which generalize swing American options and will see that in the binomial Cox–Ross–Rubinstein (CRR) market the option price $V^*$ is equal to the value of the multiple stopping Dynkin game with discounted payoffs under the unique martingale measure and provide a dynamical programming algorithm which allows to compute both this value and a corresponding perfect hedge. We will discuss also hedging with risk for swing game options. In real market conditions a writer of an option may not be willing for various reasons to tie in a hedging portfolio the full initial capital required for a perfect hedge. In this case the seller is ready to accept a risk that the portfolio value may become lower than the payoff obligation which requires to add money from other sources. In the setup here the writer is allowed to add money to his portfolio only at moments when the contract is exercised. The shortfall risk is defined as the expectation with respect to the market probability measure of the total sum that the seller added from other sources. It turns out that for any initial capital $x < V^*$ there exists a hedge which minimizes the shortfall risk and this hedge can be computed by a dynamical programming algorithm.

## 9.2.2 *Setup*

As in Lecture 5 we consider here the CRR market model on the probability space $\Omega = \{1, -1\}^N$ which is the set of finite sequences $\omega = (\omega_1, \omega_2, ..., \omega_N)$; $\omega_i \in \{1, -1\}$ with the product probability $P = \{p, 1 - p\}^N$, $p > 0$. The CRR binomial model of a financial market consists of a savings account $B_n$ with an interest rate $r$ which without loss of generality (by discounting) we assume to be zero, i.e., $B_n = B_0 > 0$ for all $n \geq 0$, and of a stock whose price at time n equals $S_n = S_0 \prod_{i=1}^{n} (1 + \rho_i)$, $S_0 > 0$ where $\rho_i(\omega_1, \omega_2, ..., \omega_N) = \frac{a+b}{2} + \frac{b-a}{2}\omega_i$ and $-1 < a < 0 < b$. Thus $\rho_i$, $i = 1, ..., N$ form a sequence of independent identically distributed random variables on the probability space $(\Omega, P)$ taking values $b$ and $a$ with probabilities $p$ and $1 - p$. Recall, that the unique martingale measure is given by $P^* = \{p^*, 1 - p^*\}^N$ where $p^* = \frac{a}{a-b}$.

We consider a swing option of the game type which has the $i$-th payoff, $i \geq 1$ having the form

$$R^{(i)}(m, n) = Y_i(m)\mathbb{I}_{m<n} + Z_i(n)\mathbb{I}_{n \leq m}, \quad \forall m, n$$

where $Y_i(n), Z_i(n)$ are $\mathcal{F}_n$-adapted and $0 \leq Z_i(n) \leq Y_i(n) < \infty$. Thus for any $i, n$ there exist functions $f_n^{(i)}, g_n^{(i)} : \{a, b\}^n \to \mathbb{R}_+$ such that

$$Z_i(n) = f_n^{(i)}(\rho_1, ..., \rho_n), \quad Y_i(n) = g_n^{(i)}(\rho_1, ..., \rho_n).$$

For any $1 \leq i \leq L - 1$ let $C_i$ be the set of all pairs $((a_1, ..., a_i), (d_1, ..., d_i)) \in \{0, ..., N\}^i \times \{0, 1\}^i$ such that $a_{j+1} \geq N \wedge (a_j + 1)$ for any $j < i$. Such sequences represent the history of payoffs up to the $i$-th one in the following way. If $a_j = k$ and $d_j = 1$ then the seller canceled the $j$-th claim at the moment $k$ and if $d_j = 0$ then the buyer exercised the $j$-th claim at the moment $k$ (maybe together with the seller). For $n \geq 1$ denote by $\mathcal{T}_{nN}$ the set of all stopping times with respect to the filtration $\{\mathcal{F}_n\}_{n=0}^{N}$ with values from $n$ to $N$.

**Definition 9.2.1.** A stopping strategy is a sequence $s = (s_1, ..., s_L)$ such that $s_1 \in \mathcal{T}_{0N}$ is a stopping time and for $i > 1$, $s_i : C_{i-1} \to \mathcal{T}_{0N}$ is a map which satisfies $s_i((a_1, ..., a_{i-1}), (d_1, ..., d_{i-1})) \in \mathcal{T}_{N \wedge (1+a_{i-1}), N}$.

In other words for the $i$-th payoff both the seller and the buyer choose stopping times taking into account the history of payoffs so far. Denote by $\mathcal{S}$ the set of all stopping strategies and define the map $F : \mathcal{S} \times \mathcal{S} \to \mathcal{T}_{0,N}^L \times \mathcal{T}_{0N}^L$ by $F(s, b) = ((\sigma_1, ..., \sigma_L), (\tau_1, ..., \tau_L))$ where $\sigma_1 = s_1$, $\tau_1 = b_1$ and for $i > 1$,

$$\sigma_i = s_i((\sigma_1 \wedge \tau_1, ..., \sigma_{i-1} \wedge \tau_{i-1}), (\mathbb{I}_{\sigma_1 < \tau_1}, ..., \mathbb{I}_{\sigma_{i-1} < \tau_{i-1}})) \text{ and}$$

$$\tau_i = b_i((\sigma_1 \wedge \tau_1, ..., \sigma_{i-1} \wedge \tau_{i-1}), (\mathbb{I}_{\sigma_1 < \tau_1}, ..., \mathbb{I}_{\sigma_{i-1} < \tau_{i-1}})).$$

Set

$$c_k(s,b) = \sum_{i=1}^{L} \mathbb{I}_{\sigma_i \wedge \tau_i \leq k}$$

which is a random variable equal to the number of payoffs until the moment $k$.

For swing options the notion of a self-financing portfolio involves not only allocation of capital between stocks and the bank account but also payoffs at exercise times. At the time $k$ the writer's decision how much money to invest in stocks (while depositing the remaining money into a bank account) depends not only on his present portfolio value but also on the current claim. Denote by $\Xi$ the set of functions on the (finite) probability space $\Omega$.

**Definition 9.2.2.** A portfolio strategy with an initial capital $x > 0$ is a pair $\pi = (x, \gamma)$ where $\gamma : \{0, ..., N-1\} \times \{1, ..., L\} \times \mathbb{R} \to \Xi$ is a map such that $\gamma(k,i,y)$ is an $\mathcal{F}_k$-measurable random variable which represents the number of stocks which the seller buy at the moment $k$ provided that the current claim has the number $i$ and the present portfolio value is $y$. At the same time the sum $y - \gamma(k,i,y)S_k$ is deposited to the bank account of the portfolio. We call a portfolio strategy $\pi = (x, \gamma)$ *admissible* if for any $y \geq 0$,

$$-\frac{y}{S_k b} \leq \gamma(k,i,y) \leq -\frac{y}{S_k a}.$$

For any $y \geq 0$ set $K(y) = [-\frac{y}{b}, -\frac{y}{a}]$ so that the above condition can be written as $S_k\gamma(k,i,y) \in K(y)$.

Notice that if the portfolio value at the moment $k$ is $y \geq 0$ then the portfolio value at the moment $k+1$ before the payoffs (if there are any payoffs at this time) is given by $y + \gamma(k,i,y)S_k(\frac{S_{k+1}}{S_k} - 1)$ where $i$ is the number of the next payoff. In view of independency of $\frac{S_{k+1}}{S_k} - 1$ and $\gamma(k,i,y)S_k$ we conclude that the above inequality is equivalent to the inequality $y + \gamma(k,i,y)S_k(\frac{S_{k+1}}{S_k} - 1) \geq 0$, i.e., the portfolio value at the moment $k+1$ before the payoffs is nonnegative. Denote by $\mathcal{A}(x)$ the set of all *admissible* portfolio strategies with an initial capital $x > 0$ and set $\mathcal{A} = \bigcup_{x>0} \mathcal{A}(x)$. Let $\pi = (x, \gamma)$ be a portfolio strategy and $s, b \in \mathcal{S}$. Set $((\sigma_1, ..., \sigma_L), (\tau_1, ..., \tau_L)) = F(s,b)$ and $c_k = c_k(s,b)$. The portfolio value at the moment $k$ after the payoffs (if there are any payoffs at this moment) is

given by
$$V_0^{(\pi,s,b)} = x - R^{(1)}(\sigma_1,\tau_1)\mathbb{I}_{\sigma_1 \wedge \tau_1 = 0} \text{ and for } k > 0,$$
$$V_k^{(\pi,s,b)} = V_{k-1}^{(\pi,s,b)} + \mathbb{I}_{c_{k-1} < L}\left(\gamma(k-1,c_{k-1}+1,V_{k-1}^{(\pi,s,b)})(S_k - S_{k-1}) - \sum_{i=1}^{L} R^{(i)}(\sigma_i,\tau_i)\mathbb{I}_{\sigma_i \wedge \tau_i = k}\right).$$

**Definition 9.2.3.** A hedge is a pair $(\pi,s)$ which consists of a portfolio strategy and a stopping strategy such that $V_k^{(\pi,s,b)} \geq 0$ for any $b \in \mathcal{S}$ and $k \leq N$.

Observe that if $(\pi,s)$ is a hedge then without loss of generality we can assume that $\pi$ is an admissible portfolio strategy, and so we will consider only admissible portfolio strategies. As usual, the option price $V^*$ is defined as the infimum of $V \geq 0$ such that there exists a hedge with an initial capital $V$.

### 9.2.3 Results

The following theorem from [15] provides a dynamical programming algorithm for computation of both the option price and the corresponding hedge.

**Theorem 9.2.1.** *Denote by $E^*$ the expectation with respect to the unique martingale measure $P^*$. For any $n \leq N$ set*
$$Y_n^{(1)} = Y_L(n), \quad Z_n^{(1)} = Z_L(n) \text{ and } V_n^{(1)} = \min_{\sigma \in \Gamma_n} \max_{\tau \in \Gamma_n} E^*(R^{(L)}(\sigma,\tau)|\mathcal{F}_n)$$
*and for $1 < k \leq L$,*
$$Y_n^{(k)} = Y_{L-k+1}(n) + E^*(V_{(n+1)\wedge N}^{(k-1)}|\mathcal{F}_n),$$
$$Z_n^{(k)} = Z_{L-k+1}(n) + E^*(V_{(n+1)\wedge N}^{(k-1)}|\mathcal{F}_n) \text{ and}$$
$$V_n^{(k)} = \min_{\sigma \in \Gamma_n} \max_{\tau \in \Gamma_n} E^*(Y_\sigma^{(k)}\mathbb{I}_{\sigma < \tau} + Z_\tau^{(k)}\mathbb{I}_{\sigma \geq \tau}|\mathcal{F}_n).$$
*Then*
$$V^* = V_0^{(L)} = \min_{s \in \mathcal{S}} \max_{b \in \mathcal{S}} G(s,b)$$
*where $G(s,b) = E^* \sum_{i=1}^{L} R^{(i)}(\sigma_i,\tau_i)$ and $((\sigma_1,...,\sigma_L),(\tau_1,...,\tau_L)) = F(s,b)$. Furthermore, the stopping strategies $s^* = (s_1^*,...,s_L^*) \in \mathcal{S}$ and $b^* = (b_1^*,...,b_L^*)$ given by*
$$s_1^* = N \wedge \min\{k|X_k^{(L)} = V_k^{(L)}\}, \quad b_1^* = \min\{k|Y_k^{(L)} = V_k^{(L)}\},$$
$$s_i^*((a_1,...,a_{i-1}),(d_1,...,d_{i-1})) = N \wedge \min\{k > a_{i-1}|$$
$$Y_k^{(L-i+1)} = V_k^{(L-i+1)}\}, \quad b_i^*((a_1,...,a_{i-1}),(d_1,...,d_{i-1}))$$
$$= N \wedge \min\{k > a_{i-1}|Z_k^{(L-i+1)} = V_k^{(L-i+1)}\}, \quad i > 1$$

*satisfy*

$$G(s^*, b) \leq G(s^*, b^*) \leq G(s, b^*) \text{ for all } s, b$$

*and there exists a portfolio strategy* $\pi^* \in \mathcal{A}(V_0^{(L)})$ *such that* $(\pi^*, s^*)$ *is a hedge.*

Next, consider an option seller whose initial capital is $x$, which is less than the option price, i.e., $x < V^*$. In this case the seller must (in order to fulfill the obligation to the buyer) add money to the portfolio from other sources. In our setup the seller is allowed to add money to the portfolio only at times when the contract is exercised. We also require that after the addition of money by the seller the portfolio value becomes positive.

**Definition 9.2.4.** An infusion of capital is a map $I : \{0, ..., N\} \times \{1, ..., L\} \times \mathbb{R} \to \Xi$ such that $I(k, j, y) \geq (-y)^+$ is $\mathcal{F}_k$-measurable, $I(k, L, y) = (-y)^+$ for any $k$, and for any $j < L$, $I(N, j, y) = \left((\sum_{i=j+1}^{L} Z_i(N)) - y\right)^+$. The set of such maps will be denoted by $\mathcal{I}$.

Thus $I(k, j, y)$ is the amount that the seller adds to his portfolio after the $j$-th payoff paid at the moment $k$ and the portfolio value after this payment is $y$. When $k = N$ or $j = L$ then clearly $I(k, j, y)$ is the minimal amount which the seller should add in order to fulfill the obligation to the buyer. Observe that when $k = N$ one infusion of capital to the seller's portfolio is already sufficient in order to fulfill the obligations even if there are additional payoffs at this moment, so we conclude that at each step that the contract is exercised there is no more than one infusion of capital. An investment strategy with an initial capital $x < V^*$ is a triple $(\pi, \mathcal{I}, s) \in \mathcal{A}(x) \times I \times S$ which consists of an *admissible* portfolio strategy with an initial capital $x$, infusion of capital and a stopping strategy. Let $(\pi, I, s)$ be an investment strategy and $b \in \mathcal{S}$ be a stopping strategy for the buyer. Set $((\sigma_1, ..., \sigma_L), (\tau_1, ..., \tau_L)) = F(s, b)$ and $c_k = c_k(s, b)$. Define the stochastic processes $\{W_k^{(\pi, I, s, b)}\}_{k=0}^{N}$ and $\{V_k^{(\pi, I, s, b)}\}_{k=0}^{N}$ by

$$W_0^{(\pi, I, s, b)} = x, \quad V_0^{(\pi, I, s, b)} = x - \mathbb{I}_{\sigma_1 \wedge \tau_1 = 0}\left(R^{(1)}(\sigma_1, \tau_1) - \right.$$
$$I(0, 1, x - R^{(1)}(\sigma_1, \tau_1))) \text{ and for } k > 0,$$

$$W_k^{(\pi, I, s, b)} = V_{k-1}^{(\pi, I, s, b)} + \mathbb{I}_{c_{k-1} < L}\gamma(k-1, c_{k-1} + 1, V_{k-1}^{(\pi, I, s, b)})(S_k - S_{k-1}),$$

$$V_k^{(\pi, I, s, b)} = W_k^{(\pi, I, s, b)} - \mathbb{I}_{c_{k-1} < L}\mathbb{I}_{\sigma_{c_{k-1}+1} \wedge \tau_{c_{k-1}+1} = k} \times$$
$$\left(R^{(c_{k-1}+1)}(\sigma_{c_{k-1}+1}, \tau_{c_{k-1}+1}) + \mathbb{I}_{k=N}\sum_{i=c_{k-1}+2}^{L} Z_i(N)\right.$$
$$\left. -I(k, c_{k-1} + 1, W_k^{(\pi, I, s, b)} - R^{(c_{k-1}+1)}(\sigma_{c_{k-1}+1}, \tau_{c_{k-1}+1}))\right).$$

Observe that if the contract was not exercised at a moment $k$ then $W_k^{(\pi,I,s,b)} = V_k^{(\pi,I,s,b)}$ is the portfolio value at this moment. If the contract was exercised at a moment $k$ then $W_k^{(\pi,I,s,b)}$ and $V_k^{(\pi,I,s,b)}$ are the portfolio values before and after the payoff, respectively. Thus the total infusion of capital made by the seller is given by

$$C(\pi, I, s, b) = \sum_{i=1}^{(c_{N-1}+1)\wedge L} I(\sigma_i \wedge \tau_i, i, W_{\sigma_i \wedge \tau_i}^{(\pi,I,s,b)} - R^{(i)}(\sigma_1, \tau_i)).$$

**Definition 9.2.5.** Given an investment strategy $(\pi, I, s) \in \mathcal{A} \times \mathcal{I} \times S$ the shortfall risk for it is defined by

$$r(\pi, I, s) = \max_{b \in S} EC(\pi, I, s, b)$$

which is the maximal expectation with respect to the market probability measure $P$ of the total infusion of capital. The shortfall risk for the initial capital $x$ is defined by

$$r(x) = \inf_{(\pi,I,s) \in \mathcal{A}(x) \times \mathcal{I} \times S} r(\pi, I, s).$$

The following result from [15] asserts that for any initial capital $x$ there exists a hedge $(\pi, I, s) \in \mathcal{A}(x) \times \mathcal{I} \times S$ which minimizes the shortfall risk and both the risk and the optimal investment strategy can be obtained recurrently.

**Theorem 9.2.2.** *Define a sequence of functions* $J_k : \mathbb{R}_+ \times \{0, ..., L\} \times \{a, b\}^k \to \mathbb{R}_+$, $0 \le k \le N$ *by the following formulas*

$$J_N(y, j, u_1, ..., u_N) = \left(\sum_{i=L-j+1}^{L} f_N^{(i)}(u_1, ..., u_N) - y\right)^+, \quad j > 0,$$

$$J_k(y, 0, u_1, ..., u_k) = 0, \quad 0 \le k \le N$$

*and for* $k < N$ *and* $j > 0$,

$$J_k(y, j, u_1, ..., u_k) =$$

$$\min\Big( \inf_{z \ge (g_k^{(L-j+1)}(u_1,...,u_k)-y)^+} \inf_{\alpha \in K(y+z-g_k^{(L-j+1)}(u_1,...,u_k))}$$

$$\big(z + pJ_{k+1}(y + z - g_k^{(L-j+1)}(u_1, ..., u_k) + b\alpha, j-1, u_1, ..., u_k, b) +$$

$$(1-p)J_{k+1}(y + z - g_k^{(L-j+1)}(u_1, ..., u_k) + a\alpha, j-1, u_1, ..., u_k, a)\big),$$

$$\max\Big( \inf_{z \ge (f_k^{(L-j+1)}(u_1,...,u_k)-y)^+} \inf_{\alpha \in K(y+z-f_k^{(L-j+1)}(u_1,...,u_k))}$$

$$\big(z + pJ_{k+1}(y + z - f_k^{(L-j+1)}(u_1, ..., u_k) + b\alpha, j-1, u_1, ..., u_k, b) +$$

$$(1-p)J_{k+1}(y + z - f_k^{(L-j+1)}(u_1, ..., u_k) + a\alpha, j-, u_1, ..., u_k, a)\big),$$

$$\inf_{\alpha \in K(y)} \big(pJ_{k+1}(y + b\alpha, j, u_1, ..., u_k, b) +$$

$$(1-p)J_{k+1}(y + a\alpha, j, u_1, ..., u_k, a)\big)\Big)\Big).$$

*Then the shortfall risk for an initial capital x is given by*

$$r(x) = J_0(x, L)$$

*and there exists an investment strategy* $(\tilde{\pi}, \tilde{I}, \tilde{s}) \in \mathcal{A}(x) \times \mathcal{I} \times S$, $\tilde{\pi} = (x, \tilde{\gamma})$
*satisfying*

$$r(\tilde{\pi}, \tilde{I}, \tilde{s}) = r(x).$$

## 9.3   Model uncertainty

As we saw before in this book main questions about financial markets such
as arbitrage, completeness and pricing of contingent claims depend on the
market probabilistic model under consideration. On the other hand, the
dynamics of real financial markets can be described only approximately
by various probabilistic models. Actually, because of complexity of the real
world doable mathematical models can serve only as certain approximations
of real phenomena. Thus, it makes sense to obtain results which do not
depend or depend very little on specifics of market models.

We will describe here one of approaches in this direction which deals
with game options in discrete time and appeared in [6]. Let $N < \infty$ be an
integer (horizon) and $I = [a, b] \subset \mathbb{R}_+ = [0, \infty)$. Introduce the sample space

$$\Omega = \{(\omega_0, \omega_1, ..., \omega_N) : \omega_0 = s > 0, \omega_i > 0, |\ln \omega_i - \ln \omega_{i-1}| \in I$$
$$\text{for all } i = 1, ..., N\}.$$

The financial market consists of a savings account (or bond) $B$ which by
discounting is supposed to be identically 1 and of a risky asset $S$ (stock). We
assume that the stock price process $S = (S_0, S_1, ..., S_N)$ is the coordinate
process on $\Omega$, i.e., $S_i(\omega) = \omega_i$ when $\omega = (\omega_0, \omega_1, ..., \omega_N) \in \Omega$ and $S_k$
represents the stock price at time $k$. Let $\mathcal{F}_k = \sigma\{S_0, S_1, ..., S_k\}$, $k \leq N$
be the $\sigma$-algebra generated by the functions $S_0, S_1, ..., S_k$. We consider a
game contingent claim with the payoff of the seller to the buyer given by
the formula

$$R(k, l) = Y_k \mathbb{I}_{k<l} + Z_l \mathbb{I}_{l \leq k}$$

where $Y_k, Z_k : \Omega \to \mathbb{R}_+$ are upper semi-continuous functions on $\Omega$ measurable with respect to $\mathcal{F}_k$ and satisfying $Z_k \leq Y_k$, $k = 0, 1, ..., N$.

In this setup a trading strategy is a pair $\pi = (x, \gamma)$ where $x$ is the initial
value of the portfolio, $\gamma = (\gamma_0, \gamma_1, ..., \gamma_N)$ and each $\gamma_k$, $k = 0, 1, ..., N$ is an
$\mathcal{F}_k$-measurable function on $\Omega$. The trading strategy $\pi$ is self-financing if the

corresponding portfolio value $X_k^\pi$ at time $k$ is given for each $k = 0, 1, ..., N$ by the formula

$$X_k^\pi = x + \sum_{i=0}^{k-1} \gamma_i(S_{i+1} - S_i).$$

Denote by $\mathcal{T}_{0N}$ the set of all stopping times with respect to the filtration $\mathcal{F}_k$, $k = 0, 1, ..., N$ taking on values between 0 and $N$. A pair $(\pi, \sigma)$ of a self-financing trading strategy $\pi$ and a stopping time $\sigma$ is called an investment strategy of the seller and it is called a hedge if, in addition,

$$X_{\sigma \wedge l}^\pi \geq R(\sigma, l) \quad \text{for all} \quad l = 0, 1, ..., N.$$

As before, the seller's super-replication (super-hedging) price of the above game contingent claim is given by

$$V = \inf\{X_0^\pi : \text{there exists a stopping time } \sigma \text{ such that } (\pi, \sigma) \text{ is a hedge}\}.$$

Since we do not have any probability measure here yet, we require the above super-hedging condition for all $\omega \in \Omega$.

We say that a probability measure $P$ on $\Omega$ is a martingale law if $S = (S_0, S_1, ..., S_N)$ is a martingale with respect to it. The set $\mathcal{M}$ of all martingale laws here is not empty. Indeed, set $\Omega_b = \{\omega = (\omega_0, \omega_1, ..., \omega_N) \in \Omega : \omega_0 = s, |\ln \omega_i - \ln \omega_{i-1}| = b, i = 1, 2, ..., N\}$ and define the probability measure $P_b$ on $\Omega_b$ by

$$P_b(\omega) = \left(\frac{1 - e^{-b}}{e^b - e^{-b}}\right)^n \left(\frac{e^b - 1}{e^b - e^{-b}}\right)^{N-n}$$

where $\omega = (\omega_0, \omega_1, ..., \omega_N)$, $\omega_0 = s$, the number of $i$'s between 1 and $N$ for which $\ln \omega_i - \ln \omega_{i-1} = b$ equals $n$ and, correspondingly, the number of $i$'s between 1 and $N$ for which $\ln \omega_i - \ln \omega_{i-1} = -b$ equals $N - n$. Then, $P_b \in \mathcal{M}$. The following result is proved in [6].

**Theorem 9.3.1.** *The super-replication price $V$ for the game contingent claim above is given by*

$$V = \inf_{\sigma \in \mathcal{T}_{0N}} \sup_{P \in \mathcal{M}} \sup_{\tau \in \mathcal{T}_{0N}} E_P R(\sigma, \tau)$$

$$= \sup_{P \in \mathcal{M}} \inf_{\sigma \in \mathcal{T}_{0N}} \sup_{\tau \in \mathcal{T}_{0N}} E_P R(\sigma, \tau)$$

$$= \sup_{P \in \mathcal{M}} \sup_{\tau \in \mathcal{T}_{0N}} \inf_{\sigma \in \mathcal{T}_{0N}} E_P R(\sigma, \tau)$$

*where $E_P$ is the expectation with respect to $P$.*

The main difference of this theorem from other super-hedging results considered in this book is that now we do not have any specific market probability measure and martingale laws above are not supposed to be equivalent to each other.

# Chapter 10

# Continuous Time Case

## 10.1 Fundamental theorems of asset pricing in continuous time

### 10.1.1 *Arbitrage*

The arbitrage theory in continuous time differs somewhat from the corresponding clear-cut theory in the discrete time discussed in Lecture 6. First, as we saw in the example from Lecture 20 in order to avoid arbitrage it is necessary to allow only certain admissible and not all self-financing trading strategies. Secondly, the admissible trading strategies should also be such that the corresponding stochastic integrals appearing in the portfolio value formulas were well defined. Thirdly, even with the above restrictions the notion of arbitrage sometimes should be slightly modified to obtain appropriate results.

We consider here a general setup with $d$-stocks whose discounted price is given by a $d$-dimensional càdlàg semimartingale $S_t = (S_t^{(1)}, ..., S_t^{(d)})$, $S_t^{(i)} > 0$, $i = 1, ..., d$, $t \in [0, T]$ with respect to a filtration $\mathcal{F}_t$, $t \in [0, T]$ satisfying usual conditions where $T > 0$ is a horizon. A non-risky asset (bond or bank account) is supposed to have price 1 identically. Next, a predictable process $\gamma : \mathbb{R}_+ \times \Omega \to \mathbb{R}^d$, $\gamma(t, \omega) = \gamma_t(\omega) = (\gamma_t^{(1)}(\omega), ..., \gamma_t^{(d)}(\omega))$ is called a trading strategy. Such trading strategy is called admissible if the stochastic integral

$$\int_0^t (\gamma_u, dS_u) = \sum_{i=1}^d \int_0^t \gamma_u^{(i)} dS_u^{(i)}$$

is well defined for each $t \in [0, T]$ in the sense of Lecture 14 and there exists a constant $M$ such that

$$\int_0^t (\gamma_u, dS_u) \geq -M \quad \text{a.s. for all} \quad t \in [0, T]. \tag{10.1.1}$$

An admissible strategy $\gamma$ is called self-financing if the corresponding portfolio value at time $t$ is given by the formula

$$X_t^\gamma = X_0^\gamma + \int_0^t (\gamma_u, dS_u). \qquad (10.1.2)$$

We deal here with already discounted quantities and self-financing strategies, and so the bank account (bond) does not play a role in the portfolio pricing. By this reason we talk about a strategy $\gamma$ and not about a pair $\pi = (\beta, \gamma)$, as before. The formula (10.1.2) is compatible with the corresponding formulas in Lectures 19–21 for the price of a self-financing portfolio since if the bond price $B_t \equiv$const then the portfolio price

$$X_t = X_0 + \int_0^t \beta_u dB_u + \int_0^t \gamma_u dS_u = X_0 + \int_0^t \gamma_u dS_u.$$

Observe also that in view of the example in Lecture 20, a condition of the type (10.1.1) is necessary in order to avoid arbitrage already in the standard Black–Scholes market.

Next, set

$$M_T = \{ \int_0^T (\gamma_u, dS_u) : \gamma \quad \text{is admissible and self-financing} \}$$

which forms a convex cone of functions in the space $L^0(\Omega, \mathcal{F}, P)$ of all (equivalence classes) of random variables on the probability space $(\Omega, \mathcal{F}_T, P)$. Let

$$C = \{ g \in L^\infty(\Omega, \mathcal{F}_T, P) : g \leq f \text{ for some } f \in M_T \}$$

and $L_+ = \{ g \in L^\infty(\Omega, \mathcal{F}_T, P) : g \geq 0 \}$. The financial market is said to provide no arbitrage opportunity (NA — no arbitrage ) if $C \cap L_+ = \{0\}$ and it is said to provide no opportunity for a free lunch with vanishing risk (NFLVR — no free lunch with vanishing risk ) if $\bar{C} \cap L_+ = \{0\}$ where the closure $\bar{C}$ is taken with respect to the norm topology in $L^\infty(\Omega, \mathcal{F}_T, P)$ (i.e., $\|g\|_{L^\infty} = ess\sup |g|$). The NA property above is the same as before, i.e., there should be no admissible self-financing strategy $\gamma$ which starts with zero portfolio value and yields at time $T$ an a.s. nonnegative portfolio value $X_T^\gamma = \int_0^T (\gamma_u, dS_u) \geq 0$ which is positive with positive probability. A somewhat stronger NFLVR property says that there should be no sequence of admissible self-financing strategies $\gamma^{(n)}$, $n \geq 1$ which start with zero portfolio value and such that the negative part $(X_T^{\gamma^{(n)}})^- = \left( \int_0^T (\gamma_u^{(n)}, dS_u) \right)^-$ of the portfolio value at time $T$ tends to zero uniformly (i.e., in $L^\infty$) while $X_T^{\gamma^{(n)}} \to X_T$ a.s. as $n \to \infty$ and $P\{X_T > 0\} > 0$.

We will need also the notion of a sigma-martingale which generalizes the notion of a local martingale. An $\mathbb{R}^d$-valued semimartingale $S = \{S_t, t \in [0,T]\}$ is called a sigma-martingale if there exists a positive predictable $d$-dimensional process $\varphi = \{\varphi_t, t \in [0,T]\}$ such that the stochastic integral $\int_0^t (\varphi_u, dS_u)$, $t \in [0,T]$ exists and is a martingale. We will also call an $\mathbb{R}^n$-valued semimartingale $S = \{S_t, t \in [0,T]\}$ locally bounded if there exists a sequence of stopping times $\tau_n \uparrow \infty$ as $n \uparrow \infty$ such that $|S_t| \leq n$ when $0 \leq t \leq \tau_n$. The following result whose proof can be found in [10] is a version of the first fundamental theorem of asset pricing in continuous time financial markets.

**Theorem 10.1.1.** *Suppose that a financial market consists of $d$-stocks whose discounted prices are represented by an $\mathbb{R}^d$-valued semimartingale $S = \{S_t, t \in [0,T]\}$ and of a bond with the price equal to 1 identically. Then the following assertions are equivalent:*

*(i) The condition NFLVR holds true.*

*(ii) There exists a probability measure $Q$ equivalent to $P$ such that $S$ is a sigma-martingale under $Q$.*

*If $S$ is locally bounded then (ii) can be replaced by*

*(ii') There exists a probability measure $Q$ equivalent to $P$ such that $S$ is a local martingale under $Q$.*

*If $S$ is bounded then (ii) can be replaced by*

*(ii") There exists a probability measure $Q$ equivalent to $P$ such that $S$ is a martingale under $Q$.*

This theorem was further extended in [74] (see also [75]). Let, again, $S = \{S_t, t \in [0,T]\}$, $S_t = (S_t^{(1)}, ..., S_t^{(d)})$ be a $\mathbb{R}^d$-valued semimartingale and weaken the admissibility condition by calling a $d$-dimensional predictable and $S$-integrable process $\gamma$ allowable if for some constant $M$,

$$\int_0^t (\gamma_u, dS_u) \geq -M\left(1 + \sum_{i=0}^d S_t^{(i)}\right) \quad \text{a.s. for } 0 \leq t \leq T.$$

**Theorem 10.1.2.** *Suppose that $S$ is a $d$-dimensional semimartingale with positive components $S_t^{(i)}, i = 1, ..., d$ (which is natural if $S$ represents the prices of $d$-stocks). There exists a probability measure $Q$ on $(\Omega, \mathcal{F}_T)$ equivalent to $P$ under which $S$ is a martingale if and only if $S$ satisfies NFLVR condition with respect to allowable (in place of admissible) self-financing trading strategies.*

### 10.1.2 Completeness

Again, consider an already discounted setup with $d$-stocks whose prices are given by an $\mathbb{R}^d$-valued semimartingale $S = \{S_t, t \in [0,T]\}$, $S_t = (S_t^{(1)}, ..., S_t^{(d)})$, $S_t^{(i)} > 0$, $i = 1, ..., d$. Denote by $\mathcal{P}(P)$ the set of probability measures $Q$ on $(\Omega, \mathcal{F}_T)$ which are equivalent to $P$ and such that $S$ is a (vector) martingale under $Q$. Assume that $\mathcal{P}(P)$ is not empty and let $P^* \in \mathcal{P}(P)$. We call here a predictable $d$-dimensional process $\gamma = \{\gamma_t, t \in [0,T]\}$ an admissible trading strategy if the stochastic integral $\int_0^t (\gamma_u, dS_u)$ exists and is nonnegative a.s. for each $t \in [0,T]$. An admissible strategy $\gamma$ will be called self-financing if its portfolio value $X_t^\gamma$ at time $t$ is given by the formula (10.1.2) and $X_t^\gamma$, $t \in [0,T]$ is a martingale under $P^*$.

A contingent claim $R$ is defined here as a positive $\mathcal{F}_T$-measurable random variable such that $E_{P^*} R < \infty$. Such a claim is said to be attainable if there exists a self-financing admissible trading strategy $\gamma$ such that $X_T^\gamma = R$ in which case $\gamma$ is said to replicate $R$. The market is said to be complete if every integrable contingent claim is attainable. The proof of the following theorem, which plays the role of the second fundamental theorem of asset pricing in continuous time, can be found in [25].

**Theorem 10.1.3.** *The following statements are equivalent:*

*(a) The market is complete under $P^*$;*

*(b) $\mathcal{P}(P)$ is a singleton;*

*(c) Every martingale $M = \{M_t, t \in [0,T]\}$ under $P^*$ can be represented in the form $M_t = M_0 + \int_0^t (\gamma_u, dS_u)$, $t \in [0,T]$ for some predictable process $\gamma_u$, $u \in [0,T]$.*

## 10.2 Superhedging in continuous time

Let $S = \{S_t, t \geq 0\}$ be an $\mathbb{R}^d$-valued right-continuous semimartingale on a probability space $(\Omega, \mathcal{F}, P)$ with respect to a filtration $\mathcal{F}_t$, $t \geq 0$ satisfying the usual conditions. Denote by $\mathcal{P}(P)$ the set of all probability measures $Q$ equivalent to $P$ and such that $S$ is a local martingale with respect to $Q$. The following optional decomposition theorem was proved in [21].

**Theorem 10.2.1.** *Assume that $\mathcal{P}(P) \neq \emptyset$. Let $Y = \{Y_t, t \geq 0\}$ be a right-continuous process which is a local supermartingale with respect to any $Q \in \mathcal{P}(P)$. Then there exists a non-decreasing right-continuous adapted (optional) process $C = \{C_t, t \geq 0\}$ with $C_0 = 0$ and a predictable $d$-dimensional process $\gamma = \{\gamma_t, t \geq 0\}$ such that the stochastic integrals*

$\int_0^t (\gamma_u, dS_u)$, $t \geq 0$ *are well defined and*

$$Y_t = Y_0 + \int_0^t (\gamma_u, dS_u) - C_t, \, t \geq 0. \tag{10.2.1}$$

For locally bounded semimartingales $S$ this theorem was proved previously in [39].

Next, we consider a financial market with $d$ risky securities (stocks) whose discounted prices are represented by an $\mathbb{R}^d$-valued right-continuous semimartingale $S = \{S_t, t \in [0,T]\}$, $S_t = (S_t^{(1)}, ..., S_t^{(d)})$ where $T > 0$ is a horizon. A non-risky security $B$ (bond, bank account) has here the constant price $B_t \equiv 1$. Consider European, American and Israeli contingent claims with the payoffs $f = f_T$, $f_t$, $t \in [0,T]$ and $R(s,t) = f_s \mathbb{I}_{s<t} + g_t \mathbb{I}_{s \geq t}$, respectively. Here, $f = f_T \geq 0$ is $\mathcal{F}_T$-measurable, $f_t \geq g_t \geq 0$ are adapted processes and

$$\sup_{Q \in \mathcal{P}(P)} E_Q f < \infty \quad \text{and} \quad \sup_{Q \in \mathcal{P}(P)} E_Q \sup_{t \in [0,T]} f_t < \infty$$

where, as before, $\mathcal{P}(P) \neq \emptyset$ is the set of all local martingale measures.

Modifying slightly the proof in Lecture 7 (for details see [39]) it is possible to show that

$$Y_t^E = ess \sup_{Q \in \mathcal{P}(P)} E_Q(f|\mathcal{F}_t), \, t \in [0,T] \tag{10.2.2}$$

and

$$Y_t^A = ess \sup_{Q \in \mathcal{P}(P), \tau \in \mathcal{T}_{tT}} E_Q(f_\tau|\mathcal{F}_t), \, t \in [0,T] \tag{10.2.3}$$

are right-continuous supermartingales with respect to all $Q \in \mathcal{P}(P)$ (i.e., right hand sides have right-continuous modifications which are denoted by $Y^E$ and $Y^A$) where, as before, $\mathcal{T}_{tT}$ denotes the set of all stopping times with values between $t$ and $T$.

By the above theorem we have optional decompositions so that for all $t \in [0,T]$,

$$Y_t^E = Y_0^E + \int_0^t (\gamma_u^E, dS_u) - C_t^E, \text{ and } Y_t^A = Y_0^A + \int_0^t (\gamma_u^A, dS_u) - C_t^A, \tag{10.2.4}$$

where $\gamma^E$ and $\gamma^A$ are $d$-dimensional predictable processes such that the integrals above are well defined and $C^E$, $C^A$ are non-decreasing right-continuous adapted processes with $C_0^E = C_0^A = 0$. Set

$$X_t^E = Y_t^E + C_t^E \quad \text{and} \quad X_t^A = Y_t^A + C_t^A, \, t \in [0,T]. \tag{10.2.5}$$

Introduce trading strategies $\pi^E = (\beta^E, \gamma^E)$ and $\pi^A = (\beta^A, \gamma^A)$, where $\beta_t^E = X_t^E - \gamma_t^E S_t$ and $\beta_t^A = X_t^A - \gamma_t^A S_t$, which make portfolio values at time $t$ equal to $X_t^E$ and $X_t^A$, respectively. By (10.2.4) and (10.2.5) the strategies $\pi^E$ and $\pi^A$ are self-financing. Now,

$$X_T^E \geq Y_T^E = f \quad \text{and} \quad X_0^E = Y_0^E = \sup_{Q \in \mathcal{P}(P)} E_Q f \qquad (10.2.6)$$

which means that $\pi^E$ is a (super) hedging trading strategy with the initial capital $\sup_{Q \in \mathcal{P}(P)} E_Q f$ which provides an upper bound for fair prices (from the seller's point of view) of the above European contingent claim. We have also

$$X_0^A = Y_0^A = \sup_{Q \in \mathcal{P}(P), \tau \in \mathcal{T}_{0T}} E_Q f_\tau \quad \text{and} \quad X_t^A \geq Y_t^A \geq f_t \qquad (10.2.7)$$

where we take $\tau \equiv t$ in (10.2.3) for the last inequality. Hence, $\pi^A$ is a (super) hedging trading strategy with the initial capital $\sup_{Q \in \mathcal{P}(P), \tau \in \mathcal{T}_{0T}} E_Q f_\tau$ which is an upper bound for fair prices (from the seller's point of view) of the above American contingent claim.

In the Israeli contingent claim case for each $\sigma \in \mathcal{T}_{0T}$ we define

$$Y_t^{I,\sigma} = \mathrm{ess} \sup_{Q \in \mathcal{P}(P), \tau \in \mathcal{T}_{t,T}} E_Q(R(\sigma, \tau) | \mathcal{F}_t), \quad t \in [0, T] \qquad (10.2.8)$$

which turns out to be, again, a right-continuous supermartingale with respect to any $Q \in \mathcal{P}(P)$. Using the above optional decomposition theorem we obtain the representation

$$Y_t^{I,\sigma} = Y_0^{I,\sigma} + \int_0^t (\gamma_u^{I,\sigma}, dS_u) - C_t^{I,\sigma}, \quad t \in [0, T]$$

with a predictable $d$-dimensional process $\gamma^{I,\sigma}$ and a non-decreasing right-continuous adapted process $C^{I,\sigma}$. As before, we set

$$X_t^{I,\sigma} = Y_t^{I,\sigma} + C_t^{I,\sigma}, \ \beta_t^{I,\sigma} = X_t^{I,\sigma} - \gamma_t^{I,\sigma} S_t \quad \text{and} \quad \pi^{I,\sigma} = (\beta^{I,\sigma}, \gamma^{I,\sigma}).$$

Then

$$X_0^{I,\sigma} = Y_0^{I,\sigma} = \sup_{Q \in \mathcal{P}(P), \tau \in \mathcal{T}_{0T}} E_Q R(\sigma, \tau) \text{ and } X_t^{I,\sigma} \geq Y_t^{I,\sigma} \geq R(\sigma, t)$$

where we take $\tau \equiv t$ in (10.2.8) for the last inequality. Thus $\pi^{I,\sigma}$ is a hedging trading strategy with the initial capital $\sup_{Q \in \mathcal{P}(P), \tau \in \mathcal{T}_{0T}} E_Q R(\sigma, \tau)$. For any $\varepsilon > 0$ we can choose $\sigma_\varepsilon \in \mathcal{T}_{0T}$ such that

$$\sup_{Q \in \mathcal{P}(P), \tau \in \mathcal{T}_{0T}} E_Q R(\sigma_\varepsilon, \tau) - \varepsilon \leq \inf_{\sigma \in \mathcal{T}_{0T}} \sup_{Q \in \mathcal{P}(P), \tau \in \mathcal{T}_{0T}} E_Q R(\sigma, \tau), \qquad (10.2.9)$$

and so the right hand side of (10.2.9) is the upper bound for fair prices (from the seller's point of view) of the above Israeli contingent claim.

**Remark 10.2.1.** Above, we used only non negativity of the optional process $C_t$, $t \geq 0$. On the other hand, we can view it as a consumption process and allow this in the definition of self-financing strategies. Then the portfolio values at time $t$ will be represented by the processes $Y^E$, $Y^A$ and $Y^{I,\sigma}$ in place of $X^E$, $X^A$ and $X^{I,\sigma}$, respectively. Still, even with consumption the above strategies make hedging portfolios starting with the same initial capital.

## 10.3   Transaction costs and shortfall risk in continuous time

### 10.3.1   *Transaction costs in continuous time*

Most of the theory of pricing of contingent claims proceeds smoothly for markets without transaction costs where as many as an investor wishes changes in a portfolio can be done at no cost. The theory of such "frictionless" markets is considered as a useful approximation of real markets since usually transaction costs form a tiny fraction of prices.

Nevertheless, it turns out that already the presence of arbitrary small proportional transaction costs lead to completely different answers on basic questions of superreplication in comparison to frictionless markets situation. It was shown in [70], [52] and [24] under increasingly more general conditions that in continuous time markets with transaction costs a trivial buy-and-hold strategy usually yields a cheapest hedge for European and American contingent claims (see also [43]). The buy-and-hold strategy means that the investor buys certain amount of stock at time zero and do not make any further trade until the expiration time $T$ or the exercise time of the buyer in an American contingent claim case.

A similar result was proved under certain conditions in [6] for the case of Israeli (game) contingent claims in continuous time with proportional transaction costs. In this case the seller's trading strategy consists of two parts $\pi$ and $\sigma$, where $\pi$ is a self-financing strategy and $\sigma$ is a cancellation (stopping) time. A trivial strategy is defined here as a pair $(\pi, \sigma)$ of a buy-and-hold strategy $\pi$ and a stopping time $\sigma$ having the form $\sigma = \inf\{t \geq 0 : S_t \in U\} \wedge T$ where $S_t$ is the stock price at time $t$ represented by a positive continuous adapted process, $T > 0$ is a horizon and $U$ is a Borel set. It is shown in [6] that if payoffs are convex and Lipschitz continuous and a certain "conditional full support" property of the stock process holds true (which ensures richness of the space of paths of this process) then, again, there exists no better (cheaper) than a trivial hedging strategy. This kind of results suggests that perhaps a full (perfect) hedge is not realistic in continuous time financial markets with the presence of transaction costs and partial hedging strategies allowing some risk should be considered instead.

### 10.3.2 *Shortfall risk*

First, consider a market in continuous time but without transaction costs. For European and American contingent claims in continuous time it was shown in [22] and [56], respectively, that there exist admissible strategies which minimize the shortfall risk. Both papers rely on Komlos theorem type arguments. For discrete time financial markets we discussed this type of results in Lecture 9. In the case of Israeli contingent claims, in view of the lack of convexity of the shortfall risk function, the Komlos theorem type arguments are not available. Still, we proved in Lecture 10 the existence of shortfall risk minimizing strategies also in game contingent claims relying on a dynamical programming approach. In the continuous time situation the latter is also not applicable, and so the existence of shortfall risk minimizing strategies for continuous time Israeli contingent claims in frictionless market is currently an open problem. Nevertheless, it is possible to approximate (in the spirit of Lecture 22) both the shortfall risk and minimizing investment strategies for Israeli contingent claims by corresponding objects in a sequence of binomial CRR models (see [13]) where computations can be carried out by the dynamical programming procedure of Lecture 10. On the other hand, it turns out that in markets with proportional transaction costs which cannot fall below certain positive value (as minimal buy and sell transaction costs is the standard practice in banks and broker companies) shortfall risk minimizing strategies exist for continuous time Israeli contingent claims as well. Roughly speaking, this kind of friction in the market creates enough compactness in order to replace Komlos theorem type arguments. We will describe here the main result from [14].

 Consider a complete probability space $(\Omega, \mathcal{F}, P)$ together with a standard one-dimensional Brownian motion $\{W(t)\}_{t=0}^{\infty}$, and the filtration $\mathcal{F}_t = \sigma\{W(s) : s \leq t\}$ completed by the null sets. Our BS financial market consists of a safe asset $B$ used as numeraire, hence $B \equiv 1$, and of a risky asset $S$ whose value at a time $t$ is given by

$$S_t^{(s)} = s\exp(\kappa W_t + (\vartheta - \kappa^2/2)t), \quad s > 0, \quad t \geq 0 \qquad (10.3.1)$$

where $\kappa > 0$ is called volatility and $\vartheta \in \mathbb{R}$ is another constant. By Lecture

19 we know that for the BS model there exists a unique probability measure $P^*$ equivalent to $P$ such that the stock price process $S^{(s)}$ is a $P^*$-martingale. As we saw in Lecture 19 the restriction of the probability measure $P^*$ to the $\sigma$-algebra $\mathcal{F}_t$ satisfies

$$\frac{dP^*}{dP}|_{\mathcal{F}_t} = \exp\left(-\frac{\vartheta}{\kappa}W(t) - \frac{1}{2}\left(\frac{\vartheta}{\kappa}\right)^2 t\right). \tag{10.3.2}$$

Next, let $T < \infty$ and let $C[0,T]$ be the space of all continuous functions $f : [0,T] \to \mathbb{R}$ equipped with the uniform topology. Denote by $C_{++}[0,T] \subset C[0,T]$ the subset of all strictly positive functions. Let $F, G : C[0,T] \to C[0,T]$ be continuous functions such that for any $t \in [0,T]$ and $x, y \in C[0,T]$, $G(x)_{[0,t]} = G(y)_{[0,t]}$ and $F(x)_{[0,t]} = F(y)_{[0,t]}$ if $x_{[0,t]} = y_{[0,t]}$, where $z_{[0,t]}$ is the restriction of $z \in C[0,T]$ to the interval $[0,t]$. We assume that $F \le G$ and there exist constants $C, p > 0$ for which

$$||F(x)|| + ||G(x)|| \le C(1 + ||x||^p) \tag{10.3.3}$$

where $|| \cdot ||$ denotes the supremum norm on the space $C[0,T]$. Consider a game contingent claim with a maturity date $T < \infty$ and continuous payoffs which are given by

$$Z_t^{(s)} = [F(S^{(s)})](t) \le [G(S^{(s)})](t) = Y_t^{(s)}$$

so that the seller pays to the buyer the amount $Y_\sigma^{(s)}\mathbb{I}_{\sigma<\tau} + Z_\tau^{(s)}\mathbb{I}_{\sigma\ge\tau}$ if the former cancels at a time $\sigma$ and the latter exercises at a time $\tau$.

In this model, purchases and sales of the risky asset are subject to transactions costs which are the maximum of a constant fee and a proportional transaction cost. Namely, if the investor buys (or sells) $\beta$ stocks then his transaction costs are given by

$$g(\beta, S) := \max(\delta, \mu|\beta|S)\mathbb{I}_{\beta\ne0}$$

where $\delta > 0$, $0 < \mu < 1$ are constants and $S$ is the stock price at the moment of trade. Presence of this minimal transaction cost yields that, in order to avoid infinite transaction costs, portfolios can only be rebalanced finitely many (but a random number of) times.

Next, we define hedging and shortfall risk in the above setup. A self-financing trading strategy with an initial position $(z,y)$ is a triple $\pi = (z,y,\gamma)$ where $z$ is the cash value of the portfolio at the initial time, $y$ is the number of stocks at this moment, and $\gamma = \{\gamma_t\}_{t=0}^T$ is an adapted, left-continuous, pure jump (i.e., all discontinuities are jumps) process with finite (random) number of jumps and initial value $\gamma_0 = y$. The random variable $\gamma_t$ denotes the number of shares in the portfolio $\pi$ at time $t$ before any change is made at this time (which is the reason why we assume that the process $\gamma$ is left-continuous). Observe that at time 0 the investor has the value $z + g(y, s) - ys$ on his savings account. Thus the portfolio (cash) value of a trading strategy $\pi$ at time $t$ is given by

$$X_t^\pi = z + \int_0^t \gamma_u dS_u^{(s)} + g(y, s) - g(\gamma_t, S_t^{(s)}) - \sum_{u \in [0,t)} g(\gamma_{u+} - \gamma_u, S_u^{(s)}), \quad (10.3.4)$$

where in the last sum there are only finitely many terms which are not equal to zero. A portfolio $\pi$ will be called admissible if $X_t^\pi \geq 0$ for any $t$. An investment strategy consists of a trading strategy and a cancellation time. Thus, formally an investment strategy with initial position $(z, y)$ is a pair $(\pi, \sigma)$ such that $\pi$ is an admissible portfolio and $\sigma \leq T$ is a stopping time with respect to the Brownian filtration. From (10.3.4) it follows that for an admissible portfolio $\pi$ the stochastic process $X_t^\pi$, $t \geq 0$ is a supermartingale with respect to the martingale measure $P^*$. The set of all investment strategies with an initial position $(z, y) \in \mathbb{R}_+ \times \mathbb{R}$ will be denoted by $\mathcal{A}(T, s, z, y)$. The set of all investment strategies will be denoted by $\mathcal{A}(T, s)$, where $s$ is the initial stock price and $T$ is the maturity date.

Next, we define the shortfall risk. Denote by $\mathcal{T}_{0T}$ the set of all stopping times less than or equal to $T$. For an investment strategy $(\pi, \sigma)$ the shortfall risk is defined by

$$\hat{r}(T, s, \pi, \sigma) = \sup_{\tau \in \mathcal{T}_{0T}} E_P \left( Y_\sigma^{(s)} \mathbb{I}_{\sigma < \tau} + Z_\tau^{(s)} \mathbb{I}_{\tau \leq \sigma} - X_{\sigma \wedge \tau}^\pi \right)^+$$

which is the maximal possible expectation with respect to the probability measure $P$ of the shortfall. The shortfall risk for an initial position $(z, y)$ is given by

$$r(T, s, z, y) = \inf_{(\pi, \sigma) \in \mathcal{A}(T, s, z, y)} \hat{r}(T, s, \pi, \sigma).$$

The following theorem says that for a given initial position $(z, y)$ there exists an investment strategy which minimizes the shortfall risk.

**Theorem 10.3.1.** *Let $(z, y) \in \mathbb{R}_+ \times \mathbb{R}$ be an initial position. There exists an investment strategy (may be not unique) $(\hat{\pi}, \hat{\sigma}) \in \mathcal{A}(T, s, z, y)$ such that $r(T, s, \hat{\pi}, \hat{\sigma}) = r(T, s, z, y)$.*

## 10.4 Swing options in continuous time

We will consider here continuous time swing game options (contingent claims) following [26], which generalize swing American options, and will see that in the standard Black–Scholes market the option price $V^*$ is equal to the value of a multiple stopping Dynkin game with discounted payoffs under the unique martingale measure. It seems that multiple stopping Dynkin's games were not studied before [15] and [26].

We start with a financial market which consists of $m + 1$ assets. One asset is a risk free bond with time evolution

$$B_t = B_0 e^{rt} \quad B_0 > 0 \quad r \geq 0$$

and the other $m$ assets are stocks whose prices $S_t^1, .., S_t^m$ satisfy the stochastic differential equations

$$dS_t^i = S_t^i (\mu_i dt + \sum_{j=1}^m \kappa^{ij} dW_j(t)).$$

Here $W(t) = (W_1(t), ..., W_m(t))$, $t \geq 0$ is a standard $m$-dimensional Brownian motion with continuous paths starting at 0 with a nonsingular covariance matrix $\kappa = (\kappa^{ij})$. Let $(\Omega, \mathcal{F}, P)$ be the corresponding probability space and $\{\mathcal{F}_t\}$, $t \geq 0$ be the Brownian filtration completed by all $P$-zero sets. As we know from Lecture 18, every martingale with respect to this filtration has a continuous modification since it can be represented as a stochastic integral. All processes here will be considered up to a horizon $T > 0$.

Solving the above stochastic differential equation we obtain an explicit stock prices representation by

$$S_t^i = S_0^i \exp \left( (\mu_i - \frac{1}{2} \sum_{j=1}^m (\kappa^{ij})^2)t + \sum_{j=1}^m \kappa^{ij} W_j(t) \right)$$

where $(S_0^1, .., S_0^m)$ are initial values of the stocks. Denote by $v^t$ the transpose of any vector or matrix $v$ and by $\bar{r}$ the row vector $(r, .., r) \in \mathbb{R}^m$. Set $\theta = \kappa^{-1}(\mu - \bar{r})^t$ and

$$G_t = \exp \left( -\theta \cdot (W(t) - \frac{1}{2}\theta) \right)$$

then, as we know from Lecture 21, the probability measure $\tilde{P}$ on $\mathcal{F}_T$ given by

$$\tilde{P}(A) = E(G_T \mathbf{1}_A), \quad A \in \mathcal{F}_T$$

is equivalent to $P$ and the process $\tilde{W}(t) = W(t) + \theta t$ is an $m$-dimensional Brownian motion on $(\Omega, \mathcal{F}_T, \tilde{P}_T)$. If $\tilde{S}_t = e^{-rt} S_t$ then by the Itô formula

$$d\tilde{S}_t^i = \tilde{S}_t^i \sum_{j=1}^m \kappa^{ij} d\tilde{W}_j(t),$$

and so the discounted process $\tilde{S}_t$ is a martingale with respect to $\tilde{P}$.

We consider a swing or multiple exercise option of the game type which has the $i$-th payoff, $i = 1, ..., l$ having the form

$$R_i(s,t) = Y_i(s)\mathbf{1}_{\{s<t\}} + Z_i(t)\mathbf{1}_{\{s\geq t\}}$$

where $Y_i(t) \geq Z_i(t) \geq 0$, $i = 1, ..., l$, $t \in [0,T]$ are $\mathcal{F}_t$, $t \in [0,T]$-adapted stochastic processes having a.s. continuous paths and such that for $i = 1, 2, ..., l$,

$$E\Big( \sup_{0\leq t\leq T} Y_i(t) \Big) < \infty.$$

This means that if the option seller cancels and the option buyer exercises the $i$-claim (right) at the times $s$ and $t$, respectively, then the former pays to the latter the amount $R_i(s,t)$. The setup includes also a necessary delay time $\delta > 0$ between cancellations and exercises, and so the game swing option is determined by the triple $(X_i, Y_i, \delta)$, $i = 1, ..., l$.

Let $\mathcal{T}$ be the set of all stopping times $\sigma$ with respect to the filtration $\{\mathcal{F}_t\}_{t\geq 0}$ and let $\mathcal{T}_{s,t}$ be the set of stopping times $\sigma \in \mathcal{T}$ such that $s \leq \sigma \leq t$.

**Definition 10.4.1.** For every $t \geq 0$ a *stopping strategy* with a horizon $t$ is a function

$\mathbf{t} : \{1, ..., l\} \times [0,T) \to \mathcal{T}_{0,t}$ such that $\mathbf{t}(1,0) \in \mathcal{T}_{0,t}$ for $i = 1$ and $\mathbf{t}(i,s) \in \mathcal{T}_{s+\delta, t\wedge(s+\delta)}$ for $1 < i \leq l$ where $\delta > 0$ is the delay between successive stoppings (exercises). Furthermore, the function $\mathbf{t}(i,\rho)$ is supposed to be $\mathcal{F}$-measurable.

For every $t \geq 0$ denote by $\mathcal{S}_t$ the set of all stopping strategies with a horizon $t$. Let $F : \mathcal{S}_T \times \mathcal{S}_T \to \mathcal{T}_{0,T}^l \times \mathcal{T}_{0,T}^l$ be a function defined by

$$F(\mathbf{s},\mathbf{t}) = ((\sigma_1, ..., \sigma_l), (\tau_1, ..., \tau_l))$$

where

$$\sigma_1 = \sigma_1(\mathbf{s},\mathbf{t}) = \mathbf{s}(1,0), \tau_1 = \tau(\mathbf{s},\mathbf{t}) = \mathbf{t}(1,0)$$

and for $1 < i \leq l$,

$$\sigma_i = \sigma_i(\mathbf{s},\mathbf{t}) = \mathbf{s}(i, \sigma_{i-1} \wedge \tau_{i-1}), \ \tau_i = \tau_i(\mathbf{s},\mathbf{t}) = \mathbf{t}(i, \sigma_{i-1} \wedge \tau_{i-1}).$$

Let $\mathcal{L}$ be the set of all sequences $(t_1, .., t_i)$, $1 \leq i \leq l - 1$ such that $(t_i + \delta) \wedge T \leq t_{i+1} \leq T$. We assume that $\mathcal{L}$ contains also the empty sequence $\phi$.

**Definition 10.4.2.** A portfolio strategy is a function $\pi$ on the set $\mathcal{L}$ such that

$$\pi(t_1, ..., t_i) = \{\beta_s^\pi(t_1, ..., t_i), \gamma_{1,s}^\pi(t_1, ..., t_i), ..., \gamma_{m,s}^\pi(t_1, ..., t_i)\}.$$

Here $\beta_s^\pi(t_1, ..., t_i)$ and $\gamma_{j,s}^\pi(t_1, ..., t_i)$, $1 \le j \le m$ are progressively measurable processes with respect to $\{\mathcal{F}_s\}_{t_i \le s}$ which satisfy

$$\int_{t_i}^T \beta_s^\pi(t_1, .., t_i) ds < \infty, \quad \int_{t_i}^T (\gamma_s^\pi(t_1, .., t_i) \cdot S_s)^2 ds < \infty$$

where $\cdot$ denotes the inner product in $\mathbb{R}^m$ and for every $t_i \le s \le T$,

$$X_s^\pi(t_1, ..., t_i) = X_{t_i}^\pi(t_1, ..., t_i) + \int_{t_i}^s \beta_u^\pi(t_1, ..., t_i) dB_u + \int_{t_i}^s \gamma_u^\pi(t_1, ..., t_i) \cdot dS_u$$
$$= \beta_s^\pi(t_1, ..., t_i) B_s + \gamma_s^\pi(t_1, ..., t_i) \cdot S_s.$$

Hence, $\pi(t_1, ..., t_i)$ is a self financing portfolio strategy starting at time $t_i$ with a value process $X_s^\pi(t_1, ..., t_i)$. For any portfolio strategy $\pi$ we also require that

$$G^\pi(t_1, ..., t_i) = X_{t_i}^\pi(t_1, ..., t_{i-1}) - X_{t_i}^\pi(t_1, ..., t_i) \ge 0$$

for every $(t_1, ..., t_i) \in \mathcal{L}$ which means that there is no infusion of capital but some money can be withdrawn at exercise or cancellation times for a payment to the option buyer. Note that in the case of the empty sequence we have

$$G^\pi(\phi) = \pi_0 - X_0(\phi) \ge 0.$$

In the case $i = l$ we define

$$G^\pi(t_1, ..., t_l) = X_{t_l}^\pi(t_1, ..., t_{l-1}).$$

**Definition 10.4.3.** An investment strategy is a pair $(\pi, \mathbf{s})$ of a portfolio strategy and a stopping strategy such that if for $\mathbf{t} \in \mathcal{S}_T$,

$$F(\mathbf{s}, \mathbf{t}) = ((\sigma_1, ..., \sigma_l), (\tau_1, ..., \tau_l))$$

then

(i) $X_\rho^\pi(\phi)$ is integrable for each $0 \le \rho \le \sigma_1$ and for every $\rho', \rho$ such that $0 \le \rho' \le \rho \le \sigma_1$,

$$E(e^{-\rho t} X_\rho^\pi(\phi) | \mathcal{F}_{\rho'}) = e^{-\rho' t} X_\rho^\pi(\phi);$$

(ii) $X_\rho^\pi(\sigma_1 \wedge \tau_1, ..., \sigma_i \wedge \tau_i)$ is integrable for any $1 < i \le l-1$ and $\rho$ satisfying $\sigma_i \wedge \tau_i + \delta \le \rho' \le \rho \le \sigma_{i+1}$ and

$$E(e^{-r\rho} X_\rho^\pi(\sigma_1 \wedge \tau_1, ..., \sigma_i \wedge \tau_i) | \mathcal{F}_{\rho'}) = e^{-r\rho'} X_{\rho'}^\pi(\sigma_1 \wedge \tau_1, ..., \sigma_i \wedge \tau_i);$$

(iii) For each $1 < i \le l-1$,

$$E(e^{-r(\sigma_i \wedge \tau_i + \delta)} X_{\sigma_i \wedge \tau_i + \delta}^\pi(\sigma_1 \wedge \tau_1, ..., \sigma_i \wedge \tau_i) | \mathcal{F}_{\sigma_i \wedge \tau_i})$$
$$= e^{-r(\sigma_i \wedge \tau_i)} X_{\sigma_i \wedge \tau_i}^\pi(\sigma_1 \wedge \tau_1, ..., \sigma_i \wedge \tau_i).$$

If, in addition,

(iv) $G^\pi(\sigma_1 \wedge \tau_1, ..., \sigma_i \wedge \tau_i) \geq R_i(\sigma_i, \tau_i)$ for every $0 \leq i \leq l$, then the above investment strategy is called a hedge for the swing game option $(Y_i, Z_i, \delta)$, $1 \leq i \leq l$.

**Definition 10.4.4.** We define the *fair price* of the swing game option $(Y_i, Z_i, \delta)$, $1 \leq i \leq l$ to be the infimum of all $x$ such that there exists a hedge $(x, \pi, \mathbf{s})$ with the initial capital $x$.

Let $\mathbf{s}, \mathbf{t} \in \mathcal{S}_T$ and $F(\mathbf{s}, \mathbf{t}) = ((\sigma_1, ..., \sigma_l), (\tau_1, ..., \tau_l))$. Set

$$H(\mathbf{s}, \mathbf{t}) = \tilde{E}\left(\sum_{i=1}^{l} e^{-r\sigma_i \wedge \tau_i} R_i(\sigma_i, \tau_i)\right)$$

where $\tilde{E}$ is the expectation with respect to the martingale measure $\tilde{P}$. The following result from [26] provides the fair price (from the seller's point of view) of the above swing game option.

**Theorem 10.4.1.** *For every $0 \leq t \leq T$ set*
$$Y_t^{(1)} = e^{-rt} Y_l(t), \quad Z_t^{(1)} = e^{-rt} Z_l(t),$$
$$V_t^{(1)} = ess \sup_{\tau \in \mathcal{T}_{t,T}} ess \inf_{\sigma \in \mathcal{T}_{t,T}} \tilde{E}\left(R^{(1)}(\sigma, \tau)|\mathcal{F}_t\right)$$
*and for $1 < i \leq l$,*
$$Y_t^{(i)} = e^{-rt} Y_{l-i+1}(t) + \tilde{E}\left(V_{(t+\delta)\wedge T}^{(i-1)}|\mathcal{F}_t\right),$$
$$Z_t^{(i)} = e^{-rt} Z_{l-i+1}(t) + \tilde{E}\left(V_{(t+\delta)\wedge T}^{(i-1)}|\mathcal{F}_t\right),$$
$$\text{and} \quad V_t^{(i)} = ess \sup_{\tau \in \mathcal{T}_{t,T}} ess \inf_{\sigma \in \mathcal{T}_{t,T}} \tilde{E}\left(R^{(i)}(\sigma, \tau)|\mathcal{F}_t\right)$$
*where by the definition*
$$R^{(i)}(\sigma, \tau) = Y_\sigma^{(i)} \mathbf{1}_{\{\sigma < \tau\}} + Z_\tau^{(i)} \mathbf{1}_{\{\sigma \geq \tau\}}.$$
*Then the fair price $V^*$ for the swing game option is given by*
$$V^* = V_0^{(l)} = \inf_{\mathbf{s} \in \mathcal{S}_T} \sup_{\mathbf{t} \in \mathcal{S}_T} H(\mathbf{s}, \mathbf{t}).$$
*Furthermore, let $\mathbf{s}^*, \mathbf{t}^* \in \mathcal{S}_T$ be stopping strategies given by*
$$\mathbf{s}^*(1,0) = \inf\{0 \leq t : V_t^{(l)} = Y_t^{(l)}\} \wedge T, \quad \mathbf{t}^*(1,0) = \inf\{0 \leq t : V_t^{(l)} = Z_t^{(l)}\}$$
$$\text{and} \quad \mathbf{s}^*(i, \rho) = \inf\{t \geq \rho + \delta : V_t^{(l-i+1)} = Y_t^{(l-i+1)}\} \wedge T,$$
$$\mathbf{t}^*(i, \rho) = \inf\{t \geq \rho + \delta : V_t^{(l-i+1)} = Z_t^{(l-i+1)}\}$$
*for $1 < i \leq l$ and $\rho \in \mathcal{T}$. Then for all $\mathbf{s}, \mathbf{t} \in \mathcal{S}_T$,*
$$H(\mathbf{s}^*, \mathbf{t}) \leq H(\mathbf{s}^*, \mathbf{t}^*) \leq H(\mathbf{s}, \mathbf{t}^*)$$
*and there exists a portfolio strategy $\pi^*$ such that $(V_0^{(l)}, \pi^*, \mathbf{t}^*)$ is a hedge.*

## 10.5 Model uncertainty in continuous time

As we saw it throughout this book, pricing of contingent claims depends on the stochastic process chosen to describe the evolution of stock prices as well as on other parameters of the market model. Since this information can be obtained only via some statistical analysis of the available data, it cannot describe precisely future developments in financial markets. By this reason the study of robust or model independent hedging problems received a substantial attention recently. We will describe here an approach from [16] and refer the reader also to references there.

We consider a financial market which consists of a nonrisky security whose price $B_t$ is taken as a numeraire, and so $B_t \equiv 1$, and of a risky asset whose price at time $t$ equals $S_t$. The market is considered on a time interval $[0, T]$ where $T < \infty$, $T > 0$ is a maturity date (horizon). Denote by $\mathcal{C}^+[0, T]$ the set of all strictly positive functions $f : [0, T] \to \mathbb{R}_+ = (0, \infty)$ satisfying $f_0 = 1$. The only assumption made about the process $S = \{S_t, t \in [0, T]\}$ is that all its paths belong to $\mathcal{C}^+[0, T]$.

Denote by $\mathcal{D}[0, T]$ the space of all measurable functions $v : [0, T] \to \mathbb{R}$ with the supremum norm $\|v\| = \sup_{0 \le t \le T} |v_t|$. For a given map $G : \mathcal{D}[0, T] \to \mathbb{R}$ consider a European contingent claim with the path dependent payoff $R = G(S)$ where $S = \{S_t, t \in [0, T]\}$ is viewed as an element of $\mathcal{D}[0, T]$. The proof in [10] relies on an approximation argument which requires certain regularity of the payoff function. Let $\mathcal{D}_N[0, T]$ be the set of piecewise constant functions from $\mathcal{D}[0, T]$ having at most $N$ jumps. It is assumed that there exists a constant $L > 0$ such that

$$|G(\omega) - G(\tilde{\omega})| \le L\|\omega - \tilde{\omega}\| \quad \text{for all} \quad \omega, \tilde{\omega} \in \mathcal{D}[0, T], \qquad (10.5.1)$$

i.e., $G$ is Lipschitz continuous, and

$$|G(\omega) - G(\tilde{\omega})| \le L\|\omega\| \sum_{k=1}^{N} |\Delta t_k - \Delta \tilde{t}_k| \qquad (10.5.2)$$

where $t_k, \tilde{t}_k$, $k = 1, ..., N$ are points of jumps of $\omega$ and $\tilde{\omega}$, respectively, (and we add points with zero jumps if there are less than $N$ positive jumps) and $\Delta t_k = t_k - t_{k-1}$, $\Delta \tilde{t}_k = \tilde{t}_k - \tilde{t}_{k-1}$. The setup includes also a probability measure $\mu$ on $\mathbb{R}_+$ such that for some $p > 1$,

$$\int x^p d\mu(x) < \infty. \qquad (10.5.3)$$

Since we assume only that $S$ is a positive continuous function, the integral with respect to $S$ is not defined, in general, which is needed for describing the value of a self-financing portfolio. Let $h : [0,T] \to \mathbb{R}$ be a continuous function with finite variation. Relying on the integration by parts formula we set

$$\int_0^t h_u dS_u = h_t S_t - h_0 S_0 - \int_0^t S_u dh_u$$

where the last term in the right hand side above is the Stieltjes integral.

Next, several definitions are due.

**Definition 10.5.1.** 1) A map $\phi : A \subset \mathcal{D}[0,T] \to \mathcal{D}[0,T]$ will be called progressively measurable if for any $\omega, \tilde{\omega} \in A$,

$$\omega_u = \tilde{\omega}_u \quad \text{for all} \quad u \in [0,t] \quad \text{implies that} \quad \phi(\omega)_t = \phi(\tilde{\omega})_t;$$

2) A semi-static portfolio is a pair $\pi = (g,\gamma)$ where $g \in L^1(\mathbb{R}_+,\mu)$ and $\gamma : \mathcal{C}^+[0,T] \to \mathcal{D}[0,T]$ is a progressively measurable map of bounded variation and the corresponding discounted portfolio value is given by

$$X_t^\pi(S) = g(S_T)\mathbb{I}_{\{t=T\}} + \int_0^t \gamma_u(S)dS_u, \ t \in [0,T];$$

3) A semi-static portfolio is admissible if there exists $M > 0$ such that

$$X_t^\pi(S) \geq -M(1 + \sup_{0\leq u\leq t} S_u^p) \quad \text{for any} \quad t \in [0,T] \text{ and } S \in \mathcal{C}^+[0,T];$$

4) An admissible semi-static portfolio is called super-replicating if

$$X_T^\pi(S) \geq G(S) \quad \text{for any} \quad S \in \mathcal{C}^+[0,T];$$

5) The (minimal) super-hedging cost of $G$ is defined as

$$V(G) = \inf\{\textstyle\int g d\mu : \text{ there exists } \gamma \text{ such that}$$
$$\pi = (g,\gamma) \text{ is super-replicating}\}.$$

Next we introduce a probabilistic structure. Set $\Omega = \mathcal{C}^+[0,T]$ and let $\mathbb{S} = \{\mathbb{S}_t, t \in [0,T]\}$ be the canonical process defined by $\mathbb{S}_t(\omega) = \omega_t$ for all $\omega \in \Omega$. Let $\mathcal{F}_t = \sigma\{\mathbb{S}_u, 0 \leq u \leq t\}$, $t \in [0,T]$ be the canonical filtration and $\mathcal{F} = \mathcal{F}_T$.

**Definition 10.5.2.** A probability measure $Q$ on the space $(\Omega, \mathcal{F})$ is a martingale measure if the canonical process $\mathbb{S} = \{\mathbb{S}_t, t \in [0,T]\}$ is a local martingale with respect to $Q$ and $\mathbb{S}_0 = 1$ $Q$-a.s. Denote by $\mathcal{M}_\mu$ the set of martingale measures $Q$ such that the probability distribution of $\mathbb{S}_T$ under $Q$ is equal to $\mu$.

Observe that if

$$\int x d\mu(x) = 1 \tag{10.5.4}$$

then $\mathbb{S} = \{\mathbb{S}_t, t \in [0, T]\}$ is a true martingale under each $Q \in \mathcal{M}_\mu$. The following is the main result of [10].

**Theorem 10.5.1.** *Suppose that a European contingent claim $G$ and a probability measure $\mu$ satisfy (10.5.1)–(10.5.4). Then the minimal super-hedging cost of $G$ is given by*

$$V(G) = \sup_{Q \in \mathcal{M}_\mu} E_Q(G(\mathbb{S})). \tag{10.5.5}$$

One can view the maximizer, if exists, of the expression $\sup_{Q \in \mathcal{M}_\mu} E_Q(G(\mathbb{S}))$ as an optimal transport of the initial probability measure $\nu = \delta_1$ concentrated at 1 to the final distribution $\mu$ with the additional constraint that the connection should be a martingale which corresponds to the Kantorovich generalization of the Monge–Kantorovich mass transport problem.

# Chapter 11

# Solutions of Exercises

## 11.1 Solutions to Chapter 1

### 11.1.1 *Martingales and stopping times*

1) Clearly, $M_n$ is $\mathcal{F}_n$-measurable for any $n \geq 0$ and

$$E(M_{n+1}|\mathcal{F}_n) = M_n e^{-\frac{1}{2}\alpha_{n+1}^2 \sigma_{n+1}^2} E(e^{\alpha_{n+1}X_{n+1}})$$

where we used that $X_1, ..., X_n$ are $\mathcal{F}_n$-measurable and $X_{n+1}$ is independent of $\mathcal{F}_n$. Since $X_{n+1}$ is a centered Gaussian random variable with the variance $\sigma_{n+1}^2$ we obtain

$$\begin{aligned}
E(e^{\alpha_{n+1}X_{n+1}}) &= (2\pi\sigma_{n+1}^2)^{-\frac{1}{2}} \int_{-\infty}^{\infty} e^{\alpha_{n+1}x} e^{-\frac{x^2}{2\sigma_{n+1}^2}} dx \\
&= e^{\frac{1}{2}\alpha_{n+1}^2 \sigma_{n+1}^2},
\end{aligned}$$

and so $E(M_{n+1}|\mathcal{F}_n) = M_n$, i.e., $M_n$, $n \geq 0$ is a martingale.

2) Again, $L_n$ is $\mathcal{F}_n$-measurable and setting $Y_n = \sum_{k=1}^n X_k$ we obtain

$$\begin{aligned}
E(L_{n+1}|\mathcal{F}_n) &= E\big(Y_n^3 + 3Y_n^2 X_{n+1} + 3Y_n X_{n+1}^2 + X_{n+1}^3 \\
&\quad -3(Y_n + X_{n+1})(\sum_{k=1}^{n+1} \sigma_k^2)|\mathcal{F}_n\big) = Y_n^3 - 3Y_n \sum_{k=1}^{n+1} \sigma_k^2 \\
&\quad +3Y_n^2 EX_{n+1} + 3Y_n EX_{n+1}^2 + EX_{n+1}^3 - 3(\sum_{k=1}^{n+1} \sigma_k^2)EX_{n+1} = L_n
\end{aligned}$$

since each $X_i$ has a symmetric distribution, and so $EX_i = EX_i^3 = 0$.

3) (a) Consider $M_n = M_n(\alpha_1, ..., \alpha_n)$ as a function of parameters $\alpha_1, ..., \alpha_n$ and then

$$L_n = \frac{\partial^{3n}}{\partial^3 \alpha_1 \partial^3 \alpha_2 ... \partial^3 \alpha_n} M_n|_{\alpha_1=0,...,\alpha_n=0},$$

i.e., we differentiate $M_n$ in each parameter 3 times at zero. Since $M_n$, $n \geq 0$ is a martingale then $L_n$, $n \geq 0$ is a martingale once we justify the differentiation of $M_n$ in parameters under the conditional expectation.

(b) Let $X_1, X_2, \ldots$ be independent random variables having moment generating functions $m_k(\alpha) = E\exp(\alpha X_k)$. Let $S_n = \sum_{k=1}^{n} \alpha_k X_k$ for some constants $\alpha_k$. Then

$$M_n = \frac{e^{S_n}}{\prod_{k=1}^{n} m_k(\alpha_k)}, \quad n = 1, 2, \ldots$$

is a martingale. Indeed,

$$E(M_{n+1}|\mathcal{F}_n) = e^{S_n}\left(\prod_{k=1}^{n} m_k(\alpha_k)\right)^{-1} E e^{\alpha_{n+1} X_{n+1}} = M_n.$$

4) We have for $\mathcal{F}_n = \sigma\{X_1, \ldots, X_n\}$,

$$\{\tau_1 = n\} = \{X_n \in (a,b)\} \cap \left((\cap_{k=1}^{n-1}\{X_k \le a\}) \cup (\cup_{k=1}^{n-1}\{X_k \ge b\})\right) \in \mathcal{F}_n$$

and

$$\{\tau_2 = n\} = \{X_n \in (a,b)\} \cap (\cap_{k=1}^{n-1}\{X_k \ge b\}) \in \mathcal{F}_n.$$

The random variable $Y_n = \max_{n \le i \le 2n} X_k$ is $\mathcal{F}_{2n}$-measurable but, in general, it is not $\mathcal{F}_n$-measurable, and so $\tau_3$ is a stopping time with respect to the filtration $\mathcal{G}_n = \mathcal{F}_{2n}$, $n \ge 1$ but not, in general, with respect to the filtration $\mathcal{F}_n$, $n \ge 1$.

### 11.1.2 *Optimal stopping*

5) There is a big literature concerning various aspects of the secretary problem (usually in infinite time) and we refer the reader to [19], [18] and others for surveys. For a finite $N$ the answer to the problem is obtained via general formulas of the single player optimal stopping problem described in Lecture 3.

6) Suppose we have $N$ candidates for the secretary position which have ranks ranging from 1 to $K$. Let $R_i$ be the rank of the candidate arriving at the day $i$, $i = 1, 2, \ldots, N$ and we assume that $R_1, R_2, \ldots, R_N$ are i.i.d. random variables taking on values from 1 to $K$ (not necessarily different). We take $\sigma$-algebras $\mathcal{F}_n = \sigma\{R_1, \ldots, R_n\}$ generated by $R_1, \ldots, R_n$ and consider the following setup. If the manager of the company A accepts a candidate with a rank $R$ we view it as a payment by the company B to the company A since we interpret the rank of a new secretary as a measure of damage to profits of the rival company B. If the manager of the company B stops the flow of candidates then this costs the company B some money which we value by 1 and interpret it as an additional payment of B to A. Thus we arrive at a Dynkin game with the payoff

$$R(m, n) = (R_m + 1)\mathbb{I}_{m<n} + R_m \mathbb{I}_{m \ge n}$$

provided the manager of company B stops the flow of candidates on the day $m$ while the manager of the company A hires the candidate arriving on the day $n$. The value of this game and the optimal stopping times can be computed by means of the general formulas described in Lecture 3.

## 11.2 Solutions to Chapter 2

1) Here $P$ itself is the unique martingale measure, $B_n \equiv 1$ and as we proved in Lecture 5 the fair price of this American contingent claim is given by the formula

$$V = \max_{0 \le \tau \le 3} E(1 + \rho_\tau)^2$$

where $E$ is the expectation with respect to $P$, max is taken over (finitely many) stopping times and we take into account that $B_n \equiv 1$. By the dynamical programming (backward induction) formulas

$V_3 = (1+\rho_3)^2$, $V_n = \max\big((1+\rho_n)^2,\ E(V_{n+1}|\mathcal{F}_n)\big)$, $n = 2, 1, 0$ and $V_0 = V$.

Since $\rho_3$ is independent of $\mathcal{F}_2$ we have

$$V_2 = \max\big((1 + \rho_2)^2,\ E((1 + \rho_3)^2|\mathcal{F}_2))\big) = \max\big((1 + \rho_2)^2,\ E(1 + \rho_3)^2\big)$$
$$= \max\big((1 + \rho_2)^2,\ \tfrac{5}{4}\big).$$

Similarly,

$$V_1 = \max\big((1 + \rho_1)^2,\ E(\max((1 + \rho_2)^2,\ \tfrac{5}{4}))\big) = \max\big((1 + \rho_1)^2,\ \tfrac{7}{4}\big).$$

Finally,

$$V_0 = \max\big(1,\ E(\max\big((1 + \rho_1)^2,\ \tfrac{7}{4}\big))\big) = 2$$

which is the fair price of this American contingent claim.

In order to find the optimal stopping time $\tau^*$ of the buyer

$$\tau^* = \min\{n \ge 0 : (1 + \rho_n)^2 = V_n\}$$

we introduce the probability space $(\Omega, P)$ in the form $\Omega = \{\omega = (\omega_1, \omega_2, \omega_3) : \omega_i = 1/2 \text{ or } = -1/2\}$ so that $\rho_i(\omega) = \omega_i$ for $i = 1, 2, 3$ if $\omega = (\omega_1, \omega_2, \omega_3)$, $\rho_0 \equiv 0$ and $P(\omega) = 1/8$ for each $\omega \in \Omega$. Then $\tau^*(1/2, \omega_2, \omega_3) = 1$ for any $\omega_2$ and $\omega_3$ since $V_1(\omega) = (1 + \rho_1(\omega))^2$ when $\omega = (1/2, \omega_2, \omega_3)$. Similarly, $\tau^*(-1/2, 1/2, \omega_3) = 2$ for any $\omega_3$ since $V_2(\omega) = (1+\rho_2(\omega))^2$ when $\omega = (-1/2, 1/2, \omega_3)$. Finally, $\tau^*(-1/2, -1/2, \omega_3) = 3$ for any $\omega_3$ since always $V_3 = (1+\rho_3)^2$. We check $E(1+\rho_{\tau^*})^2 = \tfrac{1}{2} \cdot \tfrac{9}{4} + \tfrac{1}{4} \cdot \tfrac{9}{4} + \tfrac{1}{8} \cdot \tfrac{9}{4} + \tfrac{1}{8} \cdot \tfrac{1}{4} = 2$ as required.

In order to construct a hedging portfolio strategy of the seller with the initial capital equal to $V$, recall first that

$$V_n = \max_{n \leq \tau \leq N} E((1 + \rho_\tau)^2 | \mathcal{F}_n), \ n = 0, 1, 2, 3$$

which is a supermartingale and by the explicit formulas from Lecture 1 its supermartingale decomposition has the form $V_n = M_n - A_n$, $n = 0, 1, 2, 3$ where the martingale $\{M_n\}_{n=0}^3$ has the form $M_0 = V_0$ and for $1 \leq n \leq 3$,

$$M_n = V_0 + \sum_{k=1}^n (V_k - E(V_k | \mathcal{F}_{k-1})) = V_0 + \sum_{k=1}^n (V_k - EV_k)$$
$$= V_0 + \sum_{k=1}^n (\max((1 + \rho_k)^2, a_{k+1}) - a_k)$$

with $a_4 = 0$, $a_3 = \frac{5}{4}$, $a_2 = \frac{7}{4}$ and $a_1 = 2$. Here $A_n = \sum_{k=1}^n E(V_{k-1} - V_k | \mathcal{F}_{k-1})$ is a non-decreasing process but it does not play role in computations here.

Recall, that according to the prescription from Lecture 5 the required trading strategy $(\beta_k, \gamma_k)_{k=1}^3$ comes from the representation

$$M_n = M_0 + \sum_{k=1}^n \gamma_k \rho_k \prod_{i=1}^{k-1} (1 + \rho_i)$$

which will determine $\gamma_k$, $k = 1, 2, 3$ and then $\beta_k$, $k = 1, 2, 3$ will be given by $\beta_k = M_{k-1} - \gamma_k \prod_{i=1}^{k-1}(1 + \rho_i)$ where we set $\prod_{i=1}^0 = 1$. Using explicit formulas in the martingale representation lemma we obtain that $\gamma_k = 2H_k \prod_{i=1}^{k-1}(1 + \rho_i)^{-1}$ where $H_k(\omega_1, ..., \omega_{k-1}) = M_k(\omega_1, ..., \omega_{k-1}, \frac{1}{2}) - M_{k-1}(\omega_1, ..., \omega_{k-1})$. Hence, $\gamma_k = 2(\frac{9}{4} - a_k) \prod_{i=1}^{k-1}(1 + \rho_i)^{-1}$ for $k = 1, 2, 3$ where, again, $\prod_{i=1}^0 = 1$, and so,

$$\gamma_1 = \frac{1}{2}, \ \gamma_2 = (1 + \rho_1)^{-1} \ \text{and} \ \gamma_3 = 2(1 + \rho_1)^{-1}(1 + \rho_2)^{-1}.$$

The same can be obtained directly from the above formulas writing $\gamma_k = (M_k - M_{k-1})\rho_k^{-1} \prod_{i=1}^{k-1}(1 + \rho_i)$ and taking into account that for $k = 1, 2, 3$,

$$(M_k - M_{k-1})\rho_k^{-1} = \rho_k^{-1}(\max((1 + \rho_k)^2, a_{k+1}) - a_k) = 2(\frac{9}{4} - a_k)$$

which is verified directly.

2) We have here a game contingent claim in the CRR binomial market model with the horizon $N = 3$, the bond and the stock prices at time $n$ are given by $B_n = 2^n$ and $S_n = \prod_{i=1}^n (1 + \rho_i)$, respectively, where $\rho_1, \rho_2, ...$ are i.i.d. random variables such that $\rho_i = 2$ with probability $1/4$ and $\rho_i = -\frac{1}{2}$ with probability $3/4$. We have here also $B_0 = S_0 = 1$ and $\rho_0 = 0$. The payoff function is given by

$$R(m, n) = Y_m \mathbb{I}_{m<n} + Z_n \mathbb{I}_{n \leq m} = (\rho_n^2 + 1)\mathbb{I}_{m<n} + \rho_n^2 \mathbb{I}_{m \geq n}.$$

Let $P^*$ be the martingale measure (which is unique here) and $P^*\{\rho_1 = 2\} = p^*$. By the equation $E_{P^*}(\frac{1+\rho_1}{2}) = 1$ we have $3p^* + \frac{1}{2}(1-p^*) = 2$, and so $p^* = \frac{3}{5}$.

By the formulas we proved, the fair price of this game contingent claim is given by the formula

$$V = \min_{0 \leq \sigma \leq 3} \max_{0 \leq \tau \leq 3} E_{P^*}\left(\frac{R(\sigma,\tau)}{B_{\sigma \wedge \tau}}\right),$$

where $\sigma, \tau$ are stopping times, and since we have here the finite probability space we have only finitely many stopping times between 0 and 3, and so we can write min and max. On the other hand, we proved the backward induction (dynamical programming) formulas for Dynkin's game values so that if

$$V_n = \min_{n \leq \sigma \leq 3} \max_{n \leq \tau \leq 3} E_{P^*}\left(\frac{R(\sigma,\tau)}{B_{\sigma \wedge \tau}} \middle| \mathcal{F}_n\right),$$

then $V_0 = V$, $V_3 = \frac{\rho_3^2}{2^3} = \frac{\rho_3^2}{8}$ and

$$V_n = \min\left(\frac{\rho_n^2 + 1}{2^n}, \max\left(\frac{\rho_n^2}{2^n}, E_{P^*}(V_{n+1}|\mathcal{F}_n)\right)\right), \quad n = 2, 1, 0.$$

Recall that $\{\mathcal{F}_n\}$ is the filtration such that $\mathcal{F}_n$ is generated by $S_1, S_2, ..., S_n$, i.e., by $\rho_1, \rho_2, ..., \rho_n$.

Thus,

$$V_2 = \min\left(\frac{\rho_2^2 + 1}{4}, \max\left(\frac{\rho_2^2}{4}, E_{P^*}\left(\frac{\rho_3^2}{8} \middle| \mathcal{F}_2\right)\right)\right).$$

Now, by independency of $\rho_3$ from $\mathcal{F}_2$,

$$E_{P^*}\left(\frac{\rho_3^2}{8} \middle| \mathcal{F}_2\right) = E_{P^*}\left(\frac{\rho_3^2}{8}\right) = \left(\frac{1}{2} \cdot \frac{3}{5} + \frac{1}{32} \cdot \frac{2}{5}\right) = \frac{5}{16}.$$

Since $\frac{\rho_2^2+1}{4} \geq \frac{5}{16}$ then

$$V_2 = \max\left(\frac{\rho_2^2}{4}, \frac{5}{16}\right),$$

and so $V_2 = 1$ with $P^*$-probability $\frac{3}{5}$ and $V_2 = \frac{5}{16}$ with $P^*$-probability $\frac{2}{5}$ and $V_2$ is independent of $\mathcal{F}_1$ (since $\rho_2$ is independent of $\mathcal{F}_1$).

Now,

$$V_1 = \min\left(\frac{\rho_1^2 + 1}{2}, \max\left(\frac{\rho_1^2}{2}, E_{P^*}(V_2|\mathcal{F}_1)\right)\right)$$

and

$$E_{P^*}(V_2|\mathcal{F}_1) = E_{P^*}(V_2) = \frac{3}{5} + \frac{1}{8} = \frac{29}{40}.$$

Next, $\max(\frac{\rho_1^2}{2}, \frac{29}{40})$ equals either 2 with $P^*$-probability $\frac{3}{5}$ (if $\rho_1 = 2$) or equals $\frac{29}{40}$ with $P^*$-probability $\frac{2}{5}$ (if $\rho_1 = -\frac{1}{2}$). Hence,

$$V_1 = \min\left(\frac{\rho_1^2 + 1}{2}, \max(\frac{\rho_1^2}{2}, \frac{29}{40})\right)$$

and $V_1$ equals either 2 with $P^*$-probability $\frac{3}{5}$ or it equals $\frac{5}{8}$ with $P^*$-probability $\frac{2}{5}$.

Finally,

$$V_0 = \min(1, \max(0, E_{P^*}(V_1|\mathcal{F}_0)))$$

and $E_{P^*}(V_1|\mathcal{F}_0) = E_{P^*}(V_1) = 2 \cdot \frac{3}{5} + \frac{5}{8} \cdot \frac{2}{5} = \frac{29}{20}$. Hence, $V = V_0 = 1$. So, in fact, the seller can cancel at time 0, pay the penalty 1 which is the price of this derivative, as we computed, it and this is his hedging strategy with the initial capital 1.

## 11.3 Solutions to Chapter 3

1) Let $Q_0 \neq Q_1$ be two martingale measures. Set $Q_\alpha = \alpha Q_1 + (1-\alpha)Q_2$ which are probability measures for each $\alpha \in [0,1]$. Then $Q_{\alpha_1} \neq Q_{\alpha_2}$ if $\alpha_1 \neq \alpha_2$ and $Q_\alpha$ is a martingale measure for any $\alpha \in [0,1]$. Indeed, if for $\alpha_1 \geq \alpha_2$,

$$Q_{\alpha_1} = \alpha_1 Q_1 + (1 - \alpha_1)Q_2 = Q_{\alpha_2} = \alpha_2 Q_1 + (1 - \alpha_2)Q_2$$

then $(\alpha_1 - \alpha_2)Q_1 = (\alpha_1 - \alpha_2)Q_2$ and since $Q_1 \neq Q_2$ we must have $\alpha_1 = \alpha_2$. Since both $Q_1$ and $Q_2$ are equivalent to $P$ then $Q_\alpha$ is also equivalent to $P$ with the Radon–Nikodym derivative

$$\frac{dQ_\alpha}{dP} = \alpha\frac{dQ_1}{dP} + (1 - \alpha)\frac{dQ_2}{dP}.$$

Next, let $\{S_n\}_{n=0}^N$ be the discounted stock value which is a martingale with respect to both $Q_1$ and $Q_2$ and let $\{\mathcal{F}_n\}_{n=0}^N$ be the filtration generated by $\{S_n\}_{n=0}^N$. Then for any $\Gamma \in \mathcal{F}_n$,

$$\int_\Gamma E_{Q_\alpha}(S_{n+1}|\mathcal{F}_n)dQ_\alpha = \int_\Gamma S_{n+1}dQ_\alpha$$
$$= \alpha \int_\Gamma S_{n+1}dQ_1 + (1 - \alpha) \int_\Gamma S_{n+1}dQ_2 = S_n,$$

and so $E_{Q_\alpha}(S_{n+1}|\mathcal{F}_n) = S_n$ a.s. for each $n = 0, 1, ..., n - 1$, i.e., $\{S_n\}_{n=0}^N$ is a martingale with respect to $Q_\alpha$ as well.

2) Since $\rho_1, \rho_2, ...$ are i.i.d. random variables on a probability space $(\Omega, \mathcal{F}, P)$, the interest rate is zero $r = 0$, so that $B_n \equiv B_0$ for all $n$, and the stock evolution is given by $S_n = S_0 \prod_{i=1}^n (1 + \rho_i)$, $P^*$ is a martingale

measure if and only if $E_{P^*}\rho_1 = 0$. Let $A_1 = \{\rho_1 < 0\}$, $A_2 = \{0 \leq \rho_1 < a\}$ and $A_3 = \{\rho_1 \geq a\}$ which are disjoint sets (events) and we know that $P(A_1), P(A_2), P(A_3) > 0$. Define a measure $P^*$ by

$$P^*(\Gamma) = p_1^* P\{A_1 \cap \Gamma\} + p_2^* P\{A_2 \cap \Gamma\} + p_3^* P\{A_3 \cap \Gamma\}$$

where $\infty > p_1^*, p_2^*, p_3^* > 0$ and setting $b_i = P(A_i)$, $i = 1, 2, 3$ we obtain the condition $p_1^* b_1 + p_2^* b_2 + p_3^* b_3 = 1$ which ensures that $P^*$ is a probability measure. We obtain by the definition of $P^*$ that for any event $\Gamma$,

$$\max(p_1^*, p_2^*, p_3^*) P(\Gamma) \geq P^*(\Gamma) \geq \min(p_1^*, p_2^*, p_3^*) P(\Gamma),$$

and so the assumption that $\infty > p_1^*, p_2^*, p_3^* > 0$ gives that $P^*$ is equivalent to $P$. In order to ensure that $P^*$ is a martingale measure we need that

$$0 = E_{P^*}\rho_1 = p_1^* c_1 + p_2^* c_2 + p_3^* c_3$$

where $c_1 = \int_{A_1} \rho_1 dP < 0$, $c_2 = \int_{A_2} \rho_1 dP \geq 0$ and $c_3 = \int_{A_3} \rho_1 dP > 0$ since $A_1, A_2, A_3$ have positive probability. Thus we have two equations

$$p_1^* b_1 + p_2^* b_2 + p_3^* b_3 = 1 \quad \text{and} \quad p_1^* c_1 + p_2^* c_2 + p_3^* c_3 = 0$$

with the constraints $\infty > p_1^*, p_2^*, p_3^* > 0$. We obtain

$$p_1^* = 1 - p_2^* b_2 - p_3^* b_3 \quad \text{and so} \quad p_2^* = -\frac{c_1 + p_3^*(c_3 - c_1 b_3)}{c_2 - c_1 b_2}$$

and thus

$$p_1^* = 1 + b_2 \frac{c_1 + p_3^*(c_3 - c_1 b_3)}{c_2 - c_1 b_2} - p_3^* b_3 = \frac{c_2 + p_3^*(b_2 c_3 - b_3 c_2)}{c_2 - c_1 b_2}.$$

The denominators above are strictly positive since $c_1 < 0$, $c_2 \geq 0$ and $b_2 > 0$. If

$$0 < p_3^* < \frac{-c_1}{c_3 - c_1 b_3}$$

then $p_2^* > 0$ by the above formula. If $c_2 = 0$ then

$$p_1^* = \frac{p_3^* b_2 c_3}{-c_1 b_2} = \frac{p_3^* c_3}{-c_1} > 0$$

provided $p_3^* > 0$. If $c_2 > 0$ then $p_1^* > 0$ always if $b_2 c_3 - b_3 c_2 \geq 0$ while if $b_2 c_3 - b_3 c_2 < 0$ then $p_1^* > 0$ provided

$$p_3^* < \frac{c_2}{b_3 c_2 - b_2 c_3}.$$

Thus, for any positive $p_3^*$ small enough we can find positive $p_1^*$ and $p_2^*$ solving two above equations, and so they admit infinitely many solutions which yield infinitely many product martingale measures $P^* = \{p_1^*, p_2^*, p_3^*\}^{\mathbb{N}}$.

## 11.4   Solutions to Chapter 4

We have a trinomial two stage market $N = 2$ with $B_n \equiv 1$, $S_n = S_0 \prod_{k=1}^{n}(1 + \rho_k)$, $\rho_1, \rho_2, \ldots$ i.i.d., $S_0 = 1$, $\rho_i$ equals $-\frac{1}{2}$ or $\frac{1}{2}$ or $1$ each with probability $\frac{1}{3}$. We consider two payoffs $f_n = (2 - S_n)^+$-put option and $f_n = (S_n - 2)^+$-call option.

1) A probability measure $\tilde{P}$ is martingale if $E_{\tilde{P}}\rho_i = 0$, i.e., $-\frac{1}{2}p_1 + \frac{1}{2}p_2 - p_1 - p_2 + 1 = 0$ where $p_3 = 1 - p_1 - p_2$ and $p_1 = \tilde{P}\{\rho_i = -\frac{1}{2}\}$, $p_2 = \tilde{P}\{\rho_i = \frac{1}{2}\}$ and $p_3 = \tilde{P}\{\rho_i = 1\}$. Hence, $p_2 = 2 - 3p_1$ and we have constraints $0 \leq p_1 \leq 1$, $0 \leq 2 - 3p_1 \leq 1$ and $0 \leq 1 - p_1 - p_2 \leq 1$, i.e., $0 \leq 2p_1 - 1 \leq 1$. Thus we obtain the condition $\frac{2}{3} \geq p_1 \geq \frac{1}{2}$ and $p_2, p_3$ are given by the above formulas.

2) Next, we consider the European put option. Then $E_{\tilde{P}} f_2 = E_{\tilde{P}}(2 - S_2)^+$. It is easy to see that $S_2 = \frac{1}{4}$ with probability $p_1^2$, $S_2 = \frac{3}{4}$ with probability $2p_1 p_2$, $S_2 = \frac{9}{4}$ with probability $p_2^2$, $S_2 = 1$ with probability $2p_1 p_3$, $S_2 = 4$ with probability $p_3^2$ and $S_2 = 3$ with probability $2p_2 p_3$. Hence,

$$E_{\tilde{P}}(2 - S_2)^+ = \frac{7}{4}p_1^2 + \frac{5}{2}p_1 p_2 + 2p_1 p_3 = p_1\left(3 - \frac{7}{4}p_1\right)$$

using the formulas above for $p_2$ and $p_3$ for any martingale measure $\tilde{P}$. Now we have to maximize this expression taking into account the constraint $\frac{1}{2} \leq p_1 \leq \frac{2}{3}$ from above. Under these constraints the above quadratic expression has the maximum at $p_1 = \frac{2}{3}$. Then $p_2 = 0$ and $p_3 = \frac{1}{3}$. Taking $\tilde{P}$ with such $p_1, p_2, p_3$ we obtain the superhedging price $V = E_{\tilde{P}}(2 - S_2)^+ = \frac{2}{3}(3 - \frac{7}{6}) = \frac{11}{9}$. Then we can make a computation of the self-financing hedging strategy with the initial capital $V$ since $p_2 = 0$ and we have here, in fact, a binomial (complete) Cox–Ross–Rubinstein market where we can use the explicit martingale representation constructed in the corresponding lemma.

In the call option case:

$$E_{\tilde{P}} f_2 = E_{\tilde{P}}(S_2 - 2)^+ = \tfrac{1}{4}(2 - 3p_1)^2$$
$$+2(2p_1 - 1)^2 + 2(2 - 3p_1)(2p_1 - 1) = -\tfrac{7}{4}p_1^2 + 3p_1 - 1.$$

Finding the maximum of this quadratic expression in $p_1$ under the constraint $\frac{1}{2} \leq p_1 \leq \frac{2}{3}$ from above gives $p_1 = \frac{2}{3}$, and so $p_2 = 0$ and $p_3 = \frac{1}{3}$. Thus the superhedging price here is $V = E_{\tilde{P}}(S_2 - 2)^+ = -\frac{7}{9} + 2 - 1 = \frac{2}{9}$. Again we arrive at a binomial (complete) Cox–Ross–Rubinstein market and we can find a corresponding self-financing hedging trading strategy with the initial capital $V$ using the explicit martingale representation in the corresponding lemma.

3) American options case.

For each martingale measure set

$$V_n^{\tilde{P}} = \max_{n \leq \tau \leq N} E_{\tilde{P}}(f_\tau | \mathcal{F}_n)$$

where the maximum is taken over finitely many stopping times ( since we have here a finite probability space). Then, as we proved in the optimal stopping section,

$$V_n^{\tilde{P}} = \max(f_n, E_{\tilde{P}}(V_{n+1}^{\tilde{P}} | \mathcal{F}_n)), \ n = 0, 1, ..., N - 1$$

with $V_N = f_N$. Since we have here $N = 2$ then $V_2^{\tilde{P}} = f_2 = (2 - S_2)^+$ in the put option case and $= (S_2 - 2)^+$ in the call option case.

Next, we deal with the put option case. Since $S_2 = (1 + \rho_2)S_1$ with $\rho_2$ independent of $\mathcal{F}_1$ while $S_1$ is measurable with respect to $\mathcal{F}_1$ then by properties of the conditional expectation and using the above formulas for $p_1, p_2, p_3$ corresponding to a martingale measure we obtain

$$E_{\tilde{P}}(V_2^{\tilde{P}} | \mathcal{F}_1) = (2 - \frac{1}{2}S_1)^+ p_1 + (2 - \frac{3}{2}S_1)^+ (2 - 3p_1) + (2 - 2S_1)^+ (2p_1 - 1).$$

Since $f_1 = (2 - S_1)^+$, $S_0 = 1$ and so $S_1 = \frac{1}{2}$ with probability $p_1$, $S_1 = \frac{3}{2}$ with probability $(2 - 3p_1)$ and $S_1 = 2$ with probability $(2p_1 - 1)$, we obtain the following. If $S_1 = \frac{1}{2}$ then by the above

$$E_{\tilde{P}}(V_2^{\tilde{P}} | \mathcal{F}_1) = \frac{3}{2} \text{ and } f_1 = \frac{3}{2},$$

i.e., in this case $V_1^{\tilde{P}} = \frac{3}{2}$ which happens with probability $p_1$. If $S_1 = \frac{3}{2}$ then

$$E_{\tilde{P}}(V_2^{\tilde{P}} | \mathcal{F}_1) = \frac{5}{4}p_1 \text{ and } f_1 = \frac{1}{2},$$

and so $V_1^{\tilde{P}} = \max(\frac{1}{2}, \frac{5}{4}p_1) = \frac{5}{4}p_1$ since $p_1 \geq \frac{1}{2}$, which happens with probability $(2 - 3p_1)$. Finally, if $S_1 = 2$ then

$$E_{\tilde{P}}(V_2^{\tilde{P}} | \mathcal{F}_1) = p_1 \text{ and } f_1 = 0,$$

and so $V_1^{\tilde{P}} = \max(0, p_1) = p_1$ which happens with probability $(2p_1 - 1)$.

Since $f_0 = 1$ and $\mathcal{F}_0$ is the trivial $\sigma$-algebra, we obtain

$$V^{\tilde{P}} = V_0^{\tilde{P}} = \max(1, E_{\tilde{P}}(V_1^{\tilde{P}})).$$

By the above,

$$E_{\tilde{P}}(V_1^{\tilde{P}}) = \frac{3}{2}p_1 + \frac{5}{4}p_1(2 - 3p_1) + p_1(2p_1 - 1) = p_1(3 - \frac{7}{4}p_1).$$

Hence,

$$V^{\tilde{P}} = V^{\check{P}} = \max(1, p_1(3 - \frac{7}{4}p_1)) = p_1(3 - \frac{7}{4}p_1)$$

since $\frac{1}{2} \leq p_1 \leq \frac{2}{3}$. The supremum of this expression under the above constraint is $\frac{11}{9}$ for $p_1 = \frac{2}{3}$. Thus, in this case the superhedging prices of the American and European put options are equal. Now we can take $p_1 = \frac{2}{3}$, $p_2 = 0$ and $p_3 = \frac{1}{3}$, which yields a binomial Cox–Ross–Rubinstein market, and compute a self-financing hedging trading strategy with the initial capital equal to 1 in this market.

Next, we deal with American call option. We have similarly to the above,

$$E_{\tilde{P}}(V_2^{\tilde{P}}|\mathcal{F}_1) = (\frac{1}{2}S_1 - 2)^+ p_1 + (\frac{3}{2}S_1 - 2)^+(2 - 3p_1) + (2S_1 - 2)^+(2p_1 - 1)$$

and $f_1 = (S_1 - 2)^+$. Now, If $S_1 = \frac{1}{2}$ then

$$E_{\tilde{P}}(V_2^{\tilde{P}}|\mathcal{F}_1) = 0 \text{ and } f_1 = 0,$$

i.e., in this case $V_1^{\tilde{P}} = 0$ which happens with probability $p_1$. If $S_1 = \frac{3}{2}$ then

$$E_{\tilde{P}}(V_2^{\tilde{P}}|\mathcal{F}_1) = \frac{1}{4}(2 - 3p_1) + (2p_1 - 1) = \frac{5}{4}p_1 - \frac{1}{2} \text{ and } f_1 = 0,$$

and so $V_1^{\tilde{P}} = \max(0, \frac{5}{4}p_1 - \frac{1}{2}) = \frac{5}{4}p_1 - \frac{1}{2}$ since $p_1 \geq \frac{1}{2}$, which happens with probability $(2 - 3p_1)$. Finally, if $S_1 = 2$ then

$$E_{\tilde{P}}(V_2^{\tilde{P}}|\mathcal{F}_1) = (2 - 3p_1) + 2(2p_1 - 1) = p_1 \text{ and } f_1 = 0,$$

and so $V_1^{\tilde{P}} = \max(0, p_1) = p_1$ which happens with probability $(2p_1 - 1)$.
    Now $f_0 = 0$ and

$$E_{\tilde{P}}V_1^{\tilde{P}} = (2 - 3p_1)(\frac{5}{4}p_1 - \frac{1}{2}) + (2p_1 - 1)p_1 = -\frac{7}{4}p_1^2 + 3p_1 - 1.$$

Now maximizing the last expression in $p_1 \in [1/2, 2/3]$ we obtain $p_1 = \frac{2}{3}$ and the superhedging price equals $V = \frac{2}{9}$. We obtained the same result as for the European call option and this is not by chance. This is true in general since the payoff function of an American call option is a submartingale with respect to any martingale measure (check!), and so its expectation is non-decreasing function of time which means that it does not make sense to exercise such an option earlier than the expiration time (horizon) implying that both options have the same fair price (and there was no need for an additional computation for this call American option).

4) In the game option case the payoff function has the form

$$R(m,n) = (1 + (2 - S_m)^+)\mathbb{I}_{m<n} + (2 - S_n)^+\mathbb{I}_{n\leq m}$$

in the put option case and

$$R(m,n) = (1 + (S_m - 2)^+)\mathbb{I}_{m<n} + (S_n - 2)^+\mathbb{I}_{n\leq m}$$

in the call option case. Observe that there are exactly 9 stopping times between 0 and 2 in our trinomial market. For each one of them $\sigma$ considered as a cancellation time by the seller we can do computations as above of the superhedging price of the American option with the payoff function $f_n = R(\sigma, n)$ and then take the minimum over these $\sigma$'s.

## 11.5 Solutions to Chapter 5

### 11.5.1 *Shortfall minimization for European and American options via recursive relations*

We start with the European contingent claims with a horizon $N < \infty$ and a payoff $f_N \geq 0$. Recall, that our goal is to minimize the shortfall risk $E\ell(f_N - X_N^\pi)$ where $\ell \geq 0$ is a convex function with $\ell(x) = 0$ for $x \leq 0$, $X_N^\pi$ is the portfolio value at the time $N$ corresponding to a strategy $\pi$ and $E\ell(X_N) < \infty$. The stock evolution and other definitions until the introduction of the set $\mathcal{A}_n(X, \rho^{(n)})$ of possible portfolio values at the time $n+1$ provided this value was $X$ at the time n remains the same as in Lecture 10. The difference appears in the corresponding recursive formulas. We set

$$I_N(x^{(N)}, y, z) = J_N(x^{(N)}, y) = \ell(f_N(x^{(N)}) - y),$$

$x^{(N)} \in (-1,\infty)^N$, $y \geq 0$ and then for any $n < N$, $z \in (-\infty,\infty)$, $y \geq 0$ and $x^{(n)} = (x_1, ..., x_n) \in (-1,\infty)^n$ we define

$$I_n(x^{(n)}, y, z) = \int_{-1}^\infty J_{n+1}((x^{(n)}, u), y + zu\kappa_n(x^{(n)}))d\mu_{x^{(n)}}^{(n+1)}(u)$$

and

$$J_n(x^{(n)}) = \inf_{z \in G_n(x^{(n)},y)} I_n(x^{(n)}, y, z)$$

where $\kappa_n(x^{(n)}) = S_0 \prod_{k=1}^n (1 + x_k)$ and $G_n(x^{(n)}, y)$ as in Lecture 10. Similarly to Lecture 10 (and even easier) we check that $I_n(x^{(n)}, \cdot, \cdot)$ and $J_n(x^{(n)}, \cdot)$ are lower semicontinuous in the arguments denoted by dots.

Next, in the same way as in Lecture 10 we construct $\pi^* \in \Pi$ which minimizes the shortfall risk. Namely, we set $X_0^{\pi^*} = x$ and inductively

$$X_{n+1}^{\pi^*} = X_n^{\pi^*} + \lambda_n(\rho^{(n)}, X_n^{\pi^*})\Delta S_{n+1}$$

where $\lambda_n(x^{(n)}, y) = arg\min_{\gamma \in G_n(x^{(n)}, y)} I_n(x^{(n)}, y, \gamma)$ if $a_{n+1}(x^{(n)}) \neq 0$ and $b_{n+1}(x^{(n)}) \neq 0$ while $\lambda_n(x^{(n)}, y) = I_n(x^{(n)}, y, 0)$ if $a_{n+1}(x^{(n)}) = 0$ or $b_{n+1}(x^{(n)}) = 0$.

Define $\Psi_N^\pi = \ell(f_N(\rho_1, ..., \rho_N) - X_N^\pi)$ and $\Psi_n^\pi = E(\Psi_{n+1}^\pi|\mathcal{F}_n) = E(\Psi_N^\pi|\mathcal{F}_n)$. Then $J_n(\rho^{(n)}, X_n^{\pi^*}) = \Psi_n^{\pi^*}$ and $J_n(\rho^{(n)}, X_n^\pi) \leq \Psi_n^{\pi^*}$ for any $\pi \in \Pi_x$. Indeed, proving this by the backward induction we observe first that $J_N(\rho^{(N)}, X_N^\pi) = \Psi_N^\pi$. Suppose that the above holds true for all $n \geq m+1$ and prove it for $n = m$. We have

$$E(J_{m+1}(\rho^{(m+1)}, X_{m+1}^\pi)|\mathcal{F}_m) = \int_{-1}^\infty J_{m+1}((\rho^{(m)}, u), X_m^\pi$$

$$+\gamma_{m+1} u \kappa_m(\rho^{(m)}))d\mu_{\rho^{(m)}}^{(m+1)}(u) = I_m(\rho^{(m)}, X_m^\pi, \gamma_{m+1}) \geq J_m(\rho^{(m)}, X_m^\pi).$$

On the other hand, if $\gamma_{m+1} = \gamma_{m+1}^* = \lambda(\rho^{(m)}, X_m^{\pi^*}) \in G(\rho^{(m)}, X_m^{\pi^*})$ then $I_m(\rho^{(m)}, X_m^{\pi^*}, \gamma_{m+1}^*) = J_m(\rho^{(m)}, X_m^{\pi^*})$. By the above definitions and the induction hypothesis,

$$I_m(\rho^{(m)}, X_m^{\pi^*}, \gamma_{m+1}^*) = E\big(J_{m+1}(\rho^{(m+1)}, X_{m+1}^{\pi^*})|\mathcal{F}_m\big)$$
$$= E(\Psi_{m+1}^{\pi^*}|\mathcal{F}_m) = \Psi_m^{\pi^*},$$

and so $J_m(\rho^{(m)}, X_m^{\pi^*}) = \Psi_m^{\pi^*}$ completing the induction step. Then $r(\pi^*) = \Psi_0^{\pi^*} = J_0(S_0, x) \leq \Psi_0^\pi = r(\pi)$ for any $\pi \in \Pi_x$, proving that $\pi^*$ is the shortfall minimizing portfolio strategy.

For American contingent claims with a nonnegative payoff function $f_n$, $n = 0, 1, ..., N$ satisfying $\max_{0 \leq n \leq N} Ef_n < \infty$ the construction is similar. Namely, we define now $I_n$ and $J_n$ by the following recursive relations

$$I_N(x^{(N)}, y, z) = J_N(x^{(N)}, y) = \ell(f_N(x^{(N)}) - y),$$

and then for any $n < N$,

$$I_n(x^{(n)}, y, z) = \max\big(\ell(f_n(x^{(n)}) - y), \int_{-1}^\infty J_{n+1}((x^{(n)}, u), y$$

$$+zu\kappa_n(x^{(n)}))d\mu_{x^{(n)}}^{(n+1)}(u)\big)$$

and

$$J_n(x^{(n)}) = \inf_{z \in G_n(x^{(n)}, y)} I_n(x^{(n)}, y, z).$$

Next, we define $\lambda_n(x^{(n)}, y)$ and $\pi^*$ via $I_n$ in the same way as above. Now, set $\Psi_N^\pi = \ell(f_N(\rho_1, ..., \rho_N) - X_N^\pi)$ and $\Psi_n^\pi = \max\big(\ell(f_n(\rho_1, ..., \rho_n) - X_n^\pi), E(\Psi_{n+1}^\pi|\mathcal{F}_n)\big)$. In the same way as above it follows that $r(\pi^*) = \Psi_0^{\pi^*} = J_0(S_0, x) \leq \Psi_0^\pi = r(\pi)$ for any $\pi \in \Pi_x$, i.e., that $\pi^*$ is the shortfall minimizing portfolio strategy in the American options case.

## 11.5.2 *Computations*

We consider the trinomial market with $B_n \equiv 1$, $S_n = \prod_{k=1}^{n}(1+\rho_k)$, $S_0 = 1$, $\rho_1, \rho_2, \ldots$ are i.i.d. random variables taking on values $-1/2$, $1/2$ and $1$, each with probability $1/3$. The horizon is 2, the striking price is also 2 and we consider the put option with the payoff function $f_2(x^{(2)}) = (2 - (1+x_1)(1+x_2))^+$. We want to minimize the shortfall starting with the initial capital $x = \frac{11}{18}$ which is half of the superhedging price of this European put option. By the recursive formulas above

$$I_2(x^{(2)}, y, z) = \left((2 - (1+x_1)(1+x_2))^+ - y\right)^+ = J_2(x^{(2)}, y),$$

$$I_1(x_1, y, z) = \tfrac{1}{3}\left(J_2((x_1, 1), y + z(1+x_1)) + J_2((x_1, \tfrac{1}{2}), y + \tfrac{1}{2}z(1+x_1))\right.$$
$$\left. + J_2((x_1, -\tfrac{1}{2}), y - \tfrac{1}{2}z(1+x_1))\right)$$

and

$$J_1(x_1, y) = \inf_{z \in G_1(x_1, y)} I_1(x_1, y, z)$$

where $y \geq 0$ and $G_1(x_1, y) = \{z : -y(1+x_1)^{-1} \leq z \leq 2y(1+x_1)^{-1}\}$ since here $a_n = -\tfrac{1}{2}$ and $b_n = 1$ for all $n$. Setting $v = z(1+x_1)$ we obtain

$$J_1(x_1, y) = \tfrac{1}{3}\inf_{-y \leq v \leq 2y}\left(((2 - 2(1+x_1))^+ - y - v)^+\right.$$
$$+((2 - \tfrac{3}{2}(1+x_1))^+ - y - \tfrac{1}{2}v)^+ + ((2 - \tfrac{1}{2}(1+x_1))^+ - y + \tfrac{1}{2}v)^+).$$

We have to compute $J_1$ for each value of $x_1$, i.e., to consider three cases: $x_1 = 1$, $x_1 = \tfrac{1}{2}$ and $x_1 = -\tfrac{1}{2}$. In view of the positive part sign appearing in the above formula the computation requires to consider the expression in brackets with a fixed $x_1$ in several disjoint domains of the form $\alpha_1 y + \beta_1 \leq v \leq \alpha_2 y + \beta_2$ for some constants $\alpha_1, \alpha_2, \beta_1, \beta_2$ where this expression is given by linear functions in $y$ and $v$ whose infimum in $v \in [-y, 2y]$ can be found immediately. It takes some time to consider all cases if done "by hand" but it is easy to write a computer program which will allow to make such computations very fast even for higher than 2 horizons. The computations lead to the representation of $J_1$ by different linear formulas in different disjoint intervals. Namely, it turns out that

$$J_1(1, y) = \begin{cases} \tfrac{1}{3} - \tfrac{1}{2}y & \text{if } 0 \leq y \leq \tfrac{2}{3}, \\ 0 & \text{if } y > \tfrac{2}{3}, \end{cases}$$

$$J_1(\tfrac{1}{2}, y) = \begin{cases} \tfrac{5}{12} - \tfrac{1}{2}y & \text{if } 0 \leq y \leq \tfrac{5}{6}, \\ 0 & \text{if } y > \tfrac{5}{6}, \end{cases}$$

$$J_1(-\frac{1}{2}, y) = \begin{cases} \frac{4}{3} - \frac{5}{3}y & \text{if } 0 \leq y \leq \frac{1}{3}, \\ 1 - \frac{2}{3}y & \text{if } \frac{1}{3} < y \leq \frac{3}{2}, \\ 0 & \text{if } y > \frac{3}{2}. \end{cases}$$

Next, for the initial capital $x = \frac{11}{18}$ we have,

$$I_0(x, z) = \frac{1}{3}(J_1(1, \frac{11}{18} + z) + J_1(\frac{1}{2}, \frac{11}{18} + \frac{1}{2}z) + J_1(-\frac{1}{2}, \frac{11}{18} - \frac{1}{2}z))$$

and

$$J_0(x) = \inf_{-\frac{11}{18} \leq z \leq \frac{11}{9}} I_0(x, z).$$

Considering $I_0$ only for $-\frac{11}{18} \leq z \leq \frac{11}{9}$ since only this range is needed to determine $J_0$ we obtain

$$I_0(x, z) = \begin{cases} \frac{79}{324} - \frac{5}{12}z & \text{if } -\frac{11}{18} \leq z \leq \frac{1}{9}, \\ \frac{19}{81} + \frac{1}{12}z & \text{if } -\frac{1}{9} < z \leq \frac{2}{9}, \\ \frac{16}{81} + \frac{1}{9}z & \text{if } -\frac{2}{9} < z \leq \frac{5}{9}, \\ \frac{17}{162} + \frac{5}{18}z & \text{if } -\frac{5}{9} < z \leq \frac{11}{9} \end{cases}$$

and $J_0(x) = r(\pi^*) = \frac{16}{81}$. Next, we compute by the general formulas above $\lambda_1(1, y) = 2\max(-y, 2(y-1))$, $\lambda_1(\frac{1}{2}, y) = \frac{3}{2}\max(-y, 2y - \frac{5}{2})$ and

$$\lambda_1(-\frac{1}{2}, y) = \begin{cases} \min(1 - y, \frac{1}{2} + 2(1-y), 2y) & \text{if } 0 \leq y \leq \frac{3}{2}, \\ 2y - \frac{7}{2} & \text{if } y \geq \frac{3}{2}. \end{cases}$$

Finally, we obtain $\lambda_0(x) = \frac{1}{9}$ which yields the minimizing strategy $\pi^*$ by the same formulas as in Lecture 10. The computations for the European call and the American options are similar.

## 11.6    Solutions to Chapter 6

1) Let $\mathcal{F}_t = \sigma\{\xi(s), 0 \leq s \leq t\}$ then it follows that $\xi(t) - \xi(s)$ is independent of $\mathcal{F}_s$ for any $t \geq s \geq 0$ since $\mathcal{F}_s$ is generated also by all increments of the form $\xi(v) - \xi(u)$, $s \geq v \geq u \geq 0$. Hence, for $s \leq t$,

$$E(M(t)|\mathcal{F}_s) = E(M(t) - M(s)|\mathcal{F}_s) + M(s)$$
$$= E(M(t) - M(s)) + M(s) = M(s)$$

since $M(t) - M(s) = \xi(t) - \xi(s)$ is independent of $\mathcal{F}_s$. Next, we claim that $M^2(t) - EM^2(t)$, $t \geq 0$ is a martingale too. Indeed, for $s \leq t$,

$$E(M^2(t) - M^2(s)|\mathcal{F}_s) = E((M(t) - M(s))^2|\mathcal{F}_s)$$
$$+ 2M(s)E(M(t) - M(s)|\mathcal{F}_s) = E(M(t) - M(s))^2 = EM^2(t)$$
$$+ EM^2(s) - 2EM(t)M(s) = EM^2(t) - EM^2(s)$$

since $E(M(t)M(s)|\mathcal{F}_s) = M(s)E(M(t)|\mathcal{F}_s) = M^2(s)$. Thus,

$$M^2(t) = (M^2(t) - EM^2(t) + a^2) + (EM^2(t) - a^2)$$

is the required submartingale decomposition since $A(t) = EM^2(t) - a^2 \geq 0$ is a non-decreasing process. Indeed, for $s, t \geq 0$ we see as above that

$$A(t+s) - A(t) = E(M^2(t+s) - M^2(t)) = E(M(t+s) - M(t))^2 \geq 0.$$

2(a) We can write

$$\{\tau \leq t\} = (\Gamma \cap \{\tau_1 \leq t\} \cup ((\Omega \setminus \Gamma) \cap \{\tau_2 \leq t\}) \in \mathcal{F}_t.$$

The last inclusion holds true since $\Gamma \in \mathcal{F}_\sigma \subset \mathcal{F}_{\tau_1} \cap \mathcal{F}_{\tau_2}$. Indeed, if $A \in \mathcal{F}_\sigma$ then

$$A \cap \{\tau_1 \leq t\} = (A \cap \{\sigma \leq t\}) \cap \{\tau_1 \leq t\} \in \mathcal{F}_t$$

and, similarly $A \cap \{\tau_2 \leq t\} \in \mathcal{F}_t$.

2(b) If $Q_+$ denotes the set of all nonnegative rationals then

$$\{\sigma_c < t\} = \cup_{r<t, r \in Q_+} \{Z_r < c\} \cap \{\zeta \leq r\} \in \mathcal{F}_t$$

and

$$\{\sigma_c \leq t\} = \cap_{n \geq 1}\{\sigma_c < t + \frac{1}{n}\} \in \cap_{n \geq 1}\mathcal{F}_{t+\frac{1}{n}} = \mathcal{F}_t.$$

Hence, $\sigma_c$ is a stopping time and $\tilde{\sigma}_c = \lim_{n \uparrow \infty} \uparrow \sigma_{c+\frac{1}{n}}$ is a stopping time as well.

## 11.7 Solutions to Chapter 7

1) Since we consider a continuous Brownian motion, the integral can be taken path-wise in the usual Riemannian sense. The components $W_k(t)$ of $W(t) = (W_1(t), ..., W_d(t))$ are independent one-dimensional Brownian motions, and so $Z_k = \int_0^1 W_k(t)dt$, $k = 1, ..., d$ are also independent. Thus, by Proposition 7.1.1(b) we have only to show that $Z_k$, $k = 1, ..., d$ are Gaussian random variables and to find their expectations and variances which are, clearly, the same for all $k$, and so we will deal only with $Z_1$ which we denote by $Z$. Set also $W = W_1$, then

$$Z = \lim_{n \to \infty} \frac{1}{n} \sum_{j=1}^{n} W(j/n)$$

since the right hand side here is the limit of Riemannian sums. Define random vectors $X = (W(\frac{1}{n}), ..., W(\frac{n-1}{n}), W(1))$ and $Y = (W(\frac{1}{n}), W(\frac{2}{n}) -$

$W(\frac{1}{n}), ..., W(1) - W(\frac{n-1}{n}))$. By Proposition 7.1.1(b), $Y$ is a Gaussian random vector since its components are independent Gaussian random variables by the definition of the Brownian motion. But $X = AY$ where $A$ is the matrix $(a_{kl})$ with $a_{kl} = 1$ if $k \geq l$ and $a_{kl} = 0$ if $k < l$. Thus, by Proposition 7.1.1(c), $X$ is also a Gaussian random vector. Hence,

$$\frac{1}{n}\sum_{j=1}^{n} W(j/n) = \langle \xi_n, X \rangle,$$

where $\xi_n = (\frac{1}{n}, \frac{1}{n}, ..., \frac{1}{n})$, is a Gaussian random variable. By Proposition 7.1.1(e) the limit $Z$ is also a Gaussian random variable. As in the proof of Proposition 7.1.1(e) we conclude that the parameters (expectation and variance) of $Z$ are the limits as $n \to \infty$ of the corresponding parameters of $\langle \xi_n, X \rangle$. Clearly, $E\langle \xi_n, X \rangle = 0$. Next,

$$E(\langle \xi_n, X \rangle)^2 = \frac{1}{n^2}E(\sum_{j=1}^{n} W(j/n))^2$$
$$= \frac{1}{n^3}(\sum_{j=1}^{n} j + 2\sum_{l=1}^{n}\sum_{j=1}^{l-1} j) \to \frac{1}{3} \text{ as } n \to \infty$$

where we used that $EW(s)W(t) = \min(s,t)$.

2) Let $W(t)$, $t \geq 0$ be the continuous Brownian motion and we will show that at each $t \geq 0$ with probability one it is not Hölder continuous with the exponent $1/2$. Observe that $X_t = \limsup_{s \to t} |W(t) - W(s)|/|t - s|^{1/2}$ is a random variable since by continuity of the Brownian motion it suffices to take lim sup over rational numbers only. Since $X_t$ has the same distribution as $X_0$ it suffices to show that with probability one $X_0 = \infty$. Observe that $X_0$ is measurable with respect to $\cap_{t>0}\mathcal{F}_t = \mathcal{F}_0 = \{\Omega, \emptyset\}$ where $\mathcal{F}_t$, $t \geq 0$ is the Brownian filtration. Hence, $X_0$ is a constant (including infinity) with probability one. If with probability one $X_0 < C$ for some $C < \infty$ then there exists $\Omega^* \subset \Omega$ with $P(\Omega^*) = 1$ and $t_0 = t_0(\omega) > 0$, $\omega \in \Omega^*$ such that $A(t) = A(t, \omega) = |W(t, \omega)|/t^{1/2} < C$ when $\omega \in \Omega^*$ and $t < t_0(\omega)$. Then $\liminf_{t \to 0} \mathbb{I}_{[0,C)}(A(t)) = 1$ a.s. and we have by the Fatou lemma that

$$1 = E\big(\liminf_{t \to 0} \mathbb{I}_{[0,C)}(A(t))\big) \leq \liminf_{t \to 0} E(\mathbb{I}_{[0,C)}(A(t)))$$
$$= E\mathbb{I}_{[0,C)}(|W(1)|) = P\{-C < W(1) < C\} < 1,$$

where we used that by the Brownian scaling $A(t)$ and $W(1)$ have the same distribution. We arrived at a contradiction to the assumption that $C < \infty$ proving the result.

3) The problem reduces easily to the claim that if $t \geq s$ and $X$ is an $\mathcal{F}_s = \sigma\{W(u), u \leq s\}$-measurable random variable then

$$E \exp\big(X(W(t) - W(s))) - \frac{1}{2}X^2(t - s)\big) = 1.$$

Expanding the exponent into the Taylor series we have

$$E\big(\exp(X(W(t) - W(s)))|\mathcal{F}_s\big) = \sum_{n=0}^{\infty} \frac{X^n E W^n (t-s)}{n!}$$

$$= \sum_{n=0}^{\infty} \frac{X^{2n}(t-s)^n}{2^n n!} = \exp(\tfrac{1}{2}X^2(t-s))$$

and the required identity follows.

4) By Itô's formula

$$\cos W(t) = \frac{\pi}{2} - \int_0^t \sin W(s)dW(s) - \frac{1}{2}\int_0^t \cos W(s)ds.$$

Hence,

$$E \cos W(t) = \frac{\pi}{2} - \frac{1}{2}\int_0^t E \cos W(s)ds.$$

Thus, $x(t) = E \cos W(t)$ satisfies the ordinary differential equation $\frac{dx(t)}{dt} = -\frac{1}{2}x(t)$ with $x(0) = \frac{\pi}{2}$, and so $x(t) = \frac{\pi}{2}e^{-\frac{t}{2}}$.

5) Observe that for each constant $\gamma$,

$$\exp(\gamma x - \frac{1}{2}\gamma^2 t) = \sum_{n=0}^{\infty} \gamma^n H_n(t, x).$$

Hence, the martingale $X(t) = \exp(\gamma W(t) - \frac{1}{2}\gamma^2 t)$ has the representation $X(t) = \sum_{n=0}^{\infty} \gamma^n H_n(t, W(t))$. On the other hand, by the Itô formula $dX(t) = \gamma X(t)dW(t)$. Solving this stochastic differential equation by successive approximations with the initial condition $X(0) = 1$ we obtain $Y_0(t) \equiv 1$, $Y_1(t) = 1 + \int_0^t \gamma dW(t) = 1 + \gamma W(t)$, $Y_2(t) = 1 + \gamma \int_0^t (1 + \gamma W(s))dW(s) = 1 + \gamma W(t) + \gamma^2 \int_0^t W(s)dW(s),...,$

$$Y_n(t) = \sum_{k=0}^{n} \gamma^k \int_0^t \int_0^{t_1} ... \int_0^{t_{k-2}} W(t_k)dW(t_1)...dW(t_{k-1}).$$

By the general result on solutions of stochastic differential equations $Y_n \to X(t)$ as $n \to \infty$. Hence, $X(t)$ has the representation

$$X(t) = \sum_{n=0}^{\infty} \gamma^n \int_0^t \int_0^{t_1} ... \int_0^{t_{n-2}} W(t_n)dW(t_1)...dW(t_{n-1})$$

and by the uniqueness of the power series (in $\gamma$) representation we obtain the required equality.

## 11.8   Solutions to Chapter 8

1) Recall that the fair price of an European contingent claim with a $\mathcal{F}_T$-measurable payoff $f_T \geq 0$ at the time $T$ is given by

$$V = E_{P^*}(e^{-rT} f_T)$$

where $E_{P^*}$ is the expectation with respect to the martingale measure $P^*$.

(a) Let $f_T = S_T^2$ where $S_T = S_0 \exp((\mu - \frac{\sigma^2}{2})T + \sigma W_T)$ is the stock price at time $T$. We can write also $S_T = S_0 \exp((r - \frac{\sigma^2}{2})T + \sigma W_T^*)$ where $W_t^* = W_t + \frac{\mu - r}{\sigma}t$ is the Brownian motion with respect to the martingale measure $P^*$, in particular, $W_T$ with respect to $P^*$ is a normal random variable with variance $T$ and mean 0. Hence,

$$V = E_{P^*}(e^{-rT} S_T^2) = S_0 e^{(r - \sigma^2)T} E_{P^*} \exp(2\sigma W_T^*)$$

$$= S_0 e^{(r - \sigma^2)T} \frac{1}{\sqrt{2\pi T}} \int_{-\infty}^{\infty} e^{2\sigma x} e^{-\frac{x^2}{2T}} dx.$$

Now,

$$\frac{1}{\sqrt{2\pi T}} \int_{-\infty}^{\infty} e^{2\sigma x} e^{-\frac{x^2}{2T}} dx$$

$$= \frac{1}{\sqrt{2\pi T}} \int_{-\infty}^{\infty} e^{-\frac{(x - 2\sigma T)^2}{2T}} e^{2\sigma^2 T} dx = e^{2\sigma^2 T}.$$

Hence, $V = S_0 e^{(r + \sigma^2)T}$.

(b) Now let $f_T = (S_T - K)^+$ be a call option payoff. Then

$$V = E_{P^*}(e^{-rT}(S_T - K)^+) = E_{P^*}(\exp(-\sigma^2 T/2 + \sigma W_T^*) - e^{-rT}K)^+$$

$$= e^{-\sigma^2 T/2} E_{P^*}(\exp(\sigma W_T^*) - e^{(\frac{\sigma^2}{2} - r)T}K)^+$$

$$= e^{-\sigma^2 T/2} \frac{1}{\sqrt{2\pi T}} \int_{-\infty}^{\infty} (e^{\sigma x} - e^{(\frac{\sigma^2}{2} - r)T}K)^+ e^{-\frac{x^2}{2T}} dx.$$

Let $x_0$ be such that $e^{\sigma x_0} = e^{(\frac{\sigma^2}{2} - r)T}K$, i.e., $x_0 = \sigma^{-1}((\sigma^2/2 - r)T + \ln K)$. Then

$$\frac{1}{\sqrt{2\pi T}} \int_{-\infty}^{\infty} (e^{\sigma x} - e^{(\frac{\sigma^2}{2} - r)T}K)^+ e^{-\frac{x^2}{2T}} dx$$

$$= \frac{1}{\sqrt{2\pi T}} \int_{x_0}^{\infty} (e^{\sigma x} - e^{(\frac{\sigma^2}{2} - r)T}K) e^{-\frac{x^2}{2T}} dx.$$

Now,

$$\frac{1}{\sqrt{2\pi T}} \int_{x_0}^{\infty} e^{(\sigma x - \frac{x^2}{2T})} dx = \frac{1}{\sqrt{2\pi T}} \int_{x_0}^{\infty} e^{-\frac{(x - \sigma T)^2}{2T}} e^{\sigma^2 T/2} dx$$

$$= e^{\sigma^2 T/2} (1 - \Phi(\frac{x_0 - \sigma T}{\sqrt{T}})).$$

Hence,

$$V = e^{-\sigma^2 T/2} (e^{\sigma^2 T/2}(1 - \Phi(\frac{x_0 - \sigma T}{\sqrt{T}})) - e^{(\frac{\sigma^2}{2} - r)T}K(1 - \Phi(\frac{x_0}{\sqrt{T}}))$$

$$= 1 - \Phi(\frac{x_0 - \sigma T}{\sqrt{T}}) - e^{-rT}K(1 - \Phi(\frac{x_0}{\sqrt{T}})).$$

Next, for $f_T = (K - S_T)^+$ we obtain similarly

$$V = e^{-\sigma^2 T/2} \frac{1}{\sqrt{2\pi T}} \int_{-\infty}^{\infty} (e^{(\frac{\sigma^2}{2} - r)T} K - e^{\sigma x})^+ e^{-\frac{x^2}{2T}} dx$$

$$= e^{-\sigma^2 T/2} \frac{1}{\sqrt{2\pi T}} \int_{-\infty}^{x_0} (e^{(\frac{\sigma^2}{2} - r)T} K - e^{\sigma x})^+ e^{-\frac{x^2}{2T}} dx$$

$$= e^{-rT} K \Phi(\frac{x_0}{\sqrt{T}}) - \Phi(\frac{x_0 - \sigma T}{\sqrt{T}}).$$

2) We have to find the fair price and a replicating strategy for a European contingent claim with the payoff $W^2(T)$ at the time $T$ and with the bond and stock prices $B_t = e^t$ and $S_t = \exp(t + W(t))$, respectively. Then the Brownian motion $W^*$ with respect to the martingale measure $P^*$ has the form $W^*(t) = \frac{t}{2} + W(t)$. By the theorem in Lecture 20 the fair price of this contingent claim is given by

$$V = E^*(e^{-T} W^2(T)) = e^{-T} E^* (W^*(T) - \tfrac{1}{2}T)^2$$
$$= e^{-T} \left( E^*(W^*(T))^2 + \tfrac{1}{4}T^2 \right) = e^{-T} (T + \tfrac{1}{4}T^2)$$

where $E^*$ is the expectation with respect to $P^*$. In order to construct the replicating trading strategy we have to act according to the general prescription from Lecture 20. Namely, we start with the martingale

$$M_t = E^*(e^{-T} W^2(T)|\mathcal{F}_t) = e^{-T} E^* ((W^*(T) - \tfrac{1}{2}T)^2|\mathcal{F}_t)$$
$$= e^{-T} \left( E^*((W^*(T))^2|\mathcal{F}_t) - TW^*(t) + \tfrac{1}{4}T^2 \right)$$
$$= e^{-T} \left( T - t + (W^*(t))^2 - TW^*(t) + \tfrac{1}{4}T^2 \right).$$

By the Itô formula

$$dM_t = e^{-T} (2W^*(t) dW^*(t) - T dW^*(t))$$

which provides the representation of $M_t$ as the stochastic integral

$$M_t = M_0 + \int_0^t e^{-T} (2W^*(s) - T) dW^*(s).$$

Hence, by the prescription of Lecture 20 we have to define

$$\gamma_u = e^{-T+u} S_u^{-1} (2W^*(u) - T) = e^{-(T+W(u))} (2W^*(u) - T)$$
$$= e^{-(T+W(u))} (u + 2W(u) - T)$$

and

$$\beta_u = M_t - e^{-u} \gamma_u S_u$$
$$= e^{-T} \left( T - u + (W^*(u))^2 - TW^*(u) + \tfrac{1}{4}T^2 \right) - e^{-T} (u + 2W(u) - T)$$
$$= e^{-T} \left( 2(T - u) + \tfrac{1}{4}(T - u)^2 + (u - T - 2)W(u) + W^2(u) \right).$$

As arguments of Lecture 20 show $\pi = (\beta_t, \gamma_t)_{t\in[0,T]}$ will be the required replicating trading strategy with the portfolio values $X_0^\pi = M_0 = V$ and $X_T^\pi = W^2(T)$.

3) By Lecture 21 we have

$$S_t^1 = S_0^1 \exp(\tfrac{1}{2}t + W_1(t) - 2W_2(t)) \text{ and } S_t^2 = S_0^2 \exp(-5t - 2W_1(t) + 4W_2(t)).$$

In our setup $\mu = (3,5)$ is the constant vector, the interest rate $r = 2$ and the volatility matrix $\kappa = \begin{bmatrix} 1 & 2 \\ -2 & 4 \end{bmatrix}$ is constant, as well as the vector $\eta = \mu - (r,r) = (1,3)$. We can write now $\eta = (1,3) = -(1,-2) + (2,1)$ where $(1,-2) \in \kappa\mathbb{R}^2$ and $(2,1) \in (\kappa\mathbb{R}^2)^\perp$. Thus we set $v = (2,1)$ and the set $\Gamma_t$ in the arbitrage theorem in Lecture 21 will be the whole $\Omega$ since $\eta \notin \kappa\mathbb{R}^2$. According to the construction in this theorem we set $\gamma_t^1 = \frac{2e^{2t}}{S_t^1\sqrt{5}}$, $\gamma_t^2 = \frac{e^{2t}}{S_t^2\sqrt{5}}$ and $\beta_t = \tilde{X}_t^\pi - e^{-2t}(\gamma_t^1 S_t^1 + \gamma_t^2 S_t^2) = \sqrt{5}t - \frac{3}{\sqrt{5}}$ where the discounted portfolio value $\tilde{X}_t^\pi$ of the trading strategy $\pi = (\beta, \gamma)$ has the form $\tilde{X}_t^\pi = t|v| = \sqrt{5}t$ and the value at time $t$ of the portfolio itself is $X_t^\pi = \sqrt{5}e^{2t}t$. Since $X_0^\pi = 0$ and $X_t^\pi > 0$ for all $t > 0$ the strategy $\pi$, which is admissible and self-financing, provides an arbitrage opportunity.

4) The fact that $\Theta_{k+1} - \Theta_k$, $k = 0,1,\dots$ is a sequence of i.i.d. random variables is proved in the same way as in Lecture 22 taking into account that $W^{(k)}(t) = W(\Theta_k + t) - W(\Theta_k)$ is a Brownian motion independent of $\mathcal{F}_{\Theta_n}$. Next, observe that if $0 \le \sigma \le \tau$ are stopping times such that $E\tau < \infty$, then

$$E(W(\tau) - W(\sigma))^2 = E(\tau - \sigma).$$

Indeed, $W^2(t) - t$, $t \ge 0$ is a martingale, and so by the optional stopping theorem

$$E(W^2(\tau) - \tau) = E(W^2(\sigma) - \sigma) = 0.$$

Since $E(W(\tau)W(\sigma)) = E(W(\sigma)E(W(\tau)|\mathcal{F}_\sigma)) = W^2(\sigma)$, we obtain $E(W(\tau) - W(\sigma))^2 = EW^2(\tau) - EW^2(\sigma) = E(\tau - \sigma)$. In particular, for $\sigma = 0$ and $\tau = \Theta_1$ we have $1 = EW^2(\Theta_1) = E\Theta_1$, proving (a).

In order to prove (b) we write

$$E|W(n) - W(\Theta_n)|^2 = E|W(\Theta_n \vee n) - W(\Theta_n \wedge n)|^2$$
$$= E(\Theta_n \vee n - \Theta_n \wedge n) = E|\Theta_n - n|.$$

By (a), $\Theta_n = \sum_{k=1}^n (\Theta_k - \Theta_{k-1})$ is a sum of i.i.d. random variables with $E(\Theta_k - \Theta_{k-1}) = 1$. Next, observe that $W^4(t) - 6W^2(t)t + 3t^2$, $t \ge 0$ is also a martingale (such martingales are obtained by differentiating in $\alpha$ at $\alpha = 0$ of the exponential martingale $\exp(\alpha W(t) - \frac{1}{2}\alpha^2 t)$ ). Hence, by the optional stopping theorem

$$0 = E(W^4(\Theta_1) - 6W^2(\Theta_1)\Theta_1 + 3\Theta_1^2) = 1 - 6E\Theta_1 + 3E\Theta_1^2 = -5 + 3E\Theta_1^2,$$

i.e., $E\Theta_1^2 = 5/3$. Thus, by the central limit theorem $\frac{\Theta_n - n}{\sqrt{\frac{5}{3}n}}$ converges in distribution to a standard normal random variable $X$. It follows that

$$\lim_{n\to\infty} n^{-\frac{1}{2}} E|\Theta_n - n| = \sqrt{\frac{5}{3}} E|X| = \frac{\sqrt{10}}{\sqrt{3\pi}} \int_0^\infty x e^{-\frac{x^2}{2}} dx = \sqrt{\frac{10}{3\pi}}$$

and (b) follows.

In order to prove (c) we proceed as in Lecture 22. Namely, we write

$$M_n = \max_{0\le k\le n} |W(k \vee \Theta_k) - W(k \wedge \Theta_k)|$$

$$\le 2\mathbb{I}_{\max_{0\le k\le n} \sup_{k\wedge\Theta_k \le t \le k\wedge\Theta_k + D^2\sqrt{n\ln n}} |W(t)-W(k\wedge\Theta_k)| > D^2 n^{1/4}(\ln n)^{3/4}}$$

$$\times \sup_{0\le t\le n\vee\Theta_n} |W(t)|$$

$$+2\mathbb{I}_{\max_{0\le k\le n} |k\vee\Theta_k - k\wedge\Theta_k| > D^2\sqrt{n\ln n}} \sup_{0\le t\le n\vee\Theta_n} |W(t)|$$

$$+D^2 n^{1/4}(\ln n)^{3/4}$$

for any $D > 0$. Now, using the inequality (8.4.34) from Lecture 22 we obtain similarly to (8.4.45) that

$$P\Big\{ \max_{0\le k\le n} \sup_{k\wedge\Theta_k \le t \le k\wedge\Theta_k^{(n)} + D^2\sqrt{n\ln n}} |W(t) - W(\Theta_k^{(n)})|$$

$$> D^2 n^{1/4}(\ln n)^{3/4}\Big\} \le nP\Big\{ \sup_{0\le t\le D^2\sqrt{n\ln n}} |W(t)| > D^2 n^{1/4}(\ln n)^{3/4}\Big\}$$

$$\le 2n^{1-\frac{D^2}{2}} \le 2n^{-2}$$

provided we choose $D \ge \sqrt{6}$. Next,

$$P\{\max_{0\le k\le n} |k \vee \Theta_k - k \wedge \Theta_k| > D^2\sqrt{n\ln n}\} \le P\{\max_{0\le k\le n}(\Theta_k - k)$$

$$> D^2\sqrt{n\ln n}\} + P\{\max_{0\le k\le n}(k - \Theta_k) > D^2\sqrt{n\ln n}\}$$

$$\le \sum_{k=0}^n \left(P\{\Theta_k - k > D^2\sqrt{n\ln n}\} + P\{k - \Theta_k > D^2\sqrt{n\ln n}\}\right).$$

By the exponential Chebyshev inequality and the independency of $\Theta_j - \Theta_{j-1}$, $j = 1, 2, \ldots$ we obtain

$$P\{\Theta_k - k > D^2\sqrt{n\ln n}\} = P\Big\{ \exp\left(\sqrt{n^{-1}\ln n} \sum_{j=1}^n (\Theta_j - \Theta_{j-1} - 1)\right)$$

$$> n^{D^2}\Big\} \le n^{-D^2} \left(E\exp(\sqrt{n^{-1}\ln n}(\Theta_1 - 1))\right)^n$$

$$\le n^{-D^2}\Big(1 + \sum_{m=2}^\infty \frac{1}{m!}\big(\frac{\ln n}{n}\big)^{\frac{m}{2}} E|\Theta_1 - 1|^m\Big)^n.$$

Similarly to (8.4.42) we obtain from here that if $D > 0$ is large enough then

$$P\{\Theta_k - k \ge D^2\sqrt{n\ln n}\} \le Cn^{-3}$$

and, similarly,

$$P\{k - \Theta_k \geq D^2 \sqrt{n \ln n}\} \leq Cn^{-3}$$

for some constant $C > 0$ independent of $n$ and $k \leq n$. Similarly to (8.4.46) we obtain by the martingale inequality that

$$E \sup_{0 \leq t \leq n \vee \Theta_n} |W(t)|^4 \leq Cn^2$$

for some $C > 0$ independent of $n$. Now, (c) follows from the above estimates together with the Cauchy–Schwarz inequality.

# Bibliography

[1] P. Billingsley, *Probability and Measure*, 2nd ed., J. Willey, New York, 1986.

[2] M. Beiglböck, W. Schachermayer and B. Veliyev, A short proof of the Doob–Meyer theorem, Stoch. Proc. Appl. 122 (2012), 1204–1209.

[3] R. Dalang, A. Morton and W. Willinger, *Equivalent martingale measures and no-arbitrage in stochastic securities market models*, Stochastics Stoch. Reports 29 (1990), 185–201.

[4] R. Carmona and N. Touzi, *Optimal multiple stopping and valuation of swing options*, Math. Finance 18 (2008), 239–268.

[5] Y. S. Chow, H. Robbins and D. Siegmund, *Great Expectations: The Theory of Optimal Stopping*, Houghton, Boston, 1971.

[6] Ya. Dolinsky, *Hedging of game options under model uncertainty in discrete time*, Electronic Comm. Probab. 19 (2014), no.19, 1–11.

[7] R. M. Dudley, *Real Analysis and Probability 2nd ed.*, Cambridge Univ. Press, Cambridge, 2003.

[8] E. B. Dynkin, *Game variant of a problem on optimal stopping*, Soviet Math.–Doklady 10 (1969), 270–274.

[9] *Louis Bachelier's Theory of Speculation. The Origin of Modern Finance*, Translated and with commentary by M. Davis and A. Etheridge, Princeton Univ. Press, Princeton NJ, 2006.

[10] F. Delbaen and W. Schachermayer, *The Mathematics of Arbitrage*, Springer, Berlin, 2006.

[11] Ya. Dolinsky and Yu. Kifer, *Hedging with risk for game options in discrete time*, Stochastics 79 (2007), 196–195.

[12] Ya. Dolinsky and Yu. Kifer, *Another correction: Error estimates for binomial approximations of game options*, Ann. Appl. Probab. 18 (2008), 1271–1277.

[13] Ya. Dolinsky and Yu. Kifer, *Binomial approximations of shortfall risk for game options*, Ann. Appl. Probab. 18 (2008), 1737–1770.

[14] Ya. Dolinsky and Yu. Kifer, *Risk minimization for game options in markets imposing minimal transaction costs*, Advances Appl. Probab. 48 (2016), 926–946.

[15] Ya. Dolinsky, Yo. Iron and Yu. Kifer, *Perfect and partial hedging for swing*

*game options in discrete time*, Math. Finance 21 (2011), 447–474.

[16] Ya. Dolinsky and H. M. Soner, *Martingale optimal transport and robust hedging in continuous time*, Probab. Th. Relat. Fields 160 (2014), 391–427.

[17] A. G. Fakeev, *On optimal stopping rules for stochastic processes with continuous parameter*, Theory Probab. Appl. 15 (1970), 324–331.

[18] T. S. Ferguson, *Who solved the secretary problem*, Stat. Sci. 4 (1989), 282–296.

[19] P. R. Freeman, *The secretary problem and its extensions: a review*, Intern. Stat. Review 51 (1983), 189–206.

[20] A. Friedman, *Partial Differential Equations of Parabolic Type*, Dover, Mineola NY, 2008.

[21] H. Föllmer and Yu. Kabanov, *Optional decomposition theorem and Lagrange multipliers*, Finance Stoch. 2 (1998), 69–81.

[22] H. Föllmer and P. Leukert, *Efficient hedging: cost versus shortfall risk*, Finance Stoch. 4 (2000), 117–146.

[23] H. Föllmer and A. Schied, *Stochastic finance*, 2nd. ed., de Gruyter, Berlin, 2004.

[24] P. Guasoni, M. Rásonyi and W. Schachermayer, *Consistent price systems and face-lifting pricing under transaction costs*, Ann. Appl. Probab. 18 (2008), 491–520.

[25] J. M. Harrison and S. R. Pliska, *A stochastic calculus model of continuous trading: complete markets*, Stoch. Proc. Appl. 15 (1983), 313–316.

[26] Yo. Iron and Yu. Kifer, *Hedging of swing game options in continuous time*, Stochastics 83 (2011), 365–404.

[27] Yo. Iron and Yu. Kifer, *Error estimates for binomial approximations of game put options*, ISRN Probab. Stat. (2014), 1-25.

[28] N. Ikeda and S. Watanabe, *Stochastic Differential Equations and Diffusion Processes 2nd. ed.*, North-Holland, Amsterdam, 1989.

[29] A. Jakubowski, *An almost sure approximation for the predictable process in the Doob–Meyer decomposition theorem*, in: Séminaire de Probabilités XXXVIII, Lecture Notes in Math. 1857, Springer, Berlin, 2005, p.p.158–164.

[30] E. Jouini and H. Kallal, *Martingales and arbitrage in securities markets with transaction costs*, J. Econom. Theory 66 (1995), 178–197.

[31] R. Jarrow and P. Protter, *A short history of stochastic integration and mathematical finance: The early years 1880-1970*, IMS Lecture Notes—Monograph Series 45 (2004), 75–91.

[32] O. Kallenberg, *Foundations of Modern Probability*, Springer, New York, 1997.

[33] J. Kiefer, *On the derivations in the Skorokhod–Strassen approximation scheme*, Z. Wahrsch. Verv. Gebiete 13 (1969), 321–332.

[34] Yu. Kifer, *Optimal stopping in games with continuous time*, Theory Probab. Appl. 16 (1971), 545–550.

[35] Yu. Kifer, *Game options*, Finance Stochast. 4(2000), 443–463.

[36] Yu. Kifer, *Error estimates for binomial approximations of game options*, Ann. Appl. Probab. 16 (2006), 984–1033.

[37] Yu. Kifer, *Hedging of game options in discrete markets with transaction costs*, Stochastics 85 (2013), 667–681.

[38] J. Komlós, *A generalization of a problem of Steinhaus*, Acta Math. Acad. Sci. Hungar. 18 (1967), 217–229.

[39] D. O. Kramkov, *Optional decomposition of supermartingales and hedging contingent claims in incomplete security markets*, Probab. Th. Relat. Fields 105 (1996), 459–479.

[40] N. V. Krylov, *Control of Markov processes and W-spaces*, Math. USSR Izvestija 5 (1971), 233–266.

[41] N. Krylov, *A simple proof of a result of A. Novikov* , arXiv 0207013, 2002.

[42] J. Kallsen and C. Kühn, *Convertible bonds: financial derivatives of game type*, in: *Exotic Option Pricing and Advanced Lévy Models*, A. Kyprianou, W. Schoutens and P. Wilmott eds., Wiley, Chichester, 2005.

[43] Y. Kabanov and M. Safarian, *Markets with transaction costs*, Springer, Berlin, 2010.

[44] Y. Kabanov and C. Stricker, *A teacher's note on no-arbitrage criteria*, Séminare de Probab. (Strasbourg) 35, Lecture Notes in Math. 1775, Springer 2001, p.p. 149–152.

[45] Y. Kabanov and C. Stricker, *The Harrison–Pliska arbitrage pricing theorem under transaction costs*, J. Math. Econom. 35 (2001), 185–196.

[46] I. Karatzas and S. E. Shreve, *Brownian Motion and Stochastic Calculus*, 2nd ed., Springer, New York, 1991.

[47] I. Karatzas and S. E. Shreve, *Methods of Mathematical Finance*, Springer, New York, 1998.

[48] I. Karatzas and I.-M. Zamfirescu, *Game approach to the optimal stopping problem*, Stochastics 77 (2005), 401–435.

[49] D. Lamberton and B. Lapeyre, *Introduction to Stochastic Calculus Applied to Finance*, Chapman& Hall/CRC, London, 2008.

[50] D. Lamberton and L. C. G. Rogers, *Optimal stopping and embedding*, J. Appl. Probab. 37 (2000), 1143–1148.

[51] J. P. Lepeltier and M. A. Maingueneau, *Le jeu de Dynkin en theorie generale sans l'hypothese de Mokobodski*, Stochastics 13 (1984), 25–44.

[52] S. Levental and A. Skorokhod, *On the possibility of hedging options in the presence of transaction costs*, Ann. Appl. Probab. 7 (1997), 410–443.

[53] R. Liptser and A. N. Shiryaev, *Statistics of Random processes. I. General Theory*, 2nd ed., Springer, Berlin, 2001.

[54] H. P. McKean, *Stochastic Integrals*, Academic Press, New York, 1969.

[55] P. A. Meyer, *Probability and Potential*, Blaisdell, Waltham, MA, 1966.

[56] S. Mulinacci, *The efficient hedging for American options*, Finance Stoch. 15 (2011), 365–397.

[57] M. Musiela and M. Rutkowski, *Martingale Methods in Financial Modelling*, 2nd. ed., Springer, Berlin, 2005.

[58] J. Neveu, *Mathematical Foundations of the Calculus of Probability*, Holden-Day, San Francisco, 1965.

[59] J. Neveu, *Discrete Parameter Martingales*, North-Holland, Amsterdam/American Elsevier, New York, 1975.

[60] B. Øksendal, *Stochastic Differential Equations*, Springer, Berlin, 6th ed., 2003.

[61] M. Ondreját and J. Seidler, *On existence of progressively measurable modifications*, Electron. Comm. Probab. 18 (2013), no.20, 1–6.

[62] G. Peskir and A. Shiryaev, *Optimal Stopping and Free-Boundary Problem*, Birkhäuser, Basel, 2006.

[63] A. Roux, *Pricing and hedging game options in currency models with proportional transaction costs*, Int. J. Theor. Appl. Finance 19 (2016), 1650043.

[64] A. Roux, K. Tokarz and T. Zastawniak, *Options under proportional transaction costs: An algorithmic approach to pricing and hedging*, Acta Appl. Math. 103 (2008), 201–219.

[65] A. Roux and T. Zastawniak, *American options under proportional transaction costs: Pricing, hedging and stopping algorithms for long and short positions*, Acta Appl. Math. 106 (2009), 199–228.

[66] W. Schachermayer, *A Hilbert space proof of the fundamental theorem of asset pricing in finite discrete time*, Insurance 11 (1992), 249–257.

[67] A. N. Shiryaev, *Probability*, Springer, New York, 1984.

[68] A. N. Shiryaev, *Essentials of Stochastic Finance*, World Scientific, Singapore, 1999.

[69] S. Shreve, *Stochastic Calculus and Finance I, II*, Springer, New York, 2004.

[70] H. M. Soner, S. E. Shreve and J. Cvitanić, *There is no nontrivial hedging portfolio for option pricing with transaction costs*, Ann. Appl. Probab. 5 (1995), 327–355.

[71] J. B. Walsh, *The rate of convergence of the binomial tree scheme*, Finance Stochast. 7 (2003), 337–361.

[72] R. Williams, *Introduction to the Mathematics of Finance*, AMS, Providence, 2006.

[73] J. M. Xia and J. A. Yan, *A new look at some basic concepts in arbitrage pricing theory*, Sci. in China, Series A: Math., v.46, No.6 (2003).

[74] J. A. Yan, *A new look at the fundamental theorem of asset pricing*, J. Korean Math. Soc. 35 (1998), 659–673.

[75] J. A. Yan, *Introduction to Stochastic Finance*, Springer and Sci. Press, Singapore, 2018.

# Index

CPSIA information can be obtained
at www.ICGtesting.com
Printed in the USA
BVHW042023211219
567403BV00006B/8/P